Comparative anatomy of the vertebrates

Comparative anatomy of the vertebrates

George C. Kent, Jr., Ph.D.

Alumni Professor of Zoology
Louisiana State University
Baton Rouge, Louisiana

With 362 illustrations

Second edition

Saint Louis

The C. V. Mosby Company

1969

Preface

Emphasis in this edition continues to be on basic patterns of vertebrate structure. I have repeatedly expressed or implied the concept that recent patterns are modifications of older patterns, that the adult is a modification of the embryo, that individual differences exist, and that structure and development are broadly determined by inheritance and adaptively modified by natural selection. This approach evokes spontaneous examination of the concept of organic evolution, which, devoid of theories of *how* or dogma of *why*, may be reduced to the axiom that the organisms of the world have been changing and that the present is a reflection of the past.

More than two dozen new illustrations have been added, a considerable number have been redrawn, and a few have been discarded. Most of the new art work was executed by Joan Landry. Fig. 14-29 was drawn by M. S. Salih, University of Baghdad; Fig. 14-23 was sketched initially by A. Ting, St. Edward's University (Texas); and Figs. 3-4, 3-25, and 3-26 were prepared by Sky Bishop. I am indebted to J. T. Bagnara, University of Arizona, for advice on pigment cell terminology and for use of Fig. 5-10.

In the selected readings list more than half the entries are new, and these were chosen partly to encourage sampling of the recent literature. The list continues to cite comprehensive works with extensive bibliographies. A short introductory chapter on the endoskeleton has been derived by extracting topics previously included in other chapters; therefore, the additional chapter has not increased the total number of pages on the skeleton. The sequence of material in some chapters has been rearranged with the inclusion of additional topics such as those on cervical vascular monitors and pit receptors in Chapter 16. Of the new topics, I especially hope that the section in Chapter 3 entitled "Words That Trigger Ideas" will prove stimulating to alert students, even though it will surely evoke dissonant opinions.

The preface to the previous edition documented my indebtedness to many individuals. Their kindness and that of subsequent colleagues and students continue to contribute immeasurably to this work.

George C. Kent, Jr.

Contents

6
Introduction to the skeleton and heterotopic bones, 124

7
Vertebrae, ribs, and sternum, 129

8
The vertebrate skull, 149

9
The appendicular skeleton, 187

10
Muscles, 213

11
Digestive system, 235

12
Respiratory system, 252

13
Circulatory system, 275

14
Urinogenital system, 311

15
Nervous system, 344

16
Sense organs, 376

17
Endocrine organs, 393

Appendix. Abridged classification of the vertebrates, 413

Literature cited and selected readings, 417

Comparative anatomy of the vertebrates

The vertebrate body

A study of comparative vertebrate anatomy is, in a sense, a study of history. It is the history of the struggle of vertebrate animals for compatibility with an ever-changing environment. It is the history of the extermination of the unfit and the invasion of a new territory by those best equipped for combat. It is a study of history, just as is the study of man's conquests, his political fortunes, and his social evolution.

The study of vertebrates is, by definition, a study of man, although not of man alone. It leads to better understanding of man's past and to an assay of his present state. As for predicting the future, a most important, though often neglected, objective of history, the biologist can predict that neither the earth nor that which grows on the earth will remain unchanged. The prediction is based in part on the fact that there has been a succession of animals and plants on the earth and that the species of today are not the same species that would have been seen 300 million years ago. On the basis of probability, it can be predicted that they will be still more different tomorrow. Since there has been a succession of species, and since all life seems to have come from preexisting life for quite some time, logic tells us that the species have been changing. This phenomenon is the premise of the discipline.

The discipline has an important func-tion. It is not that of promoting the premise. It is, instead, one of continually seeking additional facts, of finding new inter-relationships, of periodically reevaluating our tentative conclusions, and of drawing new ones. When comparative anatomy ceases to be a search for the truth, it will have surrendered its status as a science and will have become a body of meaningless facts. The facts are important, but their meaning is immeasurably more so. To the study of no discipline is the dictum of *Proverbs*, ". . . in all thy getting, get under-standing," more applicable.

The premise that the vertebrate species have been changing is strengthened by the observation that all vertebrates, past and present, are constituted in accordance with a basic architectural pattern. It is the purpose of this book to assist the reader to discover this pattern and to show in what direction, in successively more recent generations, the pattern has been modified. The modifications appear in all the body systems—integumentary, skeletal, muscular, digestive, respiratory, circulatory, urogenital, nervous, and en-docrine.

GENERAL BODY PLAN

The body is typically divided into a head, trunk, and tail. Paired appendages are associated with the trunk in all but a few vertebrates. A neck develops between

1

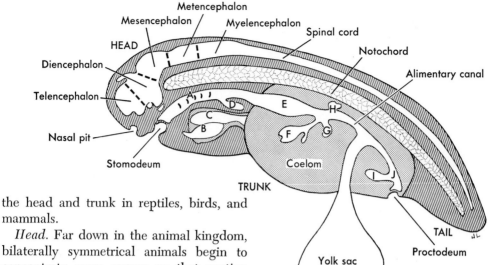

Fig. 1-1. Sagittal section of vertebrate embryo illustrating basic pattern of vertebrate structure. **A,** Third visceral arch of pharynx, lying between the second and third visceral slits; **B** and **C,** ventricle and atrium of heart; **D,** diverticulum, giving rise to the lung in tetrapods and swim bladder in fishes; **E,** stomach; **F,** liver bud and gallbladder below; **G,** ventral pancreatic bud; **H,** dorsal pancreatic bud; **I,** urinary bladder of tetrapods; **J,** cloaca. The stomodeum is separated from the digestive tract by a thin oral plate. The proctodeum is separated from the cloaca, **J,** by a cloacal membrane. The brain has five major subdivisions: telencephalon and diencephalon (forebrain), mesencephalon (midbrain), and metencephalon and myelencephalon (hindbrain).

the head and trunk in reptiles, birds, and mammals.

Head. Far down in the animal kingdom, bilaterally symmetrical animals begin to concentrate sense organs on that portion of the body which first enters a new milieu. Many invertebrates "creep up" on their environment. An earthworm goes a short distance forward, tests the new environment carefully with sense organs at the cephalic end, and, if the environment is friendly, proceeds deeper and deeper into the area ahead. If the environment appears inimical to the welfare of the earthworm, it withdraws. Thus, an advantage is gained by increasing the number of testing devices at the anterior end of the body. Concentration of sense organs in the cephalic region of vertebrates is accompanied by an increase in the size of the brain. Cartilage and bone develop for the protection of brain and sense organs. The mouth and its associated jaws and the respiratory mechanisms add to the complexity of the anterior end. The concentration of numerous mechanisms in the anterior end of the body is known as **cephalization.**

Trunk (Figs. 1-1 and 1-2). The trunk is the part containing the body cavity (**coelom**). Surrounding the cavity is the body wall, covered by skin on its outer surface and by **parietal peritoneum** on its inner surface. In addition to skin and parietal peritoneum, the body wall consists chiefly of muscle, vertebral column, and ribs and sternum when present. The body wall must be incised in order to expose the visceral organs. The latter are covered by the **visceral peritoneum,** which is continuous with the parietal peritoneum via **dorsal** and **ventral mesenteries.** Those few visceral organs that do not have mesenteries lie against the body wall just external to the parietal peritoneum. In such a location an organ is said to be **retroperitoneal.**

Most of the digestive organs are suspended within the coelom. The heart oc-

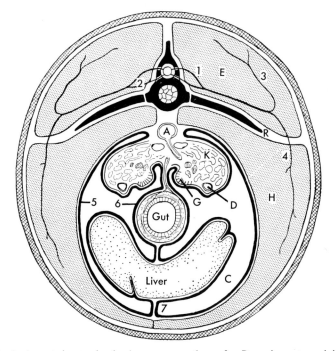

Fig. 1-2. Typical vertebrate body in cross section. **A,** Dorsal aorta, giving off renal artery to kidney; **C,** coelom, lying between **5** and **6**; **D,** kidney duct; **E,** epaxial muscle; **G,** gonadal ridge; **H,** hypaxial muscle and body wall; **K,** kidney (nephros); **R,** rib or transverse process of vertebra. **1,** Dorsal root of spinal nerve; **2,** ventral root; **3,** dorsal ramus of spinal nerve; **4,** ventral ramus; **5,** parietal peritoneum; **6,** visceral peritoneum; **7,** ventral mesentery. A remnant of the notochord lies within the centrum of a vertebra (immediately dorsal to **A**). The spinal cord (dorsal, hollow nervous system) lies above the centrum surrounded by the neural arch.

cupies the **pericardial cavity,** an exclusive subdivision of the coelom. Lungs, when present, occupy a portion of the coelom. Kidneys and their ducts, on the contrary, are retroperitoneal, but the urinary bladder usually bulges into the coelom. Gonads and their ducts arise in embryos as retroperitoneal structures but often succeed in bulging into the coelomic cavity or even in developing a dorsal mesentery during later life. Passing caudad in the roof of the coelomic cavity is the large dorsal aorta from which pass, via mesenteries when necessary, arteries supplying the various organs. The major venous channels of the trunk utilize the body wall or the mesenteries to course forward toward the heart.

The neck (**cervix**) is a narrow anterior

Fig. 1-3. Locomotion in a fish.

elongation of the body wall of the trunk in which the vertebrae are different from trunk vertebrae. Fishes have no cervical (neck) vertebrae, and amphibians have only one. The coelomic cavity does not extend into the cervical region. In addition to the vertebral column, the neck consists primarily of muscle, nerves, and elongated tubes (esophagus, long arteries, veins, and trachea) connecting the head and the trunk.

Tail. The tail is an extension of the body wall behind the caudal opening of the digestive tract. Like the neck, it contains no coelom. In fish and tailed amphibians it is used chiefly for locomotion (Fig. 1-3). The tail of all vertebrates consists almost

exclusively of a caudal continuation of the body wall muscles for power, of the nervous system for innervation, of the vertebral column for muscle support, of the caudal artery for arterial blood, and of the caudal vein for vascular drainage. Frogs and toads exhibit a locomotor tail prior to meta-

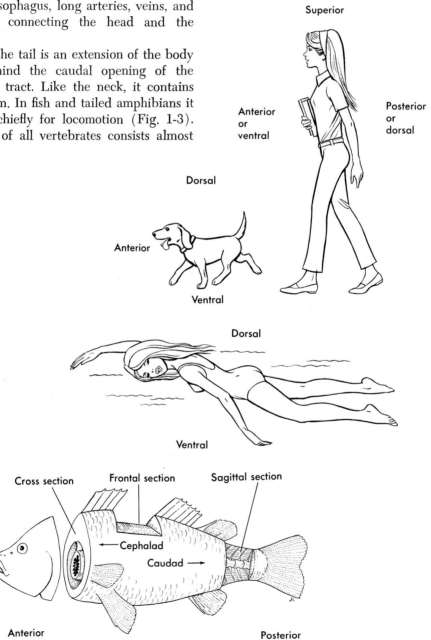

Fig. 1-4. Terms of direction and position and planes of sectioning of the vertebrate body.

morphosis when, as water-dwelling larvae, they need it most. Reptiles and mammals exhibit tails that are not always important as locomotor organs. Modern birds have reduced the tail region to a nubbin, but the earliest known birds had long tails. Mammals exhibit elongated, prehensile tails (monkeys), foreshortened tails (hamsters), flyswatters (cattle), balancers (squirrels), and organs of defense (porcupines), all of which help keep these animals from coming to an unnecessary end. Even man exhibits a tail in early embryonic life (Fig. 1-10). Vestiges may be observed on any human skeleton as the lowest three or four caudal (coccygeal) vertebrae.

Appendages. The vertebrates are subdivided into two groups, **Pisces** and **Tetrapoda,** according to the structure of the paired appendages. Pisces usually exhibit pectoral and pelvic fins. Tetrapods (amphibians and all higher vertebrates) typically exhibit jointed limbs. The internal skeleton of limbs varies only in minor details in highly specialized appendages such as the wings of birds, the flippers of whales, and the hands of man.

Bilateral symmetry. Vertebrates exhibit three principal body axes: an anteroposterior (longitudinal) axis, a dorsoventral axis, and a left-right axis. With reference to the first two, structures found at one end of each axis are different from those at the other end. The third axis terminates in identical structures on each side. Thus the head differs from the tail and the dorsum differs from the venter, but right and left sides are mirror images of each other. An animal with this arrangement of body parts exhibits bilateral symmetry.

It is sometimes convenient to discuss parts of the vertebrate body with reference to three principal anatomical planes. Two body axes establish an anatomical plane. The **transverse plane** is established by the left-right and the dorsoventral axes. A cut in this plane is a cross section (Fig. 1-4). The **frontal plane** is established by the left-right and longitudinal axes. A cut in this plane is a frontal section. The **sagittal plane** is established by the longitudinal and dorsoventral axes. A cut in this plane is a sagittal section. Sections parallel to the sagittal plane are parasagittal sections. Acquainting oneself with these concepts constitutes a simple exercise in anatomy and logic.

VERTEBRATE CHARACTERISTICS: THE BIG FOUR

We may now consider the four definitive structural characteristics of vertebrates: (1) the presence of a notochord, at least in the embryo, (2) the presence of a pharynx with pouches or slits in its wall, at least in the embryo, (3) the occurrence of a dorsal, tubular nervous system, and (4) the development of vertebrae. These are the "big four" vertebrate characteristics. The first three are chordate characteristics and are found also in protochordates. Other features typically associated with vertebrate anatomy, but not necessarily unique among vertebrates, will be discussed as satellite characteristics.

Notochord

The notochord at its peak of development is a flexible, longitudinal endoskeletal rod of turgid living cells located immediately ventral to the central nervous system and dorsal to the alimentary canal (Fig. 1-1). The location is explained partly by its origin from the roof of the primitive gut as illustrated in Figs. 4-6 and 4-8. In embryonic vertebrates the notochord begins under the midbrain and extends to the tip of the tail. During embryonic life the part of the notochord in the head becomes embedded in the floor of the skull, and the part in the trunk and tail becomes surrounded by cartilaginous or bony rings called **vertebrae** (Fig. 7-6). The vertebrae usually compress the notochord, often interrupt it metamerically, and may even completely obliterate it.

In most fishes and aquatic amphibians the adult notochord is constricted within each vertebra (Figs. 7-5 and 7-7), although there are living gnathostome fishes in which it is not constricted (Fig. 7-8). In frogs and toads the notochord is practically obliterated at metamorphosis.

Fossil reptiles that gave rise to later reptiles, birds, and mammals had complete adult notochords that were constricted within each vertebra. This remains true in "living fossil" reptiles like *Sphenodon*. However, in modern reptiles and mammals the notochord is almost obliterated. A vestige remains in mammals within the intervertebral discs separating successive centra. The vestige consists of a soft spherical mass of connective tissue called the **pulpy nucleus.** Modern birds lack even this vestige.

In protochordates and agnathans the notochord has a different fate. It continues to grow as the animal grows and never becomes surrounded by vertebrae. There-fore, it remains throughout life as the chief axial skeleton. Also, in amphioxus the notochord extends forward beyond the midbrain, and even beyond the anterior end of the brain (Fig. 2-4). In urochordates the notochord is confined to the tail, and it disappears at metamorphosis when the tail is resorbed (Fig. 2-2).

It is apparent that the notochord has been disappearing as an adult structure in recent vertebrates. But the development of a notochord in every vertebrate embryo—even the embryo of man—is a reminder that all vertebrates are built in accordance with a basic architectural plan.

The notochord is surrounded by one or more sheaths apparently secreted by the outer layer of notochord cells. The outermost sheath (**elastica externa**) has elastic fibers in its matrix. The sheaths are thick

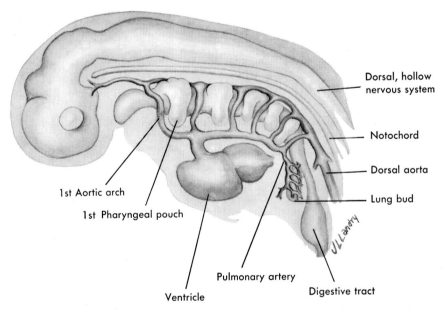

Fig. 1-5. Basic pattern of pharyngeal architecture as exemplified by a composite verte-brate embryo. The notochord lies ventral to the nervous system and extends forward as far as the midbrain. A series of pharyngeal pouches have evaginated from the lateral walls of the digestive tract. Six aortic arches connect the heart and ventral aorta with the dorsal aorta. (Typically, the first aortic arch disappears before the sixth one has formed.) The anterior end of the dorsal, hollow nervous system is enlarging to form the brain. Although a lung does not form in all vertebrates, it is an ancient structure and is repre-sented by a swim bladder in most fishes.

in those forms in which the notochord plays an important skeletal role throughout life.

Pharynx

The pharynx is the region of the alimentary canal exhibiting pharyngeal pouches in the embryo (Fig. 1-5). The columns of tissue lying between the pouches, as well as those in front of the first pouch and behind the last are pharyngeal arches (Figs. 1-6 and 1-7). The pouches may rupture through to the exterior to form pharyngeal slits. These slits may remain throughout life, or they may be temporary. If they remain throughout life, as is the case in animals that have gills, the adult pharynx is that part of the alimentary canal having the slits. If the slits are temporary, as in animals that finally breathe by lungs, the adult pharynx is that part of the alimentary canal connecting the oral cavity and esophagus.

Pharyngeal pouches and slits. The basic plan of the vertebrate pharynx is illustrated in all vertebrate embryos. A series of paired pharyngeal (or visceral) pouches arise as diverticula of the pharyngeal endoderm (Figs. 1-5 to 1-7). The pouches invade the pharyngeal wall and grow toward the surface of the animal. Simultaneously, an **ectodermal groove** grows toward each pharyngeal pouch (Figs. 1-6 and 1-7). Soon, only a thin **branchial plate** of ectoderm and endoderm separates the ectodermal groove from the pharyngeal pouch. When the branchial plate ruptures, as it usually does, a passageway is formed between the pharyngeal lumen and the exterior. This embryonic passageway is a pharyngeal (or visceral) slit. The slits may be temporary or permanent.

The pharyngeal slits are permanent as long as the animal lives in water and breathes by gills (Fig. 1-8). In dogfish

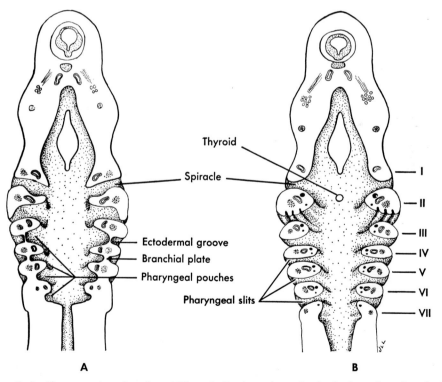

Fig. 1-6. Pharyngeal arches (**I** to **VII**) and slits in embryonic shark, frontal section, looking down onto floor of pharynx. **A,** Earlier stage. **B,** Appearance at later stage.

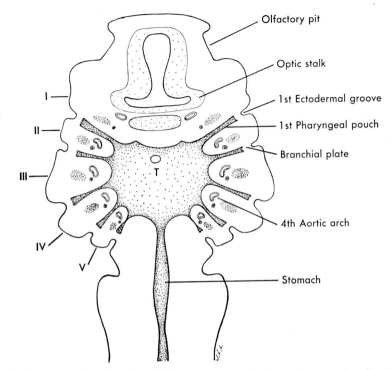

Fig. 1-7. Frontal section of embryonic frog pharynx, showing pharyngeal arches (**I** to **V**), pharyngeal pouches, and ectodermal grooves. **T,** Thyroid evagination.

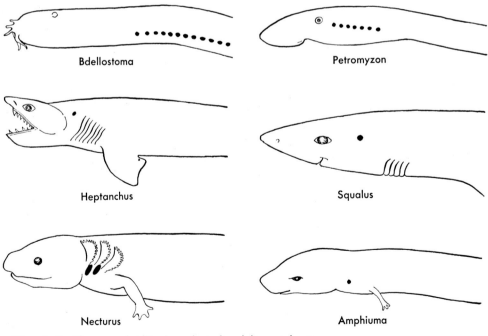

Fig. 1-8. Pharyngeal slits in selected adult vertebrates.

shark embryos six pharyngeal pouches form and all six rupture to form slits. In the walls of the last five pouches highly vascularized gill surfaces develop and the adult pouches are therefore **gill pouches or gill chambers.** The opening to the exterior is an **external gill slit.** In the wall of the first embryonic pouch an abortive gill surface (**pseudobranch**) develops and the passageway to the exterior becomes the **spiracle.** In some fishes more than six pharyngeal pouches develop in the embryo. For example, *Heptanchus (hept =* seven), another shark, has eight embryonic pouches, the last seven of which become gill pouches. Agnathans may have as many as fifteen pouches. Modern bony fish usually have five embryonic pouches, the last four of which are associated with gills.

The pharyngeal slits that form in the embryo are temporary if the animal is going to live on land and breathe by means of lungs. In reptiles, birds, and mammals no gill surfaces develop in the pharyngeal pouches, and the pharyngeal slits are transitory. Of the five pouches that develop in chicks, only three rupture to the exterior, and these close again. Only one or two of the more anterior pouches of mammals may rupture. Embryologists tell us that the cervical fistulas occasionally seen in newborn human infants (Fig. 1-9) are usually the result of the failure of the second pharyngeal slit to close. In amphibians, gill slits are present in the aquatic larvae, but at metamorphosis, when the animal ceases to respire by gills, the slits typically close. Some tailed amphibians do not metamorphose fully, and in these one or more slits may remain open (Fig. 1-8, *Necturus, Amphiuma*). Of the six pharyngeal pouches that form in frog embryos, four give rise to functional gill slits in tadpoles. These slits close permanently when the tadpole metamorphoses into a frog.

Although the pharyngeal pouches of tetrapods do not give rise to respiratory structures or to permanent slits, the first one becomes the eustachian tube and mid-

Fig. 1-9. Cervical fistula resulting from persistent pharyngeal slit.

dle ear cavity, and the second in mammals becomes the pouch of the palatine tonsils. The endodermal walls of one or more pouches give rise to thymus tissue in all vertebrates, and in terrestrial forms the walls of several pouches give rise to parathyroid glands.

The formation of at least transient pharyngeal pouches and slits in all vertebrate embryos supports the concept that all vertebrates are related, through eons of time, to ancestral stock from which they inherited common genetic codes.

Pharyngeal arches. Each embryonic pharyngeal pouch or adult pharyngeal slit is separated from the next by a column of tissue called a pharyngeal (visceral) arch (Figs. 1-6, 1-7, and 1-10). Each pharyngeal arch typically contains certain basic structures, including (1) a visceral cartilage or its replacing bone, or preskeletal mesenchyme (Fig. 8-1), (2) branchiomeric muscle (Fig. 10-10), (3) branches of certain cranial nerves, and (4) a blood vessel (aortic arch, Fig. 1-5), which, at least in the embryo, directly connects the ventral and dorsal aortas. These basic components are also found in front of the first pouch and directly behind the last. Therefore, a pharyngeal arch may be more fully characterized as one of a series of columns of pharyngeal tissue located between successive pharyngeal pouches or slits, as well as in front of the first pouch or slit and

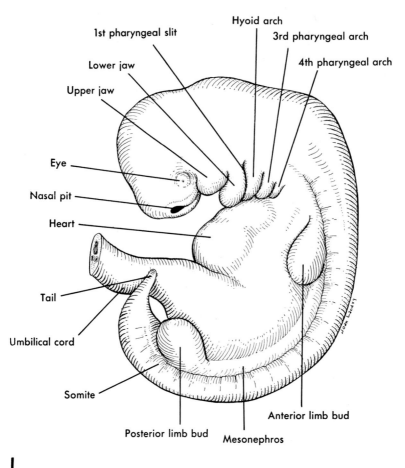

Fig. 1-10. Typical vertebrate embryo at 5 mm. stage. Drawn from a human embryo approximately 4½ weeks after fertilization.

immediately behind the last, covered externally by ectoderm and internally by endoderm, and containing branchiomeric muscle, cranial nerves, an aortic arch, and elements of the pharyngeal skeleton.

The upper and lower jaws and associated muscles, nerves, and vessels in front of the first pouch or slit constitute the **mandibular arch.** The second, or **hyoid arch,** is directly behind the first pouch or slit. The remaining pharyngeal arches are referred to only by number. Dogfish sharks have seven pharyngeal arches (Fig. 1-6, *B*). Tetrapods usually have four or five complete arches and parts of a sixth.

The boundaries of a visceral arch can be determined by inspection from the exterior when there are ectodermal grooves or pharyngeal slits to serve as landmarks, and then only. After the grooves disappear or the slits close, the components of the arches become reoriented. In most tetrapods, therefore, pharyngeal arches are anatomical entities in embryos only.

It is evident that the primitive vertebrate pharynx was a device for filtering food out of a respiratory water stream and that in lower vertebrates it still is. The structural and functional alterations that occurred in the pouches, skeleton, muscles, vessels, and nerves of the pharynx as vertebrates shifted phylogenetically from gills

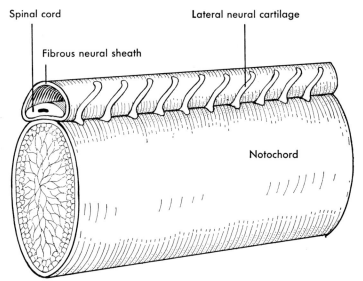

Fig. 1-11. Lateral neural cartilages of the lamprey.

to lungs is one of the fascinating chapters of vertebrate evolution. These adaptive changes are documented in the ontogeny of tetrapods and will be described in later chapters.

Dorsal, hollow central nervous system

The central nervous system in vertebrates consists of a brain and a spinal cord and contains a central cavity. Dorsal, hollow central nervous systems are found only among chordate animals. Their location and the presence of a cavity result from the fact that the central nervous system arises as a longitudinal neural plate in the dorsal ectoderm, which subsequently folds over and becomes enclosed within the dorsal body wall, to form a hollow tube (Figs. 4-7 and 15-3).

Nerves connect the central nervous system and the various organs of the body. The nerves, along with associated ganglia and plexuses, constitute the peripheral nervous system. The spinal nerves of most vertebrates are metameric, arising at the level of each body segment and passing to the skin and muscles of that segment and to the viscera. The metameric arrangement of spinal nerves in vertebrates is evident in Fig. 15-4. Ten cranial nerves arise from the brain in fish and amphibians and twelve in reptiles, birds, and mammals. Evidence will be presented later indicating that the extra two nerves in higher vertebrates are spinal nerves that became "trapped" within the skull.

Vertebrae

The vertebral column, often inappropriately called "the spine," is the salient characteristic of vertebrate animals. Vertebrae are independent, axial skeletal elements that provide more rigidity than does a notochord, at the same time permitting a considerable degree of flexibility (Figs. 7-10 and 7-20). Nevertheless, successive vertebrae sometimes ankylose, to form an inflexible column as in some fish, turtles, and birds (Fig. 7-16).

A vertebra typically consists of a centrum surrounding the notochord, of various processes for muscle attachment and for restricting the flexibility of the column, and of a neural arch protecting the spinal cord (Figs. 7-1, 7-3, and 7-7). In the tail a hemal arch, resembling an inverted neural arch, often protects the caudal blood vessels. Intervertebral elements are frequently interposed between successive vertebrae (Fig. 7-2).

Agnathans lack typical vertebrae. Indeed, whether they have vertebrae at all is a matter of definition. In lampreys a series of paired lateral neural cartilages are perched upon the notochord lateral to the spinal cord (Fig. 1-11). These cartilages are reminiscent of neural arches, but whether they are rudimentary vertebrae that fail to develop, vestigial vertebrae from a hypothetical ancestor that had a typical vertebral column, or whether they are entirely different structures is not known. In hagfishes lateral neural cartilages are limited to the tail.

SATELLITE CHARACTERISTICS
Skin

The integument of vertebrates consists of an outer epidermis and an underlying dermis (Fig. 5-4). Embedded within the skin are many varieties of structures, including bony armor, spines, scales, feathers, hairs, nails, horns, hoofs, and claws. Many types of glands develop from the skin and open on the surface (Fig. 5-1). The skin is modified locally to form transparent membranes such as the conjunctiva of the eye, mucous membranes of the lips, respiratory surfaces of the gills, and light organs in deep water fishes.

Metamerism

Serial repetition of body structures in the longitudinal axis is known as metamerism. In earthworms metamerism is apparent at a glance, the successive body segments being clearly delineated externally as well as internally. In crayfish the metamerism is obvious in the caudal region, but in the cephalothorax it is obscured by the carapace dorsally. Internally, the metamerism of earthworms and crayfish is expressed in many systems.

Vertebrates, too, exhibit a basic metamerism that is clearly expressed in embryos and is retained in numerous adult systems. Yet no external evidence of metamerism is seen because the skin is not metameric. If, however, the integument is stripped from the body, a series of identical muscle segments is revealed in most vertebrates except birds and mammals (Fig. 10-3). And only a glance at the embryo (Fig. 4-7) of a bird or mammal will reveal the fundamental metamerism of these vertebrates. The serial arrangement of adult ribs, vertebrae, spinal nerves, kidney tubules, and many blood vessels and muscles is an expression of the fundamental metamerism of vertebrates.

Respiratory mechanisms

Chordates carry on respiration by means of highly vascularized membranes derived from the pharyngeal wall (gills) or floor (lungs). Internal gills are situated in gill pouches opening to the exterior via gill slits. Only in protochordates do the slits open into an atrium surrounding the pharynx, with an unpaired atriopore leading to the exterior. The external gills of larval fish and amphibians typically develop as outgrowths of the surface of a pharyngeal arch and wave about in the water (Fig. 12-6).

Lungs arise from a midventral evagination of the pharyngeal floor. The resulting endodermal sac pushes into the coelomic cavity but remains connected with the pharynx by an air tube.

In addition to gills and lungs, vertebrates sometimes carry on respiration by other devices such as skin, buccopharyngeal lining, and (during embryonic life) special extraembryonic membranes. These organs will be discussed in later chapters.

Coelom

Like many other animals, vertebrates have a coelomic cavity between the body wall (somatopleure) and digestive tube (splanchnopleure). The coelom is subdivided in fishes, amphibians, and many reptiles into a pericardial cavity housing the heart and a pleuroperitoneal cavity housing most of the other viscera, including the lungs (Fig. 1-12). The pericardial and pleuroperitoneal cavities are separated by a membranous transverse septum. In some reptiles and in birds and mammals

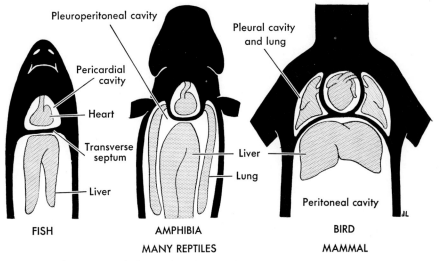

Fig. 1-12. Subdivisions of the coelom in vertebrates.

the lungs become isolated in pleural cavities that are subdivisions of the pleuroperitoneal cavity. The transverse septum is then supplemented by other septa, which may be muscular. In some male mammals, outpocketings of the coelom house the testes, and these scrotal cavities must be counted as a fourth subdivision of the coelom.

Digestive organs

The alimentary tract exhibits a series of specialized regions for the acquisition, processing, temporary storage, digestion, and absorption of food and for the elimination of the unabsorbed residue. Typical are the oral cavity, the pharynx, which is also respiratory, the esophagus, which is as long as the neck, the stomach, and the intestine. The last is often coiled, thereby increasing the absorptive area without increasing the body length. The tract exhibits a number of embryonic or adult ceca or diverticula, including the liver and pancreas.

The terminal segment of the digestive tract in all but a few vertebrates is the cloaca, which opens to the exterior via the vent. Mammals develop an embryonic cloaca, but it is later subdivided by partitions so that adults lack a cloaca. The

intestine then opens independently to the exterior via an anus.

Urinogenital organs

The urinary and reproductive organs of vertebrates are closely interrelated. Their ducts arise in adjacent regions of the mesoderm, and the two systems use common passageways.

The kidneys are the chief organs for regulating the elimination of excess water. They also assist in maintaining an appropriate electrolyte balance. In primitive vertebrates fluid and metabolic wastes were removed from the coelomic cavity by microscopic kidney tubules resembling somewhat the nephridia of earthworms. In most vertebrates, however, the substances to be excreted are collected by the tubules directly from the blood. The tubules transmit the fluids to a pair of longitudinal ducts that pass caudad and empty into the cloaca or urinary bladder.

The reproductive organs include gonads, ducts, and accessory organs such as shell glands, storage chambers, and copulatory mechanisms. Early in development all vertebrate embryos have gonadal and duct primordia for both sexes. When the animal is genetically constituted to become a female, the gonad primordia develop into

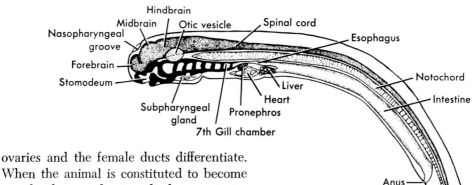

Fig. 1-13. Ammocoete larva, early stage.

ovaries and the female ducts differentiate. When the animal is constituted to become a male, the gonad primordia become testes and the male ducts grow larger. The duct system associated with the opposite sex usually disappears. Spermatozoa pass from the testes via sperm ducts, but the eggs are usually extruded into the coelomic cavity before entering the female tract. Cyclostomes lack reproductive ducts, and the sperm and eggs pass into the coelomic cavity and then to the exterior via pores in the caudal body wall.

Circulatory system

Whole blood is confined to arteries, veins, capillaries, and sinusoids. A single, ventral, multichambered, muscular pumping organ, the heart, is located under or caudal to the pharynx. Blood flows from the heart in a ventral aorta and then through aortic arches into the dorsal aorta. The latter conducts the blood caudad. It is to the circulatory system laden with vital oxygen that vertebrates owe their moment-to-moment existence. Vertebrates also have a lymphatic circulatory system.

Although almost all adult vertebrates have blood cells containing the red respiratory pigment hemoglobin, at least three species of fishes have no cells resembling erythrocytes and no demonstrable blood pigment. These are *Champsocephalus gunnari, Pseudochaenichthys georgianus,* and *Chaenocephalus acertus.* They are native to the waters of South Georgia Island in the South Atlantic.

Sense organs

Vertebrates have a wide variety of sense organs that inform them about the con-

stantly changing environment in which they live and about conditions within the body. Certain specialized sense organs are characteristic of vertebrates. These and other receptors will be discussed in Chapter 16.

LARVAL LAMPREY (Figs. 1-13 and 1-14)

Some of the structural features that characterize vertebrates have been discussed. We may now turn to a brief description of the ammocoete, a larval cyclostome that embodies many of the basic architectural features of vertebrates in simple form.*

External features. The body is divided into a head, trunk, and a tail that commences at the anus. Median dorsal and ventral fins extend nearly the length of the trunk and tail. The metameric muscles of the trunk and tail are visible through the skin. In older larvae the most anterior structure of the head is the oral hood, the side walls and roof of which encompass a chamber, the buccal funnel (stomodeum). The funnel is open anteriorly and ventrally. Guarding the mouth are numer-

---

Ammocoetes was once a generic name applied to larval lampreys when they were erroneously believed to be adult animals. After their true status as larvae was discovered, the term "ammocoete" was retained to designate the larva. The ammocoete phase lasts from two to six years.

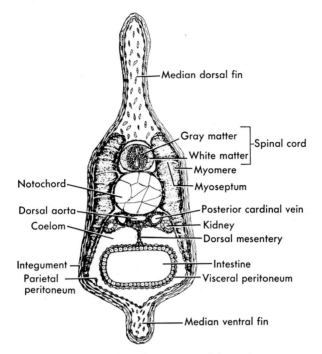

Fig. 1-14. Cross section of a larval cyclostome caudal to pharynx.

ous fleshy, finger-like oral papillae. On the surface of the head just above the anterior end of the brain a nasopharyngeal groove denotes the location of the single external naris. Several series of pores occur on the sides of the head. These are openings of neuromast organs. Seven gill slits occur on each side in the pharyngeal region.

In young larvae the undeveloped oral hood lies under the anterior end of the brain. The buccal funnel is smaller and opens directly ventrad, and the oral papillae are shorter and thicker. The tail region of young larvae is short.

Integument (Fig. 5-11). The epidermis is stratified and richly supplied with unicellular glands. The dermis is thin and consists chiefly of connective tissue. No hardened structures of any kind develop in the skin.

Skeleton. The notochord is the sole axial skeletal structure at this stage. The tapering anterior end underlies the hindbrain. It passes caudad to terminate at the tip of the tail. The notochord is surrounded

by a thick inner fibrous sheath and an outer elastic sheath.

Central nervous system. The central nervous system consists of the brain and spinal cord. The brain consists of three vesicles— the prosencephalon, or forebrain; the mesencephalon, or midbrain; and the rhombencephalon, or hindbrain. The latter merges imperceptibly with the spinal cord, which continues to the tip of the tail.

Special sense organs. From the external naris a nasal canal slants downward and backward to open into a median olfactory sac. The latter is probably derived from a pair of fused sacs, since paired olfactory nerves pass to the forebrain. From the olfactory sac the nasal canal continues downward and backward, ending blindly in a nasopharyngeal sac.

The paired eyes are covered by opaque skin and are of little use to the larva. Two median eyes extend upward toward the skin from the roof of the forebrain. These are the pineal and parapineal bodies. At their distal ends just under the skin are a simple lens and receptor cells. Nerve fibers connect the median and paired eyes with the forebrain.

Otic vesicles lie lateral to the hindbrain above the first gill chamber. They will later develop into a simple membranous labyrinth, or inner ear. A neuromast system opens by pores on the side of the head and is connected to the brain by afferent nerves.

Body wall musculature. The body wall musculature consists of a series of identical muscle segments (myomeres) extending the entire length of the body and separated by connective tissue partitions (myosepta) to which the muscle fibers attach. The myomeres constitute the chief mass of the body wall.

Coelom. The pericardial cavity containing the heart is broadly open to the main coelom in ammocoetes, although it becomes a separate cavity in the adult lamprey. The coelom also contains the esophagus, intestine, and liver.

Digestive tract. The buccal cavity terminates at the velum, which consists of two fleshy folds suspended like curtains between the buccal cavity and pharynx. The velum causes a current of water to pass into the pharynx. The pharynx is that part of the digestive tract exhibiting gill slits. At its caudal termination the pharynx is continuous with the esophagus, a short muscular tube of small diameter leading directly into the uncoiled intestine. There is no stomach. From the anteroventral floor of the intestine a large diverticulum, the liver, extends forward to a position immediately behind the heart. Embedded within the liver is a gallbladder. The liver undergoes fatty degeneration, and the gallbladder disappears during metamorphosis. The intestine terminates at the anus. There is no cloaca. Food particles in the incurrent stream of water are ensnared by mucus secreted by the pharyngeal epithelium. The food moves caudad in the pharyngeal chamber, into the esophagus, and thence to the intestine.

A complicated subpharyngeal gland lies below the pharynx and opens into it by a short duct (Fig. 17-10). Many of the cells of the subpharyngeal gland are ciliated, and some accumulate iodine. The subpharyngeal gland is therefore a primitive thyroid gland. The significance of its complicated structure, much of which degenerates at metamorphosis, is unknown.

Respiratory mechanisms. The pharyngeal wall exhibits seven gill pouches. Each pouch opens to the exterior via a small gill slit.

Circulatory system. The heart lies in the pericardial cavity just caudal to the pharynx. It has a sinus venosus, atrium, and ventricle. Blood is pumped forward by the ventricle into the ventral aorta. This vessel is unpaired near the heart but is paired farther cephalad. From the ventral aorta, blood passes via afferent branchial arteries to the capillary beds of the gill. Oxygenated blood then passes via efferent branchial arteries into the dorsal aorta, which courses caudad. The aorta supplies branches to all the organs of the body.

Venous blood returns from the head via a pair of large anterior cardinal veins and from the trunk and tail by two large posterior cardinal veins lying lateral to the dorsal aorta. Anterior and posterior cardinal veins flow into common cardinal veins that course mediad, to empty into the sinus venosus. Blood returns from the intestine via a ventral intestinal vein that ends in the capillaries of the liver. The ventral intestinal vein and its tributaries therefore constitute a hepatic portal system. From the capillaries of the liver, blood proceeds forward in a hepatic vein to the sinus venosus. The sinus venosus enters the atrium, which leads to the ventricle. There is no renal portal system.

Urogenital organs. A primitive kidney embedded in a mass of fat occurs retroperitoneally just behind the head region. It consists initially of three to six tubules, each of which drains the coelom by a peritoneal funnel, collecting liquid wastes secreted by vascular tufts (glomeruli). Subsequently, additional tubules are added farther back. From each kidney a duct passes caudad. The ducts terminate in a median urogenital papilla that opens to the exterior behind the anus. Medial to the kidneys lies an elongated gonadal ridge, which is paired in early larvae but unpaired in older ones. The ridge is the rudiment of the future unpaired testis or ovary.

Chapter summary

1. Vertebrates have a notochord, a dorsal, hollow nervous system, and a pharynx with pharyngeal pouches in the embryo. In addition, except for agnathans, they have cartilaginous or bony vertebrae consisting of centra, arches, and processes.

2. The vertebrate body is divided into head, trunk (elongated anteriorly as a neck in amniotes), and tail. The tail is characteristic of all vertebrate embryos and appears to have functioned primitively for locomotion in water. Two pairs of lateral appendages (fins in fishes, limbs in tetrapods) are typical.

3. The notochord is the most primitive skeletal structure of vertebrates. In agnathans it is the sole axial skeleton and is surrounded by thick sheaths capped by neural cartilages. In other fishes and in primitive tetrapods the notochord becomes surrounded by cartilaginous or bony vertebrae, and is constricted within each centrum. In modern tetrapods the notochord is usually obliterated during embryonic life except for intervertebral vestiges.

4. The nervous system consists of a central nervous system (brain and spinal cord) and of a peripheral nervous system composed of nerves and associated ganglia and plexuses. The central nervous system is dorsal and hollow because it typically arises as a dorsal ectodermal groove that sinks into the dorsal body wall to form a tube. There is a wide array of general and special sense organs.

5. The pharynx exhibits a series of pharyngeal pouches that rupture to the exterior to form temporary or permanent pharyngeal slits in most species. The slits serve as gill slits in fishes and most aquatic amphibians. They are evanescent in other tetrapods.

6. In front of and behind each embryonic pharyngeal pouch or adult gill slit lies a pharyngeal arch containing a skeletal element for support, branchiomeric muscle for moving the arch, nerves for the innervation of the muscles and sense organs, and an aortic arch connecting the ventral and dorsal aortas. Pharyngeal arches are readily distinguishable as long as ectodermal grooves or pharyngeal slits remain. They are not identifiable externally after the pharyngeal slits close.

7. Vertebrates are metameric with respect to numerous systems. They have a two-layered skin. They respire by gills or lungs, sometimes assisted by an array of accessory respiratory organs, including the skin and, in embryos, extraembryonic membranes.

8. There are at least two coelomic chambers (pericardial and pleuroperitoneal cavities) and sometimes four (pleural, pericardial, peritoneal, and scrotal cavities).

9. The digestive tract consists of oral cavity, pharynx, esophagus, stomach, intestines, cloaca or its derivatives, and numerous small or large glandular or nonglandular diverticula.

10. The urinogenital system includes kidneys (nephroi), gonads, ducts, and accessory reproductive organs. Kidneys typically consist of vascular glomeruli and of microscopic tubules that collect and transport kidney secretions to a pair of longitudinal ducts.

11. The blood circulatory system includes a ventral, multichambered, muscular heart, arteries, veins, capillaries, and, in restricted locations, sinusoids. A lymphatic circulatory system is also present.

12. The larval lamprey (ammocoete) has been described. This organism illustrates many of the basic architectural features of vertebrates in a primitive state.

Chapter
2
Protochordates

No discussion of the vertebrates would be complete without mention of their closest relatives, the protochordates. There are two subphyla of protochordates: the **Urochordata** (or **Tunicata**), which include sea squirts (Fig. 2-1) and certain other forms, and the **Cephalochordata,** or amphioxus (Fig. 2-3). The sessile sea squirt, dully squirting water out of its excurrent siphon, seems a far cry from even the simple little amphioxus, not to mention the lowest jawless fishes. Yet, protochordates and vertebrates have three characteristics in common—characteristics shared by no other organisms: dorsal, hollow nervous system, pharyngeal slits, and notochord. These three morphological traits occur in the embryos, at least, even if one or more are not always retained in the adults. The sharing of these characteristics impels us to conclude that vertebrates are more closely related to protochordates than to any other group of living organisms.

UROCHORDATES

The name Urochordata (*uro* = tail) indicates that the notochord is confined to the tail. The synonym Tunicata refers to the tough tunic that surrounds the living organism. The tunic contains a cellulose-like substance, tunicin. It is delicate in appearance, often beautifully colored, and usually transparent. All tunicates live in the sea, and many are colonial. They ex-

hibit a free-swimming larval stage, but many adults are sessile. There are three groups of tunicates—larvaceans, thaliaceans, and sea squirts.

Larvaceans are tiny, transparent tunicates that do not metamorphose, but remain larval throughout life. They live near the surface of the sea among the plankton, which is their food. They have a single

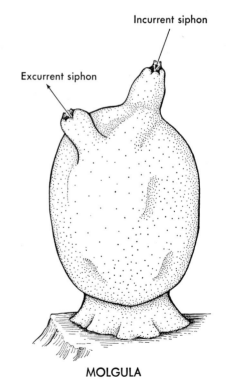

Fig. 2-1. Sea squirt.

pair of gill slits opening directly to the exterior. The larvae are hermaphroditic, but the sperm mature before the eggs so that there is no self-fertilization. The condition in which larval organisms attain sexual maturity and reproduce in the larval state is known as **paedogenesis.**

Thaliaceans are cylindrical tunicates lacking a tail. They have, therefore, another method of locomotion. Through the transparent tunic may be seen bands of muscle tissue. When these contract, a stream of water is emitted from the excurrent siphon at the end opposite the mouth, and this propels the animal forward. Thus, the lack of a tail appears to be no handicap. But since a notochord in tunicates is confined to the tail, thaliaceans have no notochord! Whether the notochord has been lost or whether the ancestors never had one is not known. Sometimes individuals unite to form a linear colony.

Sea squirts

Adult sea squirts are usually sessile, lack a notochord, and have no dorsal nervous system (Fig. 2-2). Because of this, for a long time sea squirts were not recognized as chordates. Not until the free-swimming larva was discovered was it found that these animals exhibit a notochord and

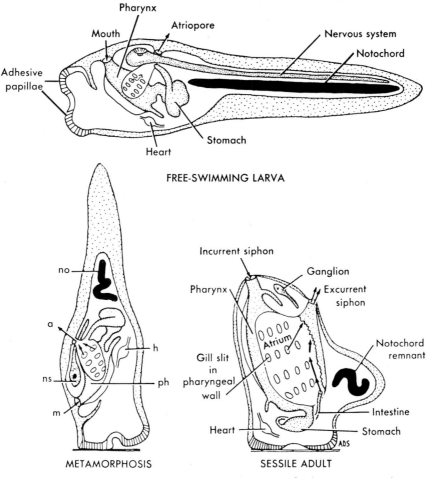

Fig. 2-2. Metamorphosis in a sea squirt. **a,** Atriopore; **h,** heart; **m,** mouth; **no,** notochord; **ns,** nervous system; **ph,** pharynx. Arrows indicate course of water. The atria have been opened to reveal entrance of pharyngeal slits.

dorsal nervous system prior to metamorphosis. Adult sea squirts live near the shore.

Larval stage (Fig. 2-2). The larval stage is represented by free-swimming organisms in which the notochord is confined to the tail. The larva is about ¼ inch long and resembles a very young frog tadpole. It is propelled by lashing of the tail. The nervous system lies dorsal to the notochord in the tail and extends into the trunk region. There it exhibits an enlargement containing a cavity that is probably related to the brain ventricle of higher forms. There is no head. Water enters the pharyngeal cavity and passes through gill slits into a chamber (atrium) surrounding the pharynx. The digestive tract also empties into the atrium. From the atrium, water and undigested food pass to the exterior. Blood is pumped from a ventral heart alternately backward for a few beats and then forward, the heart pausing before reversing the direction of flow. Three adhesive papillae covered by a sticky secretion are located at the forward end for attachment to other objects. At metamorphosis the papillae serve to attach the larva to the final substrate. By this time a major reorganization of the internal organs has commenced. The tail region is resorbed as in the frog tadpole. The notochord degenerates and only remnants are left. The site of attachment to the substrate becomes broader and roughened. The nervous system becomes altered in location and structure, and a sessile adult sea squirt has evolved.

Adult stage. Several genera of sea squirts are known to marine zoologists. *Ciona*, a long, slender, yellow-green, polyp-like form, occurs as a solitary organism off the coast of southern California. *Styela*, also solitary, is a brown form that has a short stalk for attachment. *Ascidia ceratodes* has green blood cells, vanadium being substituted for iron in the pigment. Other ascidians may be orange, white, red, or colorless, and many are transparent.

Digestive tract. In a typical adult sea squirt, water laden with minute organic and inorganic matter enters the organism through an incurrent siphon, which terminates at a membrane, the velum. Cilia in the pharyngeal wall propel an incurrent stream past the velum and into the pharynx, the largest organ of the body. Here the food particles are ensnared in mucus secreted by the endostyle, a groove in the pharyngeal floor. The food-bearing mucus is then moved by papillae and ciliary action to the stomach. From the stomach a short intestine empties into the atrium.

Respiratory system. The pharynx is the chief respiratory organ. Its highly vascular walls are perforated by rows of gill slits that open into the atrium. Oxygen and carbon dioxide are exchanged between the water and the blood vessels of the pharynx. Water passes through the gill slits into the atrium. From the atrium the water is discharged along with waste products of the digestive tract through the excurrent siphon. The forceful discharge of water when the animal is irritated inspired the descriptive name "sea squirt."

Notochord. No internal skeletal system exists in adult sea squirts, the notochord having degenerated during metamorphosis.

Nervous system. A single, solid, elongated ganglion, a remnant of the trunk ganglion of the larva, lies near the mouth. Nerve strands pass from the ganglion to all parts of the body. There is no central nervous system other than the ganglion, and no special sense organs have been described.

A neural gland with a ciliated duct passing to the pharynx lies close to the neural ganglion. It is of nervous tissue origin and may be equivalent to the neural component of the pituitary of vertebrates.

Reproductive system. Sea squirts are hermaphroditic. The gonads lie in the intestinal convolution, and a duct passes from each gonad to the atrium. In colonies sea squirts arise asexually by budding from a single parent. Each individual then reproduces sexually.

Circulatory system. A single, elongated heart lies near the pharynx. Arising from each end is a vessel. Blood is propelled first into one vessel for several pulsations and then into the other. Most of the vascular channels are large sinuses.

CEPHALOCHORDATES
Amphioxus

Amphioxus means "sharp at both ends." Any member of the subphylum Cephalochordata may be called an amphioxus or lancelet (little spear), but the correct generic name for the lancelet commonly studied in the laboratory is *Branchiostoma*. *Asymmetron* is the only other genus in the subphylum.

Amphioxus occurs along sandy beaches throughout most of the globe. The animals lie on their backs or sides in shallow water just beyond the shore or bury themselves in the sand with only the mouth region protruding. They swim by jerky movements in which the tail is drawn toward the head with lightning speed, causing the body to assume as S shape, after which they straighten up. They repeat the maneuver for several seconds, making little headway, and then fall over to their side as though exhausted. These same movements permit them to burrow into soft, wet sand with surprising speed. Adults vary from less than 1 inch to more than 3 inches in length, the largest being *Branchiostoma californiense*, which occurs along the beaches in southern California and on the California peninsula. Off the coast of China, amphioxus is collected in quantity and sold as a table delicacy.

The living animal is semitransparent but

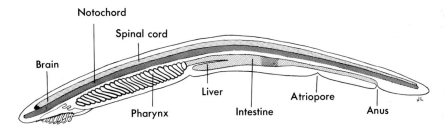

Fig. 2-3. Internal structure of a larval amphioxus.

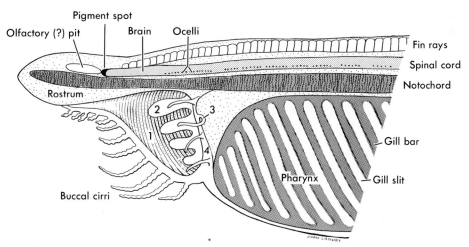

Fig. 2-4. Cephalic end of an amphioxus shown in sagittal section. **1,** Vestibule bounded by oral hood; **2,** wheel organ projecting into vestibule; **3,** velar tentacle; **4,** velum.

becomes opaque when immersed in preserving fluids. There are no paired appendages. A median dorsal fin, a caudal fin, and a median ventral fin are continuous around the tip of the tail. A pair of longitudinal ridges, the metapleural folds, project from the ventrolateral body wall along each side of the midventral line, extending from the anterior end to the atriopore. The folds may be observed with a hand lens on the intact specimen and in cross sections of the trunk. The body is practically all trunk and tail, for there is almost no cephalization. When a specimen is examined, the metameric arrangement of the body wall musculature, as seen through the skin, is a striking feature of its anatomy. In mature specimens a longitudinal row of bulging, whitish-appearing gonads is visible through the body wall ventrolaterally on each side.

The vestibule is a large chamber beneath the rostrum at the cephalic end of the body. The vestibule is bounded laterally by fleshy curtain-like folds and is open ventrally (Fig. 2-4). The rostrum and the lateral folds constitute an oral hood. The oral hood and vestibule resemble a tunnel, which leads caudad to an almost perpendicular membrane, the velum, pierced by the mouth. Fringing the anterior free border of the oral hood are tentacle-like buccal cirri, which act as sieves and also have chemoreceptive cells. Each is supported by an internal skeletal rod. A set of short, thick, finger-like projections constituting the wheel organ project forward into the vestibule from the base of the velum. The wheel organ maintains a current of water into the mouth. A third group of projections, the delicate velar tentacles, are attached to the velum around the mouth, which is always open. The tentacles, like the buccal cirri, have chemoreceptors that monitor the incurrent water stream.

Integument (Fig. 5-1, A). The skin of the amphioxus consists of a single layer of epidermal cells and a thin dermis. Interspersed among the epidermal cells are gland cells. In immature specimens the epidermal cells are ciliated, as in the larval stage of some worms. The cilia later disappear, and the epidermis secretes a cuticle resembling that of annelids. Immediately internal to the dermis is the body wall muscle.

Skeleton. The notochord is the chief skeletal element. It extends from near the anterior tip of the rostrum to the extreme tip of the tail, lying under the nerve cord. Its continuation anterior to the forwardmost tip of the brain is a condition found only in Cephalochordata (*cephalo* = head). The outermost layer of the notochord is referred to as the notochord epithelium. Surrounding the notochord is a thick noncellular sheath, produced by the surrounding mesoderm. Small fin rays lie within the dorsal and ventral fins. In the pharyngeal wall between the gill slits are fibrous rods of connective tissue reinforcing the gill bars. Similar rods form a ring about the base of the buccal cirri.

Body wall musculature. The musculature of the body wall is metameric. It consists of an uninterrupted series of <-shaped muscle segments called **myomeres,** extending from the anterior tip of the body to the tip of the tail. Each myomere is separated from the next by a connective tissue partition, the **myoseptum,** and is supplied by a spinal nerve arising from the spinal cord at the same level. As the amphioxus develops, the left side of the body is displaced forward approximately one half a somite. As a result, the corresponding myomeres of the left and right sides are not directly opposite each other. Since the myomeres are <-shaped, cross sections of the body will include several successive myomeres (Figs. 2-5 and 2-6).

Brain. The amphioxus exhibits a nervous system resembling that of vertebrates in basic structure. The brain, however, does not exhibit forebrain, midbrain, and hindbrain. Instead, two subdivisions are recognized: an anterior **prosencephalon,** containing a single enlarged ventricle, and a more posterior **deuteroencephalon.** The prosencephalon is lined with cilia and long fila-

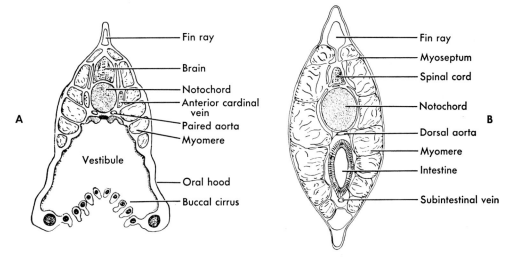

Fig. 2-5. Cross sections of an amphioxus. **A,** Anterior to mouth. **B,** Posterior to atriopore.

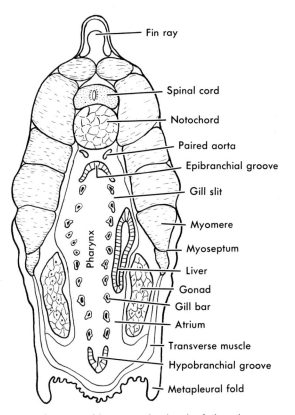

Fig. 2-6. Cross section of an amphioxus at the level of the pharynx.

mentous projections of the ependymal cells, demonstrable only with electron microscopy.

Attempts to homologize the parts of the brain of the amphioxus with those of a vertebrate have not been entirely successful. In an amphioxus the notochord extends anterior to the brain. Does this indicate the absence of a forebrain? The answer must await further neurophysiological research. Whether or not the nerves that supply the gills should be considered cranial nerves complicates the problem. If the branchial nerves are omitted, there are seven cranial nerves (including the nervus apicis or terminalis). If the branchial and oral nerves are included, there are thirty-nine. The absence of semicircular canals, eyes, lateral-line system, and foramen magnum deprives us of landmarks that would be helpful. Because of these difficulties, it is not possible to decide at this time just where the brain ends. We must be content to say that it merges imperceptibly with the spinal cord.

Spinal cord and nerves. The canal within the spinal cord is observable in cross section. It is lined by nonnervous supporting elements called **ependymal cells.** Near the end of the cord the nervous elements disappear and the cord is composed of ependymal cells alone. A similar situation occurs in vertebrates. A single membrane (meninx) surrounds the brain and spinal cord of an amphioxus.

Spinal nerves arise in a dorsal and ventral series, dorsal nerves (chiefly sensory) alternating with ventral nerves (motor) along each side of the cord. Dorsal and ventral nerves do not unite to form spinal nerves as do the dorsal and ventral roots in most vertebrates. This condition is thought to be primitive. The distribution of the spinal nerves is discussed on p. 351. Since the somites of the two sides are not exactly opposite one another, the dorsal and ventral nerves of the left and right sides do not arise directly opposite one another.

Sense organs. The relatively small size of the brain is correlated with the paucity of organs of special sense. There are no retinas, semicircular canals, or lateral-line organs. It is doubtful whether an olfactory epithelium is present. The amphioxus apparently has chemical receptors that are sensitive to bitter stimuli, though not to sweet, as judged by man's tastes. These chemoreceptors are particularly abundant on the buccal cirri and velar tentacles directly in the path of the incurrent water stream. They are also scattered on other surfaces of the body, the tail being more sensitive than the trunk. Touch receptors, which elicit withdrawal, occur over the entire body surface.

The most characteristic sense organs are the light-sensitive, pigmented **ocelli** embedded within the ventrolateral walls of the spinal cord (Fig. 2-7). Each ocellus consists of a receptor cell and a cap-like melanocyte. The melanocyte lies between the receptor cell and the incoming light rays. A conducting process extends away from the base of the receptor cell. During the early part of the twentieth century the ocelli were also referred to as Hesse cells, after the scientist who described them in 1898. The structure of the receptor cell and melanocyte has now been studied by electron microscopy and has been described.[107] The cell membrane at the apical border of the receptor cell exhibits a maze of infolded tubules extending 2.5 μ into

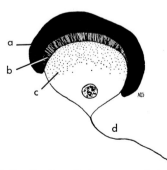

Fig. 2-7. Ocellus (light receptor) from spinal cord of an amphioxus. **a,** Melanocyte; **b,** apical border of receptor cell; **c,** receptor cell; **d,** process for conduction of impulse.

the cell. The infolded tubules exhibit expansions at their bases, below which lie many mitochondria. The melanocyte is packed with large pigment granules, each consisting of a membranous sac containing many subgranules. Occasional finger-like processes of the melanocyte extend into recesses of the receptor cell. In their microanatomy and embryogenesis, these photoreceptors resemble invertebrate receptors such as the rhabdomeres of arthropods and of many achordates, rather than the visual receptor cells of vertebrates. Thus morphology, pursued by modern techniques, continues to assist in unraveling phylogenetic relationships.

The ocelli probably assist in the orientation of the organism as it burrows into the sand. The prominent pigmented eyespot at the cephalic tip of the brain is thought not to be a photoreceptor.

Coelom and atrium. As a result of embryonic changes, the coelom is pushed aside and almost crowded out by the large atrial chamber surrounding the pharynx. A compressed remnant of the coelom occurs just internal to the body wall muscle, and other remnants are found adjacent to the gonads, around the ventral aortas, and in the metapleural folds. A discussion of the development of the coelom of the amphioxus will be found in Chapter 4 under mesoderm formation.

Digestive tract. The mouth, located in the velum, leads to the pharynx. The pharyngeal wall is perforated by gill slits varying in number in different species but exceeding sixty in adults. The slits open into a large atrium. Between each slit is an oblique gill bar containing a fibrous skeletal rod. Twice as many gill slits occur in the mature amphioxus as in the larva because each embryonic slit becomes divided into two by the downgrowth of a secondary partition, the tongue bar, from the dorsal border of the slit (Fig. 2-8). This partition reaches the ventral border of the slit and fuses with it. Thus adults have primary and secondary gill bars.

In the pharyngeal floor is a longitudinal ciliated hypobranchial groove or endostyle. In the roof is a ciliated epibranchial groove (Fig. 2-6). Anteriorly, just behind the velum, a ciliated peripharyngeal band passes dorsad to meet the epibranchial groove. The cells of these bands and grooves secrete considerable mucus. Microorganisms and other small particles of organic matter passing into the pharynx with the incurrent stream of water are entangled in the mucus, which is propelled anteriad by the cilia in the hypobranchial groove, dorsad on the peripharyngeal band or on the ciliated surfaces of the gill bars, and caudad toward the esophagus in the epibranchial groove. Large particles are

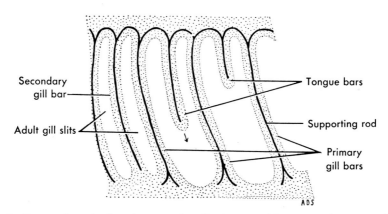

Fig. 2-8. Tongue bars in the pharyngeal wall of an amphioxus growing ventrad (arrow) to subdivide the embryonic pharyngeal slits into two adult gill slits.

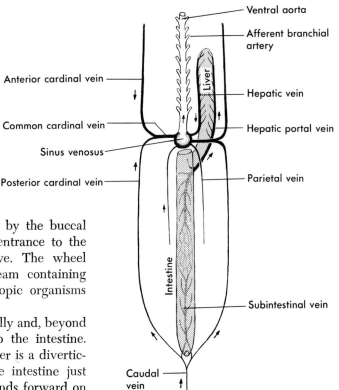

Fig. 2-9. Basic venous channels and ventral aorta of an amphioxus, dorsal view. The liver has been rotated 90° around its long axis. The hepatic portal in life is ventral to the liver and the hepatic vein is dorsal.

excluded from the pharynx by the buccal cirri, which fold over the entrance to the vestibule, acting as a sieve. The wheel organ directs a water stream containing diatoms and other microscopic organisms into the mouth.

The pharynx tapers caudally and, beyond the last gill slit, leads into the intestine. The hollow, club-shaped liver is a diverticulum from the floor of the intestine just behind the pharynx. It extends forward on the right side of the pharynx. The cells of the liver secrete a digestive enzyme similar to that of the vertebrate pancreas. The intestine terminates at the anus. As in many achordates, the entire digestive tract is ciliated.

Respiratory system. The incurrent water stream carries oxygen as well as foodstuffs to the pharynx. Relieved of microscopic food particles, the water passes between the gill bars, where it comes in contact with the respiratory epithelium en route to the atrium. The atrium opens to the outside via the atriopore.

Circulatory system. The amphioxus has the basic circulatory pattern of vertebrates. Lacking, of course, are vessels to appendages and the extensive vascular channels characteristic of the vertebrate head. The heart of an amphioxus consists only of a sinus venosus, there being no atria or ventricles. Blood is pumped by the muscular, pulsating hepatic vein that connects the liver with the sinus venosus and by the ventral aorta. These two muscular ves-

sels perform the function of the missing atria and ventricles. The remaining blood vessels have very thin walls, and histologically the arteries, veins, and capillaries are alike. Arteries are those vessels which carry blood from the sinus venosus to the gills and then supply the vascular channels within the body wall and viscera. Veins collect blood from the various vascular channels and return it to the sinus venosus.

ARTERIAL CHANNELS. The colorless blood courses forward beneath the pharynx in a muscular, contractile ventral aorta, which commences in the sinus venosus at the caudal border of the pharynx. From the

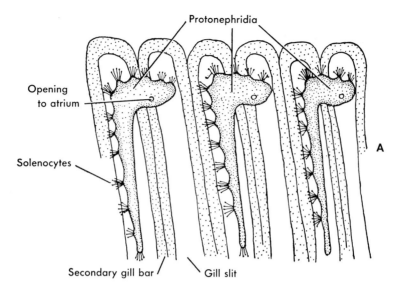

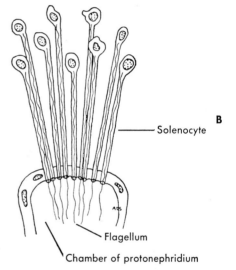

Fig. 2-10. Protonephridia, **A,** of an amphioxus, and detail of solenocytes. **B.** Solenocytes project into the coelom at their free ends and empty into the chamber of a protonephridium at their bases.

ventral aorta a series of thin-walled aortic arches (the base of which are afferent branchial arteries, as indicated in Fig. 2-9) pass up the primary gill bars, and, by branches, pass up the secondary bars on each side, where the blood is oxygenated. Before joining the dorsal aorta, the aortic arches contribute to an anastomosing network of vascular channels associated with the protonephridia. Blood from the aortic arches then enters the paired, nonmuscular, dorsal aortas. These pass caudad above the pharynx and unite just behind it to form the unpaired dorsal aorta. This distributes the blood by paired vessels to the body wall and by median vessels to the visceral organs. The dorsal aorta continues into the tail as the caudal artery.

VENOUS CHANNELS (Fig. 2-9). The venous channels are essentially identical with the basic veins of vertebrates. From the capillaries of the tail, a single caudal vein courses forward and divides into right and left posterior cardinal veins. These veins pass forward in the lateral body wall, internal to the musculature, to a point just behind the pharynx. Here the posterior cardinal veins meet two anterior cardinal veins passing caudad from the rostrum and the pharyngeal wall. Blood passes from

the union of the posterior and anterior cardinal veins into one of the two common cardinal veins, directed mediad and dorsad on each side. The common cardinal veins meet in the midline just behind the pharynx and dorsal to the intestine to form the large sinus venosus. Two parietal veins drain the dorsolateral region of the body wall caudal to the pharynx. These course forward lateral to the dorsal aorta to terminate in the sinus venosus.

Drainage from the visceral organs is by way of a single subintestinal vein arising from the caudal vein at the point of bifurcation of the posterior cardinals. The subintestinal vein passes forward along the ventral aspect of the intestine, from which it receives tributaries. It then breaks up into smaller channels midway in the intestine before reconvening ventral to the intestine to continue forward as the hepatic portal vein. This enters capillaries in the liver. From the liver a muscular, contractile hepatic vein passes to the sinus venosus.

Urogenital system. The amphioxus exhibits sexual dimorphism, that is, the sexes are separate. Mature gonads are visible through the muscle and skin of the trunk from the caudal end of the pharynx to the anus. They are partly surrounded by a remnant of the coelom. Sperm and eggs are shed into the atrial cavity, since there are no genital ducts.

Removal of fluid wastes is accomplished by protonephridia lying in dorsal groups beside the secondary gill bars (Fig. 2-10). Each protonephridium consists of a tiny chamber into which drain several clusters of solenocytes. Each solenocyte has a funnel opening into the coelom and a hollow stalk containing a flagellum. The motion of the flagellum causes a current of coelomic fluid to pass into the solenocyte and down the stalk to the protonephridial chamber. Each protonephridium opens by a single small pore into the atrial cavity. The wastes are washed away by the respiratory stream. The excretory mechanism bears a striking resemblance to the nephridia of marine annelids and to the flame cells of achordates.

Amphioxus and the vertebrates

Although an amphioxus resembles a vertebrate in many respects, differences are evident. An amphioxus has almost no cephalization; it has a notochord, but no vertebral column; it has gill slits, but in large numbers, emptying into an atrium; it has a dorsal, hollow nervous system, but one in which the brain lacks the major vertebrate subdivisions; and it has no paired sense organs. The amphioxus has a segmented musculature, but the segments extend to the anterior tip of the head; it has median fins, but no paired ones; it has a two-layered skin, but the outer layer is only one cell thick; it has a digestive tract terminating in an anus at the base of a postanal tail, but with a hollow liver diverticulum and no other endodermal evaginations; and it has aortic arches and venous channels similar to the basic channels of vertebrates, but no heart. The amphioxus is coelomate, but the coelomic cavity is greatly restricted. The coelomic fluid accumulates liquid wastes as in lower vertebrates, but the excretory protonephridia resemble those of achordates. The amphioxus may be a modified vertebrate, but more likely it is a descendant of a stem chordate that gave rise also to vertebrates (Fig. 2-13).

ENTEROPNEUSTA: INCERTAE SEDIS

The term "incertae sedis" means of uncertain status. Despite years of deliberation, not all animals have been assigned to a specific taxonomic group without dissent. Among animals of uncertain status are the balanoglossid worms, or enteropneusts, *Balanoglossus* and *Dolichoglossus* (Figs. 2-11 and 2-12).

In 1870 Karl Gegenbaur formed a new taxonomic group, Enteropneusta, incorporating balanoglossid worms. Four years later the phylum Chordata was established by Ernst Haeckel, incorporating the Urochordata, Cephalochordata, and Vertebrata. In 1884, when the chordate phylum was celebrating its tenth birthday as a taxonomic group, William Bateson added enteropneusts to the phylum chordata as the subphylum Hemichorda, meaning "half of a notochord." The addition was accepted, but not without dissent. Bateson's decision was based on three facts:

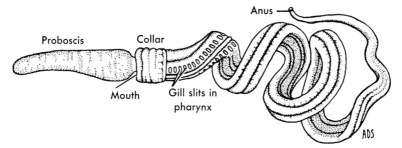

Fig. 2-11. The enteropneust *Dolichoglossus.*

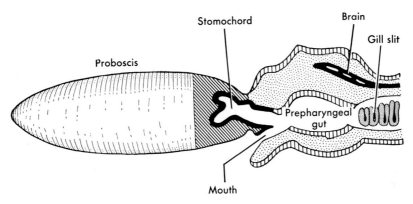

Fig. 2-12. Head of *Dolichoglossus,* sagittal section (except proboscis).

1. Enteropneusts exhibit, in addition to a ventral nerve cord, a dorsal nerve cord consisting of a groove of epithelial cells. (All epithelial cells are ciliated in enteropneusts.) In the collar region the groove sinks under the surface, forming a tube ("brain") with a continuous or discontinuous lumen. The occurrence of the dorsal nerve cord as an ectodermal groove and tube suggests that balanoglossids may share a common ancestry with vertebrates.

2. Enteropneusts exhibit a series of gill slits in the pharyngeal wall, leading directly to the exterior.

3. Enteropneusts exhibit a short diverticulum of the gut, called a **stomochord,** extending forward into the proboscis.

Bateson considered the stomochord to be homologous with the notochord of chordates. Today most investigators think otherwise and prefer to classify the enteropneusts as a phylum close to the echin-

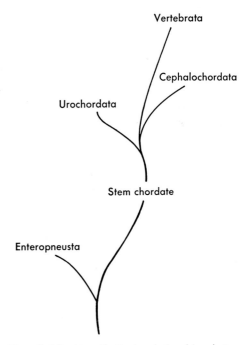

Fig. 2-13. Hypothetical relationships between protochordates and vertebrates.

oderms, since most enteropneusts exhibit a free-living larval stage (tornaria larva) resembling closely the larvae of some echinoderms. To many conservative zoologists, however, enteropneusts are still "incertae sedis."

CHORDATE RELATIONSHIPS

The exhibition of a tadpole-like larval stage in some protochordates, the occurrence of an echinoderm-like larval stage in enteropneusts, the occurrence of thaliaceans having protochordate characteristics but lacking a notochord, and the exhibition of many vertebrate characteristics in the amphioxus lead to the inference that protochordates, echinoderms, enteropneusts, and vertebrates may be distant relatives. If so, the common ancestor existed eons ago in geological time, and the sequence of evolutionary changes giving rise to these diverse forms is not at all clear at present.

Chapter summary

Protochordates are invertebrates belonging to the phylum Chordata. They typically exhibit, at least as larvae, a notochord, slits in the pharyngeal wall, and a dorsal, hollow nervous system, as do vertebrates. There are two groups of protochordates: the tunicates and the cephalochordates. Tunicates (Urochordata), which include sea squirts, permanently larval larvaceans, and notochordless thaliaceans, are characterized by the presence of a tunic that surrounds the body and a notochord confined to the tail region. Cephalochordates are represented by the amphioxus. Its structure suggests that it is a descendant of an ancestor shared with the vertebrates.

3

Parade of the vertebrates

There are 49,000 different species of animals exhibiting vertebral columns. Fortunately for Noah 30,000 of these species are fishes. Also, there are amphibians, reptiles, and mammals that share with fishes the streams, ponds, and even the seas as their permanent abode. Amphibians, especially tailed forms, are very much at home in fresh water, but saltwater amphibians are unknown. Of present-day reptiles, only sea snakes and marine turtles inhabit the seas. Other turtles and snakes remain permanently in fresh water or are land forms. Alligators and crocodiles are both aquatic and terrestrial. Only birds of all vertebrates do not occur as permanent residents of the waters of the earth, although many species are entirely dependent for food on the organisms that occur in the water. Among mammals, ~~only~~ whales, dolphins, and porpoises are permanently aquatic. +Sirenia

There is evidence that animal life originated in the primeval fresh waters which at one time covered the earth, and that primitive vertebrates in their early beginnings were water-dwelling animals with pharyngeal slits for carrying on respiratory and food-ensnaring functions. From the facts related in this book, it will become apparent that all vertebrates, including man himself, develop at first as though they were being groomed for life in the water.

The purpose of the present chapter is to parade before the reader a number of representative vertebrates that will be referred to again in later pages and to mention their probable ancestors and their nearest living relatives. The program for this parade of the vertebrates is the brief classification in the Appendix. You may find it advisable to refer to the present chapter and to the abridged classification regularly during your study of vertebrate architecture.

VERTEBRATE TAXA

Schemes of classification are devices for lumping into groups (taxa) animals that are similar. These taxa are then arranged in a phylogenetic tree. The closer together two animals are in the tree, the more closely related they are considered to be, and the more recently they both may have evolved from a common ancestor. Animals lowest in the tree generally have the greatest number of primitive traits. Those higher in the tree show greater modifications of the primitive pattern, although the primitive pattern manifests itself more or less in their embryos at least.

There are two superclasses of vertebrates, Pisces (fishes) and Tetrapoda (*tetra* = four, *pod* = foot). Chief subdivisions are classes, subclasses, superorders, orders, suborders, families, genera, and species. The following classes are generally recog-

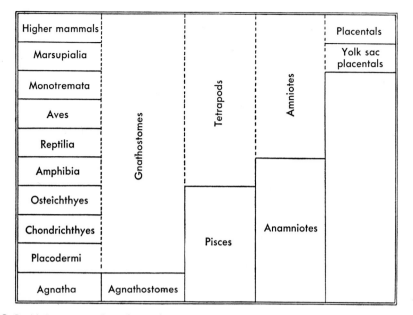

Fig. 3-1. Major categories of vertebrates.

nized, although the Agnatha are sometimes given the status of a superclass or even a subphylum.[4]

Commencing with reptiles, animals exhibit a special membrane that surrounds the developing embryo. This membrane is the **amnion,** and animals that possess it are said to be **amniotes.** Fishes and amphibians, since they do not exhibit this membrane, are **anamniotes.** Agnathans lack true jaws and hence are **agnathostomes;** all other vertebrates are **gnathostomes** (Fig. 3-1).

Since the time of Carl von Linné (Latin, Linnaeus), a Swedish naturalist, taxonomic nomenclature has been latinized by agreement among zoologists of the world. This arrangement permits zoologists of all languages to understand one another without translation when the name of any ani-

mal is employed. For instance, not every zoologist knows what "un chat," "die Katze," or "el gato" is. All these words refer to the common domestic cat, but *Felis domestica* is the taxonomic name for this species, and every zoologist, no matter what his language, knows that name. The use of the generic name *Felis* separates certain cats—tigers, lions, jaguars, domestic cats, and others—from the lynx *(Lynx).* The word *domestica* separates the domestic cat from the tiger *(Felis tigris),* lion *(F. leo),* and jaguar *(F. onca).* Use of the binomial designation (that is, the employment of two Latin names, the first designating genus and the second, species) was introduced by Linnaeus in the tenth edition (1758) of his classic book *Systema Naturae.* The generic name must always be capitalized; the specific name is never capitalized. *Felis* is seldom used by itself, for who knows what cat is being talked about? The term *domestica* is never used alone, for this name may be applied to species of many different phyla and lacks significance without the generic name. (The housefly is *Musca domestica.*) *Felis domestica* is only one animal, the domestic cat. Man is *Homo sapiens.*

---

*This class is known only by study of fossil remains.

	Dogfish	Necturus	Domestic cat
Superclass	Pisces	Tetrapoda	Tetrapoda
Class	Chondrichthyes	Amphibia	Mammalia
Subclass	Elasmobranchii		Theria
Superorder		Salientia	Ferera
Order	Selachii	Caudata	Carnivora
Suborder	Squaloidea		Fissipeda
Family		Proteidae	Felidae
Genus and species	Squalus acanthias	Necturus maculosus	Felis domestica

Accompanying is a classification of the spiny dogfish, the spotted *Necturus*, and the domestic cat.

Many Latin names have a meaning that, when translated, may be of assistance in recognizing the members of the group. Thus, Chondrichthyes means "cartilaginous fishes," Carnivora means "flesh-eaters," and *maculosus* means "spotted." The habitual use of a dictionary of Latin and Greek word roots will richly reward the science student who takes the extra time required to look up a meaning, for many recur again and again in the literature of a science. To be familiar with these roots is to have a semantic tool that ranks in importance with the scalpel and the Bunsen burner.

PISCES

Most authorities recognize four classes of fishes: **Agnatha, Placodermi, Chondrichthyes,** and **Osteichthyes.** Placoderms are all extinct. So are large numbers of Chondrichthyes and Osteichthyes.

Agnatha: ostracoderms and cyclostomes

Agnathans are especially interesting because the extinct forms, the ostracoderms, are the oldest known vertebrates. These lived in fresh water nearly 500 million years ago (Figs. 3-2, 3-3).

Ostracoderms had no jaws, and no pectoral or pelvic fins. Broad bony plates were embedded in the dermis of the head and anterior part of the trunk, and the more caudal parts exhibited smaller plates more often called **dermal scales.** The plates and scales constituted a bony armor that protected them from attacks through the skin and hence the names "armored fishes" and "ostracoderms" (*ostraco*=shell, *derm*= skin). Their survivors, the cyclostomes, have neither armor nor scales. It must be deduced, therefore, that the cyclostomes have lost the armor (that is, have undergone mutations) during the intervening ages. In addition to the bony dermal skeleton, some ostracoderms also had a bony internal skeleton. Today's cyclostomes have no bone in the skin or any place else. Thus, it may be deduced that the absence of bone in living cyclostomes is not a primitive trait. To say this in another way, the total enzymatic and hormonal complex necessary for the deposit of bone is no longer present in the Agnatha of today. Most ostracoderms had a single nostril on top of the head, and this condition remains in cyclostomes.

Adult cyclostomes (Fig. 3-4) have a long, slender, eel-like body up to 3 feet in length; hence they are sometimes called **lamprey eels.** (These are not the true eels, which are bony fishes.) Cyclostomes exhibit no pectoral or pelvic fins, no jaws, no scales, no bone, and no vertebral column. The notochord is the chief skeletal structure of the trunk (Fig. 1-11). Living cyclostomes are grouped in two orders, Myxinoidea (hagfishes) and Petromyzontia (lampreys).

Lampreys. A large buccal funnel lined with many horny denticles (Fig. 5-12) helps keep the adult lamprey, which is primarily a parasite, attached to the host fish while a tongue-like rod covered with spiny processes rasps the flesh of the victim. The nostril is located dorsally and behind the anterior tip of the head. A nasal

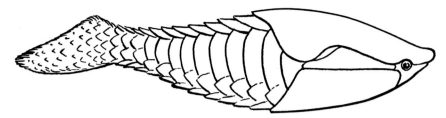

Fig. 3-2. Ostracoderm, a very ancient armored, jawless fish.

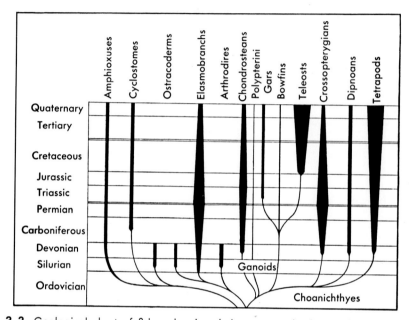

Fig. 3-3. Geological chart of fishes showing their range and relative abundance through time. (From Atwood: Comparative anatomy, ed. 2, St. Louis, 1955, The C. V. Mosby Co.)

canal passes to the olfactory sac, continues farther, and then terminates blindly as a nasohypophyseal sac. From the wall of this sac in the embryo arose the adenohypophysis, a part of the pituitary gland. The gills lie in seven gill pouches, each of which opens independently by an external gill slit leading to the exterior. Unicellular glands are numerous throughout the epidermis; hence the skin is slimy.

A well-known species is *Petromyzon marinus*, the marine lamprey. Each spring these lampreys migrate into the freshwater rivers and brooks, where they lay their eggs. In twenty to twenty-one days the small nonparasitic larvae, known as ammocoetes, emerge. The ammocoete was once thought to be an adult species. It has been described in detail in Chapter 1. After three to four years the larvae metamorphose into immature adults and migrate to the sea. There they attain sexual maturity in preparation for the journey back to the spawning places. A freshwater species lives as adults in the Great Lakes between Canada and the United States. *Lampetra* is a brook lamprey that does not migrate to the sea, spawns upon reaching sexual maturity, and promptly dies without further feeding.

Hagfishes. Hagfishes have a shallow buccal funnel lacking horny denticles. They feed on live and dead fish and also consume a large variety of small invertebrates.

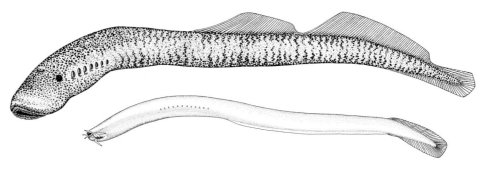

Fig. 3-4. Lamprey *Petromyzon* above; hagfish *Bdellostoma* below.

A fringe of stubby, finger-like papillae surrounds the buccal funnel. Just above the funnel is a single nostril. A nasopharyngeal canal leads from the nostril to the olfactory sac and thence to the pharyngeal cavity, carrying respiratory water to the gills. The eyes are vestigial and functionless. *Myxine glutinosa*, the Atlantic form, has six pairs of gill pouches (occasionally five or seven) opening into a common efferent duct which, in turn, opens to the outside by a single pore (Fig. 12-4). *Bdellostoma stouti*, common off the coast of California, has ten to fifteen gill pouches, each opening independently to the exterior. Unlike lampreys, hagfishes do not migrate to fresh water to spawn, and the larval stage occurs within the egg membranes. An excellent, interesting, and authoritative reference on the biology of *Myxine* is cited at the end of the book.[4]

Placodermi

Placoderms were armored fishes that were increasing in numbers in the fresh waters of the Devonian period as the ostracoderms were disappearing. With paired fins and bony jaws, they were more specialized than ostracoderms. However, the fins of some species (acanthodians, Fig. 9-8) were mere hollow spines projecting from the trunk and associated with a small web of skin, and the jaws were strange, moveable plates of bone. Another group, the arthrodires (*Coccosteus*, Fig. 3-5), exhibited a heavy bony shield in the dermis of the skin of the head and pharyngeal region and another shield in the anterior part of the trunk, the two shields meeting in a moveable joint (*arthro* = joint). The rest of the body of arthrodires was usually naked, but acanthodians were completely covered with small bony plates resembling the scales of later armored fishes. Alfred Romer considers the placoderms as an "experimental" step in the evolution of jawed fishes but perhaps not in the main line of vertebrate evolution.[145] Their dermal armor links them genetically with their predecessors, the ostracoderms.

Chondrichthyes

Chondrichthyes (Fig. 3-6) are cartilaginous fishes (*chondro* = cartilage, *ichthy* = fish) that exhibit no bone, although certain cartilaginous areas become infiltrated with deposits of lime salts that resemble bone in firmness. The ancestors of these fishes had a bony skeleton, and the absence of bone in this instance is considered a specialization, not a primitive characteristic. These fishes vary from graceful, streamlined specimens such as sharks to the much-flattened rays and skates. The most common Chondrichthyes are elasmobranchs.

The Chondrichthyes are very numerous at present but were far more so in ancient times. This is confirmed by the large number of fossil species known, some of which are identified only by a horny spine, by their otoliths (calcareous concretions developing within the inner ear), or by their teeth. Let it not be assumed, however,

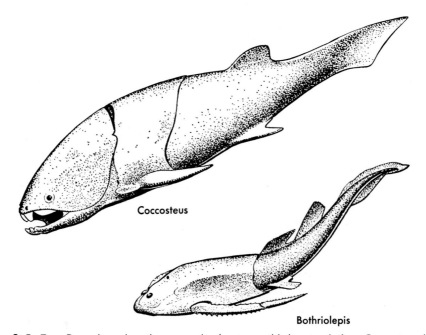

Fig. 3-5. Two Devonian placoderms, each about one-third natural size. *Coccosteus* is an arthrodire; *Bothriolepis* is an antiarch. (From Colbert: Evolution of the vertebrates, New York, 1955, John Wiley & Sons, Inc.)

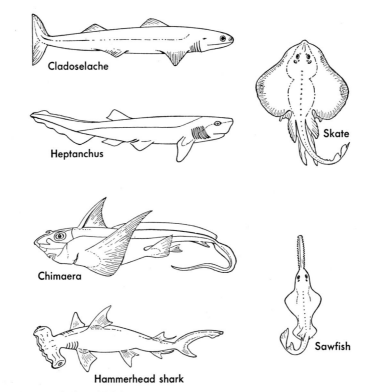

Fig. 3-6. Group of Chondrichthyes.

that fragments alone remain of these ancient hordes. Many remarkably preserved Chondrichthyes with intact viscera and muscle fibers have been uncovered within comparatively recent years, and it is likely that further discoveries will be forthcoming. Some of the ancient forms were heavily armored, as might be expected. An abridged classification of Chondrichthyes will be found on p. 414.

Elasmobranchs. The elasmobranch fishes are often divided into three orders: **Cladoselachii,** all of which are extinct; **Selachii,** or sharks; and **Batoidea,** or rays, skates, and sawfishes. Elasmobranchs differ from other Chondrichthyes in exhibiting placoid scales. The first pharyngeal slit is small, contains only a rudimentary gill (**pseudobranch**), and is called a **spiracle.** The remaining slits serve as true gill slits that are "naked," or visible, on the side of the pharynx rather than covered by an operculum. The pelvic fins of the male (claspers) are modified for transfer of sperm. Except in primitive forms such as *Cladoselache* (Fig. 3-6), the mouth is on the ventral surface rather than being terminal.

Sharks. Nearly every student of comparative anatomy becomes acquainted with sharks through laboratory dissection. Often used is the spiny dogfish of the Atlantic, *Squalus acanthias,* named for the prominent spine associated with each dorsal fin. *Squalus suckleyi* is the Pacific spiny dogfish. *Mustelus* is the "smooth dogfish" in the sense that it lacks dorsal spines. These species have a spiracle and five gill slits. Hexanchid sharks have six gill slits plus a spiracle and heptanchid sharks have seven. Other sharks include hammerheads, bonnetheads, basking sharks, and man-eating sharks. *Cladoselache,* known only as a fossil, is one of the more primitive sharks. It had seven gill slits and a spiracle.

Sharks are of interest to students of vertebrate anatomy because of the generalized structure exhibited in the arrangement of their various systems. If one were to seek a living blueprint of the vertebrate body—an architectural plan which, by

modifications here and deletions there, could serve as a pattern from which other vertebrates might be constructed—the dogfish would serve well. The anatomy of the eye of the dogfish is in all significant respects the anatomy of the human eyeball. The distribution of the nerves to the jaws, to the olfactory epithelium, to the inner ear, and to the musculature is essentially the arrangement found in man. The arrangement of the visceral skeleton, the aortic arches, the chief venous channels, and the urogenital system of the dogfish is, in essential features, identical with the arrangement of these same structures in the embryos of all higher vertebrates. Thus, knowledge of the anatomy of the dogfish constitutes a point of departure for the study of higher vertebrates culminating, if one wishes, in the study of man.

Rays, skates, and sawfishes. Rays, skates, and sawfishes are elasmobranch fishes whose bodies are more or less flattened. If one could grasp the lateral body wall of a dogfish just above the gill slits and stretch the body wall laterad to form a "wing" without interfering with the location or shape of the coelomic cavity, he would produce the body form of a ray or skate. The gill slits in rays and skates are ventral in position (Fig. 12-20), but the spiracle is on the dorsal aspect. Movement is accomplished by undulation of the winglike lateral body wall. The tail in the sting ray is modified into an organ of defense and offense. In *Raia* it becomes an electric organ capable of delivering a high-voltage electric current. Rays may attain gigantic proportions. The giant ray of tropical waters weighs nearly half a ton and has a pectoral finspread of more than 20 feet!

Sawfishes (Fig. 3-6) are elasmobranchs that do not become as flattened as rays and skates. Characteristic of sawfishes is the elongated rostral portion of the head, which bears sharp sawtooth projections along the lateral edges. As in rays, the gill slits are ventral in position.

Holocephalians. *Chimaera,* the ratfish (Fig. 3-6), is a genus of atypical Chondrichthyes which, though exhibiting numerous spines, lacks scales over most of the body and has a smooth, slippery skin. The gill slits are covered by a fleshy boneless operculum. The spiracle is closed. Instead of teeth there are hard, flat plates along the exposed surfaces of the jaws. The upper jaw, unlike that in elasmobranchs, is solidly fused with the cartilaginous brain case.

Osteichthyes

Osteichthyes (*osteo* = bone; *ichthy* = fish) have a skeleton composed partly of bone; the gill slits are covered by a bony operculum that grows from the second visceral arch; the scales are ganoid, cycloid, or ctenoid; and most bony fishes have an air-filled swim bladder arising from the esophagus (Figs. 12-7 and 14-14). In all but lungfish the cloaca is so shallow as to be practically nonexistent. There are never more than five gill apertures, and the males do not develop claspers. All Osteichthyes may be classified as either ray-finned or lobe-finned fishes.

Ray-finned fishes. Ray-finned fishes are bony fishes in which numerous horny rays develop in the dermis of the fins. The rays account for the subclass name, Actinopterygii (*actino* = ray, *pterygo* = small fin or wing). Ray-finned fishes include three subgroups: (1) chondrosteans, or cartilaginous ganoids (which have bone as well as cartilage, however), (2) holosteans, or bony ganoids, and (3) teleosts, or modern bony fishes.

Of the three groups, the lower two, referred to as **ganoid fishes** (Chondrostei and Holostei), have seen better days. Time was when these forms, well supplied with a bony internal skeleton, protected on the head by bony dermal plates, and covered over the entire body by large ganoid scales, were the dominant fishes. They are now represented by only a few forms, and not all retain typical ganoid scales.

By far the most abundant ray-finned fishes at present are the Teleostei. Since they apparently are of more recent origin

than ganoids, teleosts are called modern bony fishes.

Chondrosteans. The Chondrostei are represented today by only a small number of species, among which are the sturgeon *(Acipenser)*, and the paddlefish, or spoonbill *(Polyodon)* (Fig. 3-7). Although ancient chondrosteans exhibited a bony endoskeleton, the more recent sturgeons and paddlefishes have an endoskeleton composed primarily of cartilage. Bony plates of dermal origin overlie the cartilaginous braincase. The presence of an internal cartilaginous skeleton coupled with a bony dermal covering in the head region accounts for the name Chondrostei or "cartilage-bone" fishes. Sturgeons and paddlefish have for the most part lost the ganoid scales, except in restricted regions.

Polypterus and *Calamoichthys* are aberrant genera found in Africa. For a long time these were classified with lobe-finned fishes because a fleshy basal lobe occurs on the pectoral fins. However, typical dermal fin rays occur in the lobes, whereas true lobe-fins lack this condition. Like ancient

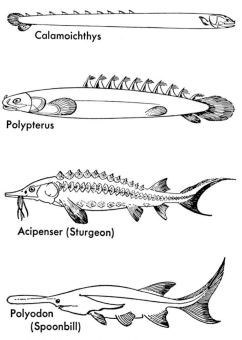

Calamoichthys

Polypterus

Acipenser (Sturgeon)

Polyodon (Spoonbill)

Fig. 3-7. Group of chondrosteans.

chondrosteans, they exhibit typical ganoid scales and a well-ossified internal skeleton.

Holosteans. Holosteans (Fig. 3-8) had a well-ossified internal skeleton. This accounts for the name Holostei (*holo* = complete, *osteo* = bone). Like other ancient fish, they exhibited prominent bony plates in the skin of the head.

Only two holosteans have survived until today—gars *(Lepidosteus)* and bowfins *(Amia)*. Both are freshwater forms, as were their more abundant ancestors. Gars are covered with typical ganoid scales. In bow-

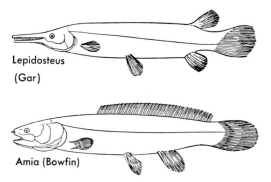

Fig. 3-8. The sole living holosteans.

fins,* ganoid scales are limited to the head, the ganoin on the surface of the scale has been lost, and the trunk and tail have cycloid scales like those of more recent bony fishes.

Teleosts. Unless the beginning student of zoology is particularly interested in ichthyology (the study of fishes), he may gain little notion of the enormous variety of teleosts (Fig. 3-9). The more than 20,000 species of teleosts (about 90% of all living fishes) have displaced ostracoderms, placoderms, chondrosteans, and holosteans, which have disappeared or are disappearing, perhaps in that order.

There are long, slim teleosts without paired appendages; short, fat ones with sails; transparent ones; fish that stand on their tails; fish with both eyes on the same side of the head; fish that carry lanterns; fish that climb trees; fish that carry their eggs in their mouths; fish that appear to be smoking pipes; fish that possess peri-

*The bowfin is also known as a choupique, cypress trout, mudfish, grindle, blackfish, and beavertail, according to the locality.

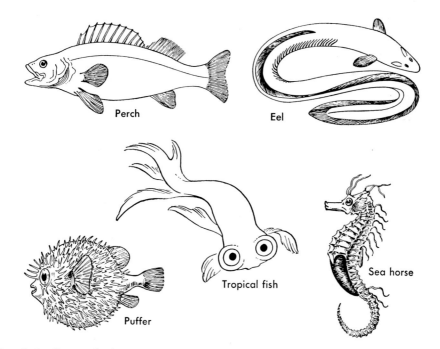

Fig. 3-9. Group of teleosts.

scopical eyes; and hundreds of additional bizarre genera. They inhabit the abyssal depths far out from the continental shelf, and they cavort in modest brooks. They run the gamut of colors, although a relatively small number of pigments assisted by myriads of light-dispersing crystals are responsible for all hues.

The skeleton of bony fishes is completely ossified. The cycloid or ctenoid scales are round, thin, pliable, and overlapping and are relatively modern. The pelvic fins are often situated far forward near the pectoral fins. There is no spiracle. These are only a few of the characteristics of teleosts. With the exception of the ganoid fishes already listed and the lobe-fins, which will be discussed next, any fish caught on a hook, viewed in an aquarium, or observed in the market place is a teleost.

Lobe-finned fishes. Lobe-finned fishes have a fleshy lobe constituting the basal part of their paired fins, a condition not found in ray-finned fishes. For this reason they are placed in the subclass Sarcopterygii (*sarc* = flesh, *pter* = fin or wing). They are sometimes called **Choanichthyes** because most of them have internal nares (choanae) that open into the oral cavity. There are two orders of lobe-fins, the Crossopterygii and the Dipnoi, or lungfishes. These two groups were differentiated by the start of the Devonian period.

Crossopterygii (Fig. 3-10). With the exception of *Latimeria*, crossopterygians are all yesterday's animals. *Latimeria*, which lacks internal nares, was found off the coast of South Africa in 1938. Prior to that date, it was supposed that crossopterygians had disappeared completely. Crossopterygians are of special interest because of their resemblance to amphibians. The skeletal elements within the fin lobe resembled the proximal skeletal elements of tetrapod limbs (Fig. 9-29). The crossopterygian skull was similar to that of the earliest amphibians (Figs. 8-20 and 8-26). Freshwater crossopterygians were capable of clambering onto the muddy banks of their native streams and migrating short distances from one body of water to another because they used their air bladders as lungs. And many had internal nares, although these may not have been used for breathing. Because of these and other morphological traits, one group of crossopterygians, the freshwater, carnivorous rhipidistians, are thought to have been the stem from which the earliest amphibians evolved.

Dipnoi. Only three living genera of Dipnoi (true lungfishes) are known. These are *Protopterus* from Africa, *Neoceratodus* (Fig. 3-10) from Australia, and *Lepidosiren* from Brazil. During the wet season these animals enjoy life in their freshwater habitats and respire by means of gills. But when the sun dries the streams, they dig deep burrows in the moist, muddy banks and spend the dry, hot season in a state of lowered metabolism (aestivation✱). This state minimizes water loss. When necessary, lungfishes may employ their swim bladders as accessory respiratory organs.

Certain structures are more specialized than among the crossopterygians. The skeleton of the fins is not reminiscent of that of a primitive tetrapod; there are peculiar bony plates instead of teeth; the swim bladder is supplied by a branch of the sixth aortic arch as in amphibians, instead of being supplied from the dorsal

---

✱The Australian form does not aestivate.

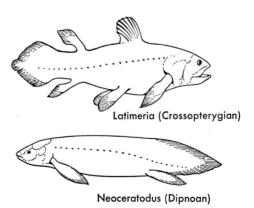

Latimeria (Crossopterygian)

Neoceratodus (Dipnoan)

Fig. 3-10. Lobe-finned fishes.

aorta as in crossopterygians; the atrium of the heart has become partially divided into two chambers as in amphibians; and external gills occur in larval stages. They have no spiracle. These amphibian-like features led early workers to conclude that lungfishes gave rise to amphibians. This view is no longer held (Fig. 3-11).

TETRAPODA

Tetrapods are chordates that ordinarily have four legs. However, snakes, legless lizards, and burrowing amphibians have no legs, and in other tetrapods one pair of legs may be absent or modified as hands, wings, or paddles. Regardless of superficial differences, legs, wings, arms, and flippers are all built according to the same basic pattern. Employing limbs, tetrapods swim, crawl, walk, run, hop, dig, climb, or fly from one environment to another and thus avoid their enemies or seek food and mates. The four classes of tetrapods are amphibians, reptiles, birds, and mammals.

Amphibia

In numerous respects amphibians resemble fishes. They respire by gills early in their lives and in some cases throughout life. The eggs have insufficient protection to prevent desiccation in air and so are usually laid in water, where, as larvae, development usually continues. Most amphibians have air bladders, which, as in some fishes, serve as lungs. Like fishes, aquatic amphibians have a canal system for reception of low-frequency vibratory stimuli from water. Resemblance

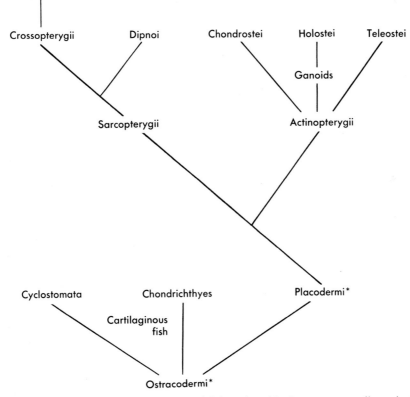

Fig. 3-11. Theoretical phylogenetic lines of fishes. Asterisked taxa are totally extinct. The distribution of fishes in time is given in Fig. 3-3.

to fish is seen in many of the systems of the body.

On the other hand, amphibians resemble reptiles in many respects. Many are well-adapted to a land habitat. They exhibit a middle ear cavity for receiving vibratory stimuli from the air. The nostrils lead to the oral cavity. The skeleton is adapted for locomotion on land. Because of such similarities, amphibians and reptiles were once classified together. Even today, students of amphibians and reptiles are allied and call themselves herpetologists.

Fish, amphibians, and reptiles are all ectothermic (cold-blooded), the temperature of their bodies being roughly the same as that of their surroundings. Most of their heat is absorbed from solar radiation. When the temperature becomes sufficiently low, ectotherms become sluggish or immobile.

Modern amphibians have some characteristics found in few other tetrapods. Almost all have a free-living larval stage with gills. The integument, which may serve as an accessory respiratory organ, is devoid of any kind of scales except in caecilians, and is moistened by ubiquitous mucous glands. Usually lacking are the thick layers of cornified epidermal cells and horny integumentary devices such as claws, which are characteristic of reptiles, birds, and mammals. Many skeletal features are diagnostic of amphibians. Among these are the presence of one sacral and one cervical vertebra. The skull exhibits some distinct amphibian features.

Modern amphibians belong to the superorder Salientia. They are probably descendants of the earliest known tetrapods, which have been placed in the superorder **Labyrinthodontia.** Labyrinthodonts are therefore considered "stem amphibians." Modern amphibians occur in three orders: Caudata (tailed amphibians), Anura (frogs and toads), and Apoda (caecilians, or burrowing amphibians).

Caudata (Urodela). Tailed amphibians (urodeles) are more generalized than frogs. They resemble somewhat the stem amphibians from which they are descended (Fig. 3-15, A). The larvae breathe by gills, exhibit gill slits, a long, keeled tail, and a lateral-line system, and often lack the hind pair of appendages. At metamorphosis posterior appendages usually develop, lungs appear, the external gills are usually resorbed, and the gills slits usually close. The streamlined, fish-like form of the trunk is not altered, the postanal tail remains, and the lateral-line canal system is often retained. Thus tailed amphibians undergo less profound alterations in structure and habits at metamorphosis than do frogs, which emerge as highly specialized, tail-less, hopping adults.

Some urodeles retain external gills throughout life, even though lungs develop. Adult animals that retain some larval traits are said to be neotenic.* A few urodeles (Plethodontidae) lose their gills but do not develop lungs.

The seven families of urodeles in the United States and Canada are listed in Table 3-1 and illustrated in Figs. 3-12 and 3-13.

Proteidae. NECTURUS. The mud puppy is the only genus of Proteidae in the United States. Its generalized amphibian anatomy has made it a classical subject of laboratory study. As a 1-inch larva, *Necturus* has tiny imperfect tetrapod appendages, but the toes at the distal ends of the hind limbs are scarcely recognizable. The belly is distended by the presence of the yolk sac, from which the embryo had been receiving nourishment prior to hatching. The tail is keeled and has dorsal and ventral fins. The larva respires by means of three pairs of external gills and exhibits three pairs of open gill slits. As the larva grows, the toes become more pronounced, the skin darkens, and the yolk no longer distends the venter. At approximately 5 years of age *Necturus* attains the length of 8 inches and is sexually mature. A pair of functional

---

*Neoteny is the prolonged retention of larval characters. When the larval trait retained is external gills, the animal may also be said to be "perennibranchiate."

Table 3-1.   Distribution of gills, gill slits, and lungs among adult urodeles

Family	Number of pairs of gills	Number of pairs of slits	Lungs
Proteidae, mud puppies	3	2	Yes
Amphiumidae, Congo eels	0	1	Yes
Cryptobranchidae, hellbenders	0	1	Yes
Salamandridae, newts	0*	0	Yes
Ambystomatidae, blunt-mouthed salamanders	0*	0	Yes
Plethodontidae, lungless salamanders	0*	0	No
Sirenidae, sirens	3	3 to 1	Yes

*Occasionally perennibranchiate (see footnote at bottom of p. 42).

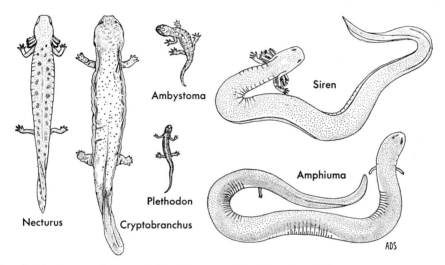

Fig. 3-12. Representatives of six of the seven families of urodeles.

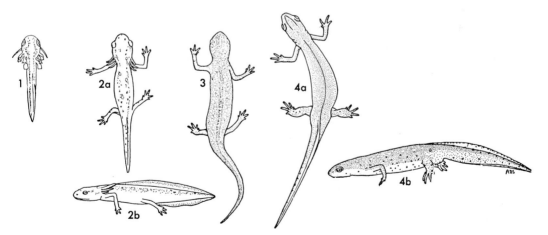

Fig. 3-13. Life history of the salamander *Notophthalmus* (formerly *Triturus*). **1,** Newly hatched larva (7 mm.); **2a,** fully formed larva (30 mm.); **2b,** fully formed larva, lateral view; **3,** red eft (70 mm.); **4a,** male newt (95 mm.); **4b,** male newt, lateral view.

lungs has developed, but the adult retains the three pairs of external gills and two pairs of gill slits. The tail is still keeled and has fins. Thus *Necturus* reaches sexual maturity without profound metamorphosis, retaining numerous larval traits.

Other tailed amphibians (Table 3-1) occasionally retain larval traits. In the case of *Ambystoma* it has been shown that if the neotenic individuals are injected with thyroid gland extract, they will complete their normal metamorphosis, lose their gills, close their gill slits, and become normal adults. *Necturus*, however, cannot be induced to discard its external gills by administration of thyroxin. One or more intracellular enzymes necessary to enable thyroxin to bring about its effect are either lacking in *Necturus* or are blocked by other metabolic substances.

There are six species or subspecies of *Necturus* in the United States and Canada, distributed in the eastern and western tributaries of the vast Mississippi River system, in the tributaries of the Great Lakes, and in numerous other large freshwater lakes and streams eastward to the Atlantic seaboard and southward to the Gulf of Mexico. *Necturus maculosus* is the most widespread of the species.

Amphiumidae. AMPHIUMA. This amphibian is a large, aquatic, eel-like urodele, which attains a length of 40 inches. One pair of gill slits remains in the adult (Fig. 1-8), which breathes entirely by means of lungs. Anterior and posterior appendages are very small, out of proportion to the length of the body, and entirely inadequate for bearing weight. *Amphiuma* (the Congo eel), the only genus in the family, occurs in the coastal regions of the United States bordering the Gulf of Mexico and the Atlantic Ocean as far north as Virginia. It is also found in parts of Missouri. There are two subspecies, *Amphiuma means means* (which has two toes) and *Amphiuma means tridactylum* (which has three toes).

Cryptobranchidae. CRYPTOBRANCHUS. This family of aquatic urodeles also has only one genus in the United States and Canada. *Cryptobranchus* looks more ferocious than other urodeles because of its broad, flattened head, its compressed tail with a thin, deep dorsal keel, the longitudinal wrinkles on its chin, and a wrinkled, fleshy fold that occurs along the entire length of the trunk on each side. *Cryptobranchus* attains a maximum length of 27 inches. One gill slit usually remains on each side, concealed beneath folds of skin in *C. alleganiensis* and exposed in *C. bishopi*.

Salamandridae. NOTOPHTHALMUS. Salamandridae are the true salamanders. There are eight species but only one genus of this family in the United States and Canada. *Notophthalmus* (formerly *Triturus*) *viridescens*, the spotted or common newt, typically exhibits three distinct phases of postembryonic life—larva, eft, and newt (Fig. 3-13). The **larva**, with gills and gill slits, lives in water. After several months gills and slits are lost, hind legs appear, and the animal, now an **eft**, emerges from the water to commence a term of residence on land, often ascending mountains to 4,000-foot altitudes. The skin becomes heavily cornified as in land animals, and the skin glands are relatively functionless. The body gradually assumes a bright orange-red color, and a series of black-bordered red spots develop along the dorsolateral aspect. The sojourn on land lasts one to three years, depending upon the locality. The land phase ends as the eft approaches sexual maturity under the stimulus of gonadotropic hormones from the pituitary gland. As the time for mating draws closer, the efts commence mass migrations down the hills, through the lowlands and meadows, toward the freshwater ponds. The migration is a manifestation of the water drive brought about by another pituitary hormone, prolactin. The skin loses its brilliant color and by the time the animals enter the ponds, many have assumed the adult coloration, olive green on the back and light yellow ventrally. Mucous glands in the skin, until now quiescent, become greatly enlarged and actively secrete. The tail changes from round to laterally compressed and once again develops a dorsal and ventral fin. The animal is now a sexually mature water-phase individual known as a **newt.**

On Long Island, New York, where the terrestrial habitat is apparently not so suitable as in other localities, the eft stage is of short duration. In this locality some larvae may never leave the water, retaining vestigial gills throughout their lives.

Notophthalmus, which spends part of its life on land with structures adapted to a terrestrial habitat and part of its life in water with structures adapted to an aquatic existence, illustrates the physiological effects of endocrines on the habits and structure of animals. Here, indeed, is justification for the term "amphibian," an animal that exhibits both (*amphi* = both) types of life (*bio* = life)—terrestrial and aquatic.

Ambystomatidae. AMBYSTOMA. Members of this family occur in three genera in the United States and Canada, the best known being *Ambystoma.* They resemble *Notophthalmus* in body form but may be differentiated by their costal grooves. Ambystomatidae are terrestrial urodeles except temporarily during the breeding season, when migration to nearby ponds occurs. Several species sometimes retain external gills and remain permanently aquatic after reaching sexual maturity. These perennibranchiate members will discard the gills and achieve the normal adult state when supplied with exogenous thyroid gland extract.

Plethodontidae. PLETHODON and DESMOGNATHUS. This family includes numerous genera, some terrestrial and others primarily aquatic. Adults lack both gills and lungs, respiration being carried on entirely through the moist skin. Although *Desmognathus* is terrestrial, it is usually found in the immediate vicinity of a spring, pond, or small stream, but *Plethodon* occurs at considerable distances from bodies of water. Plethodonts lay their eggs on land in moist places such as the undersides of logs or attached to rocks in damp caves. The larvae

hatch with legs already formed and may never enter water. In such instances larval gills are of no advantage and are discarded a few days after hatching.

Sirenidae. SIREN and PSEUDOBRANCHUS. The only two genera in this family are found in muck, or swamplands. They are perennibranchiates that never develop hind limbs. *Siren* usually has three gill slits, but not all of these may be patent. *Pseudobranchus* retains one open gill slit.

Anura. Adult frogs and toads are fully metamorphosed, tailless amphibians in which the caudal vertebrae are fused into one elongated bone, the urostyle. Like other amphibians, they exhibit an aquatic larval stage lacking paired appendages. As metamorphosis approaches, the larvae sprout anterior and posterior appendages, lose their gills, close the pharyngeal slits, resorb the tail, and develop extensive appendicular muscles. The lateral-line system disappears. The anuran thereafter breathes by lungs assisted by the skin and lives on land or in the water.

A few anurans do not live near water and cannot deposit their eggs in water. But what would happen if larvae, equipped with external gills, no legs, and a fish-like body for swimming, were to hatch from their jelly envelopes only to find they were not in water? The species would almost surely perish. But the species survive, thanks to an adaptive mutation. All development through the larval stage occurs while the young are still within the jelly envelopes. Upon hatching, a fully metamorphosed miniature adult emerges to assume life on land. Thus, these frogs (robber frogs and others) skip a free-living aquatic tadpole stage and thereby continue to exist. (Remarkable as it may seem, this is not unusual among tetrapods. Reptiles, birds, and mammals do this regularly!) The South American tree frog does not lay her eggs near the water either. Instead, she carries around the developing eggs in a brood pouch under the skin of her back. When the young have metamorphosed, the skin of the back of the mother splits, and

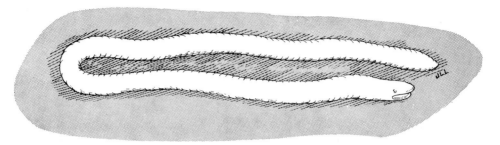

Fig. 3-14. Legless amphibian. The annulate structure is a specialization for burrowing.

fully metamorphosed young frogs hop out.

There are seven families of anurans in the United States and Canada, the most common being Bufonidae (toads), Hylidae (tree toads), and Ranidae (frogs). *Bufo* may be distinguished from *Rana* by the warty skin, broad waist, short hind limbs, and elevated parotoid glands located behind the eyes (Fig. 5-14). Tree toads have suction cups on the ends of their digits, which aid in climbing. In one family of anurans the male has a stubby tail, used as a copulatory organ. In other families the eardrum may be absent, the upper jaw may be toothless, the pupils may be horizontal or vertical, and there may be an extra digit near the thumb or big toe.

The immediate ancestors of frogs and toads are thought to be represented by an extinct order of Salientia, the **Proanura**. They had skulls very much like those of modern anurans. However, they had long ribs and a tail, and their hind limbs had not become modified for jumping.

Apoda. Apoda (Fig. 3-14), or caecilians, are limbless amphibians, which, except for a few aquatic forms, live in burrows in the earth. Their eyes are small and are sometimes buried beneath the bones of the skull. They have minute scales buried in the skin, vestiges of the dermal scales of ancestral amphibians. They may have as many as 250 vertebrae! Caecilians have a foreshortened tail, the vent being almost terminal in position, as in anurans. They lay large eggs laden with much yolk. The larval stage occurs in the egg envelops in

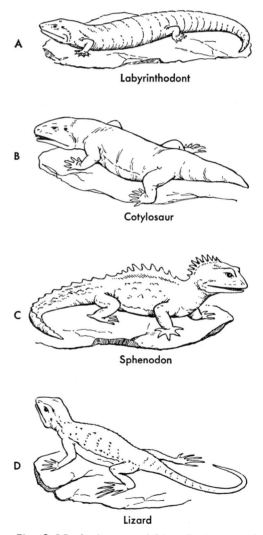

A, Labyrinthodont

B, Cotylosaur

C, Sphenodon

D, Lizard

Fig. 3-15. A, Stem amphibian. **B,** Stem reptile or reptile-like amphibian (*Seymouria*). **C,** "Living fossil" reptile (*Sphenodon*). **D,** Modern lizard (*Anolis*). The stem amphibian and lizard are separated by 300 million years.

species that lay their eggs on land. One aquatic genus gives birth to its young. Apoda occur primarily in the tropics, where there are many species.

Labyrinthodontia. Labyrinthodonts (Fig. 3-15) are the oldest known tetrapods. They were abundant at the time our present coal deposits were being formed, about 300 million years ago. They resembled salamanders and early reptiles, having an elongated, muscular body, which probably sagged onto the ground, and a strong, muscular tail. The largest were the size of crocodiles. Although they retained bony scales where the belly came in contact with the earth, the lateral and dorsal portions were often devoid of scales, or the scales were minute.

Some labyrinthodonts were probably terrestrial, but the majority were freshwater forms. They are thought to have arisen from freshwater rhipidistian crossopterygian fish. Not only were crossopterygians the ancestors of a large number of extinct and modern amphibians, but as nearly as can be determined, one branch also gave rise to reptiles.

Reptilia

From the ancient labyrinthodont amphibians there arose, if we read correctly Nature's handwriting in the earth, a group of tetrapods destined to be named, 300 million years later, the cotylosaurs (Fig. 3-15). These were the first, or stem, reptiles (*repere* = to crawl). From them developed a distinguished and hoary group of descendants, the members of the class Reptilia.

The cotylosaurs have vanished. Gone with them are their descendants the dinosaurs, those ruling reptiles that held court for 150 million years when oil for modern diesels was being formed from rotting vegetation. Gone, too, are the flying reptiles (pterosaurs) and the aquatic, viviparous ichthyosaurs (Fig. 9-19), with paddles for limbs. Vanished are hundreds upon hundreds of mighty or weak, great or small, fast or lumbering reptiles, which

slowly bowed out of the vertebrate parade while mammals were first raising their weak, small voices in the forests and at the edges of the bogs. Remaining with us even today are a few successful cotylosaur descendants: the turtles, an ancient group, which thus far has doggedly persevered in the struggle for existence; the lizards, more recent additions to reptilian society; the snakes, which are lizards deprived of their appendages through some fortuitous circumstance; and the crocodilians (crocodiles and alligators). These few forms mirror the effect of mutations and the ravages of a changing environment upon cotylosaurs and their descendants. Birds and mammals represent still other mutations of reptilian chromosomes, but they have become so different from the cotylosaur type that they are no longer classified with reptiles.

Reptiles made significant advances over amphibians and fishes in the acquisition of three special extraembryonic membranes, which have emancipated them and their descendants (birds and mammals) from the necessity of laying eggs in water.

1. The **amnion** (Fig. 4-11) is a membranous sac filled with watery, salty, amniotic fluid. The embryo develops in this fluid just as the embryos of fishes and amphibians develop in the pond or sea. The fluid is secreted by the cells of the amnion. To state the situation fancifully, instead of the mother going to the water to deposit her eggs, the developing embryo secretes its own pond around itself! Because of the amnion, reptiles, birds, and mammals are called **amniotes.**

2. In the process of establishing the amnion another membrane, the *chorion,* is produced. The chorion lies against the eggshell or, in the case of euviviparous amniotes, against the lining of the mother's uterus.

3. From the hindgut sprouts a sac, the **allantois,** which comes to lie closely applied to the chorion (Fig. 4-11, *B*).

Allantois and chorion usually become so intimately associated that the resulting membrane is often referred to as the **chorioallantoic membrane.** The vascularized chorioallantoic membrane lying against the porous eggshell (or against the uterine lining of the mother) takes the place of larval gills for respiration. Kidney wastes collect in the allantoic sac and the fluids are largely reabsorbed by the allantoic vessels, thereby conserving water. Because they have these extraembryonic membranes, reptiles found themselves able, and indeed forced, to lay their eggs upon the land. Their young hatch fully formed, skipping the larval stage, ready at once to seek their food from the terrestrial environment. Not only were reptiles and their descendants liberated from the necessity of returning to water to lay their eggs, but those that dwell in water must emerge onto the land to do so. Otherwise, they must give birth to their young, as do some marine snakes.

Reptiles are covered with a layer of dead, cornified epidermal scales, often underlaid by bony dermal plates. Always, the skin is thick and dry, with few glands. This condition results in the conservation of water, an advantage to animals that would otherwise tend to dry out by evaporation in air. The ends of the digits are supplied with claws. A newer type of kidney, the metanephros, has come into existence and is found thereafter in birds and mammals. The ventricle of the heart is partially or completely divided into right and left ventricles. In modern reptiles and birds the two primitive occipital condyles have fused to form a single occipital condyle beneath the foramen magnum. The pelvic appendages articulate with two sacral vertebrae instead of with one as in amphibians. Like amphibians but unlike birds and mammals, reptiles are ectothermic.

These then are reptiles: scaly, clawed, mostly terrestrial tetrapods lacking feathers and hair, which (except for a few viviparous forms) lay large, shell-covered, yolk-laden eggs on land, the embryos of which are surrounded by an amnion, and the young of which are hatched fully formed (Fig. 3-16).

For purposes of classification the skull and teeth are primarily utilized as criteria to separate the groups. This is effective because the skeleton is evolutionarily con-

Fig. 3-16. *Anolis* emerging from egg. (Courtesy Carolina Biological Supply Co., Burlington, N. C.)

servative, and the skeleton and teeth are more often preserved as fossils. The temporal region of the skull (behind the eyes) is accorded special consideration for classification purposes (Fig. 8-27). An abridged classification of reptiles will be found on p. 415. There are numerous extinct reptilian orders not indicated in the classification.

Cotylosauria (Fig. 3-15, *B*). Cotylosaurs are the stem reptiles from which subsequent reptilian stock arose. The skull resembled that of labyrinthodonts.

Chelonia. These ancient reptiles, the turtles, probably have remained relatively unchanged for 175 million years. Turtles are identified by their armored shell of bony plates. There are no teeth, and the jaws are beak-like. A reduction or loss of many of the body wall muscles has occurred. The ribs and the vertebral column of the trunk are solidly fused with the shell.

Rhynchocephalia. Sphenodon (Fig. 3-15, *C*) is an ancient lizard found only in New Zealand. It has been called a living fossil because it is the only remaining member of an entire rather primitive order. One would expect it to have disappeared millions of years ago, giving way before the onslaught of more modern forms. The factor that has preserved it thus far may have been its lack of competition and isolation on the islands. The teeth, unlike those in true lizards, are not embedded in sockets.

Squamata. Squamata are the modern lizards and snakes. There are occasional limbless lizards such as the glass "snake," which, when captured by the tail, breaks off the end of the tail (**autotomy**) and then regenerates a new one. The horned "toad" of the western deserts is a lizard. So, too, are chameleons that change color to blend with their background.

Snakes have lost both pairs of limbs and both girdles, although a few retain a trace of the pelvic girdle. They are thought to have developed from an ancient lizard-like reptile. The two dentary bones of the lower jaw do not fuse in the midline anteriorly (at what would be the chin). This is a distinguishing skull character of snakes. The epidermis of the skin is periodically shed, usually in one piece. The process is known as molting, or ecdysis.

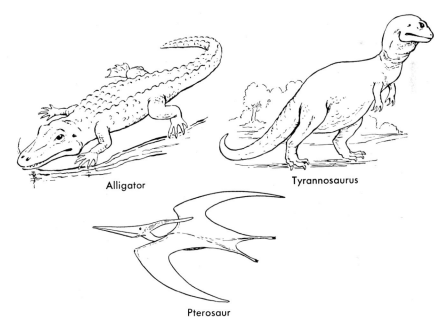

Alligator

Tyrannosaurus

Pterosaur

Fig. 3-17. Representative archosaurs.

Archosauria. Archosaurs (Fig. 3-17) were the dominant land vertebrates in the period known as the Age of Reptiles. Included among archosaurs are crocodilians, extinct flying pterosaurs, and dinosaurs. Though dinosaur means "fearful reptile," most were neither as fearful nor as massive as *Tyrannosaurus,* which was the end of a line.

Crocodiles and alligators are large amphibious reptiles with bony plates underlying the dry, leathery skin of the back (Fig. 5-16). They have a long secondary palate which separates the nasal passageway from the buccal cavity all the way back to the pharynx. The teeth are in sockets. The heart exhibits a completely divided ventricle. Crocodiles may be differentiated from alligators on the basis of the shape of the snout, which is slender in crocodiles and broader in alligators; by the fact that the nasal bones form a septum dividing the nasal aperture in alligators; and by the fact that the fourth tooth of the lower jaw in crocodiles usually fits into a lateral notch in the upper jaw and can be seen when the mouth is closed, whereas in alligators this tooth fits into a medial notch. Crocodiles are common in the subtropical and tropical waters of Asia, Africa, the Americas, and Australia. Alligators occur in the warm southern region of North

Cynognathus

Fig. 3-18. Synapsid (mammal-like) reptile about the size of a large dog. (From Colbert: Evolution of the vertebrates, New York, 1955, John Wiley & Sons, Inc.)

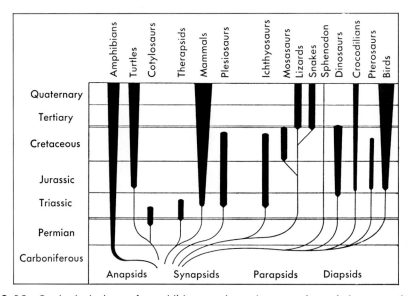

Fig. 3-19. Geological chart of amphibians and amniotes to show their range through time. (From Atwood: Comparative anatomy, ed. 2, St. Louis, 1955, The C. V. Mosby Co.)

America *(Alligator mississippiensis)* and in China *(Alligator sinensis).* Another family member, easily mistaken for an alligator, lives in South America. It is a caiman. According to Cope, alligators are of much more recent origin than crocodiles and are not known as fossils.[128]

Synapsida. Synapsida are reptiles, long extinct, from which mammals are believed to have emerged after hundreds of millions of years of chromosomal changes (Figs. 3-18 and 3-19). They arose early during reptilian evolution. Later synapsids had two occipital condyles, a mammalian

secondary palate, and a dentition consisting of incisors, canines, and grinding molars—all characteristic of mammals. Their limbs were like those of mammals and capable of generating considerable speed. The dentary was the largest bone in the lower jaw, presaging the mandible of mammals. Like other reptiles, they had only one bone in the middle ear, a tiny braincase (the part of the skull immediately surrounding the brain), and a parietal (parapineal) foramen. The skull was intermediate between that of the reptiles and mammals. Nothing is known of the skin.

Fig. 3-20. A, *Archaeopteryx,* the earliest known bird of the Jurassic period. **B,** A pigeon, for comparison. How many skeletal changes can you find? (From Colbert: Evolution of the vertebrates, New York, 1955, John Wiley & Sons, Inc.)

Aves

Excellent evidence indicates that birds, that is, vertebrates with feathers, arose from an archosaurian reptile probably related to the reptilian order Saurischia (p. 415). Saurischia included small and large dinosaurs. Saurischian dinosaurs could stand and run on their hind limbs, freeing the forelimbs for subsequent modification for flight. They had a long tail and a fairly extensive set of teeth. Birds lost some reptilian traits such as tail and teeth but retained others such as scales on the legs and feet, a single occipital condyle, and a modified diapsid skull. Many internal systems of birds are reminiscent of reptiles.

Archaeornithes. Few fossil links between reptiles and birds have been found. There are two fossil genera that we call birds because they have feathers; but most of their other characteristics were reptilian. These fossil birds are *Archaeopteryx* (Fig. 3-20) and *Archaeornis.* Three fossil specimens of *Archaeopteryx* (a single feather in one case) and only a single specimen of *Archaeornis* have been recovered. These came from slate deposits in Bavaria (Germany). They have been placed in the subclass Archaeornithes (*arch* = ancient, *ornis, ornitho* = bird), the oldest birds on record. Whether they were in the direct line to modern birds is not known. They were about the size of a crow, had long reptilian tails with a row of feathers attached along the lateral borders, and had teeth. Their bones were solid instead of hollow as in modern birds. The wings were feebly developed, and the breastbone was small, indicating weak flight muscles. Probably these early birds were not capable of sustained flight.

Neornithes. Neornithes (*neo* = new), more recent in origin than Archaeornithes, are made up of three groups. **Odontognaths** are all extinct. **Paleognaths** are ratites, incapable of flight. **Neognaths** are carinates, the flying forms.

Odontognaths (*odonto* = tooth, *gnatho* = jaw) are birds discovered in marine deposits in Kansas (U.S.A.) They had many features of modern birds, including a foreshortened tail and hollow bones. *Hesperonis* (western bird) had teeth and vestigial wings. More than a hundred specimens have been uncovered. *Ichthyornis* had powerful wings, but the tooth-bearing jaws originally ascribed to *Ichthyornis* are now known to have belonged to a small dinosaur.[157]

Ratites. Ratites are a comparatively small group of flightless birds (although one group, the tinamous, are fliers) that face extinction unless steps are taken to protect them. Many are known only as fossils. They have small, incompetent wings but powerful leg muscles that permit them to run agilely. Among extant ratites are the ostrich, kiwi, emu, rhea, and cassowary. One extinct wingless form from New Zealand, the moa, ranged up to 13 feet tall and laid eggs more than 1 foot long and 9 inches in diameter! Another, the Madagascar elephant bird, was 10 feet tall and weighed 800 pounds.

Carinates. Structural modifications have adapted most modern birds for sustained flight. The bipedal condition of their reptilian precursors freed the forelimbs to evolve into wings. The sternum has become greatly enlarged and deeply keeled (*carina* = keel), thereby providing an extensive surface for attachment of the massive breast muscles that furnish power for flight. By increasing the surface area of the forelimbs, feathers provide an airfoil. By insulating the body, feathers conserve energy, which is an advantage to homoiotherms (animals that maintain a fairly constant body temperature). (Homoiothermy in an egg-layer necessitates incubation of the developing embryo by the parent.) Numerous modifications have resulted in reduced body weight and increased buoyancy. The earlier reptilian tail has been reduced to a stump called the **uropygium** (*pyg* = rump); the skull bones have become lightweight, and teeth have been lost; the large intestine has been shortened, and the urinary bladder has been lost; and many bones are hollow and contain diverticula of the lungs. The unique arrangement of the air ducts and lungs are described in later chapters. There are about twenty-eight orders and 8,600 species of carinates. A few, such as the great auk and penguin, have lost the ability to fly.

Homing instinct. Many birds exhibit remarkable powers of returning to their nests when released at great distances over unfamiliar territory. This phenomenon has been put to the service of man in the instance of the homing, or carrier, pigeon *Columba livia.* The bird utilizes obscure, but not unfathomable, environmental clues at each step of the journey, which serve to direct it closer to known territory. Among the clues that have been suggested are the direction of the sun, moon, or stars (although birds can find their way on cloudy days and moonless nights) and magnetic forces in the atmosphere of which man may be unaware. Birds have been thrown off their courses by radio waves emanating from broadcasting towers and by natural magnetic phenomena. It has also been suggested that birds may have some neural mechanism that enables them to record subconsciously the turns made in the journey away from the nest, even when the animal is being transported in a cage to the site of release, and that this mechanism then guides them in their return by being utilized in a reverse pattern, left turns on the outward journey being converted into right turns on the home flight and vice versa. The theory is tenable neurologically. However, birds placed on phonograph turntables that have revolved continuously en route and those that have been chloroformed have still been able to find their way back to home territory.

Migration. Many birds are annual migrants, passing part of a year in one geographical region and the remainder elsewhere, sometimes migrating great distances. The Arctic tern spends several months above the Arctic Circle and the remaining months in the Antarctic, traveling 22,000 miles round trip each year! During migration, birds move in mass flights,

often at night and at an elevation of approximately 2,000 feet. Those destined for the same geographical location may pass over approximately the same routes (flyways) year after year. A great flyway is located directly over the Gulf of Mexico between the Yucatan peninsula and the Gulf Coast of the United States. Other flyways are located overland between Central America and the United States and from the West Indies via the Florida peninsula. Migration is associated in part with mating. The physiological state necessary to predispose a bird for migration is under hormonal control. However, the immediate stimulus that causes a bird to depart at any given date is related to climatic conditions. Not all birds are migratory, and all parts of the United States and Canada have permanent resident species as well as migrant ones.

Mammalia

Mammals, descendants of synapsid reptiles, are unique in the exhibition of mammary glands (*mamma* = breast), which provide nourishment for the helpless young, and in having hair on at least part of the body. Distinguishing modern mammals from other vertebrates, too, is the mandible, composed of a single dentary bone on each side of the mandibular symphysis; the occurrence of three bones in the middle ear cavity; the presence of a muscular diaphragm separating a thoracic from an abdominal cavity; the presence (in most mammals) of sweat glands; the division of the embryonic cloaca into urogenital and intestinal passageways in all but the lowest orders; heterodont dentition (except in toothed whales); the development of two sets of teeth (milk teeth and a permanent set); the presence of marrow within the bones; circular, biconcave enucleate red blood cells; the loss of the right fourth aortic arch, so that the great artery leaving the heart turns to the left; the development of a pinna, or sound-collecting lobe, accessory to the outer ear; the specialization of the larynx, which permits greater range of vocalization than in any other animal; and the extensive development of the cerebral cortex.

Like birds, mammals are homoiothermic. Like amphibians and synapsid precursors, they have two occipital condyles.

The lowest mammals (Prototheria; *proto* = early, *theria* = beasts) lay large reptilian eggs. The remaining mammals (Theria) produce tiny eggs just visible to the naked eye, and these are retained in the uterus of the mother until birth. Newly born mammals are the most helpless of all offspring, being dependent on the mother for relatively long periods of time. At first they receive nourishment from her mammary glands; later, under her tutelage, they learn to seek food and to evade the enemy—the two chief problems of the young of all vertebrates.

Mammals, especially because of limb structure, have been able to achieve a greater diversity of habitats than have any other group of vertebrates. They burrow in the ground, hop across the deserts, race over the plains, swing from the branches of trees, propel themselves through the air in true flight, and swim at great depths in the oceans—each activity made possible by modifications of body structure. The structure of the teeth is related to their feeding habits. Herbivores eat plant life (chiefly grasses), which they crop with their incisors; carnivores prefer meat,

which they kill and tear with their enlarged canine teeth; and omnivores feed on both plants and animals. Some, but not all, insectivorous mammals are toothless.

Prototheria. Prototheria are the most reptile-like of mammals. There is only one order, the egg-laying monotremes.

Monotremata. There are two families of monotremes, represented by the spiny anteater, *Echidna*, and the duck-billed platypus, *Ornithorhynchus* (bird-nose), both occurring primarily in Australia (Fig. 3-21). These animals lay eggs containing a large amount of yolk and resembling those of reptiles and birds. The mammary glands are modified sweat glands and secrete a sweat-like milk, which accumulates on tufts of hairs within a temporary pouch on the abdomen. No teats occur. The egg of the spiny anteater is incubated within the pouch, and that of the duckbill is incubated in a nest. The newly hatched duckbill is picked up by the mother with her mouth and placed in the pouch, where it remains until old enough to live independently. In the pouch the young lick the "milk" exuded onto the surface of the hairs.

Echidna has a long, prehensile tongue used to capture insects. The name "spiny" anteater comes from the fact that, intermingled with the hairs over much of the body, are modified spiny hairs. The duck-like bill of *Ornithorhynchus* consists of

EGG-LAYING MAMMAL
Platypus

Fig. 3-21. Duck-billed platypus, a monotreme.

elongated upper jaw bones covered by horny skin. The feet are webbed, and the animal spends most of its time in the water. The nest is built in a tunnel on the bank of a stream. The eggs, often two in number, are nearly round, about ¾ inch in diameter, and covered with a white pliable shell.

In addition to laying eggs, monotremes are reptile-like in other respects. They retain a cloaca with a vent to the exterior (*mono* = one; *trema* = hole). They have a ventral mesentery extending the length of the abdominal cavity and may exhibit a functional abdominal vein. The testes are located within the abdomen. The outer ear has no pinna. The malleus and incus of the middle ear are larger than in other mammals, resembling the articular and quadrate bones of reptiles. The brain, like that of reptiles, lacks the great transverse fiber tract (corpus callosum) connecting the two cerebral hemispheres in other mammals. They are homoiothermic, but the body temperature is less stable than that of higher mammals, varying as much as 13° C.

Theria. Mammals other than monotremes give birth to living young. The eggs have only enough yolk to provide energy for the earliest mitotic divisions and none to permit growth of the embryo. Nourishment during intrauterine life must therefore be supplied from the body of the mother. Except in marsupials, this is accomplished by means of a highly vascular **placenta** composed of the chorioallantoic membrane of the fetus and the specialized uterine lining of the mother.

Despite the fact that no yolk mass occurs in typical mammalian eggs, embryonic mammals develop a yolk sac—a reminder of their reptilian ancestry. In marsupials the yolk sac serves as the fetal part of the placenta (**yolk-sac placenta**).

Monotremes, marsupials, and primitive placental mammals may have evolved independently from the first reptile-like mammals. For this reason marsupials are not regarded as descendants of mono-tremes, and the placental mammals are not necessarily descendants of marsupials.

Marsupialia. Marsupials (Fig. 3-22) are mammals usually exhibiting a permanent abdominal pouch of muscle and skin called the **marsupium.** In this pouch the immature, helpless young are transported, protected, incubated, and fed after birth until they are old enough to require suckling no longer. Within the pouch are teats. The marsupium is not to be confused with the temporary pouch of female monotremes, which arises only in connection with gestation. The marsupium opens forward in some marsupials and caudad in others. The walls of the pouch are supported by two slender marsupial bones that attach to the pelvic girdle and diverge as they project forward.

At birth the young are more immature than those of any other mammal. For example, the gestation period of the opposum is twelve and one-half days, and in kangaroos eleven days. It is not known whether

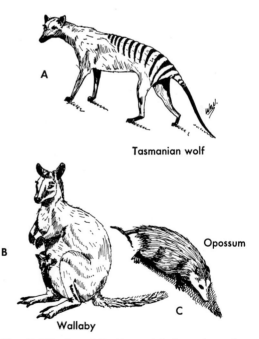

Fig. 3-22. A and **B**, Marsupials from Australia. **C**, The only native American marsupial (*Didelphis virginiana*).

A

Tasmanian wolf

B

Wallaby

Opossum

C

this is related to the fact that marsupials have a yolk-sac placenta. After the young are born, the mother places each one in the pouch and against a teat, which she forces into the mouth and part way down the esophagus of the young. Then by muscular contractions milk is pumped into the esophagus until the young have developed the necessary muscles for suckling.

The opossum *Didelphis* is the only native marsupial in North America. Another genus of marsupials occurs in Central and South America. All native Australian mammals except bats are marsupials or monotremes. Among Australian marsupials are the kangaroo, Tasmanian wolf, bandicoot rabbit, wallaby, wombat, Australian anteater, and phalangers. Australian marsupials resemble many common mammals in surprising details. There are wolf-, fox-, bear-, rabbit-, and cat-like marsupials. Some resemble flying squirrels, and there are also marsupial moles. The occurrence of no native mammal higher than a marsupial in Australia and the resemblance of Australian marsupials to many higher mammals raise interesting questions regarding how this came about.

Why should there be no nonflying mammal higher than a marsupial in Australia? It is widely accepted that Australia was once connected by a land bridge with the mainland of Asia and that subsequently this bridge was covered with water. It is theorized that mammals other than monotremes and marsupials had not yet reached Australia when the isolation of that continent occurred. Higher mammals could not thereafter have spread to Australia because of the water barrier. As a result, only marsupials and monotremes would occur in Australia. Thus, geographical isolation could account for the absence of other mammals on that island. Whether it does account for their absence is not known.

PARALLELISM. Why should Australian marsupials resemble placental mammals such as wolves, foxes, and bears? Surely Nature has shown herself to be so versatile that we can hardly believe she exhausted her power to create new types! Can it be that mammals in Australia and those elsewhere have proceeded along somewhat parallel evolutionary lines because they had identical genetic codes derived from common ancestors? If so, here would be an instance of parallel evolution. Parallel evolution is a term used to denote evolution that is occurring, or has occurred, in two **isolated** groups of **originally related** animals, wherein the results of the evolutionary changes in the two groups continue to be similar. The two groups must be isolated by geographical barriers, by physiological barriers, by differences in mating seasons, by psychological barriers, or by some other factor, so that there can be no interbreeding.

It could be that similar environmental conditions in Australia and elsewhere, rather than similar heredity, resulted in similar adaptive modifications in the isolated groups. The resemblances discussed above would then be instances of convergent evolution (p. 59), rather than of parallelism.

Insectivora. Among the most ancient of placentals and among those exhibiting the most primitive mammalian characteristics are the insectivores (Fig. 3-23). Although at one time abundant, they are represented today chiefly by a small group of retiring, often subterranean survivors: the burrowing moles, the shrews, and the spiny European hedgehogs. Insectivores subsist primarily on a diet of insects, worms, mollusks, and other small invertebrates, although they also devour small vertebrates.

Insectivores are regarded as closely related to the primitive placentals from which mammals arose. Among their primitive characteristics are a flat-footed (plantigrade) gait, five toes, smooth cerebral hemispheres, small, sharp, pointed teeth with the incisors, canines, and premolars poorly differentiated (Fig. 11-4), a large allantois and large yolk sac in the embryo, and often a shallow cloaca. The testes are retained in the abdominal cavity in some genera (a primitive trait), and they never descend fully into scrotal sacs in any genus.

Moles have foreshortened but very stout anterior limbs, with forefeet that are broad and more than twice as large as the hind feet, an adaptation for digging (Figs. 3-23, 9-13). The neck is short and the shoulder muscles are so powerful that the head and trunk seem to merge. Their tiny eyes are practically useless, but a delicate sense of smell locates distant food, and the sensitive tip of the elongated snout tells them when they have encountered it.

Shrews superficially resemble mice. They are shy, busy little fighters with a keen sense of hearing. They too have an elongated, bewhiskered, sensitive snout, but their anterior limbs are not modified as in moles. The teeth are tipped with brown coloration, and the incisors are long and

Fig. 3-23. The mole, an insectivore.

Fig. 3-24. Tree shrew. (Courtesy Delta Regional Primate Research Center, Covington, La.)

recurved. The eastern long-tailed shrew is the smallest mammal in the world.

The tree shrew (Fig. 3-24) is classified by some authorities in the order Insectivora, and by others as the most primitive member of the order Primates. It has traits that bridge the gap between the two orders.

Dermoptera. Dermoptera are insectivorous mammals about the size of a cat. They exhibit a broad, muscular fold of skin (**patagium**) stretched between the neck, forelimbs, hind limbs, and tail. There is only one genus, *Galeopithecus,* the flying lemur of Burma, Siam, the Malay Peninsula, the East Indies, and the Philippines. Although referred to as flying lemurs, these mammals do not exhibit true flight, and they are not lemurs. They are actually gliding mammals.

Dermoptera were once classified among insectivores because they resemble the latter in several respects, especially in brain structure. They have also been classified with the Chiroptera, although the patagium in flying lemurs is not so well developed nor are the fingers elongated.

Chiroptera. Bats are the only vertebrates excepting pterosaurs and birds that have achieved true flight. This is possible because of the well-developed patagium (wing) stretching between the neck, limbs, tail, and the four greatly elongated, clawless fingers. The thumbs project from the anterior margins of the wings and bear claws used for hanging from rafters or ledges in caves. The five digits of the hind limbs all bear claws. Teats (usually two) are limited to the thoracic body wall. The pectoral muscles are strong and the sternum is keeled, although not as much as in birds. All bones are slender, and those of the arm and hand are greatly elongated (Fig. 9-18). Bats have large pinnas (external ear lobes) and the face and head glands are unusually numerous and enlarged, causing bats to have a bizarre appearance.

Bats constitute a large order, but are not too well known with respect to morphology and physiology. Their power to avoid obstacles while in flight at night is remarkable. They emit high-pitched, vibratory impulses inaudible to the human ear, which, upon striking an obstacle and returning to the animal, are utilized in the avoiding response. Bats may be insectivorous, frugivorous (fruit-eaters), or sanguinivorous (subsisting on the blood of other mammals). Vampire bats have received considerable attention because of sanguinivorous habits. Incisor teeth occur on the upper jaw only, and there is only one pair. The incisors are razor-sharp and point toward one another so that they slit the skin of prey. As blood oozes from the wound, the vampire licks it up and departs without awakening the sleeping victim, which is often a domestic animal. Associated with the vampire habit of taking only fluid nourishment is the very small lumen of the esophagus, through which no solid food could possibly pass.

ADAPTATIONS AND LAMARCKISM. The adaptations of the teeth and esophagus and the fluid diet inspire the interesting speculative question whether the small lumen in the esophagus forced bats to adopt a fluid diet of blood (the only available fluid containing all the nourishment needed by a mammal) or whether a fluid diet during many generations was responsible for a deterioration of the esophagus. If the former is the case, it is fortunate that vampire bats hit upon blood as a source of nourishment; otherwise, they could not have survived reduction in the size of the lumen. If, as believed by Lamarck, deterioration of the esophagus is attributable to disuse of this part, modern genetics has thus far been unable to fathom the mechanism whereby the change became hereditary. The possibility also remains that sanguiniferous bats were already sanguiniferous before chance mutations altered the esophagus. If so, decrease in size of the esophageal lumen would have had no deleterious effect on these bats. Further discussion of Lamarckism will be found on p. 63.

CONVERGENCE. The evolution of "flying" lemurs (Dermoptera), "flying" squirrels (Rodentia), "flying" phalangers (Marsupialia), and bats, all of which develop a patagium, may at first glance seem to be instances of parallel evolution. It is possible that the flying phalanger and the flying lemur are products of parallel evolution. They almost certainly came from common stock and are not too far removed, perhaps, from that stock. However, the development of the patagium in flying squirrels and in bats, both of which are probably far removed from common stock, suggests that the patagium in most cases is not a result of parallel evolution as herein defined. Perhaps the condition is comparable to the development of flippers in whales and fins in fish; to the development of the "hand wing" of bats and the "arm wing" of birds; or, to go further afield, to the development of the vertebrate eye with its lens and retina in relation to the development of the eye of a squid with similar structures.

The term **convergent evolution** has been applied to the situation wherein two or more groups of animals arrive at a similar evolutionary trait when the groups cannot be considered to have come from common ancestors other than very remote ones. Convergent evolution implies that unrelated species **long dissimilar** may, through adaptive mutations, approach closer and closer to one another in superficial resemblances. In parallel evolution the groups are **initially related** and **are diverging,** while developing similar traits. In convergent evolution the groups are **initially unrelated,** or only remotely related, and approach one another in superficial resemblances by developing similar traits.

Primates. The Primates (pronounced Prī-mā'-tēz), a group of primarily arboreal mammals, include Prosimii (lemurs and tarsioids) and Anthropoidea (monkeys, apes, and man). They arose 60 million years ago. The predominant structural feature of primates is the modification of the fingers so that the thumb can be made to touch the ends of the other four fingers of the same hand. The big toe is also opposable in most primates, although not in man. The digits are typically provided with flattened nails rather than claws. Very often there is a prehensile tail, which serves as a strong, agile, organ for grasping, supplementing the fingers. The cerebral hemispheres of the brain of primates are larger than in any other mammal. Despite specializations of the hands, feet, and brain, several rather primitive structural features are exhibited by primates— a fact responsible for their classification below certain other mammalian orders. Among these primitive features (which lead one to suspect that primates may have arisen from an insectivore-like precursor) are the flat-footed gait, the occurrence of five digits rather than a reduced number, the retention of a large clavicle, and the retention in many primates of the primitive centrale, a bone between the proximal and distal rows of wrist bones.

The teats are thoracic, and there is usually only one pair. In some lemurs, which usually produce more than one offspring at a time, there is an extra pair of teats in the region of the groin. One male lemur has a teat on each shoulder. Primates, including man, have hair over most of the body.

PROSIMIANS. Lemurs, of which the largest is the size of a cat, receive their name (*lemures* = ghosts) from the nocturnal habit of swinging silently through the trees while most anthropoids, including man, are asleep. The foramen magnum is so situated that the long axis of the head is in line with the long axis of the body, as in most other mammals but unlike the condition in tarsioids and anthropoids. They have a long tail, which is not prehensile. The second toe has a sharp claw-like nail instead of a flattened one. The placenta is nondeciduate; that is, the membranes of the embryo do not root themselves into the uterine lining of the mother, and hence there is no trauma of the uterus at birth. This is a

Fig. 3-25. *Tarsius,* a small primate.

primitive condition. Many fossil genera have been recovered.

Tarsius (Fig. 3-25) is a small prosimian the size of a rat, living in the East Indies and the Philippines. Tarsioids, like lemurs, are arboreal and nocturnal. They have enormous eyes specialized for night vision, which are close together and directed forward so that there is an overlap in the fields of vision of the left and right eyes, as in anthropoids. The head is more nearly balanced transversely on the end of the vertebral column. Both the second and third toes have sharp rather than flattened nails. The placenta is deciduate. In general, tarsioids resemble anthropoids more than they resemble lemurs.

ANTHROPOIDS. This suborder is divided into three superfamilies: Ceboidea, New World monkeys; Cercopithecoidea, Old World monkeys; and Hominoidea, apes and man. In anthropoids the head is at right angles to the long axis of the vertebral column; the eyes are directed forward and are close together; the cerebral hemispheres are greatly developed; there is a

bony external auditory meatus in all but the ceboids; there are thirty-two teeth in the permanent set; the placenta is deciduate; and there is typically only one offspring born at a time.

Ceboids are the monkeys of South America and the small, squirrel-like marmosets of the same region. The best-known South American monkeys are *Cebus* (the capuchin), *Ateles* (the spider monkey), and *Alouatta* (the howler monkey). The latter are named for their loud, screeching cries, which are aided by a greatly enlarged hyoid apparatus and larynx, giving the appearance of a thyroid tumor (Fig. 12-18). The ceboids were once referred to as the platyrrhines (*platy* = flat, *rhino* = nostril) because in most of them the nostrils are separated by a broad, flattened, cartilaginous septum.

Cercopithecoids are the monkeys of the Old World. Best known are baboons, mandrills, and the macaque or rhesus monkey (*Macaca*) from which the symbol Rh, designating blood groups, was derived. This group was at one time referred to as the catarrhines (*kata* = down), since the nostrils open downward and lie close together, with a narrow nasal septum, as in man.

Hominoids include the apes and man. The best-known apes include the gibbon, the orangutan, the gorilla, and the chimpanzee. The last is probably the most educable animal in the world, next to man.

Man, both fossil and modern, belongs to the family Hominidae. Since the discovery of Neanderthal man in Germany in 1856, fossil bony parts belonging to hominids have been found in considerable numbers. With each new discovery the problem arises whether to include the new member in existing species, or to erect new species, or even new genera. For example, Neanderthal man is sometimes classified as a subspecies of modern man (*Homo sapiens neanderthalensis*), sometimes placed in a separate species (*Homo neanderthalensis*). The rules for classifying other vertebrates are applied to the taxonomy of man. The

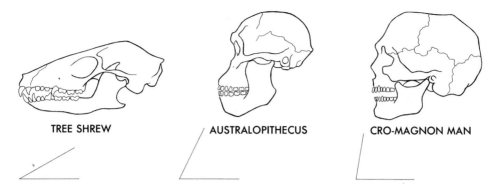

Fig. 3-26. Three theoretical steps in the evolution of the skull of man. The facial angles are shown beneath each skull.

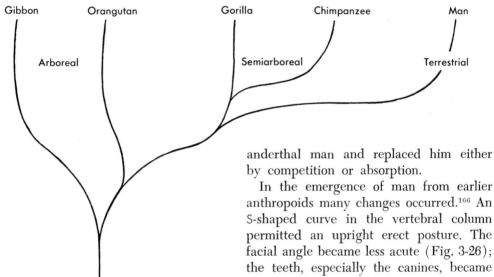

Fig. 3-27. Theoretical relationships of the hominoids.

oldest known hominids were discovered in South Africa and are about 2 million years old. They are usually assigned to the genus *Australopithecus* (*australo* = southern, *pithecos* = ape). They used simple bone tools. More recent hominids, more than a half million years old, have been classified as *Homo erectus*. They fashioned tools and used fire. Modern man, *Homo sapiens*, who added atomic energy to his array of tools, occupied Europe with Ne-

anderthal man and replaced him either by competition or absorption.

In the emergence of man from earlier anthropoids many changes occurred.[166] An S-shaped curve in the vertebral column permitted an upright erect posture. The facial angle became less acute (Fig. 3-26); the teeth, especially the canines, became smaller; the frontal lobes of the cerebral hemispheres enlarged, resulting in an enlarged braincase and a more prominent forehead; the eyebrow ridges became reduced; the nose became more prominent; the tail became confined to embryonic stages; the arms became shorter; a metatarsal arch developed in the otherwise flat feet; and the big toe moved in line with the other toes and ceased to be opposable. Articulate speech appeared, and many other changes occurred. The theoretical relationships of the hominoids is diagrammed in Fig. 3-27.

It is to the massive development of the frontal lobes of the cerebral hemispheres, to the opposable thumb, and to articulate speech that man owes his present dominance in the animal kingdom. With his fingers he can construct instruments of offense and defense and machines to lighten his burden, and with them he can scrawl symbols that convey experiences and techniques to the corners of the earth and to generations unborn. With his voice he can communicate with contemporary fellow creatures and exchange ideas with delicate shades of meaning. With his brain he can associate sensory stimuli that he currently receives with those recalled from earlier experiences, and after meditation he can elect a mode of action that he calls "intelligent." With his brain, too, he can enjoy the esthetic beauty of the imponderable universe, search for ultimate truth, and dream of a Utopia, which, but for his present residual animal nature, could be his heritage.

Edentata. Edentates are chiefly South American forms that have diverged in several respects from the basic mammalian pattern exemplified by insectivores. They may be grouped into two categories: hairy edentates, such as sloths and South American anteaters, and armored edentates, best known of which is the armadillo. Several giant armored fossil edentates are known. Teeth may be absent, as in sloths and anteaters (hence the name "edentate"), but they are more often present, as in armadillos and numerous fossil species. Incisors and canines are usually absent, and the molars lack enamel, although evidence of an enamel organ has been found. The digits are clawed. The vertebrae of edentates exhibit additional articular processes, from which is derived the name Xenarthra (*xen* = guest, *arthro* = joint), by which they are sometimes known. Exactly where to classify these animals is problematical. They are perhaps more closely related to the next order than to any other.

Pholidota. Pholidota is another insectivorous group of mammals containing one genus, *Manis*, the pangolin or scaly anteater (Fig. 5-34). Pangolins have scales resembling those of lizards. Between the scales grow scattered hairs. The scales have been considered by some investigators to represent clumps of agglutinated hairs. Because they lack teeth, pangolins at one time were included in the order Edentata.

Lagomorpha. Lagomorpha is an order established to include rabbits and hares. They differ from rodents in the exhibition of two pairs of incisors in the upper jaw, a small pair lying immediately back of, rather than alongside of, a much larger pair. The anterior pair of incisors, like those of rodents, are long, sharp, deeply embedded, and continue to grow throughout life. The smaller, posterior pair lack the sharp cutting edge. Rabbits differ from hares in having comparatively shorter ears and legs, in producing naked young, and in inhabiting burrows of their own construction. There are also other differences.

Rodentia. Rodentia is an order of gnawing animals containing many common and widely distributed families. Characteristic of rodents are the single pair of upper and lower incisor teeth. These are long, curved, and covered with enamel on their outer surfaces, providing a chisel-like edge for gnawing. The incisors have a persistent pulp and hence continue to grow throughout life. Since canine teeth are absent, there is a diastema, or stretch of jaw devoid of teeth, behind the incisors (Fig. 11-4). The intestinal cecum of rodents is very long and coiled, an adaptation to a vegetarian (cellulose) diet. The digestive enzymes of rodents are not able to digest cellulose, but harbored in the intestinal tract are cellulose-digesting microorganisms, especially bacteria, which convert cellulose into simpler carbohydrates. The testes are abdominal in location but may be lowered into the scrotal sacs.

Among common rodents are porcupines, in which the hairs are modified to form needle-like quills. Contrary to popular belief, the quills are not discharged through the air. Rodents include woodchucks, beavers, pocket gophers, marmots, chipmunks, squirrels, and the ubiquitous rats and mice. The coypu was introduced into the state of Louisiana in 1937 from Argentina and has successfully established itself in the marshes. The fur (nutria) has economic value. Rodents not established in the wild in the United States of America but utilized in research laboratories or on fur farms include the cavy (guinea pig), hamster, and chinchilla.

Cetacea. Whales, dolphins, and porpoises are moderately large or massive aquatic mammals attaining up to 100 feet in length and 150 tons in weight. Their body structure is highly adapted to an aquatic habitat. The tail has a horizontal terminal fin (fluke). A dorsal fin, devoid of internal skeletal support, sometimes occurs. The anterior limbs are paddle-like and serve chiefly as balancers. Although the fingers are not separated from the paddle, finger bones are present (Figs. 9-20 and 9-21). There is no clavicle. Posterior limbs and a girdle occur only as vestiges embedded within the trunk musculature. A few hairs, which can easily be counted, occur on the muzzle; otherwise the body is naked. Even the muzzle hairs are lacking on some whales. The nostrils (blowholes) are usually far back on the skull, where they may be united to form a single large respiratory opening. The water spout, composed of vapor formed when the animal exhales, may last from three to five minutes. The diaphram is unusually muscular. Most cetaceans have teeth, but whalebone whales (blue whales and finbacks) have, instead, a horny grating of whalebone (baleen) suspended from the roof of the mouth (Fig. 5-38). Through this grating is strained sea water containing nourishment in the form of myriads of minute aquatic organisms and other organic matter. A heavy layer of fat (blubber) under the skin assists in conserving body heat while the animal swims in the cold depths of the ocean. Because of their very great bulk, the larger species dare not approach too close to shore lest they become marooned and die, crushed by tons of their own flesh.

Cetacea are the best adapted of all marine mammals. They never leave the water. Baby whales are born in the water and hang onto one of the two inguinal teats of the mother while she swims about. All other aquatic mammals, depend on land for their breeding places (rookeries).

Mutations and Lamarckism. Any alteration of the genetic code resulting in changes in succeeding generations is known as a mutation. Loss of the olfactory nerve in whales was a mutation. If, before the mutation, whales had been relying on the sense of smell to stay well fed and safe from enemies, it is probable that, because of the mutation (loss of the olfactory nerve), whales would have become extinct. Chance mutations may have accounted for the extinction of many vertebrates, which, because of their new nature, were no longer able to live the same lives as their ancestors. On the other hand, chance loss of the olfactory nerve would not have affected whales, since they were not relying on smell any more than would a human being while swimming under water.

According to Lamarck's idea, whales would have lost the olfactory apparatus because they did not use it. More fully stated, Lamarck's doctrine asserts that when a part is employed in successive generations, it may become successively stronger, larger, or better modified for its role; when a part is neglected through many successive generations, it tends to become smaller and vestigial. The doctrine has not been acceptable to students of inheritance, for the reason that it offers no explanation of how use or disuse in any single generation may be translated into alterations of the DNA molecule in the sperm and eggs of that generation. If there is any means whereby alterations in the soma can be translated into alterations in the genetic code, that process has not yet been discovered.

Need as a basis for evolution. There is no scientific basis for the widespread misconception that, because an animal "needs" a specific structure, the structure will appear as a mutation. Populations that "need" something and do not acquire it become extinct. Their alternative is to occupy a new niche in which the need no longer exists and in which they can still compete. For example, it cannot be maintained that whales modified their forelimbs into paddles because they "needed" them. Some other explanation for the modification must be sought. Presently, the best available explanation, even if not a completely satisfying one, is to attribute the fulfillment of such "needs" to chance.

Carnivora. The carnivores, or flesh-eating mammals, are represented by terrestrial and aquatic forms. Terrestrial carnivores include cats and their kin (lions, tigers, leopards, panthers, bobcats, and lynx), dogs and related forms (wolves, foxes, and coyotes), bears and pandas, hyenas, and numerous forms economically important for their furs such as raccoons, martens, weasels, mink, otters, skunks, and badgers. Aquatic carnivores include sea lions, seals, and walruses. Adaptations for their meat-eating habits include powerful jaws and elongated canines capable of spearing and tearing flesh (Fig. 11-4). Carnivores typically have five toes (sometimes four) with strong, sharp claws. Clavicles are reduced or are completely lacking. The cecum is small, in contradistinction to that of vegetarian mammals. The cerebral cortex is extremely convoluted, and the animals are capable of considerable learning.

Aquatic carnivores (Fig. 3-28) have numerous adaptations for life in the water. The paddle-like limbs are awkward for walking on land, and much of the forelimb is included within the body wall. The hands and feet are webbed, and the finger bones have increased in number, as in cetaceans. There is a tendency for nails to disappear. In seals the hind limbs are directed caudad and are attached to the tail. Thus, movements of aquatic carnivores on land consist principally of pulling themselves awkwardly along or of flopping about. Nevertheless, the rookeries are on land and the young cannot even swim.

Tubulidentata. These are represented by a single species, the South African aardvark. This is a burrowing anteater about 6 feet in length with few teeth, relatively few coarse hairs, an elongated pig-like snout, and a long tongue for seeking out termites.

Fig. 3-28. Aquatic carnivores (sea lions). (Courtesy The American Museum of Natural History, New York, N. Y.)

Ungulates and subungulates. At this point it is necessary to introduce a group of four orders of placental mammals known collectively as ungulates (*ungula* = hoof) and subungulates. They are mostly large herbivores, capable of sustaining themselves on grasses. They all exhibit hoofs rather than nails or claws on the ends of their digits. The two higher orders, **Artiodactyla** (deer, etc.) and **Perissodactyla** (horses, etc.), walk on the tips of their toes. These exhibit the most specialized appendages encountered among tetrapods, unless birds are excepted. The two lower orders, **Proboscidea** (elephants) and **Hyracoidea** (coneys), do not walk on the tips of the toes and are subungulates. Ancestral ungulates were probably flat-footed, with five toes. Modern ungulates have no more than four toes. Reduction in the number of toes is illustrated by the familiar example of the horse, the Eocene ancestors of which had four toes in front and three behind. With successive mutations the number of functional toes was reduced to one.

The teeth of modern ungulates are characteristic. Their grinding surfaces exhibit crosswise ridges separated by deep grooves for grinding grasses (Fig. 11-5). Those teeth that lack ridges and grooves exhibit instead two or more points (cusps) on each molar tooth. Canine teeth tend to be reduced or absent. Ungulates lack a clavicle and are the only mammals exhibiting horns, although several models lack them.

One other order of mammals, the **Sirenia,** will be treated in the midst of the ungulates, since they seem to have been derived from the same stock as ungulates, although they do not exhibit hoofs.

Proboscidea. Proboscideans have a proboscis, or trunk, composed of a greatly drawn out upper lip, which is accompanied in its overgrowth by the nostrils (Fig. 11-4). They have scanty hair and thick, wrinkled skin, which earned for them the name pachyderms (*pachy* = thick, derm = skin). The incisor teeth of one or both jaws are greatly elongated to form tusks. Canine teeth are absent, and the molars are very large grinders, as in ungulates. Proboscideans are bulky animals, and the limbs are almost vertical pillars of bone and muscle. Elephants, of which there are an African and an Asiatic species, are the sole living forms. Extinct mammoths and mastodons belong to this order.

Hyracoidea. Hyracoidea, an order of subungulates, contains a single genus, *Hyrax,* popularly known as the coney. In some traits coneys resemble rodents (Fig. 3-29). They have short ears; they tend to hunch the body in a rodent-like stance when they are at rest; the muffle (upper lip) is split as in rodents; and the incisor teeth grow from a persistent pulp. Indications of a closer relationship with ungulates than with other mammals, however, are the hoof-like nails, absence of a clavicle, exhibition of ungulate-like molars and reduction of the digits of the hand to four and of the foot to three.

Sirenia. Although resembling whales, the dugong and manatee have characteristics indicating that they may have orig-

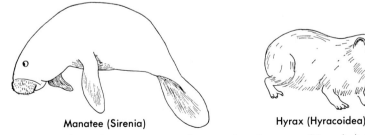

Manatee (Sirenia)   Hyrax (Hyracoidea)

Fig. 3-29. Two subungulates: a sea cow (manatee) and a coney (*Hyrax*). A third subungulate, the mastodon, is illustrated in Fig. 11-4.

inated from stock closely related to proboscideans. Sirenians are stout, clumsy-appearing, freshwater or marine mammals (sea cows) with an overgrown, wrinkled, almost pathetic-looking trunk-like snout covered with scattered, coarse bristles (Fig. 3-29). The rest of the body is naked except for a few scattered hairs. The fore-limbs are paddle-like flippers, but within the flippers the typical appendicular bones are intact. Hind limbs are completely absent, and the pelvic girdle is reduced to remnants. The tail in some cases is flattened horizontally like that of whales. Sea cows have an ungulate array of molar teeth for grinding seaweed.

Perissodactyla. Modern perissodactyls occur in three families: horses (including asses and zebras), tapirs, and rhinoceros. Many extinct forms are known. Perissodactyls walk on the hoofed tips of one, three, or occasionally four toes (Figs. 9-23 and 9-25) and are distinguished by the fact that the body weight is borne chiefly on a single digit. This is known as a **mesaxonic** foot (Fig. 9-24). Perissodactyls are usually referred to as odd-toed ungulates. Tapirs and some rhinoceros, however, have four toes on the forelimb, three on the hind limb.

Artiodactyla. Artiodactyla are ungulates in which the weight of the body is supported equally by two toes. This is known as a **paraxonic** foot (Figs. 9-24 and 9-25). Artiodactyls are usually referred to as even-toed ungulates, but at least one extinct artiodactyl had five toes on the forelimb. Artiodactyls include pigs, hippopotamuses, peccaries, camels, llamas, deer, antelope, oxen (including domestic cows), giraffes, goats, and sheep.

Artiodactyls, with the exception of pigs, hippopotamuses, and peccaries, have stomachs divided into not fewer than three chambers, sometimes four (Fig. 11-12). They bolt their food, which then passes to the rumen, the first segment of the stomach. Such animals are called **ruminants.** At their leisure, they force undigested food balls back up the esophagus and masticate this cud more thoroughly.

VARIATION, GEOGRAPHICAL ISOLATION, AND ORGANIC EVOLUTION

In this chapter we have had a glimpse of the variety of vertebrate life that exists. But variation is not restricted to differences between species, or even between isolated populations of a single species. Variation occurs within populations, and even among litter mates. Indeed, no two products of sexual reproduction are likely to be identical for the following reasons. The number of kinds of gametes or of homozygous genotypes possible with n pairs of genes is 2^n. The number of different kinds of genotypes possible is 3^n. Even if there were only 100 pairs of genes, forty-eight digits would be needed for writing the number of kinds of genotypes possible. And there are thousands and thousands of pairs of genes. In addition, there may be as many as four hundred mutational sites on a single gene.[141] The number of possible combinations of hereditary traits therefore staggers the imagination. For example, the gastrohepatic artery in a population of *Felis domestica* is sometimes 20 mm. long, sometimes 1 mm. long, or the gastric and hepatic arteries may arise independently and there is no gastrohepatic artery. The number of variations that actually occurs in the arterial channels alone in any population is such that one could identify a single shark, or a single human being, from his arterial channels. In actual practice, each of us does better than that! We can identify a single human being simply by looking at the surface of the face and head, or at the print made from his fingertips. We could, of course, recognize an individual organism by looking at the sequence of nucleotides in its DNA. Individual genetic variation, coupled with geographic isolation, seems to be the matrix out of which new species evolve.

What happens when a small group of individuals on the periphery of a population wanders sufficiently far from the other members that the isolates can establish a new colony out of contact with the original population? Such a geographically isolated group may have one of three possible fates. If they are not partially preadapted to the newer environment, they may become extinct. Or they may fuse with nearby populations of another, closely related species, establishing a zone of hybridization. Or, as a result of chance mutations affecting only one population, and by recombinations and selection pressures, a steady genetic divergence may occur, and the two populations may become reproductively isolated, should their ranges again overlap. When reproductive isolation has occurred, a new species will have been produced. The length of time required for origin of species by geographical isolation depends on many factors, but the final emergence of new species seems almost inevitable.

Although new animal species may arise other than by geographical isolation, the origin of species by mutations or recombinations arising within an existing inter-breeding population is no longer supported by most modern geneticists.

Not only have new species been appearing at different rates during different epochs of the earth's history, depending in part on changes that produce extreme modifications of climate and geography, but species are also disappearing at varying rates. Since the start of the twentieth century fifty animal species are known to have become extinct.[165] Many more species have come and gone than exist today. Because of the continued rise and demise of species, the biota on earth has been changing since the first life appeared on this planet.

The lines of evolution leading from ancient fishes to modern tetrapods may pass through the forms indicated in Fig. 3-30. Notice particularly that ancestral lines lead through very ancient forms, all of which are now extinct. A chart showing the relative abundance of the vertebrate classes through time is provided in Fig. 3-31.

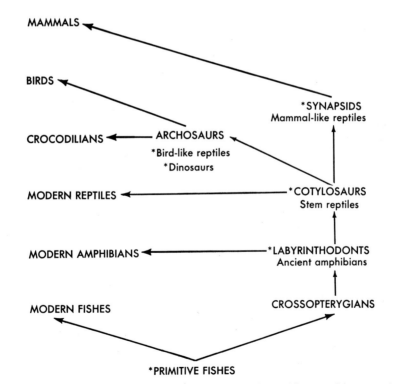

Fig. 3-30. Main lines of vertebrate evolution. Forms indicated by asterisk are extinct.

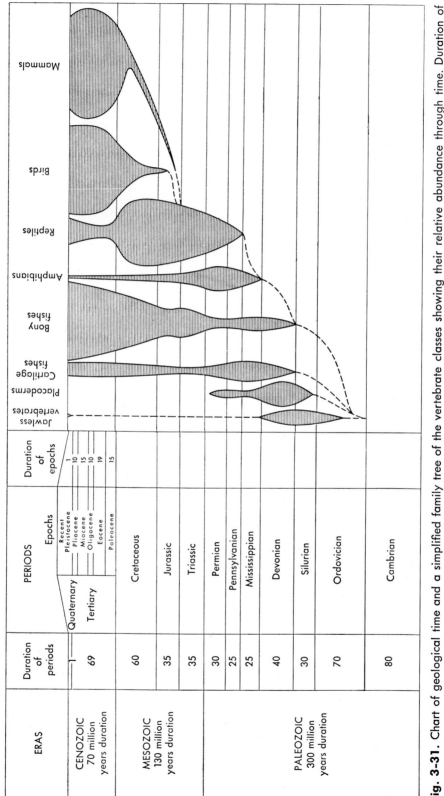

Fig. 3-31. Chart of geological time and a simplified family tree of the vertebrate classes showing their relative abundance through time. Duration of periods is in millions of years. (From Colbert: Evolution of the vertebrates, New York, 1955, John Wiley & Sons, Inc.)

WORDS THAT TRIGGER IDEAS

Words are a necessity for conceptual thought. The more words one can command, the greater the variety of ideas one may entertain. The following words stimulate thinking in relation to evolutionary concepts. You will read them, hear them, and use some of them. There will be differences of opinion with reference to some of the connotations. However, if calling attention to these abstract terms stimulates discussion, inclusion of them will have been justified.

Primitive is a relative term. It refers to the beginning, or origin. A primitive trait is one that appears in a stem ancestor from which arose an array of subsequent species, some or all of which may retain the trait. The notochord is primitive, since it occurred in the first chordates. The placoderms were primitive fish because they gave rise to an array of later fish. Ancient insectivorous mammals were primitive because they gave rise to an array of later forms with hair and mammary glands. Somewhere in phylogeny there was a primitive primate and a primitive human species. One cannot always be certain that a given structure is primitive. For example, the lateral neural cartilages of agnathans are primitive only if they reveal an original condition from which typical vertebrae later evolved.

Generalized refers to structural complexes that, at least in some of the descendants, have undergone subsequent adaptation to a variety of conditions. The hand of an insectivore was, and remains, a generalized mammalian hand. It was competent to evolve into the wing of a bat, the hoof of a horse, the flipper of a seal, and the hand of a primate. A generalized group of animals has demonstrated that it was genetically suitable for divergent evolution, that is, evolution in many directions. Labyrinthodonts were generalized tetrapods. The terms "generalized" and "primitive" come into contrast in that "generalized" connotes a state of potential adaptability

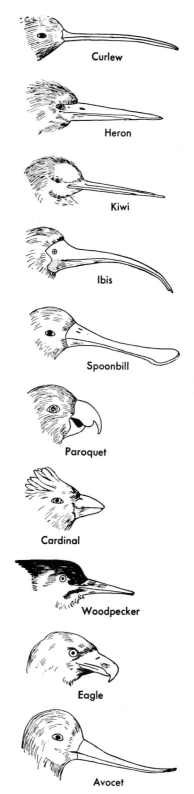

Fig. 3-32. Specializations of the beak in a few birds.

and "primitive" connotes a state of being ancestral or unchanged.

A **specialized** condition is one that represents an adaptive modification. Vertebrate wings are specializations of anterior limbs, and beaks are specialized upper and lower jaws. Beaks (Fig. 3-32) may be needle-like for extracting nectar from flowers (hummingbirds), chisel-like for drilling holes (woodpeckers), hooked for piercing and tearing captured prey (raptorial birds such as hawks and eagles), long and pointed for capturing moving fish, lizards, and other prey (herons), or recurved for extracting grubs from burrows (the female huia of New Zealand). Increased specialization connotes increased adaptation to specific conditions. The greater the specialization, the less may be the potential for further adaptive changes.

Modified connotes any state of change from a previous condition, a mutated state. If the presence of bone is a primitive trait in vertebrates, wholly cartilaginous skeletons are modifications of the condition. The modification (loss of the potential to form bone) was a specialization if it adaptively modified the animal. Modifications are not necessarily adaptive, and if they are not, they may portend the demise of the species, since any change is statistically more likely to place the animal in a less competitive state. An exception would be a modification that proved to be a preadaptation. (Preadaptations provide animals with fortuitous modifications that enable them to enter new niches or that anticipate conditions not existing at the time of the modification.)

The terms **higher** and **lower** express the relative position of major taxa in a phylogenetic tree. To the extent the tree is accurate, the terms express relative mutational distance of a given taxon from a common ancestor when compared with some other taxon. Birds and mammals evolved from reptiles, hence are said to be higher than reptiles, but mammals as a class are not usually considered higher than birds as a class, since both evolved

from an immediate common precursor. The terms may become meaningless when used to compare a species within one taxon with a species within another taxon, as when comparing a modern frog with a cotylosaur, or man with a hummingbird.

Simple is a relative term connoting a lack of complexity of component parts. A simple state is not necessarily a primitive one. The skull of *Necturus* is simple compared with that of a teleost, but it is not primitive. The primitive may also be far from simple.

The term **advanced** should connote a modification in the direction of further adaptation. Unfortunately, the word has overtones connoting progress and hence might be misconstrued. It is a matter of opinion of one species—man—whether or not a modification in another species represents progress. The phrase "more recent" may be more effective than the term "advanced."

Degenerate is another value judgement word. For example, it is sometimes applied to cyclostomes by those who think cyclostomes have lost jaw skeletons, paired appendages, bone in the skin, and other characteristics of typical vertebrates. However, the condition of the cyclostome may represent an adaptation to a semiparasitic state, and as such might better be characterized as "specialized." These agnathans may have specialized themselves right into a state of neosimplicity! To call them "degenerate" would seem to discount the value of adaptive modification. "Degenerate" would seem to be a term that should be avoided.

The words **vestigial** and **rudimentary** require explanation. A phylogenetic remnant that was better developed in an ancestor is vestigial. The forelimbs of whales are said to be vestigial, since ancestors of whales most likely were tetrapods with functional tetrapod appendages. The yolk sac of the mammalian embryo is vestigial. The term rudimentary is used in two different senses, phylogenetic and ontogenetic. In the phylogenetic sense structures that became more fully exploited in descendants are said to be rudimentary in the phylogenetic precursor. For example, the lagena of the inner ear of fishes is sometimes referred to as a rudimentary cochlea, since it evolved into a cochlea in later forms. In the ontogenetic sense, a structure that is undeveloped or not fully developed is said to be rudimentary. The muellerian duct may be considered rudimentary in most male vertebrates. It is not always possible to be certain whether a structure is rudimentary or vestigial. The pseudobranch of the shark *Squalus acanthias* is vestigial if it represents a gill that, in ancestral forms, was a full-fledged functional gill surface. However, if the pseudo-branch is not a vestige, but a potential future gill surface, then it is phylogenetically rudimentary. The majority opinion is that it is vestigial. The words are valuable stimulants to conceptual evolutionary thought, but it must be remembered that most phylogenetic relationships are theoretical.

A structure that grows smaller during the lifetime of an individual is **atrophied.** The thymus gland of mammals is atrophied after sexual maturity, and gonads atrophy after hypophysectomy.

If the foregoing thoughts trigger discussion, no matter how dissonant, the space employed in presenting them will have been well utilized.

Chapter summary

1. The vertebrate classes are Agnatha, Placodermi, Chondrichthyes, Osteichthyes, Amphibia, Reptilia, Aves, and Mammalia. The subdivisions of these classes are indicated in an abridged classification, pp. 413 to 416.

2. Pisces (fishes) include the first four classes listed in 1. The first class, Agnatha, exhibit no true jaws. The remaining vertebrates are gnathostomes.

3. Living agnathans are fishes with eel-like bodies that lack true jaws, have no paired appendages or girdles, have a skin devoid of scales, and exhibit a round, sucker-like mouth. Extinct agnathans (ostracoderms) are the oldest vertebrates known. They were armored fishes.

4. Placoderms are extinct gnathostomes covered with a bony armor.

5. Chondrichthyes are fishes with cartilaginous skeletons, five to seven naked gill slits (except Holocephali), and a covering of placoid scales. They were more abundant in ancient times. In many respects they illustrate the basic architectural plan of vertebrate construction.

6. Osteichthyes are bony fishes. They exhibit ganoid, cycloid, or ctenoid scales and have a bony operculum that covers the gill slits. The crossopterygian *Latimeria* and the ganoid fishes are the oldest living fishes in the class. Teleosts are the most modern Osteichthyes. In a few bony fishes the air sacs are used as lungs.

7. Tetrapods are vertebrates with four jointed legs or descendents thereof. Modifications of these appendages include wings (birds), flippers (seals), and arms (primates). A few tetrapods have lost one or both pairs of appendages.

8. Amphibians are tetrapods with glandular skin devoid of integumentary armor except for the minute dermal scales of caecilians. Most amphibians lay their eggs in water, and there is usually a gill-bearing larval stage. Neotenic amphibians exhibit incomplete metamorphosis.

9. Reptiles, birds, and mammals (amniotes) exhibit as embryos an amnion, chorion, and allantois. Fishes and amphibians lack these extraembryonic membranes and are anamniotes.

10. Reptiles are amniotes with dry skin, epidermal and dermal scales, and claws. They typically lay large eggs surrounded by a shell. Some, however, are viviparous.

11. Birds are amniotes with feathers. Scales persist on the legs and feet, which bear claws.

The anterior appendages are modified for flight in most birds (carinates), but in a small group (ratites) anterior appendages are usually vestigial or absent.

12. Mammals are amniotes with hair. Prototheria lay reptilian-like eggs and their mammary glands are primitive. Theria give birth to living young. Marsupials use the yolk sac as a placenta. The remaining orders have a chorioallantoic placenta.

13. There is evidence that vertebrates originated in primeval fresh waters and that they were, in their earliest beginnings, water-dwelling animals with gill slits.

14. Geographical isolation seems to account for the advent of most new species.

15. As an exercise in semantics, the following words have been examined: primitive, generalized, specialized, modified, higher, lower, simple, advanced, degenerate, vestigial, and rudimentary.

Chapter

4

Early vertebrate morphogenesis

In previous chapters reference has been made to the "basic architectural plan" upon which all vertebrates are constructed. This phrase has two implications. The first is that fishes, amphibians, reptiles, birds, and mammals all conform to a **basic pattern of anatomical structure** insofar as the various systems of the body are concerned. This uniformity of basic anatomy may be revealed by dissection of adult animals. The second implication is that there is a **uniformity of developmental processes.** This fact is revealed by a study of comparative vertebrate embryology. In order to appreciate fully the uniformity of structure of adult vertebrates, some knowledge of embryonic development is necessary.

THE VERTEBRATE EGG AND VIVIPARITY

Vertebrate egg types. Eggs of vertebrates vary in the amount of yolk they contain and in the distribution of the yolk within the egg. The eggs of mammals, except those of the egg-laying monotremes, contain very little yolk. That which they do contain is evenly distributed throughout the cytoplasm in the form of fat droplets and small yolk globules. Because of the **even distribution of the yolk,** the mammalian egg is said to be **isolecithal** (syno-

nym, **homolecithal**).* The eggs of mammals for the most part can just be seen with the naked eye. The eggs of all other vertebrates contain moderate to massive amounts of yolk that tends to be concentrated at the lower or **vegetal pole.** The relatively yolk-free cytoplasm tends to be concentrated at the **animal pole.** Eggs in which the cytoplasm and yolk tend to accumulate at opposite poles are **telolecithal** eggs.

For convenience eggs can be grouped according to the amount of yolk they contain. Many embryologists refer to eggs with very little yolk (like those of the amphioxus and mammals) as **alecithal** eggs, to eggs with moderate amounts of yolk (like those of freshwater lampreys, ganoid fishes, lungfishes, and amphibians) as **mesolecithal** eggs, and to eggs with massive amounts of yolk (like those of marine lam-

---

*The eggs of the amphioxus and *Styela*, like those of mammals, also have very little yolk, but that which they contain is not evenly distributed, as in isolecithal eggs, being concentrated centrally (**centrolecithal** eggs), as in insects and other arthropods. It becomes concentrated at one pole at the time of first cleavage. The yolk-laden eggs of teleosts are also centrolecithal at spawning but become telolecithal minutes after penetration by the sperm.

preys, elasmobranchs, teleosts, reptiles, birds, and monotremes) as **polylecithal** eggs.

Oviparity. Animals that deposit their eggs (that is, lay or spawn their eggs) are said to be **oviparous** (*ovum* = egg, *parere* = to bring forth). Most vertebrates from monotremes downward are oviparous. The fertilized egg, when laid, contains sufficient yolk to support development of the zygote into a free-living, self-nourishing organism. The baby chick develops from a yolk-laden egg and at hatching is built essentially like an adult. The frog arises from an egg with less yolk and hence has a tadpole (**larval**) stage considerably different from the metamorphosed state. When there is little yolk, as in the egg of the amphioxus, the free-living, self-nourishing state must be achieved even more quickly than in frogs. Accordingly, the larvae hatch (escape from their protective membranes) in an even more immature state than does the frog tadpole. The amphioxus hatches into an externally ciliated, free-swimming embryo eight to fifteen hours after fertilization. The mouth perforates, and larval feeding can commence twenty-eight hours later, depending on the temperature of the water. At that time the actual bulk of the larva (1 mm.) is probably little greater than that of the egg. The embryo is transparent and has only one gill slit (located on the side opposite the off-center mouth), and tissue differentiation has only just commenced. The notochord has not yet become established as an independent structure.

Viviparity. In many vertebrates the egg is retained within the mother's body during embryonic development. Living young are delivered, and the species is said to be viviparous (*vivi* = alive). The relationship between the mother and the embryo or embryos within her varies from one in which the mother provides protection and little else, as in *Squalus acanthias,* to one in which the embryo is dependent on the mother for all its nourishment, for oxygen, and for carrying away the waste products

of metabolism, as in viviparous mammals. Many intermediate degrees of dependency have evolved. The term **ovoviviparity** has been coined to designate the viviparous condition in which protection and little else is provided, and the term **euviviparity** (true viviparity) is used to designate the viviparous condition in which the embryo cannot develop independently of the nourishment provided by the mother.

Viviparity in one degree or another has evolved independently in many vertebrate groups, including those with very little yolk in the egg (mammals) and those with considerable yolk. It seems to have developed at least fifteen independent times in teleost fishes, at least ten times in lizards, and at least six times in snakes. It is found in elasmobranchs and is said to have occurred in the extinct *Ichthyosaurus.* There are viviparous urodeles and anurans that give birth to fully metamorphosed young in terrestrial environments. Birds are the only entire class of vertebrates exhibiting no viviparity. The frequency with which viviparity arose in lower vertebrates seems to have made its appearance in mammals almost inevitable.

The dogfish *Squalus acanthias* is a good example of an ovoviviparous organism. The lining of the gravid uterus is wrinkled by richly vascularized, scalloped, longitudinal folds easily observed in the laboratory. The pups are nourished by the yolk from their own yolk sac (Fig. 4-10). Biochemical and radioisotope studies have indicated that the pups receive oxygen but no appreciable nourishment or inorganic material from the mother. The pups may be removed from the mother two to three months before birth (the gestation period is twenty to twenty-two months) and will complete their development, utilizing the nourishment from their yolk sac, as long as they are confined in finger bowls or plastic tubes containing fresh sea water.

In many instances of ovoviviparity and in all instances of euviviparity certain highly vascularized tissues of the embryo lie in intimate contact with, or invade, maternal tissues, which provide the nourishment and oxygen required for development. The nourishment (hexose sugars, amino acids, and other substances) is provided either by **secretion** from maternal tissues or by **active transport** across the membranous barriers separating the bloodstreams of the embryo and mother. In teleosts the young may develop in the ovarian follicle that produced the egg, there having been no ovulation. The young of *Gambusia* develop in this way, although much of the nourishment is provided by the yolk until it is depleted. Or the embryos may develop in the lumen of the ovarian sac. Among adaptations of the embryo for development in these locations are enlargement of the embryo's pericardial sac, which lies in contact with the maternal tissues of the follicle or ovarian sac; enlargement of the embryonic gut, which lies in contact with the maternal tissues and later atrophies; and formation of villus-like projections of the embryonic rectum, protruding through the vent of the embryo into the surrounding nutrient-rich medium. In some teleosts the young develop at first in the follicular chamber and exhibit an enlarged pericardial sac, then later pass into the lumen of the ovarian sac, resorb the hypertrophied pericardial sac, and develop villi. The yolk sac and allantois, both of which are well-vascularized extraembryonic membranes, also serve as sites of absorption of nutrients in viviparous vertebrates.

The maternal tissues, under the influence of hormones, also exhibit adaptations associated with the presence of the embryos. The wall of the ovarian follicle occupied by the teleost embryo may develop vascular villi, or the walls of the ovarian sac may develop vascular folds, which separate one embryo from another or which may even project into the mouth or behind the operculum of the embryo. In *Dasyatis americana,* a euviviparous sting ray, the gravid uterus is lined with villi 2 to 3 cm. long, which produce a copious, viscous,

cream-colored secretion that nourishes the embryo. Secretions of uterine glands provide nourishment for the unimplanted blastocysts of all mammals, and perhaps for the implanted blastocyst throughout pregnancy in perissodactyls and artiodactyls. **Histotrophic nutrition** is the term applied to nutrition by **secretions** from maternal tissues, as contrasted with nutrition by substances actively transported from the bloodstream of the mother.

FERTILIZATION

Fertilization is the series of processes by which a sperm cell approaches the ovum, penetrates the egg membranes, progresses (as a male pronucleus with a haploid number of chromosomes) toward the egg pronucleus (also haploid), and fuses with the latter, thus establishing the zygote. An extended description of these activities is given by Willier and coauthors.[49] Once fertilization has occurred, the zygote commences to divide, and development of a new organism is under way.

In viviparous vertebrates fertilization must take place within the body of the female, and sperm are usually introduced by intromittent organs of the male. Fertilization must also be internal in oviparous vertebrates whenever the egg is covered by an impenetrable shell. In oviparous fishes, frogs, and toads external fertilization is the rule. The male ejects a cloud of sperm over the eggs as they are being extruded. In urodeles, however, fertilization is usually internal even though the eggs are subsequently extruded. The male deposits a sac of sperm (**spermatophore,** Fig. 4-1) on the substrate of the pond. The spermatophore is taken into the cloaca of the female by her own activity or is placed there by the male. The sperm escape from the viscous mass and migrate up the female reproductive tract to the eggs.

Number and protection of eggs as factors in survival. When eggs are retained within the female body, or when they are protected by a parent after being laid, the number of eggs produced is small and the

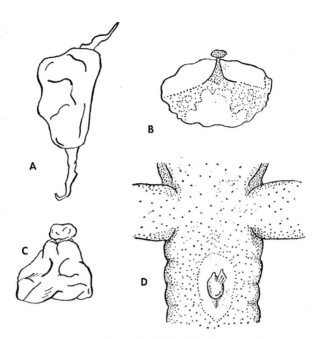

Fig. 4-1. Spermatophores. **A,** *Necturus.* **B,** *Notophthalmus.* **C,** *Ambystoma.* **D,** Cloaca of female *Desmognathus* containing spermatophore. (Redrawn from Bishop.[3])

mortality is low. When, however, the eggs are laid outside the body and are left unprotected and exposed to the environment, the mortality prior to hatching is high. A naturally high mortality rate is counterbalanced by production of large numbers of eggs. Thus, oviparous fishes and amphibians have many young (not all of which reach maturity); reptiles and birds have fewer young.

Natural selection at work. It is conceivable that an oviparous species that does not conceal or guard its eggs might, as a result of mutations, tend to produce fewer and fewer eggs. Because of the probable destruction of a percentage of the eggs, the species would tend to become more rare. The mutations resulting in smaller numbers of eggs, unaccompanied by other mutations providing for protection of the few eggs produced, would finally result in the destruction of the species. **Natural selection** is a term applied to indicate that natural phenomena inexorably determine which species will survive. Logically those fit to cope with (that is, adapted to) their

surroundings are likely to survive, and those unfit are likely to perish. To natural selection may be attributed the following facts: there are few vertebrates that lay very small numbers of eggs in unsheltered locations; there **are** vertebrates that lay small numbers of eggs in guarded locations; and there are others that lay large numbers of eggs (up to 60 million in some fishes) in unguarded locations.

The concept of natural selection by survival of the fittest (best adapted) as a factor responsible for determining the direction of evolution was proposed in 1858 simultaneously by Charles Darwin (1809 to 1882) and Alfred Russell Wallace (1823 to 1913) in a joint paper read before the *Linnean Society of London.* The title of the paper was "*On the Tendency of Species to Form Varieties; and on the Perpetuation of Varieties and Species by Natural Processes of Selection.*" Of the two authors, Darwin had by far the most evidence. In 1859 he published some of his evidence in a classic entitled *The Origin of Species,* and additional evidence was

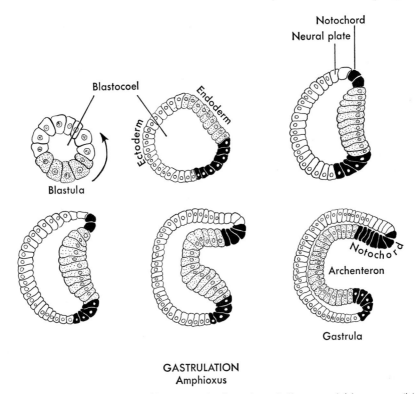

GASTRULATION
Amphioxus

Fig. 4-2. Gastrulation in an amphioxus, sagittal section. Cells around blastopore (black) are actively mitotic.

presented by him in subsequent papers. Darwin's evidence was based on comprehensive and voluminous observations of plants and animals. Natural selection is no longer regarded as the sole factor responsible for evolutionary direction, but it is recognized as one factor contributing to the rise or fall of species.

CLEAVAGE AND THE BLASTULA

The early cell divisions of the zygote are referred to as **segmentation,** or **cleavage.**[*] As a result, the zygote becomes subdivided into smaller and smaller cellular units forming (usually) a hollow spherical mass known as the **blastula,** with a wall one to several cells in thickness. Each cell of the blastula is a **blastomere;** the cavity is the **blastocoel.**

---

[*]In elasmobranch fishes, as in insects and some other invertebrates, the early cleavage stages consist of nuclear divisions alone, to be followed later by formation of cell boundaries.

The egg of the amphioxus divides rapidly and completely into two cells and then four, eight, sixteen, and so on, although not with precision, producing a blastula with cells differing only slightly in size (Fig. 4-2). The blastocoel is in the center of the blastula.

In amphibians and in many other forms whose eggs contain considerable yolk, segmentation results in a blastula composed of cells of unequal size (Fig. 4-3). At the animal pole where the protoplasm is concentrated, cell division proceeds rapidly and the cells are small. At the vegetal pole, segmentation proceeds more slowly and the cells are large and laden with yolk. The blastocoel is eccentric, being located near the animal pole.

In most species with massive amounts of yolk, segmentation is accomplished at the animal pole only, dividing the protoplasmic region into a plate of blastomeres (**blastoderm**) perched upon the unseg-

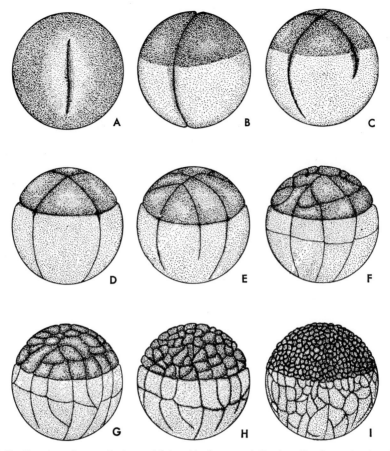

Fig. 4-3. Cleavage in a tailed amphibian (*Ambystoma*). Dark cells, the animal pole; light cells, the vegetal pole containing yolk. (After Eycleshymer; from Nelson: Comparative embryology of the vertebrates, New York, 1953, Blakiston Division, McGraw-Hill Book Co., Inc.)

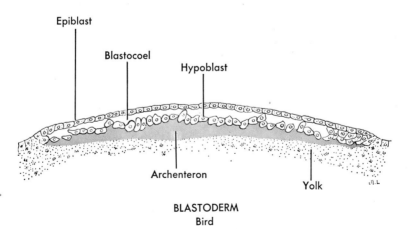

BLASTODERM
Bird

Fig. 4-4. Gastrula of bird, sagittal section. The blastoderm (the developing embryo at the animal pole) consists of epiblast and hypoblast. The latter is now the roof of the gut (archenteron). Head end is to the left.

mented yolk mass (Fig. 4-4). The blasto-coel is eccentric and occupies the animal pole. The blastoderm gives rise to the embryo and to extraembryonic membranes. One such membrane is the yolk sac, which grows down around the massive yolk, encompassing it.

The eggs of typical mammals exhibit an animal-vegetal polarity, just as do the eggs of the reptiles, even though mammalian eggs have almost no yolk. The first cleavage division typically divides the egg into two blastomeres (Fig. 4-5). One blastomere represents the animal pole; the other represents the vegetal pole. The descendants of the latter hurriedly form a nutritional membrane, the **trophoblast** (*troph* = food). The hurried production of the trophoblast is an adaptation for absorbing nourishment from the uterine fluids at the earliest possible moment, since the egg has almost no nourishment of its own. The blastocoel forms as a cleft in the trophoblast cell area. With increase in size of the trophoblast and of the blastocoel, the animal cells become an **inner cell mass.** From this mass the embryo develops. Because of the cyst-like nature of the mammalian blastula it is called a **blastocyst.**

Comparing cleavage in an amphioxus, a frog, and a bird or dogfish, it is apparent that the greater the quantity of yolk, the less can cleavage segment the entire egg mass. The quantity of yolk is but one factor determining the patterns of cleavage. Other factors are intrinsic in the cytoplasm of the egg. As a result of cleavage, all vertebrates from fish to man pass through a phase of development known as the blastula. During this stage major organ-forming areas are being established in preparation for gastrulation.

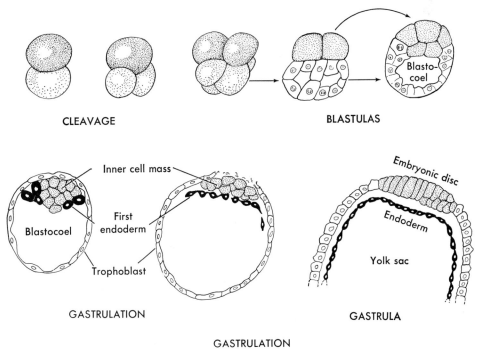

CLEAVAGE

BLASTULAS

Blastocoel

Inner cell mass

First endoderm

Blastocoel

Trophoblast

GASTRULATION

Embryonic disc

Endoderm

Yolk sac

GASTRULA

GASTRULATION
Mammal

Fig. 4-5. Cleavage, formation of blastula, and gastrulation in a mammal. At cleavage (upper left) the dark cell is the animal pole. The trophoblast is the first extraembryonic membrane to form, and it arises from the vegetal pole. The inner cell mass from the animal pole gives rise to the embryo. The yolk sac is functionless. The cavity of the yolk sac is not homologous with the blastocoel.

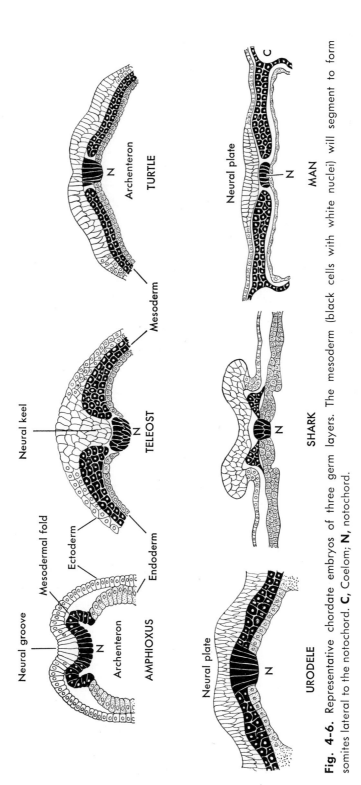

Fig. 4-6. Representative chordate embryos of three germ layers. The mesoderm (black cells with white nuclei) will segment to form somites lateral to the notochord. **C,** Coelom; **N,** notochord.

GASTRULATION

Gastrulation is the process whereby presumptive (that is, future) endoderm, mesoderm, and notochordal tissue of the blastula migrate to the interior of the embryo. Rapid cell proliferation provides cells for these migratory movements.

In an amphioxus the endodermal plate gradually folds into the blastocoel (Fig. 4-2), assisted by rapid cell division around the periphery of the site of infolding. Soon, presumptive notochord cells flow inward to become the temporary roof of the primitive gut. Then, presumptive mesoderm cells pass inward to lie in the roof of the gut at either side of the notochord. At this stage the embryo consists of an outer tube of ectoderm and an inner tube, the primitive gut (**archenteron**). The roof of the primitive gut (Fig. 4-6, amphioxus) is future notochord and mesoderm. The walls and floor are endoderm. The entrance into the primitive gut is the **blastopore.** Cells proliferated from the blastopore rim contribute to both tubes, thus effecting anteroposterior elongation of the embryo, now a **gastrula.** Further changes associated with late gastrular stages will be discussed in terms of **neuralization** (formation of the neural tube), **notogenesis** (formation of the definitive notochord), and establishment of the **mesoderm** as a third germ layer.

In the preceding paragraph we have observed gastrulation in an animal in which there is no yolk mass to impede the process. In many vertebrates, however, gastrulation is influenced by the yolk. Most amphibians manage to tuck the unwieldly yolk inside the embryo, where it contributes to the endoderm. This is accomplished partly when the smaller cells at one edge of the animal pole (Fig. 4-3) grow downward over the yolk, a process called **epiboly.** Epiboly and associated processes too complex for discussion here produce the inner tube (future endoderm) and outer tube (future ectoderm) just as effectively as in the amphioxus. The notochord lies temporarily in the roof of the archenteron, as in the amphioxus (Fig. 4-6).

Vertebrates with a massive yolk have a more difficult problem than amphibians incorporating the yolk into the body of the developing embryo. The small cells of the blastoderm do not grow downward around the massive yolk. Instead, the blastoderm splits (**delaminates**) into an upper sheet of cells (**epiblast**), which will give rise to the epidermis and neural tube, and a lower sheet (**hypoblast**), the endodermal roof of the primitive gut (Fig. 4-4). The yolk is the gut floor. Later, the endodermal sheet grows downward around the yolk to form the yolk sac.

Mammalian gastrulation, like that of reptiles, commences with a separation of the blastoderm (**inner cell mass**) into epiblast and hypoblast. Next, the endoderm of the hypoblast grows downward around a yolk that is not there (Fig. 4-5)! Development of a yolk sac in the yolkless mammals is additional evidence of a genetic relationship with egg-laying vertebrates.

FORMATION OF NEURAL TUBE AND NOTOCHORD

Synchronized with gastrulation are the processes of neuralization (formation of the dorsal, hollow nervous system), notogenesis (formation of the notochord), and formation of the mesoderm. These processes will be discussed separately.

Neuralization. In most vertebrates the dorsal, hollow nervous system (brain and spinal cord) commences as a longitudinal **neural groove** in the dorsal ectoderm (Figs 4-6 and 15-3). The groove is wider anteriorly, and this part becomes the brain (Fig. 4-7). The walls and floor of the neural groove constitute the **neural (medullary) plate.** The plate sinks into the dorsal body wall and the **neural folds** at the borders of the groove close over the latter, converting the groove into a longitudinal **neural tube.** The process of dorsal closure of the groove commences a little behind

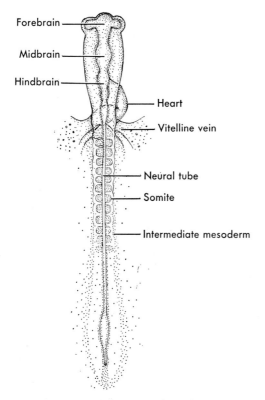

Forebrain

Midbrain

Hindbrain

Heart

Vitelline vein

Neural tube

Somite

Intermediate mesoderm

THIRTY-THREE HOUR CHICK EMBRYO

Fig. 4-7. Chick embryo of approximately thirty-three hours' incubation. The optic vesicles are beginning to evaginate from the forebrain. The dorsal mesoderm (represented by the somites) is metameric.

the anterior end of the body and sweeps cephalad and caudad. Finally, only a small opening remains at the anterior end (**anterior neuropore**) and at the posterior end (**posterior neuropore**). These openings close somewhat later.

In teleosts, ganoids, and cyclostomes, instead of forming a groove, the neural plate develops a solid, ventrally directed, keel-like ridge, which separates from, and sinks beneath, the surface ectoderm (Fig. 4-6, teleost). Eventually it develops a lumen and becomes tubular, as though it had arisen from a groove.

The enlarged anterior end of the neural tube becomes the brain. The remainder of the neural tube becomes the spinal cord.

Both the brain and cord contain a cavity, the **neurocoel.** The neurocoel of the brain is represented by the brain ventricles. The smaller cavity within the spinal cord persists as the central canal. As a result of the formation of the central nervous system from a dorsal groove (or by the process of the hollowing out of an ectodermal neural keel), the central nervous system of chordates is ectodermal, dorsally located, and tubular.

Notogenesis. In the amphioxus the notocord develops as follows: A ribbon of cells (the notochordal plate) in the roof of the archenteron commences to bulge upward toward the neural groove (Fig. 4-8, *B*). This plate finally pinches off from the archenteron to become the anterior part of the notochord (Fig. 4-8, *C*). The body of the amphioxus is elongating rapidly at this time, and the notochord shares in the process by addition of cells directly from the dorsal rim of the blastopore. At the time of hatching of the ciliated amphioxus larva (eight to fifteen hours after fertilization), the notochord is still a mere ridge in the roof of the primitive gut.

In most vertebrates the future notochord is transitorily located, throughout at least part of its extent, in the roof of the primitive gut. Its relation to the archenteron is a very old one. As an inducer of certain steps in neural tube formation, the notochord as well as the archenteric roof plays an important role.

MESODERM FORMATION

Amphioxus. In the amphioxus the first mesoderm arises from a pair of mesodermal bands located in the dorsolateral wall of the primitive gut (Fig. 4-8). These bands fold upward to establish a pair of longitudinal mesodermal folds. Soon, a short segment at the anterior end of each fold constricts off from the rest of the band, closes at the base, and forms a pair of hollow mesodermal pouches. Another pouch differentiates behind the first, then another. About the time that two mesodermal pouches have formed, the embryo

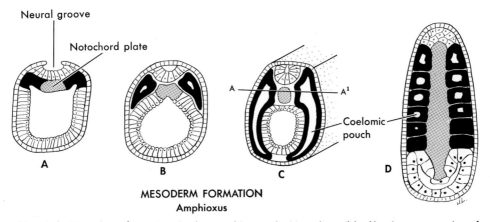

Neural groove

Notochord plate

A

B

C

D

A — — — A¹

Coelomic pouch

MESODERM FORMATION
Amphioxus

Fig. 4-8. Mesoderm formation in the amphioxus. **A,** Mesoderm (black) arises as a series of dorsolateral outpocketings from the archenteron. **B,** Each outpocketing becomes a mesodermal (coelomic) pouch. **C,** Each pouch grows ventrally until they meet to form a ventral mesentery. **D,** Early larva in frontal section (cut at A to A¹).

hatches into a free-swimming, ciliated larva. As the larva continues to elongate, the elongating mesodermal bands continue to fold upward and segment into mesodermal pouches. The pouches are hollow because, in closing at their base, each one entraps a little of the gut cavity. A frontal section of an amphioxus larva at the time that six somites have formed is seen in Fig. 4-8.

After the dorsal, hollow mesodermal pouches are established, they grow ventrally, pushing between the ectoderm and the endoderm (Fig. 4-8, *C*). These downward-growing lateral mesodermal pouches finally meet underneath the gut and fuse, establishing a temporary ventral mesentery. The outer wall of each lateral pouch lies against the ectoderm and is called the **somatic mesoderm.** Together with the ectoderm it forms the **somatopleure** (lateral body wall). The inner wall of each pouch lies against the endoderm and is called the **splanchnic mesoderm.** Together with the endoderm it constitutes the **splanchnopleure.**

The cavity between the somatic and splanchnic mesoderm is the coelom. In the amphioxus the coelom is segmented for a while because of its origin from a series of pouches (Fig. 4-8, *D*). Later, the walls of the successive pouches rupture, establishing a single long coelom on each side. Still later, the ventral mesentery ruptures, and the left and right coelomic cavities become confluent underneath the gut.

Vertebrates. Vertebrates do not go through the process of developing their initial mesoderm as dorsolateral outpocketings of the gut. Instead, the first mesoderm migrates forward from the blastopore rim, pushing between ectoderm and endoderm. These migrating cells establish the dorsal mesoderm as a metameric series of **mesodermal somites** alongside the notochord and neural tube (Fig. 4-7). Thus, the mesoderm is temporarily segmented as in the amphioxus. **Lateral mesoderm,** consisting of somatic and splanchnic sheets, also forms but is unsegmented from its first appearance. Connecting the segmented **dorsal mesoderm (epimere)** with the unsegmented **lateral mesoderm (hypomere)** is a narrow ribbon of **intermediate mesoderm (mesomere),** which is segmented in sharks and mostly unsegmented, except anteriorly, in higher forms.

Dorsal and ventral mesenteries develop (Fig. 11-15). The ventral mesentery subsequently disappears, except at the liver, where it remains as the **falciform ligament,** and at the urinary bladder of amniotes, where it becomes the **ventral mesentery of the bladder.** The fate of the dorsal, intermediate, and lateral mesoderm of vertebrates will be outlined briefly.

Fate of the dorsal mesoderm

The dorsal mesoderm of embryonic vertebrates consists of segmentally arranged mesodermal somites (Fig. 4-7). These are aligned beside the neural tube, along the entire length of the trunk and tail, and in the head in varying numbers. In the head the somites disappear without making direct contributions to the body, except in certain fishes. Each trunk and tail somite exhibits three regions—myotome, sclerotome, and dermatome—which give rise to parts of the musculature, skeleton, and skin, respectively (Fig. 4-9).

Myotome and myotomal muscle. From the myotomes of the mesodermal somites arise, **ontogenetically** or **phylogenetically,** most of the striated skeletal muscles of the body except those of the visceral arches. Muscles derived from myotomes are known as **myotomal muscles.** The most primitive myotomal muscles are the metameric mus-

cles of the body wall. Their anlagen grow downward from the myotomes into the somatopleure, pushing between the ectoderm and somatic mesoderm (Fig. 4-9). Here, in lower forms, they become arranged segmentally as **myomeres,** one myomere arising from each myotome (Fig. 10-6). Since the somites are metameric, the body wall muscles tend to be metameric. In higher vertebrates some of the lateral body wall muscles arise in situ, rather than as actual contributions from the dorsal mesoderm.

Appendicular muscles in fishes arise as buds off the body wall muscle masses. In higher forms they arise from concentrations of loosely arranged, embryonic mesoderm (**mesenchyme**) within the developing limb bud.

Certain muscles of the head are myotomal, having originated from somites either directly or by phylogenetic derivation. These are chiefly the extrinsic eyeball muscles and the muscles of the tongue.

It has been shown in tailed amphibians that unidentified inductive substances elaborated by the notochord and by other embryonic regions are necessary if the

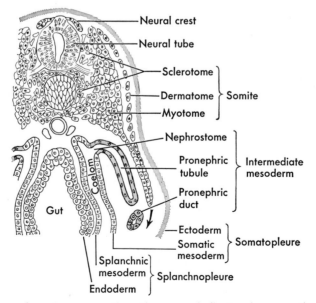

Fig. 4-9. Frog embryo in cross section. The arrow indicates downgrowth of somitic cells into the lateral body wall to contribute myotomal muscle.

myotomes are to differentiate into muscle tissue.

Dermatome and origin of the dermis. The dermatome of the mesodermal somite contributes mesenchyme that migrates toward the ectoderm and gives rise to the dermis of the skin along the dorsolateral part of the trunk only. The dermis of the more lateral and ventral parts of the trunk and that of the appendages arises from the somatic mesoderm, perhaps supplemented by mesenchyme cells from nearby neural crests. (This requires further study. Some investigators think this is true at least in amphibians.) The dermis of the head arises from head mesenchyme, which is largely of neural crest origin (neurectoderm).

Sclerotome and origin of the vertebral column and ribs. The sclerotome of the mesodermal somite gives rise to the vertebral column and to the proximal parts of each rib. Mesenchyme cells from the cephalic end of one somite and from the caudal end of the preceding somite migrate toward, and surround, the notochord and neural tube (Figs 4-9 and 7-4). These cells then deposit cartilaginous or bony vertebrae. This accounts for the intersegmental arrangement of the centra and ribs. The metameric arrangement of vertebrae and ribs is an expression of the metamerism of the mesodermal somites.

Fate of the lateral mesoderm

The lateral mesoderm is confined to the trunk and consists of somatic and splanchnic sheets (Fig. 4-9, somatopleure and splanchnopleure). Between these sheets is the coelom. The somatic mesoderm gives rise in general to the lateral and ventral body wall of the trunk, except for the epidermis and any muscle that invades it from the dorsal mesoderm. The skeleton of the lateral and ventral body wall (sternum and distal parts of the ribs and the girdles) is a product of the somatic mesoderm. The dermis is its outermost product, and the parietal peritoneum is its innermost product. The splanchnic mesoderm gives rise

to the smooth muscles and connective tissue of the digestive tract and its outpocketings and also to the heart. The visceral peritoneum is the outermost derivative of the splanchnic mesoderm. Special regions of the peritoneum contribute adrenal cortical tissue, the germinal epithelium of the gonads, and the lining of the muellerian ducts in some species.

Fate of the intermediate mesoderm

The intermediate mesoderm consists of a longitudinal ribbon of mostly unsegmented mesoderm extending the length of the trunk and lying between the mesodermal somites and the lateral mesoderm. It is sometimes referred to as "nephrogenic mesoderm" because it gives rise to the kidneys and associated ducts.

Mesenchyme of the head, pharynx, and tail

In all vertebrates evanescent mesodermal somites develop in the head anterior to the otocyst and, in the shark at least, these give rise to the eyeball muscles (Fig. 10-6). The mesenchyme dorsal to the pharyngeal arches arises, in lower forms at least, from somites in that location. Some of the mesenchyme of the pharyngeal wall migrates from trunk mesoderm and from neural crests (neurectoderm). Out of these mesenchymal masses, including head somites, trunk mesoderm, and neurectoderm, arise most of the structures of the head and pharyngeal arches.

The tail bud mesenchyme becomes organized into concentrations lateral to the notochord and neural tube. These concentrations give rise to all the structures of the tail, except nervous tissue and epidermis.

FATE OF THE ECTODERM

Stomodeum and proctodeum. During early embryonic development the oral cavity begins as a shallow median invagination of the ectoderm of the head, called the **stomodeum** (Fig. 1-1). This contributes the ectoderm of the anterior part of the oral cavity. A corresponding evagination of the embryonic foregut grows toward the stomodeum until only a thin membrane, the **oral plate**, separates the two. The oral plate finally ruptures. The anterior portion of the oral cavity is therefore ectodermal, and the posterior portion is endodermal. From the ectodermal portion arise the enamel of the teeth and the epithelial covering of the anterior portion of the oral cavity and tongue. Glands of several types (poison, salivary, mucous, etc.) arise from the ectoderm of the oral cavity. An ectodermal evagination called **Rathke's pouch** grows from the roof of the stomodeum toward the floor of the brain, contributing the adenohypophysis.

An invagination similar to the stomodeum develops in relation to the cloacal portion of the hindgut (Fig. 1-1). This caudal invagination is the **proctodeum.** When the cloacal plate separating the proctodeum from the hindgut ruptures, the endoderm of the hindgut becomes continuous with the ectoderm of the proctodeum. The proctodeum gives rise to the terminal part of the cloaca in lower vertebrates and to the anal canal (the caudalmost segment of the digestive tract) in mammals.

Epidermis and its derivatives. The ectoderm gives rise to the epidermis of the skin and its derivatives, including epidermal glands that invade the dermis but open onto the surface by ducts. The mesoderm plays a role in differentiation of skin. Crude collagen can substitute for this role, carrying differentiation all the way to cornification.

Nervous system and sense organs. The origin of the nervous system from the ectoderm has already been described. The special sense organs of vertebrates, including the olfactory epithelium, retina and lens of the eye, membranous labyrinth, neuromast system, and many of the taste buds, are ectodermal in origin. Because the neural crests are ectodermal, all neural crest derivatives are ectodermal. These include, in addition to all ganglia of the nervous system, all chromaffin tissue including the adrenal medulla, probably all pigment cells, the neurilemmal sheaths of peripheral nerve fibers (some of which, however, emerge from the neural tube in chick embryogenesis), the pia mater and arachnoid meninges (in amphibians at least), and much of the pharyngeal arch skeleton and neurocranium.

FATE OF THE ENDODERM

The endoderm gives rise to the epithelial lining of the entire alimentary canal from the stomodeum to the proctodeum. Since pharyngeal pouches arise as outpocketings of the endoderm, the derivatives of the pouches—thymus and parathyroid glands, the eustachian tube and middle ear cavity of tetrapods, and mammalian tonsils (Fig. 17-13)—have endodermal components. Arising as midventral evaginations of the pharyngeal endoderm are the thyroid gland and the lining of the respiratory tract of tetrapods from the glottis to the tips of the lungs. The swim bladders of fishes are also lined with endoderm. Caudal to the pharynx the endoderm evaginates to form the liver, gallbladder, pancreas, various cecums including the rectal gland of dogfishes, and the yolk sac and allantois. The urinary bladder of amniotes arises from the allantois and hence its lining is predominantly endodermal. Parts of the vagina and urethra of mammals are also endodermal because they are products of the embryonic cloaca.

EXTRAEMBRYONIC MEMBRANES

Yolk sac. Many embryonic vertebrates are provided with special extraembryonic membranes extending outside the body. These membranes arise early in ontogeny and perform certain services for the embryo until birth or hatching. One such membrane is the yolk sac (Figs. 4-5, 4-10, and 4-11). The yolk sac surrounds the yolk and is continuous with the embryonic gut. The sac is highly vascular, the **vitelline** (yolk) **vessels** being connected with circulatory channels within the embryo.

Yolk particles within the yolk sac are usually digested by enzymes secreted by the lining of the sac and are transported to the embryo by the vitelline veins. The yolk sac thus grows smaller as the embryo grows larger. As it shrivels, it slowly disappears through the ventral body wall, leaving a temporary scar where it was retracted. The intracoelomic remnant of the yolk sac is finally incorporated into the wall of the midgut, or it may remain as a small diverticulum. In man such a diverticulum (Meckel's diverticulum) remains in about 2% of the population. Its average position is on the ileum, 1 meter before the ileocolic valve, and its average length is about 5 cm.

An interesting modification associated with the yolk sac develops in embryonic spiny dogfishes. When the embryo is about 30 mm. long, a diverticulum of the yolk sac develops within the body close to where the sac enters the duodenum. The diverticulum grows until it occupies much of the coelomic cavity. Dogfish pups ready to be born and with the yolk sac almost completely retracted into the body cavity may be opened and the diverticulum may be easily demonstrated. It is distended with yolk, which has also spilled over into the spiral intestine and duodenum. This remaining yolk serves to nourish the new-born pup until it is able to seek its food from the environment. TeWinkel[48] has shown that the yolk sac and diverticulum are lined with rapidly beating cilia, which cause yolk platelets to pass from the extraembryonic yolk sac up the hollow yolk stalk and into the diverticulum.

Since the yolk sac in viviparous forms is highly vascularized and lies close to the maternal tissues, it often serves as a membrane for absorbing nourishment or oxygen from the parent. It functions thus in some elasmobranchs (sting ray, but not shark), in some viviparous lizards, and even in some mammals. In euviviparous lizards and in marsupials the yolk sac in association with the chorion constitutes the **choriovitelline placenta (yolk sac placenta)** lying against the uterine lining of the mother. In other mammals (Rodentia, Lagomorpha, Insectivora, and Chiroptera) this same type of placenta plays a prominent, but usually temporary, role during gestation.

Allantois. The embryos of reptiles, birds, and mammals develop a second extraembryonic membrane, the allantois. This arises from the cloaca as a large, sac-like evagination (Fig. 4-11, *B*). The allantois grows into the extraembryonic coelom, where it meets a third extraembryonic membrane, the chorion. The allantois and

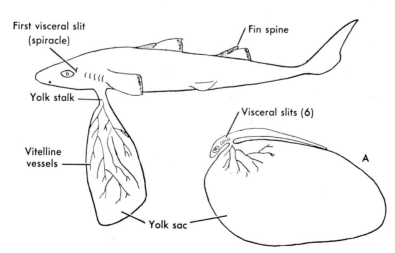

Fig. 4-10. Dogfish embryos, with their yolk sacs, removed from the uterus. **A,** Earlier developmental stage.

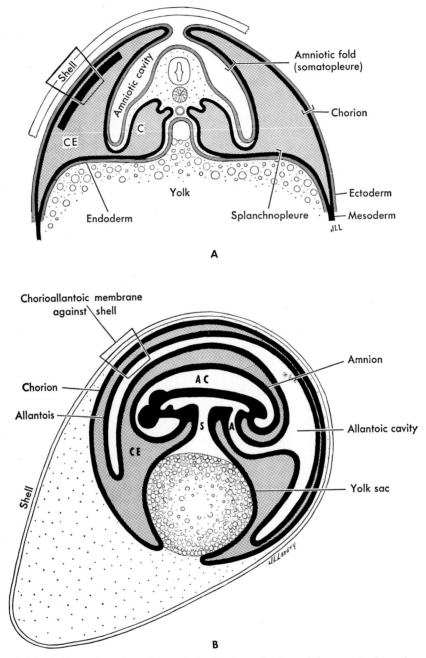

Fig. 4-11. A, Formation of amnion and chorion by upfolding of the somatopleure (amniotic fold). The box at the upper left shows the future chorioallantoic membrane lying against the shell. The black bar represents the future allantois. **B,** Later stage, after allantois has formed as an evagination from the cloaca. This drawing is based on a bird embryo. *A,* Allantoic stalk; *AC,* amniotic cavity containing amniotic fluid; *C,* intraembryonic coelom; *CE,* extraembryonic coelom; *S,* yolk stalk.

chorion together constitute a **chorioallantoic membrane.**

In oviparous reptiles, birds, and monotremes the chorioallantoic membrane comes in contact with the inner surface of the porous egg shell (Fig. 4-11, *B*). In this location, because the allantois is highly vascularized, the chorioallantoic membrane serves as a respiratory organ for exchange of oxygen and carbon dioxide between the fetal organism and the atmosphere, via the porous shell. The cavity of the allantois can serve as a repository for nitrogenous wastes from the kidney and for any excreta from the digestive tract, since the kidney ducts and large intestine empty into the cloaca. Water that collects in the allantois, excreted by the kidneys, can be reclaimed by the vessels of the allantois.

In viviparous mammals the chorioallantoic membrane comes in contact with the lining of the mother's uterus, rather than with the lining of an eggshell. In this location the membrane continues to serve as a fetal respiratory organ, exchanging oxygen and carbon dioxide with its uterine environment. The chorioallantoic membrane performs two additional functions in viviparous mammals. It serves as a site for transfer of nutrients from the mother to the young, and it serves as a site for transfer of metabolic wastes into the tissues of the mother. Thus, the chorioallantoic membrane in fetal viviparous mammals is respiratory, nutritive, and excretory in function until that time when the membrane loses intimate contact with the uterine lining. That event, normally, is during parturition. The part of the chorioallantoic membrane that lies in intimate association with the uterine lining of the mother is the **chorioallantoic placenta.** It is a more modern structure than the yolk sac placenta.

In rodents, rabbits, insectivores, bats, and man the allantoic evagination grows only part way out the umbilical cord, then dwindles to terminate as a blind sac. However, its vessels (allantoic or umbilical arteries and veins) continue toward, and

vascularize, the chorion, which functions as though the allantois were present.

The base of the allantois—that part closest to the cloaca—develops into the urinary bladder, except in birds. The part of the allantois between the distal tip of the bladder and the umbilicus may remain after birth as a **middle umbilical ligament (urachus).** The portion beyond the body wall is discarded at hatching or birth.

Amnion and chorion. The embryos of egg-laying amniotes are subject to drying and jarring, since they lie within an eggshell exposed to the air. The amnion, a fluid-filled sac surrounding the embryo, prevents drying and absorbs any shock incidental to jarring. The young of these animals, therefore, develop in fluid, just as surely as though the eggs had been laid in a pond or in the sea! Viviparous amniotes have inherited the amnion; so even man spends his first nine months in a watery fluid.

The amnion and chorion come into existence when delicate upfoldings of the somatopleure meet above the embryo and fuse (Fig. 4-11, *A*). The chorion expands to become intimately associated with the eggshell or with the lining of the maternal uterus.

Placenta. The **fetal placenta** is that modified region of the extraembryonic membranes (yolk sac, choriovitelline or chorioallantoic membranes, or the chorion alone) lying in intimate association with the female reproductive tract of viviparous vertebrates. The associated modified lining of the female tract is the **maternal placenta.** Other adaptations to viviparity have been discussed earlier in the chapter.

The anatomical relationship between the fetal and maternal tissues may be made more intimate in mammals by the presence of **chorionic villi,** which may dangle within uterine blood sinuses. In such cases part of the uterine wall is torn away at birth when the fetus disengages from the uterus. These mammals are said to ex-

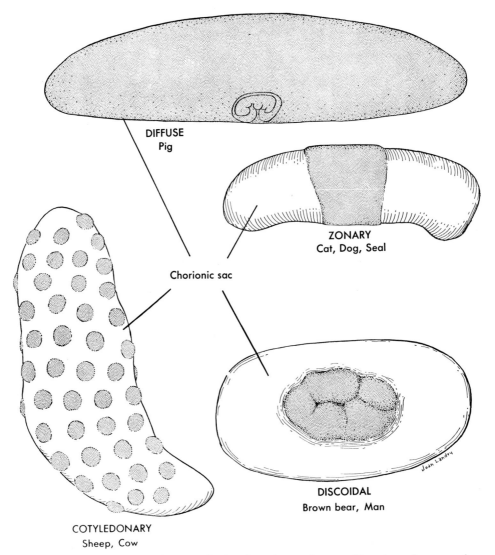

Fig. 4-12. Chorionic sacs of mammals showing placental areas. The pig embryo can be seen through the sac.

hibit **deciduate placentas** (*decidere* = to fall off and hence to be shed). In domestic cats and numerous other mammals, the fetal placenta lies in simple contact with the uterine lining, and, at birth, the fetal membranes simply peel away from the uterus without tearing it. These mammals are said to exhibit **contact placentas.** The extraembryonic membranes (plus any sloughed uterine lining) are delivered as the **afterbirth.**

Placental tissue (Fig. 4-12) may be distributed on the chorion (which becomes a chorionic sac surrounding the rest of the conceptus) in isolated patches (**cotyledonary placenta**); in a band encircling the chorion (**zonary placenta**); in a single, large, discoidal area (**discoidal placenta**); or diffusely over the entire surface of the chorion (**diffuse placenta**). The latter condition is thought to be the more primitive. Numerous articles have been published on the comparative anatomy of the vertebrate placenta.[35] The placentas of mammals, at least, are a source of gonadotropic hormones.

Chapter summary

1. Oviparous vertebrates deposit eggs that hatch in water or on land. These eggs are usually mesolecithal (containing moderate amounts of yolk) or polylecithal (containing abundant yolk). The amphioxus deposits small eggs with very little yolk (alecithal eggs) that hatch quickly into immature, free-living, ciliated larvae with poorly developed organ systems.

2. The yolk in eggs containing little yolk may be evenly distributed throughout the egg (isolecithal or homolecithal condition), or it may be centrally located (centrolecithal condition). The yolk, when not scarce, is aggregated at one pole (vegetal pole), whereas the cytoplasm is aggregated at the opposite (animal) pole. This is known as the telolecithal condition.

3. Viviparous vertebrates develop in the body of the female parent. In the ovoviviparous condition the eggs are laden with yolk, and the mother provides protection but little else. In the euviviparous condition the eggs are practically devoid of yolk, and nourishment must be supplied via the maternal tissues. In most cases of viviparity certain highly vascularized tissues of the embryo lie in intimate association with maternal tissues, which are likewise remarkably adapted for viviparity. Viviparity has evolved in every vertebrate class except birds.

4. Fertilization may be external or internal. It is always internal in viviparous forms. As a result of fertilization and cleavage, a blastula is formed.

5. Gastrulation, neuralization, notogenesis, and mesoderm formation quickly follow the establishment of a blastula.

6. Ectoderm contributes the epidermis and its derivatives, the nervous system, most of the special sense organs, and the derivatives of the stomodeum and proctodeum. Through the neural crests it also makes other contributions, including the adrenal medulla.

7. The brain and spinal cord arise typically as a longitudinal groove in the dorsal ecto-derm, which sinks beneath the surface to establish the dorsal, hollow nervous system. In teleosts, ganoids, and cyclostomes the brain and cord arise from a ventrally directed, solid keel of dorsal ectoderm, which detaches and secondarily becomes hollow.

8. Endoderm gives rise to the lining of the the digestive tract and its evaginations, to the derivatives of the pharyngeal pouches and pharyngeal floor, to the urinary bladder in amniotes, and to at least part of the cloaca or its derivatives.

9. The notochord in most forms arises partly from the roof of the archenteron.

10. Mesoderm is subdivided into the dorsal mesoderm or epimere (segmented), intermediate mesoderm or mesomere (partially segmented), and lateral mesoderm or hypomere (unsegmented, except in the amphioxus).

11. The mesodermal somites (dorsal mesoderm) are subdivided into myotome, sclerotome, and dermatome, giving rise, respectively, to somatic muscles (ontogenetically or phylogenetically), to vertebrae and the proximal parts of the ribs, and to the dermis of the dorsolateral skin.

12. The intermediate mesoderm gives rise chiefly to the kidney and its ducts, and to certain reproductive ducts.

13. The lateral mesoderm contributes somatic mesoderm to the body wall and splanchnic mesoderm to the digestive tract. The pharyngeal arch mesenchyme is derived partly from lateral mesoderm and partly from neurectoderm.

14. The coelom lies between somatic and splanchnic mesoderm. It is initially segmented in the amphioxus and unsegmented in vertebrates.

15. Extraembryonic membranes include the yolk sac (the oldest membrane) and an amnion, chorion, and allantois. The amnion and chorion were reptilian innovations for preventing desiccation. The allantois, likewise a reptilian innovation, served initially as a fetal respiratory membrane. In vivip-

arous mammals the allantois, functioning as a unit with the chorion (chorioallantoic placenta), became also a site of absorption of nutrients and elimination of metabolic wastes. The yolk sac sometimes functions as a placenta independently or in association with the chorion (choriovitelline placenta).

16. Mammals exhibit contact or deciduate placentas with villi arranged in a zonary, cotyledonary, discoidal, or diffuse pattern.

Chapter
5
Skin

The skin covers all exposed surfaces of the body, including the exposed portion of the eyeball (where it is called the **conjunctiva** and is usually transparent) and the external surface of the eardrum. It is directly continuous with the lining of all passageways opening to the surface.

The skin of all vertebrates is built in accordance with the same basic blueprint. It consists of an **epidermis,** derived from the ectoderm, and a **dermis** (**corium**), derived from the mesoderm. Variations among species involve (1) the presence or absence of bone in the dermis, (2) the relative abundance of glands in water forms, and (3) specializations of the surface layer (**stratum corneum**) of the epidermis in land-dwelling forms.

FUNCTIONS

The integument has many important functions among vertebrates, including protection of the organism from injurious external influences, exteroception (the receipt of stimuli from the external environment), respiration (especially among amphibians), regulation of body temperature in homoiotherms (warm-blooded vertebrates), regulation of the ratio of salts to water in body fluids, storage of reserve foods in homoiotherms, nourishment of mammalian young, locomotion, and diverse additional functions.

Protection. Among the most ancient protective integumentary structures of vertebrates is bony dermal armor, such as that encountered among ostracoderms and placoderms, sturgeons, and turtles. It has been modernized to form lightweight dermal scales in modern fishes. In the head region the dermal skeleton may be particularly heavy and thereby shields the brain and sense organs. Elsewhere, it helps prevent compression of soft internal organs.

Integumentary pigments serve a protective function. Fish, amphibians and many reptiles may alter the arrangement of pigment cells, and of pigment granules within the cells, to provide protective coloration. Pigment also protects vertebrates against solar radiation.

Skin glands such as those secreting substances that may be bitter, poisonous, or offensive to the nose ward off potential enemies. If the enemy lives through the first experience, it may not seek a second encounter! Scent glands may also announce the presence of a specific sex.

Pelage (fur) and plumage are defensive weapons in one respect. The bristling coat of an angry mammal and the ruffled plumage of a frightened bird make them more ominous. Claws, nails, horns, spines, barbs, needles—all are integumentary devices conferring advantages in the struggle for existence.

Exteroception. Nerve endings in the skin warn vertebrates against inimical forces in the external environment. These endings are subject to stimulation by touch, changes in pressure, extremes of heat and cold, and chemicals. Woe to the vertebrate that is oblivious to changes in the environment, for he may be destroyed by an enemy or starve to death.

Respiration. The gills are covered, in part at least, with modified skin. Amphibians carry on considerable respiration through the skin of the entire body.

Temperature control. Warm-blooded animals are faced with the necessity of excluding or conserving heat. Fur and feathers provide insulation against excessive heat in warm climates and help conserve heat in cold ones. Deposits of fat within the deep layers of the skin also serve as insulation. Sweat glands provide cooling by evaporation. When heat radiation is desirable, vasodilation of integumentary vessels occurs, and the skin becomes a radiator. When heat conservation is necessary, the vessels constrict.

Water and electrolyte balance. Although the kidneys are often thought of in relation to the elimination of salts and water, sweat glands share this responsibility in many mammals, and chloride-secreting cells often occur on the gills of marine fishes.

Food storage. In addition to serving as insulation, fat in the deep layer of the skin or under the skin of warm-blooded forms serves as reserve food. Hibernating animals accumulate subcutaneous fat prior to hibernation. Migratory birds deposit fat during the weeks preceding migration.

Nourishment of young. Mammary glands of the skin of mammals manufacture milk. Without the nourishment produced by the mother's skin, newborn mammals would starve.

Locomotion. Fins contain flexible rodlike modifications of the skin called **dermal fin rays.** Adhesive pads on the digits assist in climbing, as do claws of amniotes. The feathers on the wings and short tail of birds serve as stabilizers and provide an airfoil.

Miscellaneous. The integument has other functions not cited. Vitamin D is manufactured in some skin. Brood pouches under the skin of some fishes and amphibians protect unhatched eggs. Some skin glands in tetrapods keep the nostrils free of water and dirt. The lacrimal gland washes the conjunctiva of the mammalian eye.

• • •

When the varied roles of the skin of vertebrates are considered, it becomes evident that the integument is more than merely a convenient way to hygienically package the underlying parts.

SKIN OF THE EFT

As an introduction to the study of the integument of vertebrates, we may select a skin unencumbered by scales, feathers, or hairs. The skin of an eft—the sexually immature, land stage of the urodele *Notophthalmus*—will serve to illustrate vertebrate skin in general. The skin of the eft, like that of all vertebrates, is composed of an outer epidermis and an inner dermis (Fig. 5-1, *B*).

Unlike the epidermis of an amphioxus (Fig. 5-1, *A*), the epidermis of vertebrates is a stratified epithelium. The deepest (basal) layer consists of columnar cells undergoing mitosis to replace those lost from the surface. Proliferation from the basal layer causes older cells to be pushed outward. As they approach the surface, they manufacture a substance called **keratin,** which renders them tough and impervious to water. The cells then become flattened (squamous) and finally die. Thus, the outermost layer (stratum corneum) of the epidermis is made up of flattened, dead, cornified (keratinized) cells. The stratum corneum is constantly being shed in patches and replaced, but in some amphibians and reptiles the stratum corneum of the entire body may be shed at one time, a process called molting, or **ecdysis.** The living cells

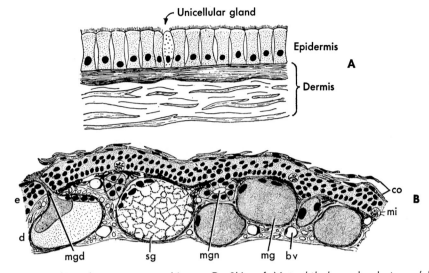

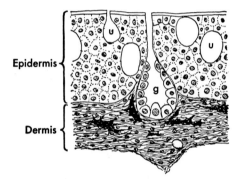

Fig. 5-1. **A,** Skin of a young amphioxus. **B,** Skin of *Notophthalmus,* land stage (eft). **co,** Cells of thin stratum corneum; **d,** dermis; **e,** epidermis; **mg,** mucous gland; **mgd,** duct of discharging mucous gland; **mgn,** new epidermal gland invading the dermis; **mi,** mitotic figure in the basal layer of the epidermis; **sg,** exhausted mucous gland; **bv,** blood vessel.

beneath the stratum corneum are intermediate in shape between columnar and squamous cells.

The dermis underlies the epidermis and is thicker. It is made up chiefly of connective tissue, blood vessels, tiny nerves, lymphatics, bases of epidermal glands, and pigment cells (melanocytes).

The integumentary glands develop from the epidermis and are **multicellular.** As the glands grow, they invade the dermis, where there is room for expansion and where they are close to the capillary beds, their source of supplies. In some regions these glands are so numerous that they constitute the chief bulk of the dermis. The lumen of each sac-like gland is drained by a duct lined with ectoderm.

SKIN OF GNATHOSTOME FISHES

The skin of fishes (Figs. 5-2 to 5-4) is built on the same basic pattern as that of the eft. Three notable differences are found, however: (1) the absence of a typical stratum corneum, (2) the presence of many **unicellular** glands in the epidermis, and (3) the presence of scales in the dermis.

Fig. 5-2. Glandular skin of the dipnoan *Protopterus.* Unicellular, **u,** and multicellular, **g,** glands in the epidermis. The dermis contains melanocytes.

The epidermis of fishes

The stratum corneum is poorly differentiated or absent in fishes. In its simplest form a stratum corneum provides a waterproof covering that impedes or prevents water loss through the skin. It is therefore an adaptation to life in air.

The absence of a stratum corneum in fishes makes it possible for them to have many functional unicellular epidermal

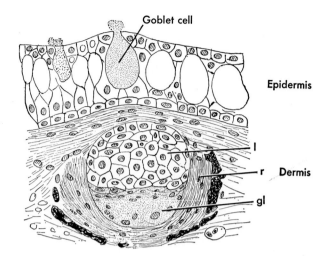

Fig. 5-3. Skin and light organ (photophore) of a luminous fish. **gl,** Luminous cells; **l,** lens cells; **r,** reflector cells (absent in some species) surrounded by melanocytes.

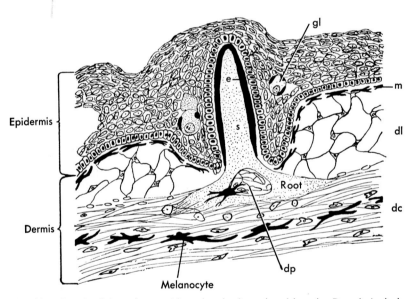

Fig. 5-4. Skin of a dogfish embryo with a developing placoid scale. Dermis includes compact layer, **dc,** and loose layer, **dl; dp,** dermal papilla within the root; **e,** enamel; **gl,** unicellular epidermal gland; **m,** melanocyte; **s,** spine of scale.

glands, since there is no layer of dead cells to prevent the secretions from reaching the surface (Figs. 5-2 to 5-4). The unicellular glands form mucous granules while still close to the basal layer. When they arrive at the surface of the skin, the cells rupture and mucus exudes onto the surface. Skin glands are most abundant in fishes that have lost the capability of developing scales.

In addition to unicellular glands, fishes have multicellular epidermal glands (Fig. 5-2), although they lack the variety found among tetrapods. Certain of the multicellular glands of fishes, especially deep water teleosts, have become modified to serve as

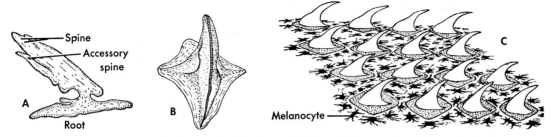

Fig. 5-5. Placoid scales from a dogfish. **A** and **B,** Two different scales dissected out of the skin. **C,** Small section of dogfish skin dehydrated and cleared. The spines are seen projecting from a root (basal plate) embedded in the dermis.

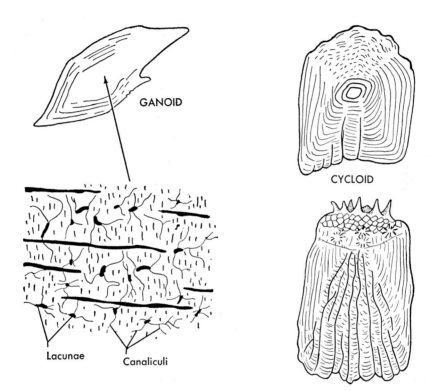

Fig. 5-6. Scales of bony fishes. Lower left, bony structure of a ganoid scale. The uppermost part of the cycloid and ctenoid scales are the free borders.

light-emitting organs, known as **photophores.** These arise in the epidermis and, by growth, invade the dermis. In one variety of photophore (Fig. 5-3) the basal part of the gland consists of luminous cells, and the more superficial part consists of mucus cells that serve as a magnifying lens. Surrounding the base of the gland in the dermis are a blood sinus and a heavy concentration of pigment cells. The light emitted by photophores is not intense and may be of many hues.

Dermal scales of fishes

A dermal scale is a scale that develops in the dermis. Five types of dermal scales have been identified among fishes. **Cosmoid** scales occurred in ancient lobe-finned fishes and are no longer present. **Placoid** scales are characteristic of elasmobranchs (Figs. 5-4 and 5-5). **Ganoid** scales predominated in chondrosteans and holosteans (the ganoid fishes), only a very few of which are alive today (Fig. 5-6). **Ctenoid** and **cycloid** scales are characteristic of modern teleosts (Fig. 5-6); however, cycloid scales also have evolved in lungfishes (Dipnoi) and on parts of the body of some holosteans (*Amia*). All dermal scales are currently considered to be modifications of the bony dermal armor of ostracoderms and placoderms. Cyclostomes and some modern bony fish such as catfish and eels have presumably lost the ability to form dermal scales, although scale anlagen appear transitorily in embryonic eels.

Although fish scales develop in the dermis, the epidermis may possibly play a role in their development,[76] and it apparently determines the direction in which teleost scales are oriented within the dermis. This has been demonstrated by reorienting the embryonic epidermis and observing the altered orientation of the dermal scales that develop later. Scales, like bone, are an important site for calcium storage in teleosts.

Placoid scales (Figs. 5-4 and 5-5). Each placoid scale exhibits a **basal plate** embedded in the dermis and a caudally directed **spine** projecting through the epidermis. The basal plate, composed of dentine-like bone, is anchored into the dermis by connective tissue fibers. The spine consists of dentine continuous with that of the basal plate. It is covered on the surface by enamel and contains a pulpy dermal core of blood vessels, nerve endings, and lymph channels. The spines give the skin a sandpaper-like texture, which can be felt by stroking the finger forward over the body. Because of its texture, sharkskin (shagreen) has been used as an abrasive.

The development of a placoid scale is illustrated in Fig. 5-4. A dermal papilla (a concentration of dermal cells pushing into the epidermis) develops at the site of each future placoid scale just under the epidermis. The papilla organizes the basal plate and, upon invading the epidermis, produces the spine. The growing spine is surrounded by cells of the basal layer of the epidermis. According to one interpretation, these epidermal cells constitute an enamel organ, which secretes the enamel. Further study is needed on the origin and structure of the enamel. The spine finally erupts though the skin.

Modified placoid scales. Tiny placoid scales, called stomodeal denticles, line the entire buccal cavity and much of the internal pharyngeal wall in some Chondrichthyes. Shark teeth are very large placoid scales. The stomodeal denticles in the vicinity of the jaws exhibit transitional stages between denticles and teeth until, on the exposed cutting surfaces of the jaws, typical teeth replace the denticles (Fig. 5-7). The teeth, in turn, are replaced by typical placoid scales in the skin bordering the jaws. Shark teeth, like other placoid scales, are not embedded in the cartilage of the jaws but are simply anchored in the dermis by connective tissue fibers. As teeth, scales, or denticles are lost, new ones replace them.

The fin spine of the spiny dogfish is a modified placoid scale lacking the broad basal plate and having an exaggerated spinous process. A layer of dermal pigment lies between the enamel and the dentine of the fin spine, darkening the spine. The posterior aspect of the spine is not provided with enamel. Fin spines develop in essentially the same manner as do typical placoid scales.

The sting (barb) of the tail of sting rays is a modified, posteriorly directed placoid scale. The ray thrusts the tail forward over its back and drives the sting forcefully into the flesh of the victim. The sting sometimes exhibits small, recurved spines along its entire length, which cause a ragged flesh wound when the barb is withdrawn. The danger to human beings wounded by a sting ray seems to be greater from

Fig. 5-7. Jaws of the Port Jackson shark showing distribution and varieties of teeth. (Courtesy Ward's Natural Science Establishment, Inc., Rochester, N. Y.)

secondary infection than from any poison secreted by associated glands.

The rostrum of the sawfish is greatly elongated (Fig. 3-6). Along its lateral borders are hypertrophied placoid scales ("sawteeth") resembling teeth, which convert the rostrum into a "saw." The covering of the saw is modified integument.

In some elasmobranchs (the basking shark, *Cetorhinus*, for example) myriads of modified placoid scales, called gill rakers, form a straining device guarding the entrance from the pharyngeal cavity into the gill chambers. This device strains out of the respiratory current large quantities of small organisms, which are passed into the digestive tract as food. (The gill rakers in the dogfish shark have a core of cartilage and are not modified placoid scales.)

Ganoid scales. Ganoid scales are composed of bone with lacunae and canaliculi (Fig. 5-6). On the surface of the bone is a hard, shiny substance called **ganoin.** Ganoid scales are typically rhomboid in shape and provide the fish with a bony armor.

Polypterus and garfishes have complete coverings of ganoid scales. Other living ganoid fishes have ganoid scales on the

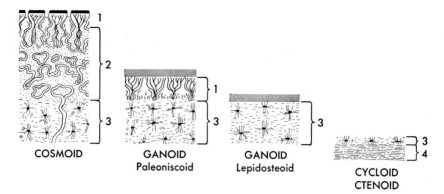

Fig. 5-8. Dermal scale structure, diagrammatical. **1,** Dentine-type bone (cosmine); **2,** spongy bone; **3,** lamellar bone; **4,** fibrous connective tissue of modern fish scale. Black surface represents enamel-like substance; gray surface represents ganoin. Layer **3** in modern scales is often acellular.

anterior part of the body and cycloid scales elsewhere.

The scales in Polypterini and *Latimeria* have a layer of dentine (a special kind of bone) between the basal layer of bone and the ganoin covering (Fig. 5-8). This is an older ganoid scale type (**paleoniscoid**). More recent ganoids have **lepidosteoid** ganoid scales.

Cycloid and ctenoid scales. Cycloid and ctenoid scales (Figs. 5-6 and 5-8) are modern. They overlap and are thin and flexible, consisting of a thin surface layer of bone that is often acellular and an underlying layer (fibrillary plate) of collagenous connective tissue. Cycloid and ctenoid scales differ from one another only in minor details, and both may occur on one fish. Ctenoid scales (*cten* = comb) have tiny tooth-like projections on their free border. As the scale grows, the epidermal covering may persist as a delicate membrane, or it may rub off completely. Among teleosts, cycloid scales occur on carp, buffalo, and similar fishes. Ctenoid scales are found only on teleosts, including perch and sunfish.

Evolutionary trends in fish scales

Ancient bony armor in the dermis of ostracoderms and placoderms consisted of broad plates of bone several layers thick. The various types of scales exhibited by more recent fishes apparently arose from armor of this kind as a result of loss of one or more layers. A theoretical line of evolution of scales is presented in Fig. 5-9.

An early stage in the evolution of dermal scales can be observed in extinct lobe-finned fishes. These exhibited small bony cosmoid scales resembling the large bony plates of armored fishes. No fish alive today has these ancient cosmoid scales.

Ganoid scales seem to be structurally altered cosmoid scales with a layer of ganoin on the surface. Cosmoid and ganoid scales were coexistent at one time, the former on lobe-finned fishes and the latter on the then existing ray-finned fishes.

Cycloid and ctenoid scales may be modified ganoid scales. They lack the hard ganoin on the surface and have only a trace of a bony layer. The scale consists chiefly of a layer of fibrous connective tissue, which makes the scale flexible.

Placoid scales are equivalent to the dentine part of cosmoid scales. The dentine is covered by enamel. Chemical analyses of ganoin, tooth enamel, and the surface material of placoid scales reveals that these differ only in the ratios of certain amino acids to one another.

Dermal fin rays

Supporting the fins of fishes are long, flexible fin rays embedded in the dermis.

The fin rays in Chondrichthyes are elongated, hair-like fibrous connective tissue rods called **ceratotrichia** (*cerato* = horn, *tricho* = hair). In bony fishes the fin rays are branched. Instead of being fibrous, each ray consists of a series of modified scales joined end to end. These rays are called **lepidotrichia** (*lepido* = scale). In some higher teleosts the fin rays have become sharp spines.

Integumentary pigments of vertebrates

The various color patterns seen in the skin of ectotherms (fishes, amphibians, and reptiles) are produced by the reflection and absorption of light by cells that contain pigment granules. Pigment cells have many long processes that ramify among the cells of the dermis immediately under the epidermis, or that may even extend into the epidermis. These processes are always present. In pigment cells known as **chromatophores** the pigment granules may at one time be dispersed throughout the cytoplasm of the cell and into the processes, and at another time they may be aggregated around the nucleus so the processes contain no pigment. The dispersal and aggregation of the pigment granules is under the regulation of neurotransmitters such as norepinephrine produced at the endings of the sympathetic nervous system and of hormones. Chromatophorotropic hormone (intermedin) from the intermediate lobe of the pituitary gland causes certain granules to disperse, thus darkening the skin. Melatonin produced by the pineal body causes certain

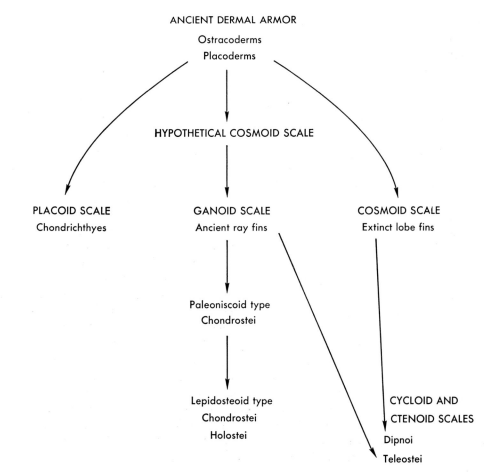

Fig. 5-9. Theoretical lines of derivation of fish scales from dermal armor.

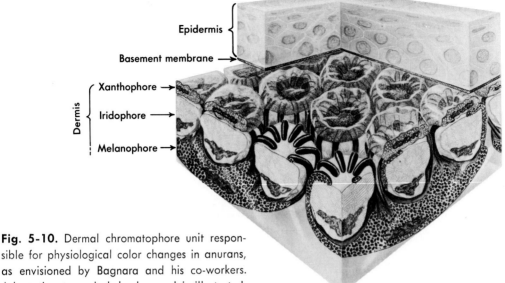

Epidermis {

Basement membrane →

Dermis {

Xanthophore →

Iridophore →

Melanophore →

Fig. 5-10. Dermal chromatophore unit responsible for physiological color changes in anurans, as envisioned by Bagnara and his co-workers. Adaptation to a dark background is illustrated. (Slightly modified from Bagnara, J. T., Taylor, J. D., and Hadley, M. E.: The Journal of Cell Biology **38**:67, 1968.)

of the granules to aggregate, thus blanching the skin.

Chromatophores are named according to the color of the contained pigment. **Melanophores** contain dark brown melanin granules (melanosomes). **Xanthophores** contain yellow, and **erythrophores** contain red granules. **Iridophores** contain a light-dispersing substance, known as guanin, which, under appropriate conditions, results in white, silvery, or iridescent skin. The colors of the skin of ectotherms result from combinations of these types of chromatophores (Fig. 5-10).

Chromatophores—pigment cells in which the granules may be aggregated or dispersed in response to nervous or hormonal stimuli—have not been demonstrated in the skin of birds or mammals. These homoiotherms do, however, have pigment cells containing melanosomes, which do not disperse or aggregate. Such cells are called **melanocytes.** Melanocytes also occur among ectotherms. Pigment cells appear to be differentiated neural crest cells that have migrated from their origin. They are, therefore, neurectodermal.

SKIN OF CYCLOSTOMES

Since absence of scales in fishes appears to be neither primitive nor typical, discussion of the skin of cyclostomes has been deferred until now.

The epidermis of cyclostomes is not especially atypical of fishes (Fig. 5-11). Unicellular glands are especially abundant. The epidermis of the buccal funnel is unusual in that it develops heavy layers of stratum corneum, which form conical, horny epidermal teeth (Fig. 5-12). Smaller cornified spines develop on the rasping tongue. The teeth and tongue spines are periodically shed, as are all cornified appendages of vertebrate skin.

The dermis of cyclostomes is unusual. It is very tough and fibrous, consisting of layer on layer of collagenous connective tissue fibers (Fig. 5-11). The structure suggests that the collagen is deposited in preparation for the differentiation of membrane bone, which, however, fails to materialize. This highly unusual condition is interpretable in terms of the history of the skin of cyclostomes. The earliest agnathans (ostracoderms) had a heavy bony dermal armor. Modern cyclostomes have lost the enzyme systems necessary for the final

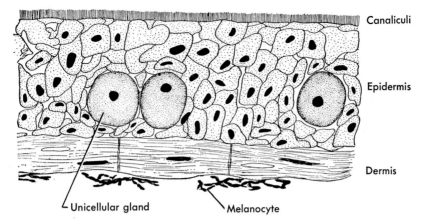

Fig. 5-11. Skin of a larval cyclostome. The striated appearance of the surface results from the presence of tiny canaliculi in the ectoplasm of the surface cells.

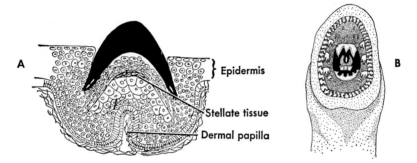

Fig. 5-12. Lamprey teeth and buccal funnel. **A,** Cornified tooth (black), a product of the epidermis. Whether the stellate tissue is the beginning of a replacement tooth is not yet known. **B,** Buccal funnel. The teeth are portrayed as white against a dark background.

steps in the differentiation of bone, but they retain the initial steps, that is, deposit of the preosseous collagenous matrix.

SKIN OF AMPHIBIANS

The skin of amphibians has been modified from that of fishes in three primary respects: (1) dermal scales have been lost in most modern forms, although they were abundant in primitive ones; (2) epidermal glands are multicellular rather than chiefly unicellular; and (3) the epidermis often exhibits a stratum corneum. These conditions, especially the presence of the dry, moisture-conserving stratum corneum, are adaptations to a semiterrestrial existence. Typical amphibian skin is shown in Fig. 5-1.

Stratum corneum. Predominantly water-dwelling urodeles have almost no stratum corneum. Predominantly land-dwelling urodeles and frogs exhibit a thin cornified layer. Efts exhibit a stratum corneum while living on land, but when they return to the ponds as sexually mature adults, the skin glands hypertrophy under the stimulus of hormones, the cornified layer is lifted off by the secretions, and the skin becomes soft and moist.

Modifications of the stratum corneum in amphibians are few. Toads exhibit a heavy stratum corneum with thickened patches, which give a warty appearance. Many species of frogs exhibit localized warty elevations of the stratum corneum. Callus-like caps develop on the tips of the fingers

Fig. 5-13. Undersurface of skin from the tail of *Necturus*, showing epidermal mucous glands (arrow) projecting into subcutaneous tissue just external to muscle.

and toes of urodeles subjected to buffeting in mountain brooks. Tadpoles have horny jaws and lips covered by rows of horny teeth, which are adaptations for feeding on vegetation during the larval stage. These larval cornified parts are lost at metamorphosis.

Glands. In amphibians such as toads, in which a thick stratum corneum is present, glands are less numerous, and the skin tends to be dry. With less cornification, multicellular glands abound. Mucous glands are more or less continually active and thus maintain the skin in a moist state. Although skin glands arise from the epidermis, they usually invade the dermis (Fig. 5-13). Serous glands usually secrete only when the animal requires the protection of the acrid, often toxic, secretion that wards off enemies (parotoid gland, Fig. 5-14). Amphibians are the lowest vertebrates in which multicellular glands are commonplace.

Dermal scales. Although dermal scales are absent in most modern amphibians, the skin of labyrinthodonts was heavily armored, like that of ancient fishes. In some cases the scales were minute. Rem-

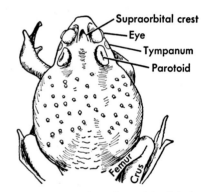

Fig. 5-14. Warty skin of toad *(Bufo)*. A serous gland, the parotoid, is seen behind the eye.

nants of armor occur today as scales of 1 to 2 mm. embedded in the dermis of some caecilians (Fig. 5-15). A few tropical toads exhibit bony dermal scales on the back. Scales embedded in the dermis will be encountered again among higher tetrapods, but only in occasional species.

Subcutaneous lymph sinuses. The relationship of the skin to the underlying muscle in frogs is unusual. Underneath the dermis of frogs lie numerous subcutaneous lymph sinuses separated from one another at intervals by connective tissue partitions

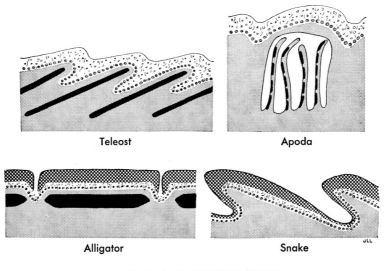

Teleost   Apoda

Alligator   Snake

DERMAL AND EPIDERMAL SCALES

Fig. 5-15. Dermal scales (black) in a fish, amphibian (caecilian), and alligator. Epidermal scales (black and white) in an alligator and snake. The epidermal scales are modifications of the stratum corneum. The gray area represents the dermis.

attaching the skin to the underlying muscle. The skin of the frog is, therefore, less intimately attached to the body than that of other vertebrates.

Looking backward and forward. The vestigial scales of caecilians permits us a backward glance into the age of dermal armor. The skin of frogs, with its thin stratum corneum, its many glands, and the absence of scales and other integumentary hard parts, is characteristic of the specializations exhibited by modern amphibians. The warty skin of toads, with its heavy cornified layer, gives us a preview of the skin of reptiles.

SKIN OF REPTILES

Reptiles in general are fully adapted to life in the dry air. Instead of returning to water to lay their eggs, as amphibians do, they must lay these eggs on land even though they live in the water. The major integumentary adaptations for terrestrial life include (1) the presence of a relatively thick stratum corneum; (2) the evolution of numerous modifications of the stratum corneum, including epidermal scales,

claws, horns, spikes, beaks, and rattles; and (3) a sparsity of skin glands. As a result of these modifications, typical reptiles have a dry, scaly skin, which retards the loss of valuable water through the surface. After all, the water problem of reptiles is not trying to unload water, as in freshwater fishes, but rather conserving the little water they are able to obtain from their environment. It is sometimes a long time between rains!

In addition to epidermal scales, these animals also retain traces of the bony dermal scales of their ancestors. These ancient vestiges will be discussed first.

Dermal scales of reptiles

The inherent potential of the dermis to develop bone has been emphasized. The earliest reptiles inherited bony dermal plates from the ancient amphibians. The latter inherited them from armored fishes. Many extinct reptiles, therefore, had large dermal scales, and it is not surprising that remnants of this bony armor still occur in living reptiles (Figs. 5-16 and 5-17).

Crocodiles and alligators have many oval

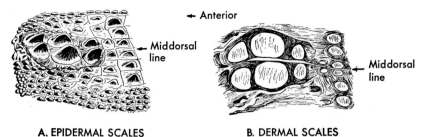

A. EPIDERMAL SCALES   **B. DERMAL SCALES**

Fig. 5-16. Alligator skin. **A,** From dorsum of neck showing epidermal scales. **B,** Same section turned upside down to show dermal bone embedded in skin beneath the tall crests. The middorsal line may be used as a reference point.

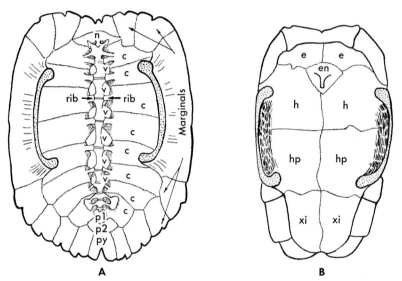

DERMAL PLATES

Fig. 5-17. Dermal plates of the carapace, **A,** and plastron, **B,** in a turtle, from an internal view. On the carapace the plates shown are, **c,** costals, united with ribs; **n,** nuchal; **p**^{1} and **p**^{2}, precaudals; **v,** vertebrals (six of the eight are labeled). Marginals encircle the periphery and include the pygal, **py.** On the plastron the plates shown are, **e,** epiplastrons; **en,** entoplastron; **h,** hyoplastrons; **hp,** hypoplastrons; **xi,** xiphiplastrons.

bony plates embedded in the dermis. Along the dorsum of the neck these plates are especially prominent (Fig. 5-16) and often exhibit crests, which give the back an awesome, prehistorical appearance. A few lizards exhibit similar, but smaller, dermal scales.

Turtles are truly armored vertebrates! The armor, or shell, consists entirely of large bony plates deposited in the dermis and arranged in specific patterns that meet in immovable sutures (Fig. 5-17). The shell covers the entire trunk of the turtle and consists of a dorsally arched **carapace** and a ventral, flattened **plastron.** The two parts are united by **lateral bridges,** which must be sawed through to expose the soft internal organs.

The dermal plates of the carapace occur in three groups, best observed on the internal surface of the shell. The following description is for *Chrysemys,* the common painted turtle. Variations occur among the species. (1) In a median row are found,

Fig. 5-18. Epidermal scales of lizards. **A,** Horned "toad." **B,** Collared lizard. (**A** courtesy Carolina Biological Supply Co., Burlington, N. C.; **B** courtesy F. W. Schmidt, naturalist photographer, La Marque, Texas.)

commencing anteriorly (Fig. 5-17, *A*), one **nuchal,** eight **vertebral,** and two **precaudal plates.** Nuchal and vertebral plates unite with vertebrae during embryonic development. (2) Immediately lateral to the foregoing series are eight **costal plates,** each united with a rib. (3) **Marginal plates** constitute a peripheral ring. The last marginal plate (**pygal plate**) is unpaired. The dermal plates of the plastron, visible on its internal surface, are (Fig. 5-17, *B*) **epiplastrons** (one pair), **entoplastron** (median), and **hyoplastrons, hypoplastrons,** and **xiphiplastrons** (all paired). Soft-shelled turtles and the leather-back sea turtle have leathery shells devoid of dermal ossification.

Epidermal scales of reptiles

The surface scales on a reptile are modifications of the stratum corneum and hence are epidermal. In reptiles the tendency to form a dry, scaly epidermis reaches an evolutionary peak. Above reptiles epidermal scales continue to be in evidence but are less widely distributed over the body. Epidermal scales are not independent elements capable of being dissected out of the skin; they are continuous with one another because of the continuity of the stratum corneum.

Development of epidermal scales is initiated by dermal papillae. These papillae are elevations of the dermis indenting the epidermis on its undersurface. The vascu-

lar beds in dermal papillae provide a continuing source of nutritives and oxygen for the production of cornified cells. The epidermis overlying a papilla becomes thickened by cell divisions. The proliferated cells move toward the surface and, in so doing, become impregnated with a scleroprotein known as **keratin.** This substance is insoluble in most solvents and shows a high sulfur content chiefly because of the amino acid cystine. When keratin is deposited in a cell, the cell is said to be keratinized or cornified, and it dies. All stratum corneum cells are cornified.

Lizards and snakes. The epidermal scales of lizards and snakes (Figs. 5-18 and 5-19) overlap one another. The enlarged scales (head shields) of the head region of snakes (Fig. 5-20, *A*) exhibit a characteristic number and arrangement and are sometimes named for the underlying dermal bones that develop within the dermal papillae. Anal shields (Fig. 5-20, *B*) also occur. The scales (or **scutes,** as they are sometimes called) that come in contact with the ground on the ventral surface of snakes provide protection from mechanical injury and furnish the friction necessary in locomotion.

Periodically, the stratum corneum of the entire body of snakes and lizards—head, trunk, tail, appendages, and even the covering of the eyeball—is shed in one piece (ecdysis, or molting). The discarded cor-

Fig. 5-19. Cottonmouth snake. **A,** Single cottonmouth scale. **B,** Scale arrangement. (Courtesy Carolina Biological Supply Co., Burlington, N. C.)

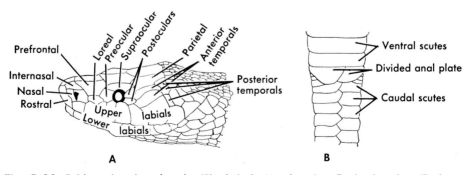

Fig. 5-20. Epidermal scales of snake *(Elaphe).* **A,** Head region. **B,** Anal region. (Redrawn from Parker.[80])

neum is left behind attached to a log or jagged rock.

Crocodiles and alligators. The epidermal scales of crocodilians (Fig. 5-16) are large and rectangular on most of the trunk and tail and are smaller on the limbs. They are not independent of one another but are part of a continuous sheet of stratum corneum, with row upon row of horny thickenings that do not overlap. Crocodilians do not shed the entire stratum corneum at one time. Instead, patches are sloughed at intervals.

Turtles. The epidermal scales on the surface of the shell in turtles are broad, quadrilateral sheets of stratum corneum and are more often referred to as scutes than as scales. These scutes lie in groups and have names (Fig. 5-21). The arrangement of the scutes **does not coincide** with that of the underlying bony dermal plates.

The scutes in the carapace of *Chrysemys* lie in three groups: **neural** (five median scutes), **costal** (four lateral pairs), and **marginal** (twenty-five scutes). The marginal scutes continue onto the ventral aspect of the shell. Inframarginal scutes occur on the lateral bridges. Other scutes develop on the plastron (Fig. 5-21, *B*). On those surfaces of the body not covered by the shell—limbs, tail, neck, and a few parts of the trunk—the epidermal scales are much smaller and resemble the smallest scales of alligators.

Other modifications of the stratum corneum of reptiles

In addition to epidermal scales, reptiles exhibit other modifications of the stratum corneum. **Claws** are modifications of the stratum corneum at the tips of the digits. They consist of two curved parts: a dorsal horny **unguis** and a ventral softer **subunguis**. These two parts cover the terminal bone in the digit. The unguis elongates more rapidly and folds down over the edges of the subunguis to form the claw. Claws occur first in reptiles and persist in birds and mammals. They evolved as nails in primates and as hoofs in ungulates. Reptilian claws are periodically shed. (African clawed toads do not have claws comparable to those of amniotes.)

The horny substance on the **beak** of turtles is a modification of the stratum cor-

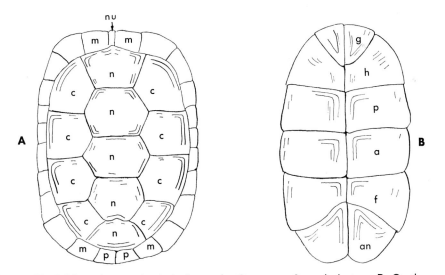

Fig. 5-21. Epidermal scutes (scales) of a turtle. Carapace, **A**, and plastron, **B**. On the carapace the scutes shown are, **c**, costals; **m**, marginals all around the periphery, including the nuchal, **(nu)**, and pygals, **(p)**; **n**, neurals. On the plastron the scutes shown are, **g**, gulars; **h**, humerals; **p**, pectorals; **a**, abdominals; **f**, femorals; **an**, anals.

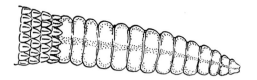

Fig. 5-22. Rattles from a rattlesnake, dorsal view.

neum. Rattlesnake **rattles** (Fig. 5-22) are rings of horny stratum corneum that remain attached to the tip of the tail after each molt. The **horns** of the horned toad (a lizard) are grotesque modifications of the stratum corneum.

Reptilian skin glands

Reptiles exhibit relatively few integumentary glands. In some snakes, scent glands open near the cloaca. Their secretion plays a role in sexual attraction. The scent left along a trail may also be of assistance in enabling snakes to congregate prior to hibernating. Male lizards exhibit femoral glands on the medial side of each hind leg. These glands secrete a substance that hardens on the skin and forms small, temporary spines said to be of assistance in holding the female during copulation. As many as four different sets of integumentary glands encircle the cloacal region in some lizards. Musk turtles exude a yellowish fluid with a musk odor from two glands on each side of the body just below the carapace. A longitudinal row of apparently nonfunctional glands occurs along the dorsum in alligators.

SKIN OF BIRDS

The term "thin-skinned" is literally true with reference to birds. The skin is a delicate membrane loosely applied to the underlying muscle, which thus retains maximum freedom for flight. Only on the legs, feet and head is the skin relatively thick and intimately attached. These regions exhibit such modifications of the stratum corneum as epidermal scales, claws, and horny coverings other than feathers. Elsewhere the body is clothed with feathers.

Skin glands are few in number. There are no dermal scales. Their loss was certainly no disadvantage in flight.

Epidermal scales, horny sheaths, and claws

Epidermal scales in birds are limited to the lower leg, foot, and base of the beak. The covering of the beak itself consists of a sheath of stratum corneum, which may be in one piece on each jaw, or which may be in several sections (compound sheath). The segments of compound sheaths have been homologized with specific reptilian scales. Horny toothlike protuberances occur as modifications of the sheath on the beaks of some birds. The sheath of the beak may be continued onto the nasal region in the form of a shield (frontal plate).

Any bony spurs on the wrist or ankle of birds, like that on the carpometacarpus of the Mexican Jaçana (Fig. 9-16) or on the tarsometatarsus of gamecocks, are covered with a sheath identical to that of the beak. (The core of the spur, like the core of the beak, consists of membrane bone.)

The comb of roosters is covered with a thick, warty stratum corneum. In the dermis lie cavernous blood sinuses that give a red color to a healthy comb.

Claws, like beaks, are diversified and adapted to many different habitats (Fig. 5-23). Although claws of birds are usually thought of as being restricted to the hind limb, one or two sharp claws may occur also at the ends of the digits of the wings (ostriches, hoatzins, geese, some swifts, and so forth). The young hoatzin uses the claws on the wings for climbing about on branches of trees. *Archaeopteryx* had three claws on each wing.

Feathers

Feathers are a remarkably complicated modification of epidermal scales and, as such, are products of the stratum corneum. There are three types of feathers on the body: contour feathers, down feathers

Fig. 5-23. Claw on the middle toe of a great blue heron. (From Atwood: Comparative anatomy, ed. 2, St. Louis, 1955, The C. V. Mosby Co.)

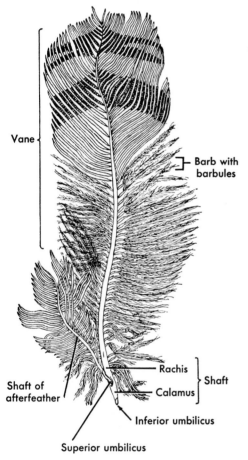

Fig. 5-24. Contour feather from a grouse.

(plumules), and hair-like feathers (filoplumes).

Contour feathers (Fig. 5-24). Contour feathers are the conspicuous feathers that give the bird its contour, or general shape. A typical contour feather consists of a horny, hollow **shaft** and two flattened **vanes.** The base of the shaft devoid of

vanes is the **calamus (quill).** The vane-bearing segment of the shaft is the **rachis.** Each vane consists of a row of **barbs,** which in turn have **barbules.** The latter have **hooklets,** which interlock with the barbules of adjacent barbs and thereby stiffen the vane (Fig. 5-25). When a contour feather is ruffled, the barbules have become unhooked. Preening reunites adjacent barbs by rehooking the hooklets and thus returns the feather to its tailored state. During preening the secretion of the uropygial gland (p. 113) is applied to the barbs. On the smaller contour feathers of the wings (**covert** feathers), the lower barbules lack hooklets, and this region of the feather is fluffy. In ostriches and some other birds all feathers are fluffy, since hooklets are absent.

Arising from a notch (**superior umbilicus**) on the shaft of a contour feather at the base of the rachis is an **afterfeather.** Usually the aftershaft is much smaller than the main shaft, but in some birds (emu and cassowary) the afterfeather is of the same length. This results in a double feather from a single calamus. At the base of the shaft an opening, the **inferior umbilicus,** leads into a cavity in the shaft, occupied by a plug of the dead dermal papilla (**pulp**). The surface of the shaft lying against the skin exhibits an umbilical groove.

Although contour feathers completely clothe most of the body, the feather follicles from which they grow are disposed in feather tracts (**pterylae**) (Fig. 5-26). A few birds, including ostriches and penguins, lack feather tracts.

Down feathers. Down feathers are small, fluffy feathers lying underneath and between the contour feathers. They may be seen by separating the contour feathers. Very young birds usually lack contour feathers and are covered at hatching by a complete coat of down. Down feathers may be ancestral to the more complicated contour feathers. They have a short calamus, with a crown of barbs arising from the free end. Hooklets are lacking. Eider-

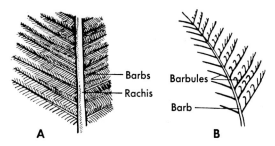

Fig. 5-25. A, Section of feather showing a series of barbs coming off the rachis. **B,** Each barb has barbules, and the barbules have hooklets. (From Atwood: Comparative Anatomy, ed. 2, St. Louis, 1955, The C. V. Mosby Co.)

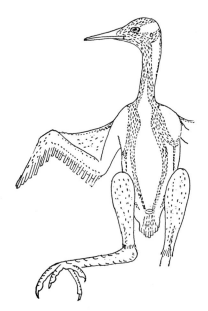

Fig. 5-26. Feather tracts.

down, used in pillows, is the down feathers of the eider duck.

Filoplumes. Filoplumes are hair-like feathers familiar to housewives who singe chickens before cooking them. Filoplumes lack vanes and consist chiefly of a thread-like shaft. They are usually scattered throughout the skin between the contour feathers. In peacocks they are very long.

Development of feathers. The future down feather at first may be identified as a minute, pimple-like elevation on the surface of the skin. Within this feather primordium lies an elongated dermal papilla (Fig. 5-27). The primordium elongates and a pit (**feather follicle**) develops around its base. The germinal layer of epidermis at the base of the follicle in the **growth zone** proliferates tall columns of epidermal cells, which push toward the distal tip of the growing feather between the dermal papilla and the epidermis (now called **periderm,** or feather sheath). These epidermal columns separate from one another, cornify, and develop into barbs with attached barbules. A growing feather still surrounded by its sheath is a **pinfeather.** When the feather sheath splits open, the fluffy barbs stretch out of their cramped quarters, and the quill elongates. When a feather is full grown, the dermal papilla in the shaft dies and becomes the pulp. The living basal portion of the papilla withdraws from the base of the shaft via

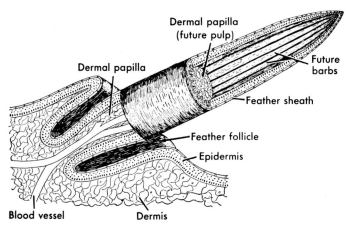

Fig. 5-27. Developing down feather.

the inferior umbilicus. Feathers that arise during successive molts develop from re-activated dermal papillae, which have already given rise to the preceding feathers. During natural molts in many birds—perhaps in all birds—feathers of the old generation are passively pushed out of the follicles by the incoming feathers.

Feather pigments. Color in bird plumage is the result of combinations of integumentary pigments found in melanocytes. The melanin granules within the barbs have an interesting ontogenetic history. **Melanoblasts** (future melanocytes) derived from neural crests migrate into the dermal papilla at the base of the developing feather and then into the epidermis of the growth zone. Here the melanoblasts mature into melanocytes with typical branched processes that ramify among the epidermal cells of the growth zone. Electron microscopy has shown that melanin granules are actively ejected from the tips of the processes of the melanocytes into the epidermal cells that are being added to the barbs. Thus, the pigment granules are in the barbs.

There are no blue granules. When viewed under a microscope by **transmitted** light, the "blue" feather is seen to be brown, the color of the melanin granules beneath the prismatic layer. The blue color observed in **reflected** light is a dispersion phenomenon, like the blue of the sky. However, a red feather is red even when viewed in transmitted light.[74]

Integumentary glands of birds

The **uropygial gland,** a prominent swelling just above the foreshortened tail, is one of the few integumentary glands found in birds. The oily secretion that exudes from it is transferred to the feathers while the bird is preening. The outer ear canal is lined with small oil glands in some birds.

SKIN OF MAMMALS

The epidermis of mammals exhibits a thick stratum corneum (particularly thick on the palm of the hand and the sole of the foot), and a variety of multicellular integumentary glands (Fig. 5-28). Modifications of the stratum corneum include epidermal scales, hair, claws, nails and hoofs (modified claws), true horn, and other miscellaneous cornified structures. Integumentary glands are of a wider variety than in any other vertebrate group, and many, such as mammary, sweat, and sebaceous glands, are strictly mammalian.

The dermis of mammals is thicker in proportion to the epidermis than in other vertebrates. No ossification occurs directly in the dermis of mammals, except in armadillos; therefore, these are the only mammals with dermal scales. However, collagenous connective tissue, which precedes formation of dermal scales in lower vertebrates, is abundant. When treated with tannic acid (that is, when "tanned"), the dermis becomes leather. Projecting into the dermis are the bases of epidermal glands and of hairs.

The superficial fascia, underlying the dermis, is a pliable cushion of loose subcutaneous connective tissue that is more abundant in some locations than in others and in mammals with little fur. Although some fat (adipose tissue) develops in the dermis, most of the fat associated with the skin is deposited subcutaneously. Whale blubber, an extreme example of subcutaneous fat, is an adaptation to life in the icy seas.

Glands

Mammary glands. Mammary glands are modified skin glands that secrete milk. They arise in both sexes from a pair of elevated ribbons of ectodermal cells called **milk lines,** extending along the ventrolateral body wall of the fetus from the axilla to the groin (Fig. 5-29). Along the milk lines develop patches of undifferentiated mammary tissue, which invade the dermis (Fig. 5-30) and then spread under it in the superficial fascia. As development progresses, teats (nipples) develop above each patch. In most mammals

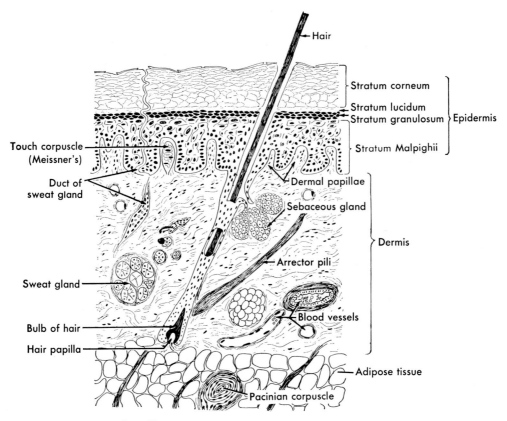

Fig. 5-28. Mammalian skin.

the duct system of both sexes develops in a circular patch beneath each nipple. In hamsters the duct system spreads only to the side.

As the female approaches sexual maturity (adolescence in primates), rising titers of female sex hormones cause the duct system to spread and branch. During pregnancy a battery of hormones in appropriate proportions causes the formation of sac-like secreting terminals (alveoli) at the ends of the duct system. In pregnant cats and many other mammals successive patches of mammary tissue spread toward each other and unite to form on each side a long, thick mass of considerable weight.

In monotremes true mammary glands do not develop. Instead, in both sexes glands resembling modified sweat glands produce a nutritious secretion, which is lapped off a convenient tuft of hairs by the young (Fig. 5-31, monotreme). Teats would probably be useless in the duckbill, since it appears doubtful whether the young, hindered by horny beaks and lacking muscular cheeks and lips, could nurse. Except during lactation, the teats of the opossum are hygienically stored in depressions within the skin.

The distribution and number of mammary glands and nipples vary with the species. A single pair of thoracic nipples occurs in apes and man. Bats also have thoracic nipples. Insectivores have one pair of thoracic and one pair of inguinal nipples. Flying lemurs have a single pair in the armpit (axillary nipples). In Cetacea nipples occur near the groin (inguinal nipples), and the baby porpoise or whale holds onto the nipple as the mother swims about in the sea. Male lemurs have a nipple on each shoulder, and nutrias have four on the back so that the babies

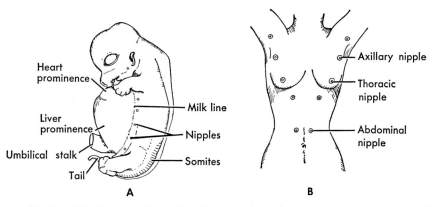

Fig. 5-29. A, Milk line and nipples in a 20 mm. pig embryo. **B,** Supernumerary nipples in man.

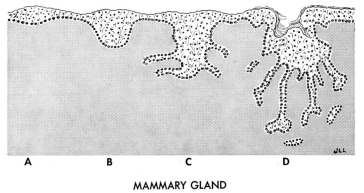

MAMMARY GLAND

Morphogenesis

Fig. 5-30. Successive stages of mammary gland development. **A,** Equivalent to a human embryo at 6 weeks; **B,** equivalent to a human embryo at 9 weeks and to mouse embryo at 16 days; **C,** intermediate stage; **D,** at birth. Gray area represents dermis.

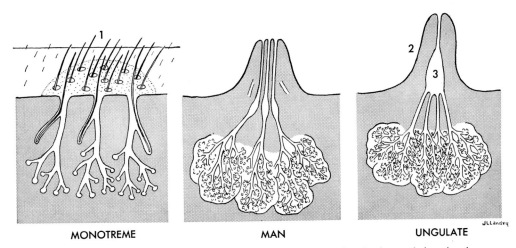

Fig. 5-31. Mammary ducts and nipples. The monotreme lacks nipples and the glands resemble modified sweat glands. Man has true teats. **1,** Hairs; **2,** false teat; **3,** cistern.

are able to ride along on the mother's back above water while nursing. In pigs, dogs, edentates, and many other mammals a series of **axillary, thoracic, abdominal,** and **inguinal** nipples is scattered all along the milk line. Supernumerary nipples may occur in any mammal, including man (Fig. 5-29, *B*). In general, there are sufficient nipples for the number of young in a litter and these are in locations adapted to the habits of the species.

In true teats (Fig. 5-31) the ducts open at the tip of the nipple as a result of the elevation of the duct-bearing area during development. In false teats the skin around the duct openings becomes elevated as the teat develops, and all ducts empty into one cistern. Suckling young are not concerned about the terminology employed, so long as the teat is provident.

Sudoriferous glands. Sweat glands (*sudor* = sweat) are long, slender, coiled tubes of epidermal cells extending deep into the dermis (Fig. 5-32, *B*). Their secretions ooze onto the surface of the skin through tortuous channels, which perforate the stratum corneum and open as pores. The glands arise in embryonic life as sim-

ple tubular ingrowths of the epidermis into the dermis. Sweat glands are widespread among mammals, but they may be absent, as in scaly anteaters and marine forms such as whales. Man, covered with less hair, has the greatest number of sweat glands per square inch of body surface. They are much restricted in distribution in many mammals and occur only on the soles of the feet (cats and mice) or on the toes, lips, ears, back, or head. The **ciliary glands,** which open into the hair sacs of the eyelashes and along the margins of the eyelids, are sudoriferous glands.

Sebaceous glands. Sebaceous glands (*sebum* = grease) secrete an oily exudate, sebum, into the hair follicle (Fig. 5-32, *B*). The oil oozes onto the surface of the skin, lubricating the stratum corneum. Fur (including human hair) glistens after brushing because of the film of oil on the hairs. Marine mammals, which are practically devoid of hair, do not have these glands. Usually several glands open in association with one hair shaft. They may open directly onto the surface of the skin in some areas.

In the outer ear canal of mammals modi-

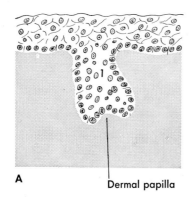

A Dermal papilla

Fig. 5-32. Successive stages in the development of hair and associated glands. **1,** Initial epidermal ingrowth into dermis indented by dermal papilla; **2,** hair follicle; **3,** developing sebaceous gland; **4,** sweat gland.

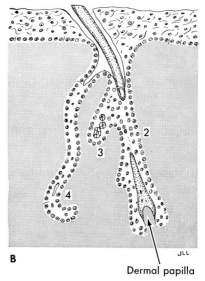

B

Dermal papilla

HAIR
Morphogenesis

fied sebaceous glands (**ceruminous glands**) secrete cerumen, a waxy grease. This wax helps trap insects that might otherwise wander deep into the canal and touch the highly sensitive eardrum. The **tarsal** (meibomian) **glands,** which secrete oil onto the exposed surface of the eyeball, are sebaceous glands situated in the dense connective tissue plate (tarsus) supporting the free edge of each eyelid.

Other integumentary glands. Mammals have a wide variety of miscellaneous skin glands opening on many surfaces of the body. The most striking of these are scent glands, which may protect the animal from foes or may attract members of the opposite sex. Scent glands occur on the feet in the goat and rhinoceros. Callus-like growths on the feet of horses appear to be remnants of similar glands. The musk gland of the musk deer is the size of a chicken's egg, embedded in the dermis of the abdominal wall. Musk is used as the basis for many perfumes. Many carnivores and rodents, including skunks, have large scent glands that open near the anus. Kangaroo rats *(Dipodomys)* have a specialized sebaceous glandular area of skin along the anterior part of the middorsal line in the most exposed area of the arched back. It seems to be primarily a scent gland. A pair of tame grisons (South American rodents) emitted such a pungent odor at all times that it was almost unthinkable to remain in a closed room with these animals for more than a few minutes. Many of the odors of the mammalian zoo are caused by scent glands, not by unhygienic conditions in the pens and cages.

Male elephants have a temporal gland that swells during the breeding season and secretes a sticky, brown fluid. At this time the male elephant is very dangerous. A gland above the eye of the peccary looks like a navel and secretes a watery fluid. The male lemur in some species has a hardened patch of spiny skin on the forearm, under which lies a gland the size of an almond. Bats have so many glands in the skin of the face and head that their features are distorted and bizarre. Glands in the skin of the external genitals of both sexes seem to be chiefly lubricatory. All of the foregoing appear to be modifications either of sebaceous or sudoriferous glands.

Hair

Hair, a modification of the stratum corneum, is found on every mammal, and nowhere else in the animal kingdom. The hair may form a dense, furry covering over practically the entire body, or it may be represented by only two bristles on the upper lip, as in some whales. Where the fur is dense, there are usually short, fine hairs (underfur) as well as longer, coarser ones. Commercial sealskin is the skin of a seal with the underfur intact.

The chief advantage of hair seems to be its insulating effect. Hairs also serve as sensitive tactile (touch) organs, since the root of each is surrounded by a basket-like network of sensory nerve endings, which, when disturbed, initiates a train of sensory impulses to the brain. Disturb a single hair on the back of your hand and note the sensation evoked. **Vibrissae** (whiskers) perform this role exclusively.

Phylogenetic origin of hair. The phylogenetic origin of hair is not known. It has been proposed[73] that a hair may be a modification of an epidermal bristle (**protothrix;** *thrix, tricho* = hair) that emerges from certain sense organs called apical pits in reptiles. These sensory pits lie at the apices of epidermal scales singly or in groups, as do hairs, and the bristles emerge from the pits to extend beyond the surface of the skin. The similarity in the arrangement of sensory pits and hairs, and the sensory function of each, supports the **protothrix theory** of the origin of hair.

Hair follicles are not distributed equidistant throughout the skin, but occur in isolated groups of two to a dozen or more. The number depends on the species. Certain monkeys exhibit groups of three, apes exhibit groups of five, and man exhibits groups of three to five. In armadillos the hairs actually lie in linear arrangement between rows of scales (Fig. 5-33). A glance at the side of your own hand, near the base of the thumb, will reveal the linear arrangement of hairs in that location. The presence of hairs between scales would seem to have been an adaptation for the receipt of tactile stimuli in a skin covered with insensitive scales.

Structure. Hairs are cornified modifications of the epidermis. They grow from hair follicles that project deep into the dermis. Each hair consists of a long shaft, a root, and a bulb. The **shaft** lies free within the hair follicle and projects above the surface of the skin. It is composed of layers of dead, keratinized cells numbering about 500,000 per linear inch. Within the hair may occur a **medulla** composed of air spaces and dermal elements. The **root** of the hair is that portion deep within the follicle where the hair has not yet separated from the surrounding epidermal cells of the follicular wall. Here the cells are becoming cornified and are dying. The **bulb** is a swelling at the base of the hair containing the dermal papilla. It is an area of rapid mitosis, which is constantly contributing new cells and making the hair longer.

Development. Hair follicles first develop as cylindrical ingrowths of the epidermis into the dermis (Fig. 5-32). Beneath the epidermal ingrowth and indenting its base, a dermal papilla organizes. With continued proliferation of epidermal cells, the hair primordium grows deeper and deeper into the dermis, nourished by vessels of the dermal papilla. When the bulb at the base of the primordium is sufficiently differentiated, cornified cells commence to make their appearance, and a hair shaft begins to rise out of the follicle.

Arrectores pilorum. Inserting on the wall of each hair follicle in the dermis is a tiny smooth muscle, the arrector pili, or "elevator of the hair" (Fig. 5-28). When the arrectores pilorum contract, the follicles and hair shafts are drawn toward a vertical position. The skin around the base of each hair is pulled into a tiny mound, causing (in man) "gooseflesh," or "chill bumps." The elevation of the hairs in many mammals causes the animal to look ferocious. It also increases the insulating capacity of the fur.

Modifications of hairs. The coarse hair on the pig's back is modified to form stiff bristles, which, because of their elasticity,

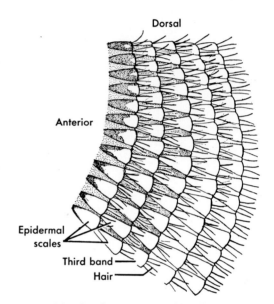

Fig. 5-33. Skin from an armadillo showing epidermal scales and hair.

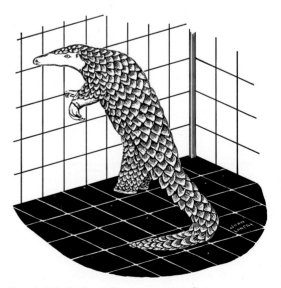

Fig. 5-34. Hair scales on a pangolin.

have been used in the manufacture of
brushes. The quills, or spines, of porcupines and spiny anteaters are rigid, needle-
like hairs. Vibrissae are long, strong, tactile
hairs. The bristles on the upper lip of
whales are vibrissae. Tufts of vibrissae occur on the wrist of cats, rodents, and some
other mammals.

Two spectacular modifications of hair
are the scales of scaly anteaters (Fig. 5-34)
and the horns of *Rhinoceros* (Fig. 5-35).
These appear to be composed of many
agglutinated hairs.

Epidermal scales of mammals

Scales such as those on the feet and
tail of rats and beavers are not uncommon among mammals. Like the superficial
scales of reptiles and birds, they are of
epidermal origin, being derivatives of the
stratum corneum. The epidermal scales of
the armadillo are also localized thickenings of the stratum corneum (Fig. 5-
33). Under these lie bony scales of dermal
origin.

Although the distribution of hairs in
groups between scaleless areas suggests a
loss of scales in mammals, it does not follow that epidermal scales of living mammals are necessarily remnants of reptilian
scales. Mammalian scales may be innovations that appeared as a mutant character
in scaleless mammalian skin. Regardless
of the relationship between reptilian and
mammalian scales, one fact stands out beyond dispute: the stratum corneum of
reptiles, birds, and mammals has in all
cases the potential of producing cornified
thickenings. These are epidermal scales.

Horns

Perissodactyls and artiodactyls have
been endowed with organs of offense and
defense—the horns on the head. The term
"horn" implies that they are constructed
of horny substance, a product of the stra-

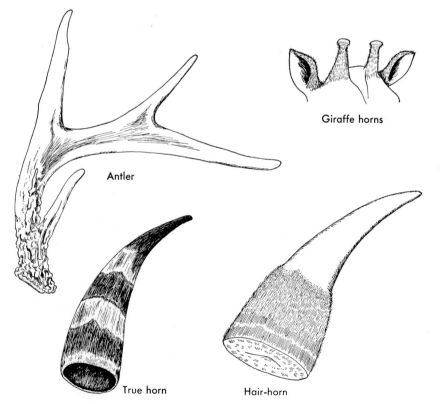

Giraffe horns

Antler

True horn

Hair-horn

Fig. 5-35. Mammalian horns and antlers.

tum corneum, but not all horns are true horns in this sense of the word. Horns may be classified in four categories: (1) hair-horns, (2) antlers, (3) true horns, and (4) giraffe horns (Fig. 5-35).

Hair-horns. Hair-horns are found in *Rhinoceros.* These are unlike other horns in that they appear to be composed of agglutinated modified hairs. They are perched upon a roughened area of the nasal bones. Some African rhinoceroses have two horns, one behind the other. Hair-horns occur on both sexes and are not shed.

Antlers. Antlers are branched structures of dermal bone attached to the frontal bone and are characteristic of the deer family. They are shed annually. When antlers are still in the formative state, they are covered on the surface with typical skin, or "velvet." This soon wears off and the naked, branched bone remains. Antlers normally occur in males only, except in reindeer and caribou.

True horns. Members of the bovine family exhibit true horns. This family includes cattle, sheep, goats, and antelopes (except the pronghorn antelope). The horn of bovines is constructed of a core of bone covered by a horny epidermal cap, which, when removed, is hollow. Bovine horns are never branched, are never shed, and occur usually in both sexes, although there are numerous exceptions to the bisexual condition. Polled cattle have lost their horns by selective breeding.

The horns of pronghorn antelopes are true horns, too. The chief difference between bovine and pronghorn horns is that the latter are branched and that the horny covering (but not the bony core) is shed annually.

Giraffe horns. The horns of giraffes are stunted antlers. They are therefore short, bony projections of the frontal bones which remain covered with unmodified skin.

Claws, nails, and hoofs

Claws, nails, and hoofs are all built on the same plan (Fig. 5-36). The last two are modifications of the more primitive claw. All three grow more or less throughout life.

The claws of mammals are identical in structure with those of reptiles and birds and represent a modification of the stratum corneum. A curved dorsal plate of cornified cells (**unguis**) and a less dense ventral plate (**subunguis**) enclose the tip of the digit.

Nails, characteristic only of primates, are claws consisting chiefly of the broad, flattened, horny unguis. The subunguis is much softer than in claws and is much reduced. As a result, the nail covers only the dorsal aspect of the finger or toe. If permitted to grow, nails become long and more claw-like.

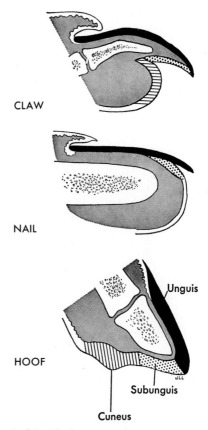

CLAW

NAIL

HOOF

Unguis

Subunguis

Cuneus

Fig. 5-36. Black area represents stratum corneum of claw, nail, and hoof. View is a sagittal section of the last phalanx.

The hoof is a modified claw characteristic of ungulates (mammals that walk on the tips of their toes). The horny unguis is U- or V-shaped instead of being flattened like a nail. It bears little or none of the animal's weight. The subunguis, also U-shaped, becomes greatly thickened. Since the subunguis consists of dead cells, the horse's shoe can be nailed into it. Lying between the arms of the U-shaped subunguis is a softer horny substance, the **cuneus**. This is similar to the callus-like pads (tori) on the foot of other mammals. The terminal bone of the digit is usually pointed when a claw is borne on it and blunt in the case of a nail or hoof.

Miscellaneous modifications of the stratum corneum

Mammals exhibit various callus-like pads that develop at sites where friction occurs. The **ischial callosities** of monkeys are thick, cornified pads of epidermis covering the ischial portion of the hip bone on which the monkey sits. Camels have **kneepads,** which absorb some of the shock when the animal kneels, preparatory to lying down. **Tori** are epidermal pads on the tips of the digits (**apical pads**) and on the palms and

soles of numerous species. The cat is able to "pussyfoot" by retracting the claws and walking silently upon the apical and interdigital pads. Tori are prominent on the hands of human embryos, but they disappear before birth. Their location is marked by friction ridges on the digits and palm (Fig. 5-37). **Corns** and **calluses** are thickenings of the stratum corneum that develop where the integument is subject to continued friction.

Toothless whales have great frayed horny sheets of oral epithelium suspended from the hard palates (Fig. 5-38). As many as 370 of these sheets of **baleen** have been counted in one whale. The apparatus serves as a massive strainer, permitting only minute foodstuffs, the sole nourishment of whalebone whales, to continue into the esophagus.

Integumentary pigments in mammals

Hair color is produced by melanin pigment of varying intensities of brown or black located between and within the hair cells. The specific distribution and density of these pigments, both granular and in solution, and the presence of air vacuoles in the medulla are responsible for all natural hair color. Gray and white hair are

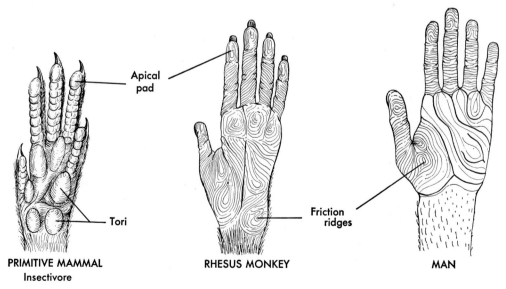

PRIMITIVE MAMMAL
Insectivore

RHESUS MONKEY

MAN

Apical pad

Tori

Friction ridges

Fig. 5-37. Tori (epidermal pads) in mammals.

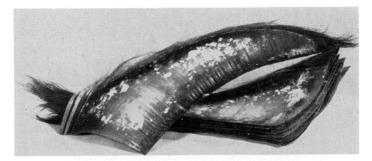

Fig. 5-38. Sheets of baleen (whalebone) from the oral cavity of a whalebone whale. The sheets may be from 2 to 12 feet long. (Courtesy General Biological Supply House, Inc., Chicago, Ill.)

the result of large numbers of air vacuoles and little pigment.

Skin color that is not attributable to blood in the capillaries of the dermis is chiefly the result of varying concentrations of melanin granules in the epidermis, especially in the basal layers. Melanocytes occur in varying numbers immediately beneath the epidermis, in the dermis. The pigment-containing processes of many of these ramify among the cells of the deeper layers of the epidermis.

Chapter summary

1. The integument of vertebrates consists of epidermis originating from ectoderm and dermis originating from mesoderm. The epidermis is a stratified epithelium. In terrestrial forms it has a surface layer (stratum corneum) of dead, keratinized (cornified) cells. Major constituents of the dermis are blood vessels, nerves, lymphatics, the bases of multicellular epidermal glands, pigment cells, and an abundance of collagenous connective tissue fibers.

2. The dermis from its earliest history has been a site for the deposit of bone on the matrix of collagenous fibers. The bone-forming potential of the dermis was expressed primitively as bony dermal armor in ostracoderms and placoderms. Later the armor took the form of scales—cosmoid at first, then placoid or ganoid, ctenoid and cycloid, the last two with very little bone. Among tetrapods this potential is expressed as bony scales or plates in the dermis of amphibians (caecilians and tropical toads), reptiles (crocodilians, lizards, and turtles), and mammals (armadillos).

3. Cosmoid scales, derived from bony dermal armor, do not occur today. They probably represent a stage in the evolution of other types of dermal scales.

4. Placoid scales are characteristic of elasmobranchs. They have a basal plate composed of dentine-like bone and a spine, which protrudes through the epidermis and is covered with an enamel-like substance. Modifications of placoid scales in elasmobranchs include teeth, denticles, some gill rakers, dorsal fin spines, barbs, and sawteeth.

5. Ganoid scales composed chiefly of bone covered by ganoin were once predominant among ray-finned fishes. They are found today only in Chondrostei and Holostei. They probably gave rise to the more modern cycloid and ctenoid scales.

6. Cycloid and ctenoid scales are characteristic of teleosts and living lungfish. They represent a specialization of dermal armor wherein most of the bone has been lost.

7. The skin of fishes is characterized by the presence of bony or fibrous dermal scales

and of many unicellular epidermal glands. The stratum corneum is poorly differentiated. In a few species dermal scales have been lost.

8. Living cyclostomes and a few other fish lack dermal scales, but ancestral forms exhibited an extensive bony dermal armor.

9. Fin rays (ceratotrichia in Chondrichthyes and lepidotrichia in Osteichthyes) are derivatives of the dermis.

10. Photophores (light organs) are modified epidermal glands of fish.

11. Chromatophores contain pigment granules, which may be aggregated or dispersed. They originate from neural crests and have been found in the skin of fishes, amphibians, and reptiles. They included xanthophores, erythrophores, melanophores, and iridophores. Melanocytes are melanin-containing pigment cells in which the granules cannot be aggregated. They are found in all vertebrate classes. Melanoblasts are undifferentiated melanocytes.

12. The skin of amphibians has many multicellular epidermal glands and is practically devoid of dermal or epidermal scales. The stratum corneum is usually absent in aquatic forms, thin in amphibious species, and thick and warty in land-dwelling toads. Caecilians and a few tropical toads exhibit dermal scales.

13. The stratum corneum of amniotes gives rise to epidermal scales, claws, nails, hoofs, horn, spines, rattles, beak and spur coverings, feathers, hairs, quills, callosities, and baleen. Epidermal scales are especially characteristic of reptiles but are present in restricted sites on birds and mammals.

14. The skin of reptiles has a thick stratum corneum, is covered with epidermal scales, and has few integumentary glands. Bony dermal scales occur in crocodilians and a few lizards. In turtles the dermal scales constitute the shell.

15. The thin skin of birds is characterized by feathers (contour, down, and filoplumes) derived phylogenetically from epidermal scales. Typical epidermal scales occur chiefly on the legs and head. Contour feathers develop from feather tracts (pterylae). Pinfeathers are developing feathers. There are no dermal scales and few skin glands.

16. The skin of mammals is characterized by the presence of hair derived from the stratum corneum. The arrangement of hair suggests that it may originally have been interscalar. Modifications of hair include hair-horns (in *Rhinoceros*), scales (of the scaly anteater), spines (in spiny anteaters), quills (of porcupines), bristles, and vibrissae. The stratum corneum is thick, multicellular epidermal glands are diverse, and epidermal scales are restricted mostly to the feet and tail, except in armadillos. Bony dermal plates occur only in armadillos.

17. Skin glands are mostly unicellular in fishes, in which they abound. They are mostly multicellular in tetrapods, being abundant in amphibians, rare in reptiles and birds, and numerous and diversified in mammals. Characteristic of mammals are mammary, sudoriferous, and sebaceous glands. All skin glands are epidermal.

18. Horns are found in some reptiles and in ungulate mammals. True horns have a bony core covered with a horny epidermal cap that is not shed, except in the case of pronghorns. Giraffe horns have a bony core covered by typical skin. Antlers are bony outgrowths from which skin has been shed. Hair-horns are probably agglutinated hair.

19. Claws are modifications of the stratum corneum, consisting of a dorsal unguis and ventral subunguis. They are characteristic of most reptiles, birds, and mammals. In the latter they may be modified as nails and hoofs.

6

Introduction to the skeleton and heterotopic bones

The skeleton or hardened tissues (*sclero* = hard) of invertebrates are most often on the surface, where they constitute a lifeless, secreted exoskeleton. The hardened tissues of vertebrates are most often underneath the surface, where they constitute a living, growing endoskeleton. As nearly as can be determined, the notochord is the oldest endoskeletal element of vertebrates. It is a special skeletal tissue that is neither bone nor cartilage and is best referred to simply as "notochord tissue." Bone and cartilage supplement the notochord and even replace it to a large extent in most adult vertebrates.

For convenience, the endoskeleton of vertebrates may be subdivided on the basis of location into three major categories of elements: (1) the **axial skeleton,** (2) the **appendicular skeleton,** and (3) **heterotopic bones.** The axial skeleton consists of the skull (including the pharyngeal skeleton), the notochord, vertebral column, and ribs; and, in tetrapods, the sternum. The appendicular skeleton consists of pectoral and pelvic girdles, the skeleton of the paired fins or limbs, and, in fishes, the skeleton of the median (unpaired) fins. (Median fins are characteristic of fishes

that use them for stabilizers.) Heterotopic bones are miscellaneous elements that develop in association with certain organs.

BONE

Bone is a very ancient tissue. It was abundant in ostracoderms and placoderms, the earliest known vertebrates. In vertebrates of all classes from fish to man bone is supplemented in its skeletal role by cartilage. However, bone has been almost abandoned in favor of cartilage by coelacanths and chondrosteans. Chondrichthyes have abandoned bone completely, except for the dentine of their scales and teeth, and cyclostomes have no bone whatsoever.

The hardness of bone is a result of the presence of much interstitial substance. This consists of a matrix of collagenous fibrils, in the meshes of which are deposited hydroxyapatite crystals. These inorganic crystals composed of calcium, phosphate, and carbonate are deposited under the influence of the bone cells. The bone cells lie in tiny pools of fluid known as lacunae (*lacuna* = lake; Fig. 6-1). The lacunae contain dissolved salts, some of which precipitate to form crystals. They also contain

Fig. 6-1. Section of bone to show lacunae (black) and canaliculi (canals radiating from the lacunae). Not all bone exhibits haversian systems (lacunae in concentric rings). (From Bevelander: Essentials of histology, ed. 5, St. Louis, 1965, The C. V. Mosby Co.)

salts that have resulted from dissolution of crystals, since bone deposit, dissolution, and reconstruction (reshaping of the bone) is a continual process. Connecting the lacunae are tiny, interanastomosing fluid-filled canals (canaliculi). In some bone, but not all, the lacunae are arranged in concentric haversian systems. Some of the bone in fishes is acellular; that is, the bone cells do not become incorporated within the bone, but retreat as bone is deposited. The process of bone formation is known as osteogenesis.

Preskeletal blastema; membrane bone and replacement bone. Before bone or cartilage can be deposited, a preskeletal concentration of embryonic mesenchyme cells, known as a blastema, must develop. The mesenchyme cells that contribute to most preskeletal blastemas are of mesodermal origin. However, the mesenchyme that gives rise to the skeleton of the head and pharyngeal arches migrates from neural crests and is referred to as **mesectoderm.** Once the blastema of the future bone or cartilage has aggregated, bone or cartilage may then be deposited. Whether bone is deposited, or whether the initial deposit is cartilage depends on the species,

the location of the blastema, and the age of the individual. And, of course, the capacity to deposit either one depends on the presence of appropriate enzyme systems.

Bone deposited directly in mesenchyme is **membrane bone.** The process of formation of membrane bone is known as **intramembranous ossification.** This process gives rise to certain bones of the skull and pectoral girdle, to bone that arises ontogenetically in the dermis of the skin (dermal bone), to the vertebrae of teleosts and urodeles, and to a few heterotopic bones. Also, periosteal bone (bone deposited by the periosteum) is membrane bone.

Bone deposited in preexisting cartilage is **replacement bone.** The cartilage is removed, and bone is deposited where the cartilage previously existed. The process is known as **endochondral ossification.**

The process of endochondral and intramembranous ossification is the same. It consists of the deposit of hydroxyapatite crystals on a matrix of collagenous fibrils. However, in endochondral ossification, cartilage must first be removed before bone may be constructed (Fig. 6-2).

Dermal bone. The dermis of the skin of

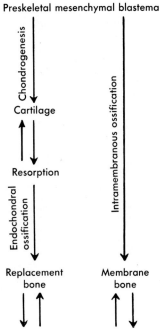

Preskeletal mesenchymal blastema

Fig. 6-2. Fate of preskeletal mesenchyme. Remodeling decelerates after sexual maturity.

vertebrates has an ancient and persistent potential to form bone. This bone arises by ossification within dermal membranes and hence is membrane bone. The earliest vertebrates (ostracoderms and placoderms) had so much bone in the skin that they are referred to as armored fishes. The ganoid fishes were also armored. They are mostly extinct, but a few are still around. Modern fishes have almost lost the ability to deposit bone in their skin. Membrane bone that has its origin in the skin is called **dermal bone.**

The turtle shell is dermal bone, and the skin of crocodilians has some bony plates (Fig. 5-16). Even in armadillos bone develops in the skin. In fact, bone develops in the skin of vertebrates of all classes except birds.

Somewhere in geological time and along pathways leading from armored agnathans, some of the bony scales and plates that previously had developed in the skin began to form under the skin. As a result,

there arose bones apparently derived **phylogenetically** from dermal armor but with an **embryonic** origin from deeper blastemas. These bones are superficial to the cartilaginous endoskeletal elements (Fig. 8-16).

Among bones with a dermal history are the membrane bones of the skull such as nasals, vomers, maxillas, and many others. In recognition of their historical derivation from skin, the membrane bones of the skull are called **dermatocranial bones.** Collectively, they are referred to as the **dermatocranium.** The membrane bones of the pectoral girdles appear also to have had a dermal history. Only those membrane bones that appear to have been derived phylogenetically from the skin may be called dermal bones.

CARTILAGE

Cartilage resembles bone in that the cells lie in pools of fluid surrounded by an interstitial matrix. The matrix of cartilage consists of chondromucoid, a glycoprotein. Unlike bone, cartilage has no canaliculi demonstrable by light microscopy and no blood vessels penetrate it, except those enroute to other organs. Also, cartilage is a deep tissue. There is no such thing as "dermal cartilage."

Cartilage is deposited within a prechondral mesenchymal blastema. The process of cartilage formation is known as **chondrogenesis.** Once cartilage has formed, it may remain throughout life, or it may be resorbed and replaced by bone. In a few locations the cartilage may be resorbed and not replaced, as in parts of the embryonic pterygoquadrate and Meckel's cartilages of birds and mammals. Some cartilages such as Meckel's are extremely primitive; others, such as those that give rise to the tympanic bulla of mammals, are relatively recent additions to the vertebrate skeleton.

Hyaline (*hyal* = clear) cartilage is a translucent cartilage found in many locations in the organism. In vertebrate embryos it is abundant, constituting a tem-

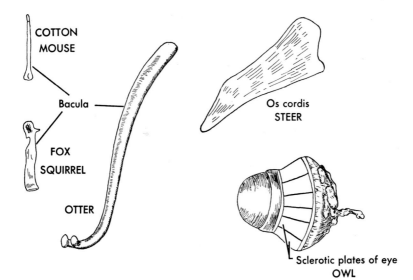

Fig. 6-3. Representative heterotopic bones.

porary skeleton to be replaced later by bone. Cartilage may have calcium salts deposited within the interstitial substance. This **calcified cartilage** is often mistaken for bone. The skeleton of sharks contains much calcified cartilage. Cartilage with thick, dense collagenous bundles in the interstitial matrix is **fibrocartilage.** The intervertebral discs of mammals are fibrocartilage. **Elastic cartilage** contains elastic fibers. In mammals it occurs in the pinna of the ear, in the walls of the outer ear canal, in the epiglottis, and elsewhere.

No cartilage has been demonstrated in ostracoderms or placoderms. Of course, this does not prove that they had none, but cartilage seems to be a relatively recent tissue, possibly an adaptation to embryonic life. If so, cartilaginous fishes have lost the genetic code necessary for ossification.

HETEROTOPIC BONES

In addition to the usual bones comprising the axial and appendicular skeletons, miscellaneous heterotopic bones also develop (Fig. 6-3). These bones are usually missing from routine skeletal preparations. Among heterotopic bones are the **os cordis** embedded in the interventricular

septum of the heart in deer and in bovines and the **rostral bone** in the pig's snout. A **baculum** (**os priapi** or **os penis**) is embedded between the spongy bodies in the penis of bats, rodents, marsupials, carnivores, insectivores, whales, and lower primates. An **os clitoridis** is embedded in the female penis (clitoris) in otters, several rodents, and rabbits.

Osseous tissue forms in the wall of the gizzard in at least one species of dove. The syrinx of birds often develops an internal skeletal element, the **pessulus.** At least one species of bat has intrinsic bone in the tongue. Bone develops in the gular pouch of a South American lizard, in the muscular diaphragm of camels, and in the upper eyelid of crocodilians (**adlacrimal,** or **palpebral,** bone). (A similar plate of connective tissue, the **tarsus,** develops in man.) The hand of the mole has a **falciform** (scythe-shaped) bone, an adaptation for digging. **Epipubic** (**marsupial**) bones are embedded in the ventral body wall of monotremes and marsupials.

Sesamoid bones (nodular bones, so called because someone was reminded of sesame seeds!) are common in the tendons of tetrapod appendages. Best known among these are the relatively large **pisiform** bone of the hand and the **patella** (kneecap). Other heterotopic bones not listed here also occur. Differentiation of osseous tissue in many of these sites is not particularly remarkable considering the skeleton-producing potential of mesoderm, especially where stress occurs.

Chapter summary

1. The endoskeleton of adult bony vertebrates consists of replacement bones that never were part of the skin, of membrane bones that are ontogenetic or phylogenetic derivatives of the skin, of membrane bones that never were part of the skin, of some cartilage that may later ossify, and of some cartilage that does not ossify. These elements are united by fibrous connective tissue ligaments to form the functional framework of the bony vertebrate body.

2. Bone is not found in cyclostomes or in Chondrichthyes.

3. Cartilage has not been described in ostracoderms or placoderms.

4. The endoskeletal elements of vertebrates consist of the following:

   A. Axial skeleton
      (1) Skull (including pharyngeal skeleton)
      (2) Notochord and vertebral column
      (3) Ribs
      (4) Sternum (in tetrapods)
   B. Appendicular skeleton
      (1) Pectoral girdles
      (2) Pelvic girdles
      (3) Skeleton of paired fins or limbs
      (4) Skeleton of median fins (in fishes)
   C. Heterotopic bones

5. The chief endoskeletal elements may be classified histologically as follows:

   A. Notochord tissue—chiefly embryonic in higher forms
   B. Bone—origin is from mesoderm or, in the head and pharynx, mesectoderm; chief histological features are bone cells, interstitial substance, canaliculi, and lacunae
      (1) Replacement bone—arises by endochondral ossification
      (2) Membrane bone—arises by intramembranous ossification
         (a) Dermal bone—derived ontogenetically or phylogenetically from skin
         (b) Other membrane bone—not of dermal derivation
   C. Cartilage—origin is from mesoderm or, in the head and pharynx, mesectoderm; chief histological features are cells, interstitial substance, canaliculi, and absence of nutritive vascular channels.
      (1) Hyaline cartilage—chief embryonic skeleton
      (2) Calcified cartilage—prominent in Chondrichthyes
      (3) Fibrocartilage
      (4) Elastic cartilage
   D. Ligaments—fibrous tissue uniting components into a system

Vertebrae, ribs, and sternum

The vertebral column and ribs constitute a major part of the axial skeleton of vertebrates, along with the skull. To these must be added the sternum in tetrapods. These structures typically consist partly of cartilage and partly of bone. In the preceding chapter cartilage and bone were introduced as tissues, and their morphogenesis was outlined briefly.

VERTEBRAL COLUMN

The vertebral column is composed of a series of essentially similar cartilaginous or bony vertebrae extending from the skull to the tip of the tail. More than one morphological variety of vertebrae is found in every column. Trunk vertebrae differ from those of the tail. In tetrapods, trunk vertebrae are further modified when associated with ribs (thoracic vertebrae), with the pelvic girdle (sacral vertebrae), or when in the neck (cervical vertebrae). Most vertebrae exhibit a **centrum** (**body**), one or two **arches**, and certain **processes** (Fig. 7-1).

Basic structure of vertebrae

Centra. The centra occupy the position immediately beneath the neural tube occupied during embryonic life by the notochord alone. During both ontogeny and phylogeny the developing centra tend in-

creasingly to encroach upon the notochord, "squeezing" it and partially or completely obliterating it within the centrum. The fate of the notochord in various vertebrates has been discussed in Chapter 1, and should be reviewed.

Often an **intercentrum,** or **intervertebral disc,** is interposed between successive centra (Figs. 7-2 and 7-9). The interposed element may be fused with the cephalic or caudal end of the centrum, thereby altering the shape of the ends, which may be convex, concave, flat, or a combination of these (Fig. 7-2).

Amphicelous centra (*amphi* = both, *coel* = hollow) are hollow at both ends. These are thought to be a rather ancient type. They are found in most fishes, in tailed amphibians, and among primitive amniotes. The cavity between two vertebrae is occupied partly or wholly by the notochord, which is expanded between centra and constricted within each centrum. **Procelous centra** (*pro* = in front) are concave anteriorly only, since at the caudal surface the concavity is occupied by an attached intervertebral disc. **Opisthocelous** centra (*opistho* = at the back) are concave posteriorly only, the intervertebral disc being attached at the anterior end. They are found in the neck in some large ungulates. **Acelous** centra (also called **amphiplatyan;**

129

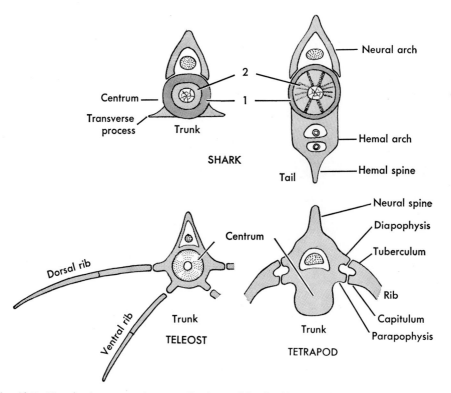

Fig. 7-1. Vertebral structure in a cartilaginous fish (shark), in a teleost, and in a tetrapod. The notochord is seen within the centrum in the shark. The hole occupied by the notochord in life is seen in the centrum of the teleost. **1,** Notochord sheath invaded by cartilage; **2,** notochord invaded by cartilage. The cartilage is sometimes deposited in radiating patterns, as shown in the tail vertebra.

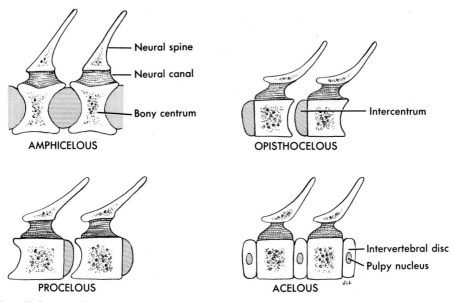

Fig. 7-2. Vertebral types based on shape of articular surfaces and shown in sagittal views. The pulpy nucleus is a remnant of the notochord.

platy = flat) are flat on both ends. They are common in mammals, in which a more or less flattened intervertebral disc lies between centra. In the neck of birds the caudal ends of the centra are shaped like a saddle (**heterocelous** centra).

Arches. Perched upon the centrum is a neural arch (Figs. 7-1 and 7-7). The successive neural arches and associated ligaments enclose a long **vertebral (neural) canal,** within which lies the spinal cord.

Inverted underneath the centrum may be a **hemal arch** (Figs. 7-1, 7-3, and 7-7). It is found in the tail of fishes, tailed amphibians, and many amniotes, including some mammals. In amniotes it is often called a "chevron bone." Hemal arches enclose the caudal artery and vein in a hemal canal.

Processes. The processes (apophyses) of the vertebrae are projections from the arches and centra. They are mostly paired and occur with more or less regularity. Some of the processes provide for increased rigidity of the column, some prevent excessive torsion, some articulate with ribs, and some serve as the attachment for muscles.

Zygapophyses are processes projecting from the neural arch. There are usually four. A pair of **prezygapophyses** (Fig. 7-3) arise from the base of the neural arch and project cephalad. The articular surfaces at their ends are directed upward and inward. A paid of **postzygapophyses** arise from the neural arch and project caudad. Their articular surfaces are directed downward and outward, articulating with the prezygapophyses just behind them. Zygapophyses do not occur among fishes.

Diapophyses (Fig. 7-1) are **transverse processes** attached primitively to the base of the neural arch (sometimes to the centrum) and extending laterad. Each articulates with the dorsal head of two-headed ribs.

Parapophyses (Fig. 7-1) are transverse processes articulating with the ventral head of two-headed ribs. If either head is reduced, the corresponding diapophyses or parapophyses may be absent.

Basapophyses are ventrolateral processes of the centrum. They articulate with the hemal arch when the latter is present. In some cases they meet ventrally to become the hemal arch.

Hypapophyses (Fig. 7-3) are prominent midventral (unpaired) projections of the centrum of certain vertebrae of reptiles, birds, and mammals. They provide attachment for muscles.

Pleurapophyses (Fig. 7-10) are transverse processes having short ribs fused with them at their tips.

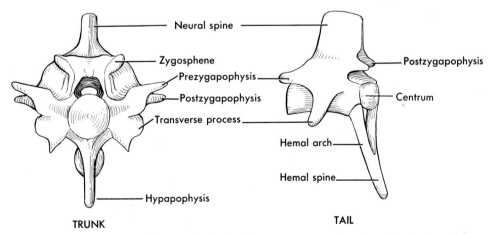

Fig. 7-3. Reptilian vertebrae (python) showing processes and arches. The trunk vertebra is shown from a cephalic view.

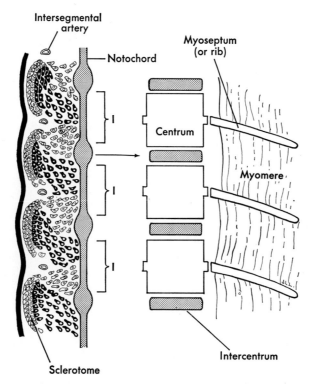

Fig. 7-4. Formation of tetrapod vertebrae by migration of the less dense anterior part of sclerotome (light cells) and the more dense posterior part of the preceding sclerotome (dark cells) to an intersegmental position, **I**, surrounding the notochord. The notochord is thereby compressed within each centrum. The intercentrum (intervertebral disc) may contain a remnant of the notochord. The myosepta and ribs are intersegmental.

Morphogenesis of vertebrae

A typical vertebra composed of a centrum and neural arch arises from concentrations of mesenchymal cells that stream out of the sclerotome of the mesodermal somites and surround the notochord and the neural tube (Fig. 7-4). The cells aggregate intersegmentally, being contributed by the sclerotomes of two successive somites. When sufficient mesenchyme cells have aggregated in the area of the future centrum and neural arch, they typically deposit a cartilaginous model of the future bony vertebra. The cartilage is later replaced by bone, except in Chondrichthyes, in which the vertebrae remain cartilaginous throughout life. In many teleosts and in tailed amphibians bone is laid down directly in the mesenchyme and the cartilage stage is omitted.

The notochord and its sheath are sometimes invaded by mesenchyme cells that deposit cartilage within these structures. Centra composed chiefly of a cartilaginous notochord are called chordal centra (Fig. 7-1, shark).

Secretion of cartilage by mesenchyme cells surrounding the spinal cord is induced in part by a cartilage-promoting factor elaborated by the cord.* Spinal cord from a chick with seventeen to eighteen somites placed in tissue culture along with mesodermal somite cells induces cartilage formation by the somite cells. Notochord **sheath** cells in chicks also induce cartilage in mesenchyme cells derived from a somite

---

*It has been hypothesized that somite cells may be unable to synthesize chondroitin sulfate, a building stone of cartilage, until they come under the influence of the cartilage-promoting factor.

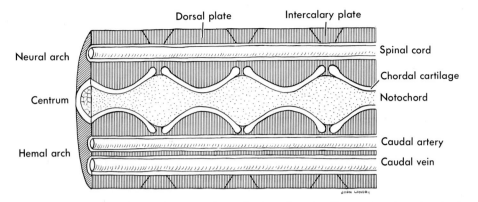

Fig. 7-5. Caudal vertebrae of *Squalus* in sagittal section. Cartilage other than that in the notochord sheath (chordal cartilage) is cross-hatched. The notochord is constricted within each centrum.

but only in cells contiguous with the sheath and not at a distance, as in the case of spinal cord. In **adult** urodeles in which the tails are regenerating after being severed, the cartilage-promoting factor of the cord (confined to the motor column but produced the length of the cord) induces the formation of cartilaginous vertebrae, which, later in regeneration, are replaced by bone.[61] This is especially interesting, since during **embryonic** development in urodeles the bony vertebrae form directly from mesenchyme, no cartilaginous stage intervening. The protein synthesis sequence thus differs in the adult compared with the embryo, an example of the sequential changes that occur as ontogeny progresses.

Vertebral column of fishes

The vertebral column of a fish is composed of **abdominal** (trunk) and **caudal** (tail) vertebrae, the latter exhibiting hemal arches. In Agnatha and in some aberrant bony fishes centra are lacking. The centra of most, but not all, fishes are the primitive amphicelous type. The notochord persists and is usually constricted within each centrum and expanded between centra.

Chondrichthyes. The vertebral column of a dogfish is fairly typical of elasmobranchs (Fig. 7-5). Like the rest of the skeleton, it lacks bone. The centrum de-

velops chiefly from the notochord and its sheath following invasion by mesenchyme cells that deposit cartilage. Contributing to the centrum is cartilage from perichordal mesenchyme. Once the centrum has formed, lime salts are deposited in the cartilage, giving rise to calcified cartilage. Anteriorly the column is fused with the occipital region of the skull. There are no articular joints between the centra. Diplospondyly (two vertebrae for each body segment) is seen in the tail and caudal region of the trunk.

The neural arch in a dogfish consists of a **dorsal plate** and an **intercalary plate** on each side of the spinal cord. Alongside of the caudal artery and vein are similar hemal plates.

The vertebral column turns upward at the end and most of the caudal fin lies ventral to the upturned column, which may even extend into a dorsal lobe. This is a **heterocercal** tail. It is thought to be an ancient condition.

In Holocephali, instead of a single centrum per body segment, the notochord sheath is converted into many calcified cartilaginous rings per segment (Fig. 7-6). Nevertheless, only one pair of dorsal plates and one pair of intercalary plates comprise the neural arches in each segment.

Teleosts. Modern bony fishes have well-ossified amphicelous vertebrae (Fig. 7-7).

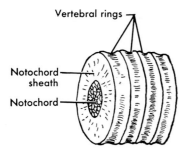

Fig. 7-6. Calcified cartilaginous notochordal rings typical of some Holocephali. The rings are more numerous than the somites. The neural and hemal arches are not illustrated.

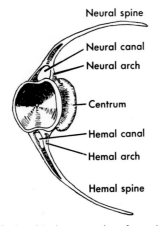

Fig. 7-7. Amphicelous vertebra from the tail of a teleost. Note tiny hole in the centrum, occupied in life by a constricted notochord.

The notochord is strongly constricted within each centrum. Between centra the notochord sheath forms strong intervertebral ligaments. The bony centra ossify directly from perichordal concentrations of mesenchyme and thus skip a cartilaginous stage. Neural arches develop and, in the tail, hemal arches. The arches usually unite with the centrum, but in some species they remain separate from the centrum throughout life. The neural spines are often greatly elongated, and successive ones may unite in the posterior region of the trunk and in the tail to form a delicate longitudinal bony rod paralleling the vertebral column. In lower teleosts blunt neural spines may be surmounted by pointed supraneural bones. Characteristic of teleost vertebrae are a variety of processes that protrude from the centra and arches and articulate with similar processes on adjacent vertebrae. For the most part the processes are not comparable with those of tetrapods.

Aberrant vertebrae. Except for teleosts, bony fishes display a gamut of vertebral types. Gars, for example, have opisthocelous centra. In lungfish and sturgeons (Fig. 7-8) no complete centra form, and the notochord is present and unconstricted throughout the length of the body. Some cartilage deposited within the notochord sheath provides rigidity. The neural arch consists of two cartilages on each side in each body segment. Lying against the notochord on each side ventrally in each segment are additional cartilages. It is uncertain whether this condition is an arrested embryonic state, a primitive condition, or an extreme specialization.

Diplospondyly. Although the majority of vertebrates develop only one centrum per body segment (that is, per embryonic mesodermal somite), species exist in which there is a duplication of elements per somite. Neural arches alone are duplicated in some locations, centra and arches in others. This condition, called diplospondyly, has already been pointed out in elasmobranchs, sturgeons, and Dipnoi. It occurs in the tail of *Amia*, where two amphicelous discs (a centrum and intercentrum) surround the notochord in each body segment (Fig. 7-9). Only the centrum bears neural and hemal arches. Duplication of neural arches is common in lizards.

Diplospondyly is reminiscent of the rachitomous vertebrae of crossopterygian fishes and early labyrinthodonts, to be described later. What advantage diplospondyly confers on fishes, if any, is not clear.

Agnatha. Aside from the lateral neural cartilages illustrated in Fig. 1-11, there are no serially arranged skeletal elements associated with the notochord of agnathans. A fibrous connective tissue tunnel for the neural tube, notochord, and blood

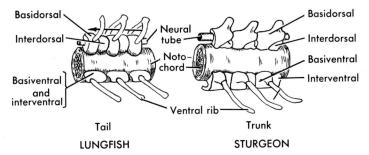

Fig. 7-8. Abortive vertebral components in adult *Neoceratodus* (lungfish) and *Acipenser* (sturgeon). The notochord sheath is invaded by cartilage. Arrow indicates canal occupied by a longitudinal ligament.

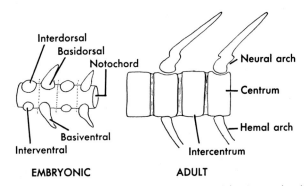

Fig. 7-9. Diplospondyly in tail of *Amia*. Basidorsals and basiventrals give rise to centra. Interdorsals and interventrals give rise to intercentra. A centrum and intercentrum develop for each body segment.

vessels affords an enclosure that takes the place of vertebrae. Caudally, the lateral neural cartilages of lampreys become irregular, and fuse to form a continuous covering of the tip of the notochord. In hagfish the lateral neural cartilages are limited to the tail.

Lateral neural cartilages may be abortive neural arches, they may be primitive elements that have not evolved further, or they may bear no genetic relationship to vertebrae.

Vertebral column of tetrapods

With terrestrial life, movements of the head became increasingly independent of movements of the vertebral column. Facilitating mobility of the head was reduction in length of the ribs on the anterior trunk vertebrae, which were then called **cervical** vertebrae. Amphibians, with little

independent movement of the head, exhibit a single cervical vertebrae (Fig. 7-10). Amniotes develop flexible, often elongated necks and have many cervical vertebrae.

Also related to the assumption of life on land were the advent of tetrapod limbs and consequent modification of the last one or several trunk vertebrae (**sacral**) for articulation with the pelvic girdle (Figs. 7-10 and 7-11). Sacral vertebrae bear short, stout transverse processes that brace the pelvic girdle against the vertebral column, in crawling, jumping, or walking. The sacral vertebrae often ankylose with one another to form a **sacrum**. They may also ankylose with adjacent trunk and caudal vertebrae to form a **synsacrum** (Fig. 7-16). Amphibians have a single sacral vertebra. The number is not large in any class.

Primitively, **caudal** vertebrae may have numbered fifty or more, but in modern

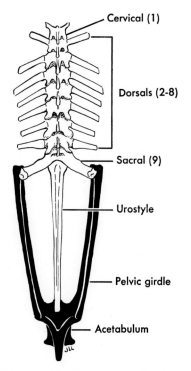

Fig. 7-10. Vertebral column and pelvic girdle (black) of a frog. The transverse processes of the vertebrae are pleurapophyses since they include short ribs.

tetrapods the number is much reduced, and terminal vertebrae consist of small cylindrical centra only (Fig. 7-17). The caudal vertebrae in apes and man are called **coccygeal** vertebrae.

All tetrapods therefore bear cervical, trunk (dorsal), sacral, and caudal vertebrae. In many reptiles and in all birds and mammals, trunk vertebrae are further differentiated into **thoracic** (located anteriorly and articulating with independent elongated ribs) and **lumbar** vertebrae (in which ribs are reduced or absent). Regional differentiation of the vertebral column constitutes only one of the many changes imposed on the vertebrate body in connection with emergence from water to land.

Amphibia. Amphibians have cervical (one), dorsal (variable number), sacral (one), and caudal vertebrae, the latter being absent in anurans. Hemal arches occur only in urodeles. Intercentra are found in ancient amphibians but have been lost in modern forms.

Necturus, typical of less specialized uro-

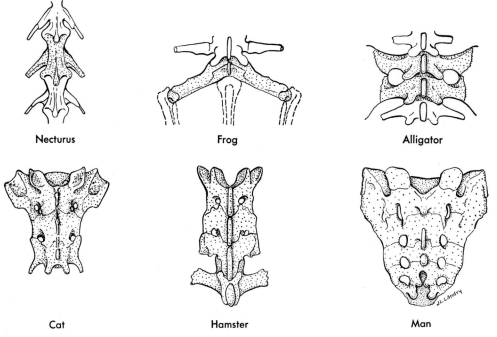

Fig. 7-11. Sacral vertebrae (stippled) of selected vertebrates, dorsal views. The sacral vertebrae ankylose to form a sacrum in the mammals illustrated.

deles, has amphicelous vertebrae, and the notochord persists within the adult centrum. Some of the more specialized urodeles have opisthocelous vertebrae, which may obliterate the notochord. Necturi have forty-three or more vertebrae. Some urodeles have as many as 100.

Anurans have nine procelous vertebrae and a urostyle. With few exceptions this is the shortest column known (Fig. 7-10). The urostyle appears to represent a number of coalesced ancestral vertebrae. In some species the cranial end of the urostyle exhibits one or more coalesced vertebrae with vestigial arches, transverse processes, and nerve foramina. Embryonically, the urostyle develops from a continous cartilaginous rod that surrounds the notochord. Only in a few anurans do remnants of the notochord remain after metamorphosis.

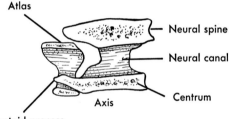

Fig. 7-12. First two cervical vertebrae of cat, sagittal sections. Much of the centrum of the atlas has become attached to the axis as an odontoid process.

Caecilians may have as many as 250 well-ossified vertebrae of the primitive amphicelous type. The notochord persists between centra.

Atlas and axis. In reptiles, birds, and mammals the anterior two cervical vertebrae are specialized. The first cervical vertebra (**atlas**) is ring-like (Fig. 7-12), since most of its centrum has become attached to the next vertebra. Articulating with the skull in a condyloid joint, the atlas provides a cradle in which the skull may rock, as in nodding "yes." Attached to the second vertebra (**axis**) is an **odontoid process**, which represents the detached centrum of the atlas. The odontoid process projects forward to rest upon the floor of the atlas (Fig. 7-12).

Since most reptiles and birds have a single occipital condyle, the atlas in these forms bears a single anterior articular surface. In mammals the occipital condyles are paired, and the anterior articular surface of the atlas is split into bilateral facets.

In numerous extinct and modern reptiles and in an occasional mammal, an additional neural arch known as the **proatlas** is interposed between atlas and skull (Fig. 7-13).

Reptilia. Reptiles exhibit well-ossified cervical, dorsal, sacral (two in all living forms), and caudal vertebrae. Dorsal vertebrae are differentiated as thoracic and lumbar vertebrae in crocodilians and most

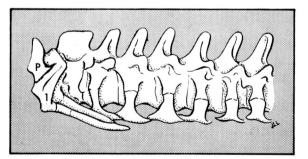

Fig. 7-13. The eight cervical vertebrae, eight cervical ribs, and proatlas, **P,** of an alligator, left lateral view. **1,** Atlas and attached first rib. Immediately behind the first rib is the rib of the axis.

lizards, since long ribs are confined to the anterior trunk region in these forms (Fig. 7-20). All types of centra occur among reptiles.

The centra of early reptiles were amphicelous, accommodating remnants of the notochord, and there were intercentra between the centra (Fig. 7-19, Cotylosaur). These conditions occur today only in the ancient *Sphenodon* and in a few lizards such as the geckos of the tropics. Most modern reptiles have procelous vertebrae and no intercentra. The alligator has eight cervical, eleven thoracic, five lumbar, two sacral (fused), and forty caudal vertebrae. Each vertebra consists of a centrum and neural arch. Chevron bones separated from the centra occur in the tail.

Snakes may have as many as 500 vertebrae, the largest number on record. The dorsal vertebrae are not divided into thoracic and lumbar vertebrae, since all exhibit long ribs. Many lizards, when captured by the tail, break it proximal to the point of capture and scurry away. Such autotomy is implemented by a zone of soft tissue placed transversely through each centrum and neural arch (the location being at the level of a myoseptum), at which point the break occurs. This seems to be an instance of diplospondyly. In addition to the usual apophyses, a special pair (**zygosphenes**, Fig. 7-3) occurs in lizards and snakes and fits into matching concavities (**zygantra**) on the preceding vertebra.

The vertebral column in turtles (eight cervical, ten dorsal, two sacral, and sixteen to thirty caudal vertebrae) is remarkable because of the extensive fusion between elements. The neural arches of the dorsal, the sacral, and the first caudal vertebrae become fused with one another and united with the vertebral plates of the carapace (Fig. 5-17).

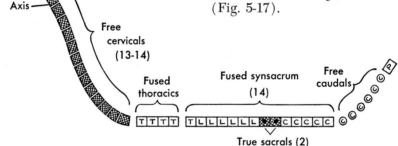

Fig. 7-14. Vertebral column of pigeon, diagrammatical. **T,** Thoracic; **L,** lumbar; **C,** caudal; **P,** pygostyle.

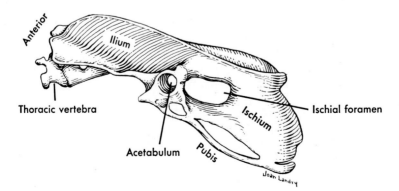

Fig. 7-15. Synsacrum and pelvic girdle of the guinea hen, left lateral view.

Aves. Like turtles, birds exhibit extensive ankylosis between vertebrae (Fig. 7-14). The last thoracic, all the lumbar, the two sacral (three in the ostrich), and the first few caudal vertebrae unite to form one adult bone, the synsacrum (Figs. 7-15, 7-16 and 7-24). The synsacrum, in turn, becomes more or less fused with the pelvic girdle. The transverse processes of the vertebrae of the synsacrum may be seen in a ventral view (Fig. 7-16). The specific number of vertebrae incorporated in the synsacrum varies in different species; thirteen to eighteen or more is common. The synsacrum provides a rigid framework for the bipedal gait. Thoracic vertebrae anterior to the synsacrum unite more or less completely in all species. Fusion includes the successive neural arches and other processes.

Posterior to the synsacrum lie several free caudal vertebrae (usually five to seven) and a **pygostyle**, constituting the skeleton of the stumpy tail (Fig. 7-24). The pygostyle begins in the embryo as several independent caudal cartilages, probably vestiges of earlier tail vertebrae. Chevron bones occur in association with caudal vertebrae in some species.

The cervical region is unusually flexible. Many birds are capable of turning their heads almost backward, of lowering them considerably below the level of their feet, and of bending their necks into sinuous curves. The specialization is especially useful in feeding and is made possible by the heterocelous vertebrae. The caudal ends are saddle shaped, with a convexity in the right-left axis and a concavity in the dorsoventral axis. The cephalic end of the next centrum is shaped to accommodate this configuration. The number of cervical vertebrae varies; the long neck of the swan has twenty-five. The hypapophyses of the cervical region may be simple or recurved spines and are sometimes fused with successive ones.

Although the free vertebrae in most birds are heterocelous, parrots, penguins, and some others have opisthocelous ones. The oldest birds had amphicelous vertebrae, which suggests the presence of notochord remnants. In modern birds the notochord is obliterated during embryonic life. The vertebrae of many birds have pneumatic foramina, which admit air sacs to a cavity within the centrum.

Mammalia. The vertebral columns of mammals are fairly uniform with respect to the number of neck and trunk vertebrae (Table 7-1). All are acelous except the cervical vertebrae of ungulates, which are opisthocelous. Perhaps the ball-and-socket arrangement in the neck of these herbivores facilitates grazing.

There are almost always seven cervical vertebrae. This is as true in the stubby, rigid neck of whales as in the elongated neck of the tallest giraffe. The sole exceptions are the edentates with six, eight, or nine and the manatee with six. In moles several cervical vertebrae ankylose, perhaps strengthening the neck for burrowing. In cetaceans and armadillos, which have no external evidence of a neck region, all cervical vertebrae are foreshortened and are more or less fused.

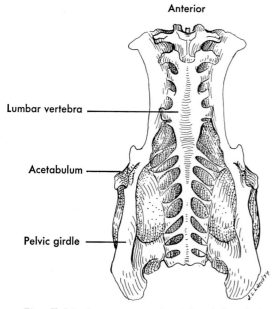

Anterior

Lumbar vertebra

Acetabulum

Pelvic girdle

Fig. 7-16. Synsacrum and pelvic girdle of the guinea hen, ventral view.

Dorsal (thoracic and lumbar) vertebrae usually number about nineteen. Ancestral mammals had about twenty-seven vertebrae anterior to the first sacral vertebra, including the cervical vertebrae.

The sacrum is composed of ankylosed sacral vertebrae, usually 2 to 5 (Fig. 7-11). A synsacrum incorporating up to thirteen vertebrae occurs in edentates, but most of these are dorsal vertebrae. A sacrum is lacking entirely in whales, a condition correlated with reduction of the pelvic girdle.

Fig. 7-17. Complete set of tail vertebrae from a hamster, left lateral view.

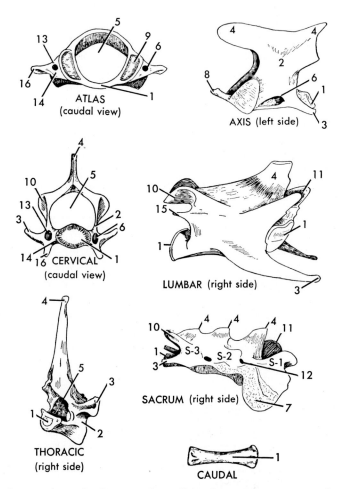

Fig. 7-18. Cat vertebrae. **1,** Centrum; **2,** pedicle; **3,** transverse process; **4,** neural spine; **5,** neural canal; **6,** transverse foramen; **7,** site of articulation with ilium; **8,** odontoid process; **9,** articular facet for axis; **10,** postzygapophysis; **11,** prezygapophysis; **12,** intervertebral foramen; **13,** diapophysis; **14,** parapophysis; **15,** accessory process; **16,** cervical rib with two heads, one fused with a diapophysis, the other fused with a parapophysis; **S-1, S-2, S-3,** sacral vertebrae.

Caudal vertebrae (Figs. 7-17 and 7-18) vary most in number. One scaly anteater has fifty. In man they are reduced to three to five, and these may ankylose to form a **coccyx.** As many as eight caudal vertebrae develop in the human embryo, however. Chevron bones occur in the tail of marsupials, monotremes, sirenians, cetaceans, carnivores (including cats), rodents, and others, but they are usually lost in the preparation of mounted skeletons. Toward the end of the tail the neural arches, chevrons, and all processes become progressively shorter until finally the last several caudals consist of centra only.

Between successive vertebrae lie **intervertebral discs** corresponding in position to the intercentra of reptiles. Anterior to the sacrum, each disc consists of fibrocartilage with a soft center, the **pulpy nucleus,** a remnant of the notochord (Fig. 7-2). Sacral and coccygeal intervertebral discs ossify.

Prezygapophyses and postzygapophyses are found on practically all the vertebrae. Prezygapophyses, especially in the thoracic and lumbar regions, often bear small metapophyses (called mammillary processes in man). A hypapophysis occurs on the lumbars in some mammals.

Evolution of vertebrae

When the cells from the embryonic somite migrate toward the notochord and neural tube preparatory to forming vertebrae, these cells aggregate into a less dense anterior mesenchymal concentration and a more dense posterior concentration. This has been interpreted as indicating that the earliest vertebral column consisted of **two** sets of elements (two bony rings) at each body segment. The occurrence of diplospondyly in so many fishes, and the presence of centra and intercentra (or intervertebral discs) in primitive tetrapods have been considered additional evidence for the hypothesis of two rings. Is the duplication of vertebral elements in each segment a primitive condition? Or is this a recent specialization? Scholars are aligned on both sides of the question.

The vertebral column of the earliest tetrapods did not consist of a single ring at each level, and it did not consist of two rings at each level. The "vertebra" of the earliest known amphibians (Fig. 7-19, A) consisted of a **hypocentrum** (a large anterior, median, wedge-shaped element that was incomplete dorsally), and two **pleurocentra** (smaller, intersegmental, posterodorsal elements). A similar condition occurred in crossopterygians. A vertebra of this kind has been called **rachitomous** (*rachi* = backbone, *tom* = cut; a vertebra of several pieces). All later tetrapod vertebrae are probably modifications of the rachitomous type. Successive changes leading to modern amniotes appear to have been characterized by progressive increase in size of the pleurocentrum. The rachitomous vertebral type was also modified in other directions (Fig. 7-19, B_2). At present

Table 7-1. Vertebrae in selected mammals

	Cervical	Thoracic	Lumbar	Sacral	Caudal
Horse	7	18 to 20	6	5	15 to 21
Opossum	7	13	6	2	19 to 35
Hamster	7	13	6	4	13 to 14
Sheep	7	13	6 to 7	4	16 to 18
Cat	7	13	7	3	18 to 25
Dog	7	12 to 13	7	3	19 to 23
Rabbit	7	12	7	4	16+
Man	7	12	5	5	3 to 5
Bat	7	11	5	5	9
Sperm whale	7	11	8	0	24

it is not clear whether the centrum of modern amphibians represents a pleurocentrum or hypocentrum.

In nearly all modern vertebrates, each vertebra commences development at several loci surrounding the spinal cord and notochord. As the separate anlagen enlarge, they may remain independent or may unite. The typical adult vertebra is, therefore, a **composite structure.** The findings of embryology, comparative anatomy, and paleontology all point to the same conclusion. An adult vertebra composed of a single centrum and neural arch for each body segment is a specialized condition. Primitively there were several skeletal elements (though not necessarily two rings) for each body segment.

RIBS

Ribs are long or short, metamerically displosed skeletal structures articulating medially with vertebrae and extending laterad into the body wall. The proximal ends of ribs are, like vertebrae, derivatives of the sclerotomes of the somites. Distally,

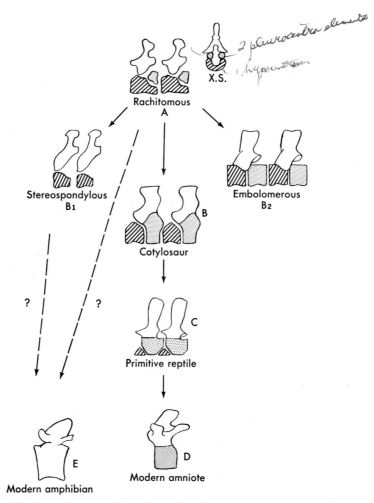

Fig. 7-19. Modifications of tetrapod vertebrae leading to modern amniotes. The rachitomous type (shown also in cross section, **X.S.**) occurred in crossopterygians and in the earliest amphibians. **B₁** and **B₂** are from labyrinthodonts. Whether the modern amphibian centrum represents a hypocentrum (diagonal lines) or a pleurocentrum (stippled) is not certain.

they may develop from the lateral mesoderm. They are laid down first as cartilage. In some vertebrates the cartilaginous rib is at first directly continuous with the cartilaginous transverse processes of the vertebra (diapophysis, parapophysis, or both). Later, when ossification centers are giving rise to bone within the centrum, processes, and proximal part of the rib, a joint develops, separating the rib from the transverse process or processes. Sometimes, in the case of short ribs, no joint develops, and a short rib remains attached to the vertebral process, which is then known as a pleurapophysis.

Fishes

Fishes usually exhibit ribs on every trunk vertebra. *Polypterus* and some teleosts (salmon, herring, and pike) exhibit two ribs on each side, associated with each centrum of the trunk (Fig. 7-1). A **dorsal rib** passes laterad in the horizontal septum between the dorsal and ventral musculature. A **ventral rib** arches ventrad in the body wall just external to the parietal peritoneum. Most teleosts exhibit ventral ribs only, and these are characteristic "fish" ribs. Dogfishes, on the other hand, develop dorsal ribs only. Agnathans lack ribs altogether, probably for the same reason that they lack centra, whatever the reason may be.

The ventral ribs in some fishes occur in the tail, meeting ventrally to form hemal arches. This condition has led to the theory that ventral ribs and hemal arches may be homologous, and for this reason ventral ribs are often called **hemal ribs.** In most teleosts, however, the transition from ventral rib to hemal arch is not so direct.

Additional small intermuscular bones resembling short ribs are frequently found in the myosepta of teleosts. These bones, along with the ribs and supraneural spines, make some edible fishes more difficult to eat.

Tetrapods

Evidence suggests that tetrapod ribs are dorsal ribs, but this has not been settled. In ancient amphibians and reptiles all vertebrae, including the atlas and many caudal vertebrae, bore free ribs. In more recent forms, ribs tend to be restricted in distribution along the column (Fig. 7-20).

The typical tetrapod rib is bicipital; that is, it exhibits two heads (Figs. 7-1 and 7-21). The dorsal head (**tuberculum**) articulates with the diapophysis of the vertebra; the ventral head (**capitulum**) articulates with the parapophysis or with a centrum. A few tetrapods (anurans, lizards, snakes, and monotremes) exhibit single-headed ribs, which result from fusion of

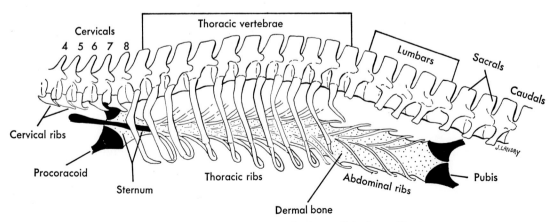

Fig. 7-20. Cervical, thoracic, and abdominal ribs (gastralia) of an alligator.

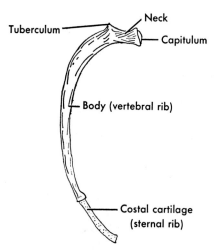

Fig. 7-21. Thoracic rib of a cat showing the two heads (tuberculum and capitulum).

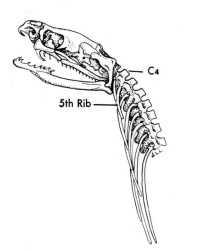

Fig. 7-22. Cervical ribs of a cobra. The first three ribs, associated with the atlas, axis, and third vertebra, are short and are hidden by the jaws, **C₄**, The neural spine of the fourth cervical vertebra.

the two heads or from disappearance of one head during development. All sacral ribs are single headed.

The ribs of amphibians are much reduced in length and no ribs reach the sternum. The ribs of a frog, fused with transverse processes, are illustrated in Fig. 7-10.

Cervical ribs. Cervical ribs are usually short, except in some reptiles (Fig. 7-22).

In most tetrapods the two heads remain fused throughout life with the diapophysis and parapophysis of the cervical vertebrae. Between the two heads may lie a **transverse foramen** (Fig. 7-18). The successive foramina provide a **vertebrarterial canal,** transmitting the vertebral artery and vein. Crocodilians and ratites always have one or more free cervical ribs, and human beings occasionally have one.

Thoracic ribs. Thoracic ribs in amniotes are greatly elongated and usually arch ventrad within the myosepta to meet the sternum. In most cases they are divided into a dorsal segment (**vertebral rib**) and a ventral segment (**sternal rib, or costal cartilage**). The sternal ribs are ossified in birds and a few mammals. A segment between the vertebral and sternal ribs occurs in some crocodilians and lizards. The last several costal cartilages may not reach the sternum, terminating instead on the preceding costal cartilage. When a costal cartilage is absent, the rib is said to be **floating.**

Thoracic ribs, the associated centra, and the sternum form a complete cage (**thoracic basket**) encompassing the thoracic viscera in amniotes. Thoracic ribs of birds and some older lizards, including *Sphenodon,* exhibit flat **uncinate processes** that overlap the next rib (Fig. 7-24). *Archaeopteryx* lacked these processes. The thoracic and lumbar ribs of turtles fuse with the carapace (Fig. 5-17, A).

Lumbar ribs. Lumbar ribs are usually short and fused with the distal tips of the transverse processes. In snakes and snakelike lizards, however, long arching ribs occur all the way to the sacrum.

Sacral and caudal ribs. Sacral ribs are short and stout. They are usually fused with the transverse processes of the sacral vertebrae, although they are separate in some reptiles. Distally, sacral ribs ankylose with the pelvic girdle and brace the latter (Figs. 7-10 and 7-11). Caudal ribs, when present, diminish in size along the base of the tail.

Gastralia (Fig. 7-20). Extinct reptiles,

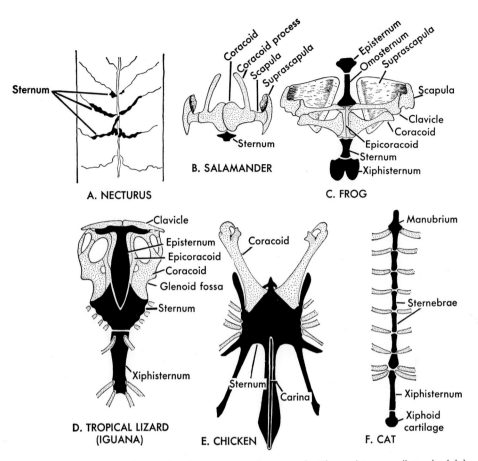

Fig. 7-23. Sternum and associated structures of tetrapods. The episternum (interclavicle) of *Iguana* is not part of the sternum.

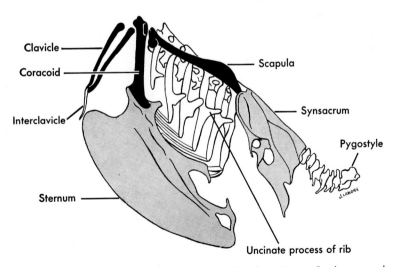

Fig. 7-24. Skeleton of trunk, tail, and pectoral girdle of a pigeon. Replacement bones of the pectoral girdle are black. Between the synsacrum and the pygostyle are free caudal vertebrae.

Sphenodon, and crocodilians exhibit a series of rib-like bones (**gastralia or abdominal ribs**) embedded in the ventral body wall near the rectus abdominis muscle, one or more in each somite. These are dermal bones, equivalent to the plastron bones of turtles. Some lizards have similar structures arising from cartilage.

STERNUM

The sternum is strictly a tetrapod structure (Fig. 7-23). It is absent completely in caecilians, in some of the simpler tailed amphibians such as *Proteus* and *Amphiuma,* and in snakes, legless lizards, and turtles. It is poorly developed in most urodeles.

The sternum of crocodiles and, when present, of lizards consists of a cartilaginous plate, often with caudal prolongations, articulating with the pectoral girdle and with a variable number of thoracic ribs (Figs. 7-20 and 7-23).

The sternum of birds is a large bony element articulating with numerous ribs (Fig. 7-24). In birds with strong muscles for flight, the sternum exhibits a midventral keel, the **carina,** for insertion of the muscles. Such birds are therefore called carinate birds. Ratites for the most part lack a carina.

The sternum in mammals is an elongated bony rod usually composed of a series of segments (**sternebrae**). The anterior sternebra is the **manubrium** (handle). The last bony segment is the xiphisternum, which bears at its caudal tip a cartilaginous **xiphoid** process. Between manubrium and xiphisternum lies the body of the sternum. In a small number of mammals (Cetacea and Sirenia) the adult sternum does not exhibit sternebrae.

Development and derivation of the sternum. The sternum arises in amniotes as a pair of mesenchymal **sternal bars** (Fig. 7-25) that later unite to form a cartilaginous model of the future sternum. Still later, ossification centers appear within the cartilage, much of which may finally be replaced by bone. The sternum of frogs also arises as paired elements, but there is considerable doubt whether the sternum of modern amphibians is homologous with that of amniotes.

Whether the sternum arose phylogenetically as a pair of ventral skeletal elements

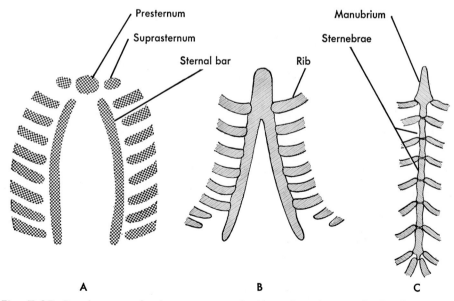

Fig. 7-25. Development of primate sternum. **A,** Mesenchymal stage. **B,** Cartilage stage, the paired sternal bars uniting with one another. **C,** Sternum of a young cat.

independent of the ribs and secondarily became associated with them, or whether it represents an evolutionary contribution from the ventral ends of ribs is not clear. The intimate relationship of the sternum with elements of the pectoral girdle (clavicle, interclavicle, and coracoids) and the presence of ossification centers sometimes giving rise to suprasternal ossicles (see next paragraph) suggests that the manubrial area, at least, may be a contribution of the pectoral girdle.

Suprasternal ossicles of mammals. In addition to the embryonic sternal bars that become the body of the sternum, three additional mesenchymal concentrations appear in many mammals. These are the median **presternal** and paired **suprasternal** (sometimes called **episternal,** but not homologous with the element of the same name in amphibians) chondrification centers (Fig. 7-25). The presternal center normally unites with the sternal bars and contributes to the manubrium. The fate of the paired suprasternal centers varies.

In some forms, which, for different reasons, are usually considered generalized, the suprasternal elements give rise to independent suprasternal cartilages or bones interposed between the clavicle and the manubrium sterni. These suprasternal ossicles are normal components of the skeleton of some insectivores, edentates, and rodents, at least. In other mammals (including marsupials and primates) the suprasternal anlagen typically unite with the manubrium along with the presternum. But even in those species in which they typically unite with the sternum, they not infrequently remain as independent bones in some individuals. Presternal ossicles are common anomalies, for instance, among all major groups of primates—lemurs, tarsiers, orangutans, chimps, monkeys, apes, and man. Since these are small bones, their presence usually goes unnoticed and they are easily lost, especially in small mammals.

The development of median presternal and paired suprasternal centers of chondrification in widely diverse mammals suggests the possibility that these are vestiges of the interclavicle and coracoids of the pectoral girdle of reptiles. Monotremes, the lowest mammals, still exhibit a median interclavicle and a pair of coracoid bones.

Chapter summary

1. The vertebral column of fishes exhibits abdominal (trunk) and caudal vertebrae. In tetrapods trunk vertebrae are subdivided into cervical, dorsal, and sacral, the transverse processes of the latter being fused with the pelvic girdle. Dorsal vertebrae are subdivided into thoracic and lumbar when long curved ribs are limited to the anterior dorsal (thoracic) vertebrae. Caudal vertebrae, usually numerous, are reduced to archless centra near the end of the series. They number few in birds and in some mammals and are absent in anurans.

2. The first two cervical vertebrae are specialized as atlas and axis in amniotes. A proatlas occurs in many reptiles and in an occasional mammal.

3. Ankylosis between vertebrae is characteristic of much of the column in turtles and birds and occurs in restricted regions in some mammals. Sacral vertebrae usually unite to form a sacrum. The sacrum of birds unites with adjacent lumbar and caudal vertebrae to form a synsacrum.

4. The notochord is prominent within the column of adult fishes and becomes reduced in modern amphibians and reptiles. It disappears as an adult structure in birds and mammals, except for the pulpy nucleus.

5. Among Agnatha, lampreys exhibit only lateral neural cartilages, two pairs per body segment, and no centrum. Hagfishes lack lateral neural cartilages in the trunk.

6. The vertebrae of Chondrichthyes are wholly

cartilaginous. Neural arches consist of a dorsal plate and an intercalary plate on each side in each body segment except caudally, where diplospondyly results in a doubling of this condition. The vertebral column consists essentially of the chondrified notochord sheath, and the notochord is intact within the sheath, although constricted within each centrum.

7. Teleost centra arise by intramembranous ossification. They consist of a centrum and an arch for each segment. The notochord is present throughout the column but is greatly constricted within each centrum.

8. Primitive vertebrae are amphicelous. These occur in extinct members of all classes as well as in most fishes, some primitive urodeles, caecilians, *Sphenodon*, and some lizards. Most anurans and most modern reptiles have procelous vertebrae. *Lepidosteus*, some higher urodeles, and (in the neck) ungulates have opisthocelous vertebrae. Birds alone have heterocelous vertebrae; most mammals, acelous.

9. Vertebral processes are numerous in fishes and for the most part are not comparable with those of tetrapods. Supraneural bones are often associated with neural arches in fishes. Tetrapod processes include zygapophyses, diapophyses, parapophyses, basapophyses, pleurapophyses, and hypapophyses. Accessory processes include zygosphenes in lizards and snakes and mammillary processes in mammals. Chevron bones of amniotes are functionally comparable with hemal arches.

10. Centra, arches, and processes arise from concentrations of mesenchyme cells that migrate from the sclerotome and surround the embryonic notochord and neural tube. Most bony vertebrae arise by endochondral ossification, but those of teleosts and modern amphibians arise by intramembranous ossification. In a few fishes the notochord sheath is invaded by mesenchyme cells that deposit cartilage within the sheath to form chordal centra.

11. The occurrence of diplospondyly in many fishes and the presence of interdorsals, interventrals, and intercentra in many primitive vertebrates suggests that the earliest vertebral column consisted of two bony rings per body segment. However, diplospondyly has also been interpreted as a specialized condition. Nevertheless, a modern adult vertebra consisting of centrum and neural arch is a specialized condition. Primitively, several elements (not necessarily two rings) contributed to the axial skeleton for each body segment.

12. Ribs occur on or articulate with nearly all vertebrae in fishes and ancient tetrapods. Those of fishes are dorsal (elasmobranchs), ventral (teleosts), or both (*Polypterus* and a few teleosts). Ventral ribs in some cases become hemal arches as they approach the tail. Agnatha lack ribs. Tetrapod ribs appear to be homologous with dorsal ribs. The typical modern tetrapod rib is bicipital.

13. Cervical ribs are usually short and fused with processes of the vertebra. Thoracic ribs are elongated and may reach the sternum. Lumbar and sacral ribs are fused with processes of the vertebrae, and sacral ribs are also fused with the pelvic girdle. Caudal ribs diminish in length along the tail. All ribs are reduced in length in amphibians.

14. Gastralia are rib-like bones (abdominal ribs) in the ventral abdominal wall of some ancient and modern reptiles.

15. The sternum is limited to tetrapods. It is absent in caecilians, some urodeles, snakes, legless lizards, and turtles. It is well developed in birds (keeled in carinates) and in mammals, in which it is often segmented. The modern amphibian sternum may not be homologous with the amniote sternum. The sternum arises as a pair of embryonic sternal bars, which later fuse.

16. Suprasternal ossicles occur widely among mammals. They may represent vestiges of the coracoid bones of reptiles. The presternal ossification center of the manubrium sterni may be a remnant of the reptilian interclavicle.

Chapter 8

The vertebrate skull

The skull of a dogfish (Fig. 8-1) serves as an excellent introduction to a study of the skulls of bony vertebrates. The dogfish skull consists of two functionally independent components, a cartilaginous **neurocranium** that surrounds the brain and certain special sense organs, and a cartilaginous **splanchnocranium** that consists of the jaws and the branchial cartilages.

Bony vertebrates develop cartilaginous neurocranial and splanchnocranial components as one step in the development of their bony skulls. The bones arise later from two sources. Some of the bones are deposited where the cartilage previously existed. These are replacement bones (Chapter 6). The remaining bones are formed by intramembranous ossification independently of the neurocranium and splanchnocranium. These membrane bones collectively constitute the **dermatocranium.**

The adult bony skull, whether of a fish or human being, is therefore a complex of three basic components, neurocranium (consisting of cartilage or of replacement bones), splanchnocranium (consisting of cartilage or replacement bones), and dermatocranium (consisting of membrane bones). These three major components are associated in a basic architectural pattern that is essentially the same in all vertebrates. That pattern and its modifications in the different classes is the subject of this chapter.

NEUROCRANIUM
Morphogenesis of cartilaginous stage*

The neurocranium of vertebrates is formed initially in cartilage, but, except in agnathans and cartilaginous fishes, the cartilage is later replaced partly or wholly by bone. All vertebrate embryos from fish to man develop a cartilaginous neurocranium soon after differentiation of the notochord. The neurocranium develops in accordance with a basic pattern incorporating the components listed in the following discussion and diagrammed in Fig. 8-2.

Notochord, parachordal, and prechordal cartilages. The neurocranium commences as a pair of parachordal and prechordal cartilages underneath the brain. Parachordal cartilages parallel the anterior end of the notochord beneath the midbrain and hindbrain. Prechordal cartilages develop anterior to the notochord underneath the forebrain. The parachordal cartilages expand across the midline toward each other and unite. In the process, the notochord and parachordal cartilages are incorporated into a single, broad, cartilaginous **basal plate.** The prechordal cartilages likewise expand and unite across the midline at their anterior ends to form an **ethmoid plate.**

---

*A neurocranium in the cartilaginous stage is often referred to as a chondrocranium, meaning "cartilaginous braincase."

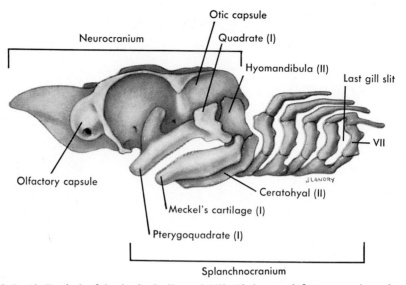

Fig. 8-1. Skull of dogfish shark. **I, II,** and **VII,** Skeleton of first, second, and seventh pharyngeal arches. The spiracle lies between the quadrate and hyomandibula. Labial cartilages, gill rakers, and gill rays are omitted.

Sense capsules. While the parachordal and prechordal cartilages are forming, cartilage also appears in two other locations: (1) as an **olfactory capsule** partially surrounding the olfactory epithelium and (2) as an **otic capsule** completely surrounding the developing inner ear (membranous labyrinth). The olfactory capsules are incomplete anteriorly since water (in fishes) or air (in tetrapods) must have access to the olfactory epithelium. The walls of the olfactory and otic capsules are perforated by foramina transmitting nerves and vascular channels.

An optic capsule forms around the retina, but the capsule does not fuse with the neurocranium. As a result, the eyeball can move independently of the rest of the skull. The capsule is usually fibrous and is known as the **sclerotic coat** of the eyeball. However, cartilaginous or bony plates form within the sclerotic coat in some vertebrates (Fig. 6-3, owl).

Completion of floor, walls, and roof. The expanding ethmoid plate fuses anteriorly with the olfactory capsules, and the expanding basal plate fuses with the otic capsules that lie lateral to the hindbrain.

The ethmoid and basal plates also expand toward one another until they meet to form a floor upon which the brain rests. Further development of the cartilaginous neurocranium involves construction of cartilaginous walls alongside of the brain and, in lower forms, of a cartilaginous roof over the brain. The cranial nerves and blood vessels are already present by this time, and the cartilage is deposited in such a manner as to leave foramina for these structures. The largest foramen is the foramen magnum in the rear wall of the neurocranium.

In the dogfish and its relatives and in lower bony fishes such as *Amia*, the brain is completely covered by a cartilaginous roof. But in more recent fishes and in tetrapods, the brain is not roofed over by cartilage, except above the foramen magnum. Therefore, in the immature skull there are soft spots (**fontanels**) containing no skeleton until such time as membrane bones form a roof over the brain.

The preceding pattern of development recurs throughout the vertebrate series and produces a cartilaginous neurocranium that protects much of the embryonic brain, the

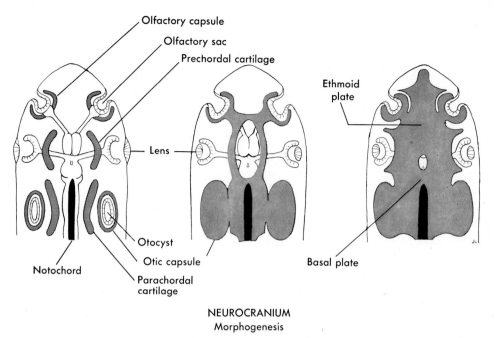

Olfactory capsule
Olfactory sac
Prechordal cartilage
Ethmoid plate
Lens
Otocyst
Otic capsule
Notochord
Parachordal cartilage
Basal plate

NEUROCRANIUM
Morphogenesis

Fig. 8-2. Initial stages in development of a cartilaginous neurocranium, diagrammatical, ventral view. At right, a cartilaginous floor underlying the brain has been completed. The notochord is incorporated into the basal plate.

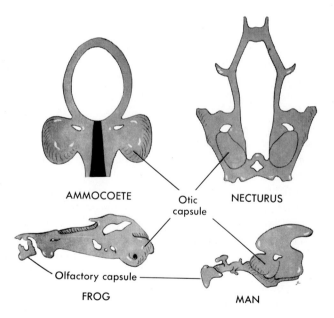

AMMOCOETE
Otic capsule
NECTURUS
Olfactory capsule
FROG
MAN

NEUROCRANIA

Fig. 8-3. Neurocrania from selected vertebrates. Man, embryonic condition; ammocoete, larval condition (lamprey); frog, after metamorphosis; *Necturus*, adult. Dorsal view of ammocoete and *Necturus*, lateral view of frog and man.

olfactory epithelia, and the membranous labyrinths (Fig. 8-3).

Adult cartilaginous neurocrania

Cyclostomes and cartilaginous fish retain a cartilaginous neurocranium throughout life. The failure to deposit bone is considered a specialization. It can be assumed on the basis of strong evidence that these fishes have a hereditary metabolic block that impedes the synthesis of some substance necessary for bone formation.

CYCLOSTOMES. The several cartilaginous components of the embryonic neurocranium just described remain in **adult** cyclostomes as more or less independent cartilages (Fig. 8-9). Failure to form a well-knit neurocranium has been regarded by some investigators as primitive and by others as a specialization. Identifiable are an olfactory capsule (a single median structure protecting the median olfactory sac), otic capsules, a basal plate, a notochord (not fused with the basal plate), and other cartilages that are not entirely homologous with those of typical vertebrates. The roof above the brain is fibrous.

Chondrichthyes. The adult neurocranium of elasmobranchs constitutes the highwater mark in the development of cartilaginous braincases. Walls are fully formed and the brain is completely roofed. The otic capsules are fused into the posterolateral wall of the braincase, and the olfactory capsules are incorporated anteriorly. The notochord is barely visible ventrally as a ridge extending forward from the foramen magnum. The hypophysis is cradled in a cartilaginous pocket, the **sella turcica.**

On the posterodorsal aspect a depression, the **endolymphatic fossa,** exhibits two pairs of perforations occupied by endolymphatic and perilymphatic ducts. These ducts contain fluids from within and around the membranous labyrinth, respectively. The two endolymphatic ducts continue to the dorsal surface of the fish, putting the endolymph of the inner ear in continuity with the sea water.

Ossification centers in the neurocranium

The cartilaginous braincase is primarily an embryonic adaptation. Except in cartilaginous fishes, it is largely replaced by bone. The process of erosion of cartilage in the embryonic neurocranium and deposit of replacement bone in its stead (**endochondral ossification**) occurs more or less simultaneously at numerous separate ossification centers. Although the specific number of such centers varies in different species, four regions of the cartilaginous neurocranium are universally involved, giving rise to replacement bones that are essentially homologous throughout the vertebrate series. These four regions—occipital, sphenoid, ethmoid and otic—will be discussed next and are illustrated in Figs. 8-4 to 8-6.

Occipital centers. The cartilage surrounding the foramen magnum typically becomes replaced by four bones. One or more ossification centers ventral to the foramen magnum produce a **basioccipital** bone underlying the hindbrain. Centers in the lateral walls of the foramen magnum produce two **exoccipital** bones. Above the foramen develops a **supraoccipital** bone. In mammals all four occipital elements may finally fuse to form a single **occipital** bone. In some vertebrates, notably modern amphibians, one or more of these elements may remain cartilaginous, although they were bony in stem amphibians.

The braincase usually articulates with the first vertebra via one or two occipital condyles located beside or below the foramen magnum. Stem amphibians exhibited a single, hollow, occipital condyle borne chiefly on the basioccipital bone but completed by the exoccipital bones. Reptiles (except those fossil forms leading to mammals) and birds still exhibit a single condyle, which may be tripartite (borne on basioccipital and exoccipital bones), or which may be borne on the basioccipital alone. Modern amphibians and mammals have diverged from the early tetrapod condition and exhibit two condyles, one on

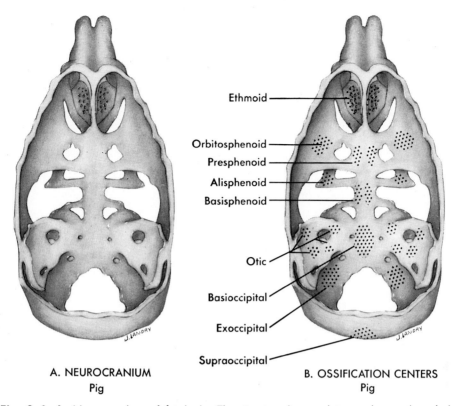

A. NEUROCRANIUM
Pig

B. OSSIFICATION CENTERS
Pig

Ethmoid

Orbitosphenoid
Presphenoid
Alisphenoid
Basisphenoid

Otic

Basioccipital

Exoccipital

Supraoccipital

Fig. 8-4. A, Neurocranium of fetal pig. The structure is complete as shown, there being no roofing cartilage above the brain. **B,** Ossification centers in typical mammalian neurocranium, based on fetal pig. The otic centers are multiple centers in the otic capsule. The ethmoid centers are interspersed among the olfactory foramina. The alisphenoid center is in the pterygoquadrate cartilage.

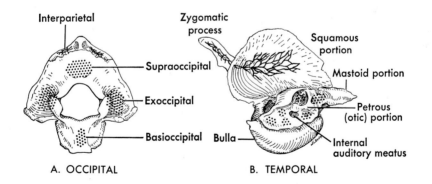

Interparietal

Zygomatic process

Squamous portion

Mastoid portion

Supraoccipital

Exoccipital

Basioccipital   Bulla

Petrous (otic) portion

Internal auditory meatus

A. OCCIPITAL

B. TEMPORAL

OSSIFICATION CENTERS

Fig. 8-5. Endochondral ossification centers (dots) in occipital area and otic capsule (petrous portion of temporal bone). Intramembranous ossification centers are shown as a black network in the interparietal, squamous, and jugal (zygomatic) regions. The centers have been superimposed on the occipital and right temporal bone of an adult cat. **A,** Caudal view. **B,** Medial view. The bulla arises from new cartilage not associated with the earlier neurocranium. The mastoid portion is an outgrowth of the petrous portion.

each exoccipital. Single occipital condyles are common among bony fishes.

Sphenoid centers. The embryonic cartilaginous neurocranium underlying the midbrain and pituitary gland ossifies to form a **basisphenoid** bone immediately anterior to the basioccipital and a **presphenoid** bone anterior to the basisphenoid. Thus, a bony platform consisting of basioccipital, basisphenoid, and presphenoid underlies the brain. The side walls above the presphenoid give rise to **orbitosphenoid** bones, which form part of the orbital wall and are perforated by the optic nerve. The side walls above the basisphenoid area (in the orbital wall behind the orbitosphenoid region) tend not to form replacement bone, and in primitive amphibians and reptiles they were often membranous. However, a **laterosphenoid (pleurosphenoid)** bone arises in this region in snakes, crocodilians, and birds.* All the sphenoid elements unite in some mammals to form a single **sphenoid bone.** Since the embryonic cartilaginous neurocranium is usually incomplete dorsally, no replacement bones lie above the brain other than the supraoccipital above the foramen magnum.

Ethmoid centers. The ethmoid region lies immediately anterior to the sphenoid complex and includes the ethmoid plate and olfactory capsules. Of the four major ossification centers in the cartilaginous embryonic neurocranium (occipital, sphenoid, ethmoid, and otic), the ethmoid more than the others tends to remain cartilaginous throughout life, particularly in tetrapods. Ossification centers that arise in this region become the **cribriform bone,** perforated by olfactory foramina, and the scroll-like **turbinal** bones of the nasal passageways of tetrapods. **Mesethmoid** bones ossify immediately anterior to the presphenoid in some mammals (carnivores, rodents, and primates) and contribute to the otherwise cartilaginous median nasal septum. In mammals the several ethmoid ossification centers may unite to form a single **ethmoid bone. (Ectethmoids,** when present, arise in part by intramembranous ossification.)

---

*The **alisphenoid** bone of mammals lies above the basisphenoid, but it does not arise from the neurocranium.

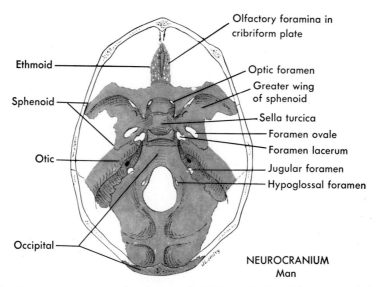

Fig. 8-6. Entire neurocranium of a human skull. The calvarium (dermatocranial roof) has been sawed off and view is looking down into skull from above. Major endochondral ossification centers are labeled at left. Immediately in front of the jugular foramen is the internal auditory meatus (black), which transmits the seventh and eighth cranial nerves.

Otic centers. The cartilage of the otic capsule may be replaced in lower tetrapods by three bony elements that completely encase the membranous labyrinth. These elements are the **prootic** in the anterior wall of the capsule, the **opisthotic** in the posterior wall, and the **epiotic** in the dorsal region. In most tetrapods one or more of these elements fuse with adjacent bones. For example, there are no opisthotic elements in frogs and most reptiles, since they have fused with the exoccipital bones. Epiotic bones fuse with nearby membrane bones in most tetrapods. In birds and mammals prootic, opisthotic, and epiotic centers all unite to form a single **petrosal (periotic) bone.** In cats and man the petrosal, in turn, unites with adjacent membrane bones to form a **temporal bone** (Fig. 8-5, *B*).

In teleost fishes two additional bones, **sphenotic** and **pterotic,** arise partly from the otic capsule, but the otic bones of fishes are not readily homologized with those of tetrapods. As many as six ossification centers have been described in the otic capsule of the human fetus.

VISCERAL SKELETON

The visceral skeleton (splanchnocranium) of vertebrates consists of a series of cartilages or bones arising in the embryonic visceral (pharyngeal) arches. They have their origin partly from neural crest cells that contribute to the mesenchyme (here called mesectoderm) of the arches. In fishes and gill-breathing amphibians the visceral skeleton consists of the skeleton of the jaws and of the gills. In tetrapods the visceral skeleton has become modified and adapted for new functions associated with life on land.

The dogfish is an excellent animal on which to begin a study of visceral skeletons because the splanchnocranium of sharks is seen in its original capacity serving as jaws and gill supports.

Visceral skeleton of dogfish

The splanchnocranium of *Squalus* (Fig. 8-1) consists of a series of seven pairs of jointed visceral cartilages, one in each visceral arch commencing with the first, and of three unpaired basibranchial cartilages in the pharyngeal floor (Fig. 8-7). The last five pairs of visceral cartilages are essentially similar; each consists of four jointed segments named (from dorsal to ventral) **pharyngobranchial, epibranchial, ceratobranchial,** and **hypobranchial.** The hypobranchial cartilages articulate ventrally

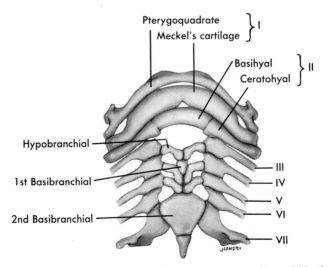

Fig. 8-7. Visceral skeleton of a dogfish shark, ventral view. **III** to **VII**, Ceratobranchial cartilages of the third to seventh pharyngeal arches.

with basibranchial cartilages. The second pair of visceral cartilages (hyoid cartilages) consist of three segments, **hyomandibular** cartilages dorsally, **ceratohyal** cartilages laterally, and an unpaired **basihyal** cartilage ventrally. In the embryo the basihyal is paired. The first pair of visceral cartilages consists of two segments on each side: the **pterygoquadrate** cartilages dorsally and **Meckel's** cartilages ventrally. The left and right pterygoquadrate cartilages meet in the dorsal midline to form the upper jaw, whereas the left and right Meckel's cartilages form the lower jaw. The joint between the pterygoquadrate and Meckel's cartilage on each side determines the location of the angle of the mouth. At the joints the upper and lower jaw cartilages are bound by strong ligaments to the hyomandibular cartilages, which suspend the jaws and gill cartilages from the otic capsule of the neurocranium. This is known as hyostylic jaw suspension (p. 157).

Projecting from the epibranchial and ceratobranchial segments of visceral cartilages III to VI are delicate cartilaginous **gill rays,** which extend into interbranchial septa and support two demibranchs. The equivalent cartilages of the hyoid arch (hyomandibular and ceratohyal cartilages) also bear gill rays but support a single demibranch in the anterior wall of the first true gill slit. The last pair of visceral cartilages (VII) do not bear a gill surface.

Short labial cartilages are attached to the jaws and lie in the marginal folds of the mouth. These occur also in some bony fishes but not in tetrapods.

Visceral skeleton of bony fishes

Bony fishes develop visceral cartilages that are later replaced wholly or in part by bone. In addition, some of the visceral cartilages become invested by membrane bone.

The pterygoquadrate cartilages of the embryo do not meet anteriorly, and only the quadrate region, at the angle of the mouth, contributes to the functional upper jaw. The remainder lies in the roof of the oral cavity and becomes associated with the braincase. Later, this part of the cartilage may be replaced by one or more bones (as many as five in teleosts) such as **epiterygoid** and **metapterygoid** bones.*

Meckel's cartilage is replaced at its posterior tip by an **articular** bone (which may also have some dermal bone contributions). The remainder of Meckel's cartilage becomes ensheathed by dermal bones, which constitute most of the adult mandible, so that the cartilage either remains as a core within the mandible or disappears. (In *Amia,* Meckel's cartilage is **not** ensheathed anteriorly, and it is replaced by **mentomeckelian bones** of endochondral origin.)

The hyoid cartilages of bony fishes give rise to a number of replacement bones. Most common are **hyomandibula, symplectic** (articulating with the quadrate and providing hyostylic jaw suspension), **interhyal, epihyal, ceratohyal,** and **hypohyal** elements (Fig. 8-8). A **basihyal (entoglossal)** de-

---

*Pterygoid, entopterygoid, and ectopterygoid bones are palatal bones of membrane origin and are not derivatives of the splanchnocranium.

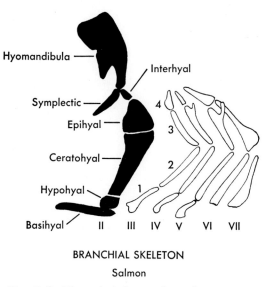

BRANCHIAL SKELETON

Salmon

Fig. 8-8. Visceral skeleton of a salmon, upper and lower jaws removed. Hyoid cartilages are shown in black. **1** to **4,** Hypobranchial, ceratobranchial, epibranchial, and pharyngobranchial elements of the third arch, respectively.

velops ventrally. Behind it may be an unpaired **urohyal,** probably a modified branchiostegal ray. The gill on the hyoid arch of bony fishes is usually rudimentary, and the hyoid skeleton is adapted to conform to movements of the operculum. It also plays a prominent supporting role in the functioning of the jaws of fishes.

The visceral skeleton behind the hyoid arch resembles the last five visceral cartilages of dogfishes except that the embryonic cartilages become replaced by bone (Fig. 8-8). Except for the last arch or two, the same four segments develop (**pharyngobranchial, epibranchial, ceratobranchial,** and **hypobranchial** elements). A series of basibranchial elements lies in the pharyngeal floor.

The last bone of the branchial series (probably a ceratobranchial) does not bear a gill, although it often bears **pharyngeal teeth.** The teeth are large ventrally and higher on the arch grow progressively smaller. The blue sucker (*Cycleptus elongatus*) has thirty-five to forty teeth on the last gill arch on each side.[54] Pharyngobranchial cartilages from the more anterior branchial

arches may likewise expand and become armed with teeth.

Amphistyly, hyostyly, and autostyly. The relationship of the pterygoquadrate cartilage and of the hyomandibular cartilage to the neurocranium and to each other exhibits interesting variants. In one arrangement the pterygoquadrate cartilage is braced against the neurocranium and so also are the hyoid cartilages via the hyomandibula. The result is amphistylic jaw suspension (*amphi* = both, *styly* = bracing), since the skeleton of both the first and second arches participates in bracing the jaws against the neurocranium. This is a rather primitive arrangement. In another modification the hyomandibula is braced against the otic capsule, and the pterygoquadrate cartilage is braced against the hyomandibula. This results in hyostylic jaw suspension, since a hyoid element braces the entire functional jaw complex against the neurocranium. This arrangement has evolved in more recent fishes and is characteristic of elasmobranchs and teleosts. In still another variant the pterygoquadrate cartilage is intimately bound to the neuro-

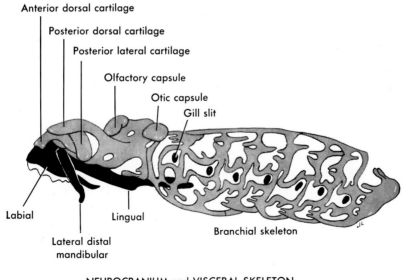

Anterior dorsal cartilage
Posterior dorsal cartilage
Posterior lateral cartilage
Olfactory capsule
Otic capsule
Gill slit
Labial
Lingual
Lateral distal mandibular
Branchial skeleton

NEUROCRANIUM and VISCERAL SKELETON
Lamprey

Fig. 8-9. Neurocranium and visceral skeleton of a lamprey. Black skeletal elements may represent vestiges of jaws. Olfactory capsule is a midline structure; otic capsules are paired.

cranium by investing dermal bones. As a result, the suspension of the jaws is said to be autostylic (*auto* = self). Autostyly occurred in the very ancient placoderms and is still found in chimaeras and in lungfish. A modern variant of autostyly is seen in tetrapods with this difference: the hyomandibula has become independent of the other hyoid elements, and is thereby available exclusively to transmit sound waves to the otic capsule, against which it is still braced. A more sophisticated terminology of the relationships between the jaws, the hyoid arch, and the otic capsule is employed by specialists.

Visceral skeleton of agnathans

The parts of the splanchnocranium of cyclostomes are not easily compared with those of higher vertebrates (Fig. 8-9). Some of the features suggest a greatly modified splanchnocranium in which the usual cartilages are no longer identifiable. For example, *Myxine* has no identifiable pterygoquadrate and Meckel's cartilage, but it does have a cartilaginous lower jaw of a different nature. This is the dental plate cartilage, which forms a V-shaped trough in the floor of the oral cavity, covered by mucosa. When it moves, the rasping teeth are moved. Beneath the dental plate is an immovable basal plate to which the muscles of the dental plate are attached. A careful study of the visceral skeleton of a hagfish[52] led to the conclusion that there is a rudimentary upper jaw fused with the neurocranium. The remainder of the visceral skeleton consists of other cartilages of unknown homology and of a fenestrated, basket-like cartilaginous framework immediately under the skin surrounding the gill slits and extending into the pharyngeal walls.

The visceral skeleton of living cyclostomes can teach us little about the evolution of the splanchnocranium from fish to man. It is interesting because of the questions it raises concerning the state of evolution (primitive, or highly modified?) of living agnathans.

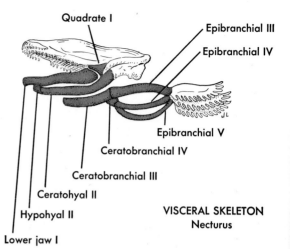

Fig. 8-10. Skull and visceral skeleton of *Necturus*. **I** to **V** are the skeletal elements of the pharyngeal arches. Not all elements derived from the pterygoquadrate cartilage are shown.

Visceral skeleton of tetrapods

In **permanently larval amphibians** such as *Necturus* and in **larval** frogs and toads, the visceral skeleton is essentially fish-like except that the number of gill-bearing arches is reduced (Fig. 8-10). Larval frogs have **six** visceral cartilages and the last four bear external gills. The adult *Necturus* has **five** visceral cartilages and the last three bear gills. There is little endochondral ossification in the visceral cartilages of either animal. However, the first visceral cartilages become invested by membrane bones.

When metamorphosis occurs in frogs, the visceral skeleton is slowly remodeled into a skeleton adapted for respiration by lungs (Fig. 8-13). The remodeled adult visceral skeleton scarcely resembles the branchial skeleton of the larva but is a direct transformation of it nevertheless.

Reptiles, birds, and mammals do not exhibit a larval state and hence never breathe by gills. Still, much of the ancestral branchial skeleton develops during embryonic life, and "metamorphosis" takes place during **embryonic** life instead of during **larval** life. The result is the same as in frogs—a visceral skeleton adapted for life on land.

In the succeeding paragraphs the modifications of the visceral skeletal elements in tetrapods will be discussed.

Pterygoquadrate cartilages. Pterygoquadrate cartilages are the embryonic upper jaw cartilages. Their fate in tetrapods is essentially the same as in bony fishes.

The quadrate portion continues to serve as the posterior tip of the upper jaw, except in mammals. It may remain as a quadrate cartilage throughout life, or it may be replaced by endochondral ossification (quadrate bone of *Necturus*, reptiles, and birds). In mammals the quadrate portion separates from the remainder of the pterygoquadrate cartilage to become the **incus** of the middle ear. The evolutionary transition from jaw to ear was a gradual one, and intermediate evolutionary steps are seen in mammal-like reptiles. It has been claimed that the bony ring (**annulus tympanicus**) to which the tympanic membrane of anurans is attached is a derivative of the embryonic pterygoquadrate cartilage.*

The more anterior part of the pterygoquadrate cartilage becomes ensheathed by one or more membrane bones (**premaxilla, maxilla, jugal,** and **quadratojugal**—Fig. 8-17). Whatever is left of this part of the cartilage lies close to the palate. It may remain cartilaginous throughout life (*Necturus*), or it may be replaced by the **epipterygoid** bone (in mammals, the **alisphenoid**).

Meckel's cartilages. Parts of the embryonic Meckel's cartilage become replaced by replacement bone, parts remain cartilaginous, and much of it is ensheathed by membrane bones (Fig. 8-11). Dry, preserved mandibles of reptiles are hollow because Meckel's cartilage has been removed. In adult birds and mammals few remnants of Meckel's cartilage remain within the mandible (Fig. 8-15).

The posterior tip of Meckel's cartilage is not ensheathed. Instead, below mammals it projects behind the ensheathed area, where, as the **articular bone,** it articulates

*If so, the annulus of anurans is not homologous with that of mammals.

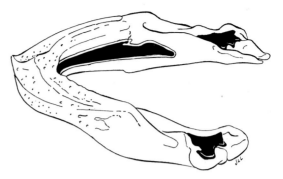

Fig. 8-11. Mandible of an adult sea turtle, from the left and above, showing core of Meckel's cartilage (black). The cartilage emerges at caudal angles to become articular surfaces.

with the quadrate of the upper jaw. Occasionally the articular portion will remain unossified as the **articular cartilage.**

In mammals the embryonic articular portion projects into the middle ear cavity and later separates from the rest of Meckel's cartilage to become the **malleus** (Fig. 8-12, *B*). The malleus still forms a joint with the quadrate (incus), but in the middle ear instead of at the tips of the jaws.

In some amphibians an additional ossification center, the mentomeckelian bone, develops in Meckel's cartilage on either side of the mandibular symphysis. This ossification center was also noted in some bony fishes. (In some species the mentomeckelian bone is a membrane bone ensheathing Meckel's cartilage instead of replacing it.)

Hyomandibular cartilages (columella or stapes). It will be recalled that the hyomandibula of sharks is interposed between the quadrate region of the upper jaw and the otic capsule containing the inner ear. Investigations have shown that the hyomandibula of tetrapods gives rise by endochondral ossification to part of the columella (stapes in mammals), which transmits sound waves from the quadrate bone (incus) to the inner ear. In perennibranchiate urodeles such as *Necturus* the hyomandibula (columella) is rudimentary.

Hyobranchial apparatus. The term hyobranchial apparatus designates collectively

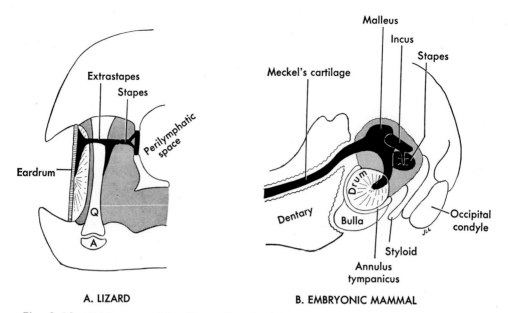

A. LIZARD   **B. EMBRYONIC MAMMAL**

Fig. 8-12. Middle ear ossicles. The malleus in the mammalian embryo is the posterior tip of Meckel's cartilage, which has become surrounded by the developing middle ear cavity (gray). **A,** Articular bone; **Q,** quadrate.

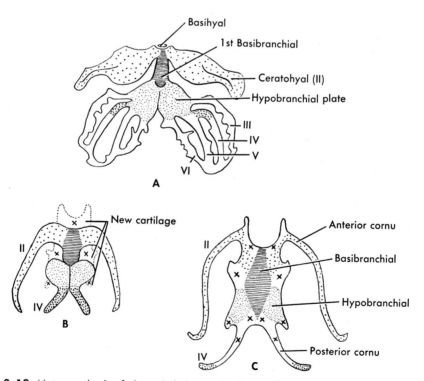

Fig. 8-13. Metamorphosis of visceral skeleton of a frog, jaws omitted. **A,** Branchial skeleton of larva. **B,** Condition in late metamorphosis. **C,** Hyoid of a young frog. Coarse and fine stipple and cross hatching indicate homologous areas. **x,** Cartilage added at metamorphosis. **II** to **VI,** Skeleton of second through sixth pharyngeal arches.

the skeletal derivatives of the hyoid arch (other than the columella) and of the remaining visceral arches in tetrapods. The term includes the various hyoids and the laryngeal skeleton.

In perennibranchiate amphibians the skeleton of arches II, III, IV, and V is much the same as in fishes, and the last three arches bear gills (Fig. 8-10). In anuran larvae the branchial skeleton is also fish-like. Cartilages III to VI unite ventrally to form a hypobranchial plate (Fig. 8-13, A). During metamorphosis the branchial skeleton is reorganized (Fig. 8-13, B and C). The hypobranchial plate contributes to the body of the adult hyoid, the ceratohyal contributes anterior horns (cornua), and the ceratobranchials of arch IV contribute posterior horns. The branchial skeleton is thereby modified for anchoring the improved tongue and for attachment of pharyngeal muscles useful in life on land.

The hyobranchial apparatus in reptiles (Fig. 8-14) consists of a body derived from the basihyal and from the basibranchial cartilages, as well as two or three slender horns. The **anterior horns** usually arise from the second pair of visceral cartilages (arch II). Other horns arise from arches III and IV. In lizards and birds the body of the hyobranchial apparatus is reduced,

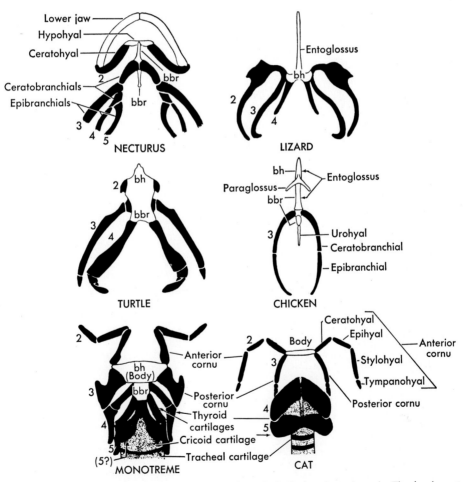

Fig. 8-14. Fate of visceral skeletal elements **2** to **5**, inclusive, in tetrapods. The basic pattern is modified in birds more than in any mammal. The pattern is almost intact in monotremes. **bh,** Basihyal; **bbr,** basibranchial.

and an **entoglossal** bone derived by endochondral ossification within the basihyal extends forward into the tongue. In snakes the entire ancestral branchial skeleton is reduced to slender vestiges.

In mammals (Fig. 8-14) the anterior horns and the associated part of the body of the hyoid arise from visceral arch II. The posterior horns and the rest of the body arise from arch III. In cats the anterior horns are the longer (**greater cornua**) and are composed of several segments. The dorsal segment (**tympanohyal**) articulates with the tympanic bulla.

In man the anterior horn is the short horn (**lesser cornu**), since a **stylohyoid ligament** replaces the three lower skeletal segments. A **styloid process** (Figs. 8-15 and 8-37) representing approximately the tympanohyal segment is fused with the temporal bone. The anterior horn is short in rabbits also. It is connected to the jugular process of the skull by a muscle having a styloid process embedded in its dorsal tendon. The hyobranchial apparatus of monotremes (Fig. 8-14) is reminiscent of the branchial skeleton of primitive tetrapods.

Laryngeal skeleton. The walls of the tetrapod larynx are supported by cartilages or replacement bones, which in many forms represent the skeleton of the last one or two visceral arches. Nearly all tetrapods have **cricoid** and **arytenoid** elements; in addition, crocodilians and mammals have **thyroid** elements. All are bilaterally disposed in the embryo, but the left and right cricoid and thyroid cartilages often unite ventrally during development. In the primitive monotremes the cartilages remain paired throughout life.

In mammals thyroid cartilages appear to be a product of the mesenchyme of arches IV and perhaps V; cricoid and arytenoid cartilages appear to be products of arch V and perhaps VI. Since the caudal end of the visceral arch series has been subject to reduction during evolution, it is not surprising that problems are encountered in relating laryngeal cartilages to specific arches. There may be instances among

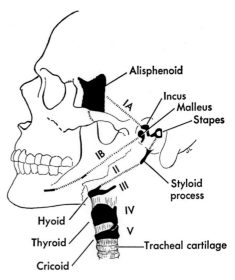

VISCERAL SKELETON

Fig. 8-15. Visceral skeleton of man. **IA,** Broken line connects derivatives of pterygoquadrate cartilage; **IB,** broken line connects vestiges and derivatives of Meckel's cartilage; **II,** broken line connects derivatives of second (hyoid) arch commencing dorsally at the stapes and terminating ventrally at the lesser horn of the hyoid; **III to V,** derivatives of third, fourth, and fifth pharyngeal arches. **III,** Greater horn of hyoid bone (illustrated also in Fig. 12-17).

modern amphibians in which the laryngeal cartilages are not visceral arch derivatives.

• • •

From the preceding discussion it is evident that the visceral skeleton is an ancient mechanism associated initially with branchial respiration. In lung-breathing tetrapods the visceral skeleton has been modified for transmission of sound (malleus, incus, and stapes), for attachment of the muscles of the modified tongue, and for support of the larynx (cricoid, thyroid, and arytenoid cartilages). These adaptations illustrate how mutations alter older structures for newer functions (Table 8-1).

DERMATOCRANIUM

The earliest vertebrates were encased in a bony armor that arose by ossification in the dermis of the skin (Fig. 3-2). In later

Table 8-1. Major skeletal derivatives of visceral arches in selected modern vertebrates

Arch	Dogfish	Teleost	Necturus	Frog	Reptile and bird	Mammal
I	Meckel's cartilage Pterygo-quadrate	Articular Quadrate Epipterygoid Metapterygoid	Articular Quadrate Cartilage in lateral roof of mouth	Articular Mentomecke-lian* Quadrate Annulus tympanicus?	Articular Quadrate Epipterygoid	Malleus Incus Alisphenoid
II	Hyomandibula Ceratohyal Basihyal	Hyomandibula Symplectic Interhyal Epihyal Ceratohyal Hypohyal Entoglossal	Rudimentary Ceratohyal Hypohyal	Columella Anterior horn Body of hyo-branchial apparatus	Columella Anterior horn Body of hyo-branchial apparatus Entoglossus	Stapes Anterior horn Styloid process Body of hyoid
III	Pharyngo-branchial Epibranchial Ceratobran-chial Hypobranchial	Pharyngo-branchial Epibranchial Ceratobran-chial Hypobranchial	Epibranchial Ceratobran-chial	Body of hyo-branchial apparatus	Second horn Body of hyo-branchial apparatus	Posterior horn Body of hyoid
IV	Branchial elements (as in III)	Branchial elements (as in III)	Branchial elements (as in III)	Posterior horn Body of hyo-branchial apparatus	Posterior horn	Thyroid cartilage
V	Branchial elements	Branchial elements	Epibranchial only	Not clearly delineated These arches may contribute to cartilages of larynx; precise homologies between laryngeal cartilages of lower vertebrates and mammals remain in doubt		Thyroid Cricoid Arytenoid
VI	Branchial elements	Branchial elements	Missing or not clearly delineated			
VII	Bears no gill Some reduction	Reduced	Missing			

*In some species.

vertebrates the dermal armor became restricted to the head and pectoral girdle area, and it began to form *under* the dermis in contact with the neurocranium and splanchnocranium (Fig. 8-16). Thus, the skull came to include dermal bones. Collectively, these dermal bones constitute the dermatocranium.

For convenience, discussion of the dermatocranium is divided as follows: (1) dermal bones roofing over the neurocranium and brain and contributing to the lateral walls of the skull, (2) dermal bones investing the pterygoquadrate cartilage, (3) membrane bones of the palates, (4) dermal bones investing Meckel's cartilage, and (5) opercular bones, not present above fishes.

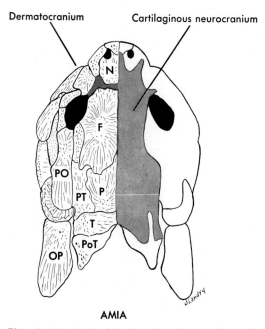

AMIA

Fig. 8-16. Skull of *Amia,* dorsal view. Dermal bones are removed on the right side to reveal underlying cartilaginous neurocranium. **F,** Frontal bones; **N,** nasal; **OP,** operculum; **P,** parietal; **PO,** postorbital; **PoT,** posttemporal; **PT,** pterotic; **T,** tabular. The bones anterior to the nasals are ethmoids. Premaxillas are not visible in this view.

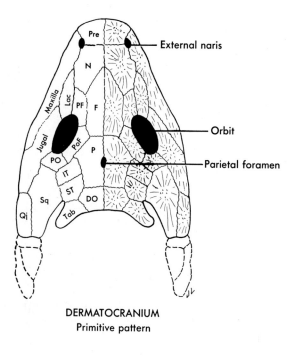

DERMATOCRANIUM
Primitive pattern

Roofing bones

The basic architectural pattern of the roofing bones of crossopterygians and ancient tetrapods is diagrammed in Fig. 8-17. A series of bones—**parietal, frontal,** and **nasal**—extended from the occipital area forward as far as the nares. The parietal bones overlay the hindbrain and the nasal bones overlay the olfactory capsule. Ancient fishes had a series of median bones that interrupted the sagittal suture, but they were lost during early tetrapod evolution. In the midline between the two parietal bones was a parietal foramen. It is still present in many fishes, amphibians, and reptiles, transmitting a third eye to a position under the skin.

Forming the walls of the orbit in the basic plan are **lacrimal, prefrontal, postfrontal, postorbital,** and **jugal** bones. Other orbital bones (**supraorbital, suborbital,** and so forth) develop in some fishes. At the posterior angle of the skull are **intertemporal, supratemporal, tabular,** and **squamosal** bones.

These roofing bones overlie the neurocranium when the neurocranium is complete above the brain (Fig. 8-16). Whenever the neurocranium is incomplete dorsally, soft spots (fontanels) can be felt in the head until the membranes under the skin have ossified (Fig. 8-18). A separate **bregmatic bone** often develops in the fontanel at the junction of the coronal and sagittal sutures (Fig. 8-19). The bone may later coalesce with the parietal and frontal, or it may remain independent. A bregmatic

Fig. 8-17. Approximation of the basic pattern from which tetrapod dermatocrania have diverged, based on a labyrinthodont. Broken lines indicate crossopterygian opercular elements lost by early tetrapods. Scale-like appearance of dermal bones shown on right. **DO,** Dermoccipital; **F,** frontal; **IT,** intertemporal; **Lac,** lacrimal; **N,** nasal; **P,** parietal; **PF,** prefrontal; **PO,** postorbital; **PoF,** postfrontal; **Pre,** premaxilla; **Sq,** squamosal; **ST,** supratemporal; **Qj,** quadratojugal; **Tab,** tabular.

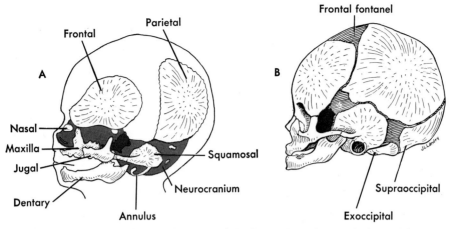

Fig. 8-18. Two stages in the development of the bony skull of man. **A,** Dermal bones are not extensively ossified; neurocranium (gray) is cartilaginous. **B,** Further ossification in roofing area and appearance of endochondral ossification centers in the occipital area of neurocranium. Black represents the greater wing of the sphenoid, which is said to develop from the pterygoquadrate cartilage.

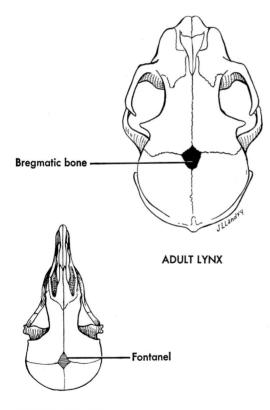

Fig. 8-19. Fontanel in a young pigeon and bregmatic bone (black) in *Lynx*. (*Lynx* redrawn from Manville.[65])

bone occasionally develops in the human skull. Paracelsus called it the **ossiculum antiepilepticum** because he supposed it prevented epilepsy.

Dermal bones investing the pterygoquadrate cartilages

The dermatocranium arches downward on the sides and front of the head to encase within its lower margin the pterygoquadrate cartilages. These dermal bones constitute most of the upper jaw (Fig. 8-17).

The most constant dermal bones of the upper jaw are the **premaxillas** and **maxillas,** both usually bearing teeth. In lower bony fishes there may be a series of maxillas (gar, Figs. 8-24 and 8-25). In teleosts the premaxillas are enlarged and the maxillas may be toothless, reduced, or removed from the upper jaw margin (carp, Fig. 8-24). In some urodeles maxillas fail to form (Fig. 8-21, *Necturus*). In birds the premaxillas are elongated and become part of the beak (Fig. 8-31). Premaxillas develop in human embryos, a discovery made by the German poet-scientist von Goethe, but they soon unite with the maxillas and lose their identity.

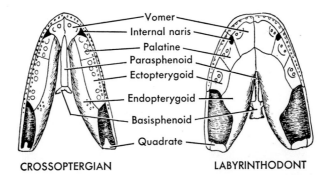

Fig. 8-20. Primary palates of a crossopterygian and a labyrinthodont. Note similarity of structure.

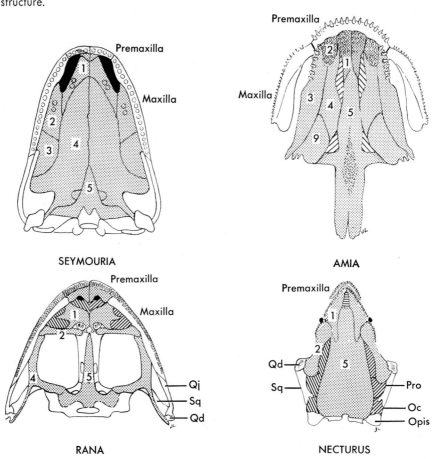

Fig. 8-21. Ventral views of primary palates of a primitive reptile (*Seymouria*), a lower bony fish (*Amia*), a modern amphibian (*Rana*), and a tailed amphibian (*Necturus*). Palatal bones (in gray) are numbered. **1**, Vomer; **2**, palatine (in *Necturus*, palatopterygoid); **3**, ectopterygoid; **4**, pterygoid; **5**, parasphenoid; **9**, epipterygoid. Cartilage is indicated by diagonal lines; internal nares are black. *Amia* lacks internal nares. **Oc**, Cartilaginous portion of otic capsule against which stapes articulates; **Opis**, opisthotic; **Pro**, prootic; **Qd**, quadrate; **Qj**, quadratojugal; **Sq**, squamosal. Note tiny teeth far back on parasphenoid of *Amia* and absence of maxillas in *Necturus*.

Jugal and **quadratojugal** bones complete the dermatocranial margin in lower tetrapods. They lack teeth. One or both may be absent in amphibians (Fig. 8-21, *Rana* and *Necturus*). With the appearance in reptiles of temporal fossae, the jugal bones become part of the zygomatic arch (Fig. 8-27).

These bones—premaxillary, maxillary, jugal, and quadratojugal—lie lateral to any remnants of the embryonic pterygoquadrate cartilages other than the quadrate portion.

Membrane bones of the palates

In cartilaginous fishes the floor of the neurocranium is also the skeletal roof of the anterior part of the pharynx. In bony vertebrates the neurocranium becomes undergirded by membrane bones, which constitute the bony **primary palate** (Figs. 8-20 and 8-21). The chief membrane bones in the primary palate are paired **vomers** (beneath the ethmoid region of the neurocranium), **palatines**, and **pterygoid** bones (including endopterygoid and ectopterygoid bones), and an unpaired **parasphenoid** (beneath the sphenoid region of the neurocranium). The pterygoid bones underlie remnants of the pterygoquadrate cartilage. Primitively, teeth occurred on most of the bones of the primary palate.

In some reptiles and in birds and mammals a **secondary (false) palate** develops in addition to the primary palate (Figs. 8-22, 8-23, 8-29 and 8-30). The secondary palate is a horizontal partition separating the pharynx into dorsal (nasal) and ventral (oral) passages. Membrane bone ossifying in the secondary palate includes the **palatine processes** of the premaxillas, maxillas, palatines, and in extremely long secondary palates (crocodilians, Fig. 8-29),

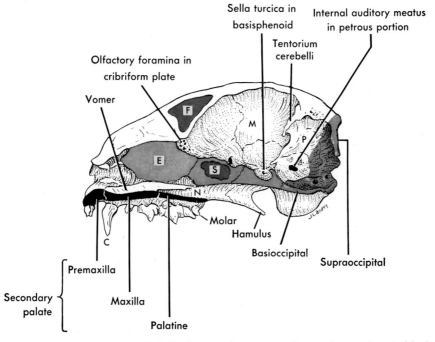

Fig. 8-22. Sagittal section, cat skull, showing bony part of secondary palate in black. **C,** Canine tooth; **E,** mesethmoid (perpendicular plate of ethmoid); **F,** frontal sinus in frontal bone; **M,** middle cranial fossa housing cerebral hemispheres; **N,** nasal passageway; **P,** posterior cranial fossa housing cerebellum; **S,** sphenoidal sinus in presphenoid bone. The teeth to left of the canine are incisors; those to the right are premolars except the last, which is a molar. The olfactory bulb of the brain lies against the cribriform plate.

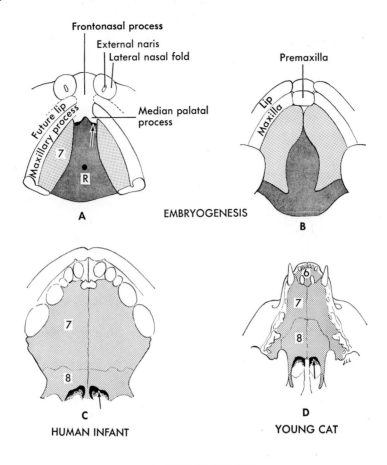

SECONDARY PALATES

Fig. 8-23. A to **C,** Formation of secondary palate in man. **D,** Secondary palate of a young cat for comparison. In **A** (fetus approximately 18 weeks old) the palatine processes of the maxilla, **7,** are growing toward the midline, forming a secondary roof (stippled) in the oral cavity. Dark gray is the primary roof containing Rathke's pouch, **R,** which gives rise to the adenohypophysis. In **B** the palatine processes of the maxillas have met anteriorly. Failure of closure will result in a cleft palate. In **C** the palatine processes of the maxillas, **7,** have met in the midline and palatine processes of the palatine bones, **8,** have also formed. In **D, 6** indicates the palatine processes of the premaxillas. Arrows indicate nasal passageway.

of the pterygoid bones. In mammals the pterygoid area of the secondary palate fails to ossify and is therefore a "soft" (membranous) palate. Palatine processes arise as horizontal shelves of bone that grow toward one another in the roof of the oral cavity (Fig. 8-23, *A* and *B*). Failure of the palatine processes to meet results in a **cleft palate.** Cleft palates are normal traits in most reptiles and in birds, and are congenital abnormalities in mammals.

Dermal bones investing Meckel's cartilages

The embryonic Meckel's cartilages become ensheathed by dermal bones. In primitive vertebrates the number of ensheathing elements was large (Table 8-2 and Fig. 8-32). In modern vertebrates the number has been reduced. Teleost and modern amphibian mandibles usually have not more than three ensheathing bones on each side, and mammals have only the

Table 8-2. Dermal bones investing Meckel's cartilage

Fishes			Tetrapods				
			Primitive	Modern			
Primitive	Crossop-terygians	Modern	Labyrintho-donts	Reptiles and birds	Amphibians	Mammals	
Dentary	Dentary	Dentary†	Dentary	Dentary	Dentary	Dentary	
Angular	Angular	Angular‡	Angular	Angular	Angular¶		
Surangular	Surangular		Surangular	Surangular			
Infradentary*	Splenial		Splenial	Splenial	Splenial¶		
Infradentary	Coronoid		Coronoid	Coronoid			
Infradentary	Prearticular	Derm-articular§	Prearticular	Coronoid			
Infradentary			Intercoronoid				
Infradentary			Precoronoid				
Infradentary			Postsplenial				

Primitive forms had a greater number of bones than modern ones. Reptiles have retained more of the primitive elements than other modern tetrapods.
*Variable number.
†Dentary incorporates mentomeckelian in some teleosts.
‡May be absent.
§May include articular of cartilage origin.
¶Part of an angulosplenial in frog.

dentary. Reptiles retain a larger number of primitive elements than other living tetrapods. The mandible of young birds is reptilian, but in adults the bones are united and are indistinguishable.

Opercular elements

The operculum is a flap of tissue that arises as an outgrowth of the hyoid arch and extends backward over the gill slits. It is membranous in Holocephali and absent in elasmobranchs. In bony fishes it is stiffened by flat plates of dermal bone. Most constant are a large **opercular** and smaller **preopercular, subopercular,** and **interopercular** bones (Fig. 8-24).

In many bony fishes a series of bony **branchiostegal rays** (seven in perch) develops in a caudally directed flap (branchiostegal membrane) of the operculum. No vestiges of opercular elements occur in tetrapods.

Reduction in number of membrane bones during phylogeny

The number of individual membrane bones has tended to be reduced during phylogeny. With reference to Fig. 3-30, p.

67, any major group at the end of an arrow has fewer membrane bones than does the group preceding it in the vertebrate line. Labyrinthodonts had fewer than the crossopterygians, cotylosaurs had fewer than the labyrinthodonts, modern reptiles have fewer than the cotylosaurs, and mammals have fewer than the mammal-like reptiles. Modern amphibians have fewer than their ancestors, the labyrinthodonts. This generalization does not imply that modern reptiles have fewer membrane bones than modern amphibians, since modern reptiles were not derived from modern amphibians.

The reduction is a result of the fusion of adjacent ossification centers to form composite bones and reduction in the number of intramembranous ossification centers. In frogs the frontal and parietal bones are represented by a single **frontoparietal.** In man the left and right frontal bones of the fetus (Fig. 8-18) unite at about the eighth year of life to form a single frontal bone; in lower primates these unite only in old age. Reduction in the number of membrane bones in the mandible (Fig. 8-32) illustrates this trend.

Membrane bones may unite with re-

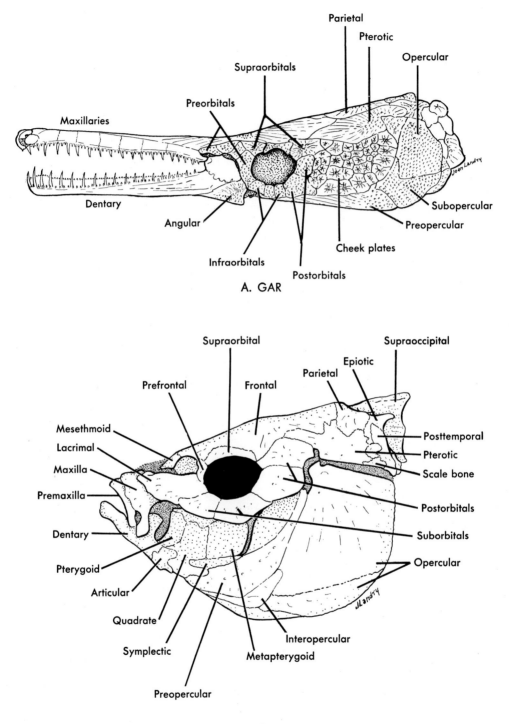

Fig. 8-24. Skull of modern fish (carp) and skull of more ancient fish (gar) for contrast. **A,** Note the scale-like nature of dermal bones in the gar and the series of maxillary bones. **B,** Dark stipple represents unossified cartilage.

placement bones, thereby giving rise to a single bone with a dual history. Postfrontal and supratemporal bones may unite with replacement bones of the otic capsule to form a **sphenotic** and a **pterotic** bone, respectively. The squamosal in mammals unites with otic elements to contribute to a **temporal** bone (Fig. 8-35). The mammalian interparietal may unite with the supraoccipital. Unions such as these result in reducing the number of individual bony components in the skulls of more recent tetrapods. Teleost fish did not participate in this reduction process.

BONY SKULLS OF THE VARIOUS CLASSES*
Fishes

Chondrostei and holostei. The skulls of these older bony fish (ganoid fishes) tend to be flattened as contrasted with the skulls of modern bony fish. The dermal bones are sculptured, scale-like, superficial in location, and covered with ganoin in *Polypterus* and in the gar (Figs. 8-24 and 8-25). In the other ganoids the dermal bones of the skull have lost their ganoin.

The neurocranium is well developed above the brain. Only in *Polypterus* is it extensively ossified. In the other living ganoids (*Amia*, sturgeon, spoonbill, and gar) ossification of the neurocranium is partial at best. The only traces of ossification in the neurocranium of sturgeons are in the otic capsule (opisthotic and prootic) and in the orbitosphenoid region. The elongated rostrum of the spoonbill is a cartilaginous extension of the neurocra-

---

*Other terms applied to bones mentioned in this chapter and not noted elsewhere include ectopterygoid (transpalatine), postparietal (dermoccipital or dermal supraoccipital), supratemporal (suprasquamosal or supramastoid), surangular (supraangular), vomer (prevomer below mammals), tabular (epiotic), and ectethmoid (lateral ethmoid). In many cases the terms are not strictly synonymous. The tetrapod septomaxillary which is occasionally present is of dermal origin and not homologous with the replacement bone of the same name in fishes. For a more complete synonymy of teleost bones, consult Harrington.[59]

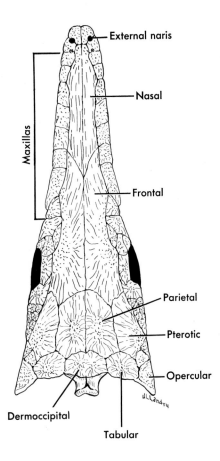

GAR

Fig. 8-25. Skull of gar, dorsal view, showing scale-like nature of dermatocranial bones.

nium. The neurocranium can be revealed by stripping away the overlying dermal bones (Fig. 8-16).

Teleostei. The teleost skull (Fig. 8-24, *B*) is highly specialized—more so than the skulls of early tetrapods. It is laterally compressed, arched, and vault-like (tropibasic) as contrasted with the broad, flat platybasic skulls of primitive fishes and early tetrapods. The neurocranium is incomplete above the brain, but the remainder is well ossified.

The dermal bones no longer resemble scales, having lost all trace of ganoin and having sunk deeper into the head. Their surface is smooth, rather than sculptured as in lower fishes. Strict homologies between the many dermal bones of teleosts and those of tetrapods cannot be established. The eyeball muscles occupy two enlarged cavities (myodomes) on each side of the head, rather than inserting on the orbital wall as in other vertebrates. Myodomes also occurred in ancestral Osteichthyes. Jaw teeth tend to be confined to a few bones or only one, the maxilla. The visceral skeleton has been discussed on p. 156.

Amphibians

Historical perspective. The skulls of crossopterygian fishes and those of the first amphibians (labyrinthodonts) were very similar (Figs. 8-20 and 8-26), strengthening the theory that tetrapods are descendants of crossopterygian-like ancestors. The skulls were platybasic, and the dermal bones were similar in arrangement and were sculptured by the imprint of the overlying dermis. The neurocranium was extensively ossified, except for the olfactory capsule area. A single occipital condyle was borne chiefly on the basioccipital but was completed by the two exoccipital bones. Internal nares pierced the primary palate, and most of the palatal bones bore teeth (Fig. 8-20). There were vacuities between the pterygoid bones and the parasphenoid, and these vacuities became exaggerated in some later amphibians (Fig. 8-21, *Rana*).

Certain modifications occurred in the transition from crossopterygian to labyrinthodont skull. The operculum was lost. The distance between the eyes and the tip of the snout increased and thus provided more facial area in tetrapods. The pineal foramen came to lie between the parietal bones instead of between the frontals. The hyomandibula became surrounded by the first pharyngeal pouch, to become the columella of the middle ear, and the wall of the otic capsule where the columella articulated became a membrane-covered fenestra (**fenestra ovalis**) for transmitting sound waves.

The skull of recent amphibians has become further modified. Ossification has been reduced both in the dermatocranium and in the neurocranium so that there are fewer bones and much more cartilage in the modern amphibian skull. Two occipital condyles are now present, one on each exoccipital bone. Palatal vacuities have increased in size, especially in the anurans, and the pineal foramen has been lost. However, the modern amphibian skull remains markedly platybasic. The visceral skeleton has been discussed earlier.

Anura. Reduction of intramembranous ossification in the roofing area has resulted in loss of prefrontal, lacrimal, postfrontal, postorbital, supratemporal, tabular, and postparietal bones and has exposed the prootic bone (otic capsule) to dorsal view (Fig. 8-28). Frontal and parietal bones

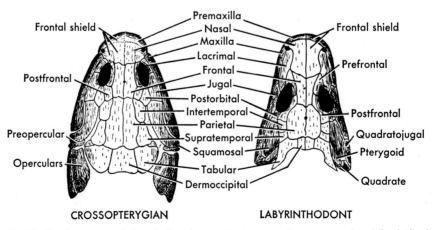

Fig. 8-26. Roofing area of the skulls of an extinct crossopterygian and a labyrinthodont. Note similarity of structure.

unite to form a single frontoparietal, which ossifies just prior to the end of metamorphosis. In the palate (Fig. 8-21) parasphenoid and vomer bones persist, ectopterygoid bones have been lost, the palatine is reduced to a transverse splinter bracing the jaws against the neurocranium anteriorly, and the pterygoid is reduced to a slender tripartite bone bracing the jaws posteriorly. Reduction in palatal elements produced an exceptionally large palatal vacuity, a convenient arrangement permitting the eyeballs to be retracted into the roof of the oral cavity!

The neurocranium is complete dorsally except for several fenestra underneath the frontoparietal bone (Fig. 8-3), and it remains mostly cartilaginous throughout life. The only replacement bones in the neurocranium are two exoccipitals, a sphenethmoid, and a prootic, which ossify at metamorphosis. (In *Triprion petasatus*, a tree frog, the ethmoid bone has a dual origin from endochondral and intramembranous ossification centers.[70])

Depending on the species, Meckel's cartilages are either replaced or ensheathed at the mandibular symphysis by mentomeckelian bones. Posteriorly, Meckel's cartilages either remain unossified or are replaced by articular bones. The bones ensheathing Meckel's cartilage have been reduced to a slim dentary and an angulosplenial.

The pterygoquadrate cartilages persist as cartilaginous bars above the pterygoid bone and as quadrate cartilages or bones at the ends of the upper jaws. The annulus tympanicus, a cartilaginous ring to which the eardrum is attached, is also thought to be a derivative of the pterygoquadrate cartilage. Dermal bones of the upper jaw approximate the primitive pattern, since there are premaxillary, maxillary, and quadratojugal bones.

Urodela. In some respects the urodele skull is more generalized than that of anurans. A slightly larger number of dermal bones remains in the roofing area. For example, prefrontals are sometimes present,

and the frontals and parietals remain separate. However, nasals fail to form in perennibranchiates. Splenials and angulars of the lower jaw also remain separate.

On the other hand, the dermal investment of the pterygoquadrate cartilage is highly modified, since many centers fail to develop. Jugals and quadratojugals are missing and, in perennibranchiates, maxillas are missing. The absence of these elements may be related to the quasi-larval status of perennibranchiates.

In the palatal area considerable cartilage remains. The palatine bone is either small or unossified and the complement of pterygoids is reduced. (In *Necturus*, Fig. 8-21, the palatine is fused with the pterygoid to form a palatopterygoid bone.) The parasphenoid in urodeles is expansive.

The pterygoquadrate cartilages have the same fate as in anurans, remaining cartilaginous in the palatal area and sometimes being replaced by a quadrate bone posteriorly. No mentomeckelian bones form in association with Meckel's cartilage, and the articular end of this cartilage is usually unossified.

Small fenestra usually remain in the neurocranium dorsally, and so it is not entirely complete above the brain. In perennibranchiates, however, the fenestrum is large. The middle ear cavity and eardrum usually fail to develop so that the adult columella is reduced or absent.

Apoda. The skulls of caecilians are typical amphibian skulls with extensive reduction of dermal elements, but they have somewhat more endochondral ossification than in other amphibians.

Reptiles

Among generalized features transmitted from labyrinthodonts to modern reptiles via the cotylosaurs are the single occipital condyle, retention of numerous dermal bones, and extensive ossification of the neurocranium. As far as these features are concerned, the modern reptilian skull is less modified than the modern amphibian skull.

Among specializations in reptilian skulls, although not universal among the species, are the tropibasic condition, presence of temporal fossae, some reduction of roofing elements in the temporal region, loss of the pineal foramen except in *Sphenodon* and many lizards, streptostylism in the quadrate bone, formation of a turbinal element in the nasal passageway, a tendency toward formation of a secondary palate, appearance of numerous vacuities, and increased prominence of the dentary bone. The neurocranium is incomplete dorsally. Most of these modifications have been transmitted to birds or to mammals or to both.

Temporal fossae. One of the major structural changes that has overtaken tetrapod skulls is the development of one or two vacuities, or fossae, bounded by bony arches in the temporal region (Fig. 8-27). Stem reptiles did not exhibit temporal fossae, and so their skulls are **anapsid** (*an* = without, *apsid* = arch). Today, only turtles lack temporal fossae, and cotylosaurs and turtles are placed together in the subclass Anapsida.

Temporal fossae first appeared in the early descendants of stem reptiles. Some descendants (Synapsida, the extinct mammal-like reptiles) developed a single temporal fossa between the postorbital, squamosal, and jugal bones, the last forming most of the underlying **zygomatic arch.** This is a **synapsid** skull. It was transmitted to mammals.

Some reptiles developed a **superior** and an **inferior** fossa. The inferior fossa corresponds to the fossa of synapsids. When two fossae are present (*Sphenodon* and crocodilians), there are two lateral temporal arches, hence the term **diapsid** skull. The superior temporal arch (or supratemporal arcade, as it is sometimes called) lies between the superior and inferior fossae. The external auditory meatus (outer ear canal) passes under cover of the arcade to terminate at the tympanic membrane stretched across a foramen recessed within the infratemporal fossa. The first birds arose from diapsid reptiles.

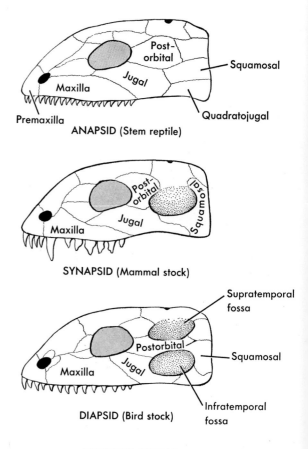

ANAPSID (Stem reptile)

SYNAPSID (Mammal stock)

DIAPSID (Bird stock)

TEMPORAL FOSSAE

Fig. 8-27. Temporal fossae in reptiles leading to birds and to mammals. The postorbital bone in the diapsid skull constitutes the superior temporal arch. The jugal and squamosal bones contribute to the zygomatic arch in the synapsid skull.

If current interpretations are correct, lizards and snakes have modified diapsid skulls, lizards having lost the inferior arch, and snakes having lost both arches.

A few extinct reptiles (ichthyosaurs, for example) exhibited one dorsally located temporal fossa, which may or may not have been equivalent to the superior temporal fossa of diapsids. This was a **parapsid** condition.

Roofing elements (Fig. 8-28). The dermatocranial roof is generalized. In the alligator skull, which is a reasonable ex-

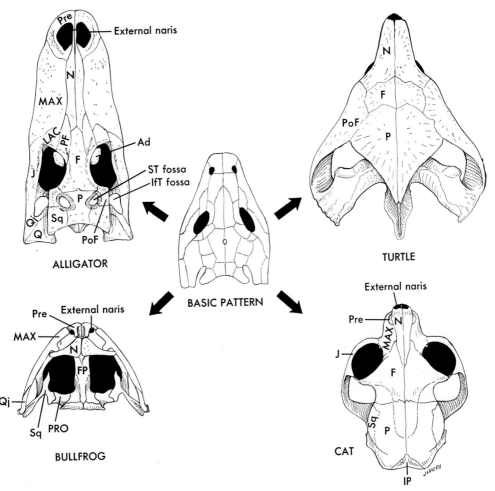

ROOFING BONES

Fig. 8-28. Roofing area and associated bones in selected vertebrates, dorsal views. The basic pattern represents a labyrinthodont. The turtle is an alligator snapping turtle, *Macroclemys temminckii.* **F,** Frontal; **LAC,** lacrimal; **N,** nasal; **P,** parietal; **PF,** prefrontal; **PoF,** postfrontal; **Pre,** premaxilla; **Sq,** squamosal; **St,** supratemporal fossa; **Qi,** quadratojugal; **Ad,** adlacrimal; **FP,** frontoparietal; **IfT,** infratemporal fossa; **IP,** interparietal; **J,** jugal; **MAX,** maxilla; **PRO,** prootic; **Q,** quadrate. As a study aid you may wish to color homologous bones on the different skulls with the same colors.

ample, are paired nasal, lacrimal, prefrontal, postfrontal, and squamosal bones. The unpaired frontal and parietal bones are paired in the embryo. The otic capsule is buried beneath the dermatocranial elements, as in early tetrapods. In this generalized reptilian roof there are more bones than in any modern amphibian. The adlacrimal (palpebral) of the eyelid of crocodilians is not typical of reptiles.

The temporal region of the turtle skull is an enigma. It has no temporal fossae, which suggests a primitive condition with complete roofing. Yet there has been considerable excavation at the rear. Supratemporal, tabular, and postparietal bones are missing, and parietal and squamosal bones have receded, leaving a wide gap in the temporal region. The postfrontal incorporates both the postfrontal and postorbital, and a single prefrontal takes the place of nasal, lacrimal, and prefrontal bones. It is possible that the temporal region as it now exists may be a restored secondary roof, but this is conjectural, and it cannot be said with certainty that the turtle skull is a true anapsid condition directly inherited from the cotylosaurs.

Primary and secondary palates. The primary palate reflects the basic tetrapod pattern, usually consisting of vomer, palatine, pterygoid, and ectopterygoid bones and a parasphenoid, the latter reduced to a rostrum attached to the basisphenoid.

Development of a secondary (false) palate occurs first in reptiles, and in crocodilians it develops to an extreme state (Fig. 8-29). Palatine processes of the premaxillary, maxillary, palatine, and pterygoid bones all unite in crocodilians to form a

very long, completely bony secondary palate with the internal nares far to the rear. Dorsal to the palate is the nasal passageway; ventral to it is the oral cavity.

In most reptiles the palatine processes of one or more of the bones listed in the preceding paragraph fail to meet, and the palate is therefore incomplete. In Fig. 8-30 are shown secondary palates with varying degrees of completeness. An incomplete palate is normal for most reptiles. When it occurs in an alligator or a mammal, the animal is said to have a "cleft" palate.

Upper jaw. The pterygoquadrate cartilages remain as quadrate bones and in *Sphenodon*, lizards, and snakes as epipterygoid bones. The investing bones of the pterygoquadrate cartilages are those of the basic pattern (premaxillas, maxillas, jugals, and quadratojugals), except in lizards and

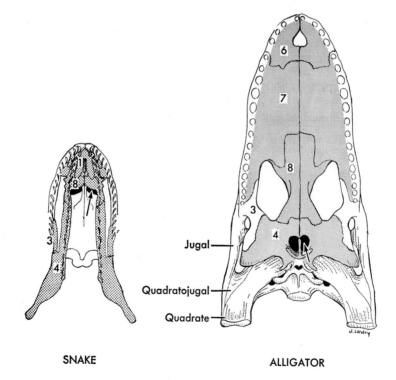

Jugal ——

Quadratojugal ——

Quadrate ——

J. Landry

SNAKE   ALLIGATOR

SECONDARY PALATES

Fig. 8-29. Secondary palates (gray) of two reptiles. Compare location of internal nares (arrows) in alligator with position in snake *(Natrix)* and turtles (Fig. 8-30). **1,** Vomer; **3,** ectopterygoid; **4,** pterygoid; **6,** palatine process of premaxilla; **7,** palatine process of maxilla; **8,** palatine process of palatine.

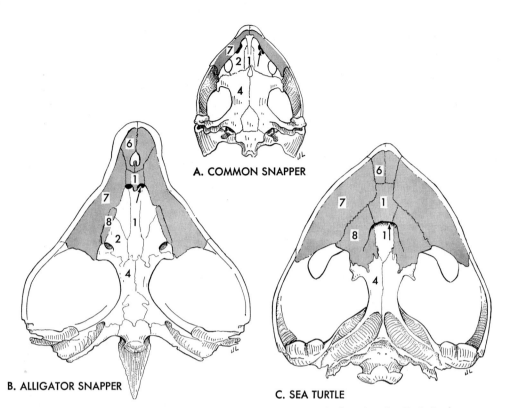

A. COMMON SNAPPER

B. ALLIGATOR SNAPPER

C. SEA TURTLE

Fig. 8-30. Species differences in the secondary palate (gray) of turtles. **A,** *Chelydra serpentina.* **B,** *Macroclemys temminckii.* **C,** *Lepidochelys olivacea* Ridley. The posterior part of the quadrate and the squamosal and supraoccipital regions have been omitted from **C.** In **A,** only the maxilla, **7,** participates in formation of the secondary palate. In **B,** the premaxillas, **6,** also participate, and the palatines, **2,** make a small contribution, **8.** In **C,** all three bones make major contributions. **1,** Vomer; **2,** palatine bone of primary palate; **4,** pterygoid; **6,** palatine process of premaxilla; **7,** palatine process of maxilla; **8,** palatine process of palatine bone. Arrows indicate position of internal nares.

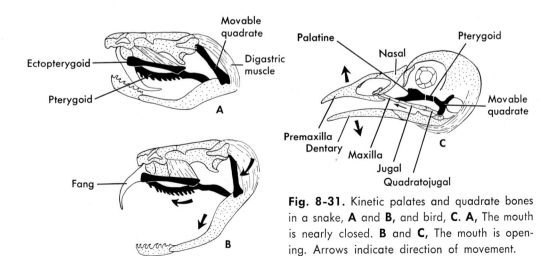

Fig. 8-31. Kinetic palates and quadrate bones in a snake, **A** and **B,** and bird, **C. A,** The mouth is nearly closed. **B** and **C,** The mouth is opening. Arrows indicate direction of movement.

snakes, in which loss of the bony arch between the upper and lower temporal fossae eliminated the jugal and quadratojugal bones.

The upper jaw and palate of *Sphenodon*, lizards, and snakes are movable as a unit, independently of the rest of the skull. The condition is referred to as **kinetism.** In addition, the quadrate is independently movable because of loss of the quadratojugal, with which the quadrate was ankylosed. The condition is known as **streptostylism.** Because of kinetism and streptostylism, snakes can swallow objects larger than their own head (Fig. 8-31).

Kinetism was present in crossopterygians and in early amphibians leading to reptiles.

It has been transmitted to birds but has been lost in modern amphibians and in mammals. Streptostylism in the reptilian precursors of mammals facilitated the transition of the quadrate from a bone of the jaw to a bone (incus) of the middle ear.

Lower jaw. The reptilian mandible preserves the basic tetrapod plan of numerous dermal bones. Typical is the lower jaw of crocodilians, which exhibits dentary, angular, surangular, splenial, and coronoid bones. These ensheath Meckel's cartilage. The posterior element of the mandible is the articular, replacing Meckel's cartilage and articulating with the quadrate. A large **external mandibular vacuity** occurs

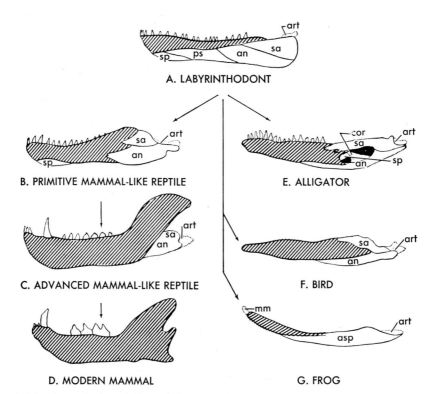

Fig. 8-32. Theoretical evolution of the mandible. **A** to **D,** Probable successive stages of evolution of the lower jaw of mammals. The dentary (oblique lines) became increasingly larger, and other elements were reduced and finally lost. **E** to **G,** Lower jaws of amphibian, reptile, and young bird for comparison with basic pattern. In the modern reptile the pattern has been modified less than in the frog. All elements are dermal bones except those in dotted outline, which are derivatives of Meckel's cartilage. **an,** Angular; **art,** articular; **asp,** angulosplenial; **cor,** coronoid; **mm,** mentomeckelian; **ps,** postsplenial; **sa,** surangular; **sp,** splenial.

between the angular, surangular, and dentary in crocodilians. It has been transmitted to birds. A smaller **internal mandibular vacuity** occurs in crocodilians behind the coronoid. The two halves of the mandible are usually firmly ankylosed, but in snakes they are united by an elastic ligament, which aids in opening the mouth widely.

In fossil reptiles leading to mammals the dentary became increasingly prominent. A **coronoid process** was developed that extended upward toward the temporal region and served for the attachment of the massive temporal muscle. Meanwhile, the other dermal bones of the mandible became reduced, presaging the loss in mammals of all mandibular bones, except the dentary (Fig. 8-32).

Neurocranium. The neurocranium ossifies almost completely, except the nasal capsule. A new development is formation of a cartilaginous **turbinal fold** in each nasal canal. This fold increases the area of the nasal epithelium and thus provides more moisture to inhaled air. (It will be recalled that reptiles were the first animals to completely free themselves of a water habitat.)

The otic capsule gives rise to prootic and opisthotic bones, although the latter may fuse with the exoccipitals, as in alligators. In turtles the otic complex is exposed to dorsal view because of erosion of dermal elements in the temporal region. Four occipital bones ossify, and the single occipital condyle inherited from labyrinthodonts may be located on the basioccipital (crocodilians) or on the basioccipital and exoccipitals (turtles). Basisphenoid and presphenoid bones ossify in the neurocranial floor. A replacing pleurosphenoid is found in the rear of the interorbital septum in snakes and crocodilians; otherwise the interorbital septum of modern reptiles is membranous.

Visceral skeleton. A columella, often consisting of a stapes and extrastapes (Fig. 8-12), occurs in the middle ear. The remainder of the visceral skeleton (hyobranchial apparatus) has been discussed on p. 161.

Genetic relationships of reptiles based on skull structure. If the skull is a reliable criterion for ascertaining genetic relationships, reptiles are related about as follows: Stem reptiles (cotylosaurs) were derived, along with modern amphibians, from labyrinthodont amphibians. Modifications of cotylosaur types produced diapsid and synapsid reptiles. From early diapsids emerged crocodilians, lizards, snakes, and

Table 8-3. Skulls of early tetrapods contrasted with those of modern amphibians and reptiles with reference to a few selected characteristics

	Early tetrapods	Modern reptiles	Modern amphibians
Neurocranium	Well ossified	Well ossified	Mostly cartilage
	One condyle	One condyle	Two condyles
	Platybasic	Tropibasic	Platybasic
Primary palate	Complete complement of dermal bones	Relatively complete	Reduced complement
	Parasphenoid small	Small	Large in urodeles
	Vacuity small	Small	Large in anurans
	Internal nares lateral	Medial	Lateral
Secondary palate	None	Partial or complete	None
Dermal roofing bones	Complete complement	Some reduction	Extensive reduction
Pineal foramen	Present	Present in some	Absent
Marginal bones	Complete complement	Usually complete	Reduced complement
Bones ensheathing Meckel's cartilage	Numerous	Numerous	Reduced complement

many species of extinct reptiles as well as birds. From synapsid reptiles emerged mammals. Evidence from other systems corroborates these general conclusions.

In Table 8-3 the skulls of early tetrapods, living reptiles, and living amphibians are contrasted with respect to a few traits. The primitive pattern is better expressed in modern reptiles than in modern amphibians.

Birds

The bird skull is essentially reptilian in structure. Reptilian dermal ossification centers develop, the secondary palate is incomplete, and a single occipital condyle occurs. Reptilian vacuities and fossae are prominent, and the neurocranium is well ossified. The reptilian complement of premaxillary, maxillary, jugal, and quadratojugal bones is present.

Modifications of the reptilian condition in bird skulls are partly associated with flight, altered feeding habits, and increased brain size. The dermal bones are very thin, and the sutures between them in adult skulls other than ratite skulls are obliterated. Although the skull is diapsid, the bony arch between the superior and inferior fossa has been lost along with other bones, which decreases the flying load. The premaxillary and dentary (and sometimes the maxillary and nasal) bones are elongated to form the beak, which is necessary for feeding (Fig. 8-31, C). The roof of the skull is domed to accommodate the much larger brain of birds.

In the roofing area the pineal foramen is lost, and the lacrimal bones are pierced by a nasolacrimal duct, which drains fluid from the surface of the eyeball into the nasal cavity. (The eyeball is subject to excessive drying as a result of flying conditions.) The enlarged brain results in the frontal and parietal bones arching downward alongside of the brain. Prefrontal and postfrontal bones have been lost. As in diapsid reptiles, the jugal and quadratojugal bones are part of the zygomatic arch (Fig. 8-31, C).

A partial secondary palate is present. No ectopterygoid develops in the primary palate, and the parasphenoid extends forward as a rostrum of the basioccipital. The palatine bones often have a movable articulation with the neurocranium, in which case the palate participates in movements of the upper jaw.

The embryonic pterygoquadrate cartilage forms a quadrate, which is streptostylic. When the quadrate is drawn forward, the zygomatic arch is pushed anteriorly, elevating the upper jaw and associated bones (Fig. 8-31).

The lower jaw has a reptilian complement of dermal bones and an articular bone replaces Meckel's cartilage posteriorly. Teeth, however, are missing in modern birds.

The floor of the neurocranium consists of the basioccipital, basisphenoid, and (usually) presphenoid. Laterally, the neurocranium gives rise to pleurosphenoid and orbitosphenoid bones. Four occipital bones usually ossify, and prootic and opisthotic bones develop in the otic capsule. The olfactory capsule forms ethmoid bones and two turbinal cartilages (instead of one, as in reptiles). Sclerotic bones ossify in the sclerotic coat of the eyeball (Fig. 6-3).

Birds, like reptiles, have a columella in the middle ear. The remainder of the visceral skeleton (hyobranchial apparatus) is illustrated in Fig. 8-14.

Mammals

The mammalian skull is derived from a synapsid reptile type. The temporal fossa is bounded ventrally by the zygomatic arch, which may be incomplete, especially in insectivores and edentates. The quadratojugal is missing from the arch, which therefore consists chiefly of the jugal (in mammals usually called malar or zygomatic). The skull has become increasingly domed as the cerebral hemispheres have expanded. As a result, the frontal and parietal bones extend downward beside the brain. Two occipital condyles occur, a condition inherited from synapsid reptiles.

A complete secondary palate is present.

Characteristic of mammals is inclusion of the articular and quadrate bones into the middle ear cavity, thereby providing mammals with three ear ossicles—the **malleus** (articular), **incus** (quadrate), and **stapes** (columella or hyomandibula). A **tympanic bulla** (large in some species and small in others) ossifies from new cartilage deposited around the middle ear cavity. Removal of the quadrate and articular from the jaws and expansion of the dentary during synapsid evolution resulted in a new articulation for the mandible, the dentary now articulating against the squamosal part of the temporal bone.

Dermatocranium. Roofing elements usually remaining in the mammalian dermatocranium are nasal, lacrimal, frontal, parietal, interparietal, and squamosal. The lacrimal becomes part of the anterior orbital wall and is perforated by a nasolacrimal canal. The squamosal may be an independent bone, as in rabbits, or it may become part of the temporal bone, as in cats and man.

The mammalian interparietal (not found in all species) arises from paired ossification centers and is not homologous with the median interparietal of the primitive pattern. *Homo erectus* had an interparietal bone, and this trait has recurred in Mongolians and related populations such as Peruvian Indians, where it is called the Inca bone (Fig. 8-33). The interparietal still occurs with high frequency among living populations in Peru.

The frontal bone often houses an air-filled sinus (Fig. 8-22). In sheep and goats that butt heads as part of a mating ritual, the frontal air sinus extends into the horns, which are outgrowths of the frontal bones. These bovines come together at speeds up to 35 miles per hour and the sinus, with its bony internal braces, helps to attenuate the compression imposed on the brain by the impact. The entire skull is so shaped that the compression wave is transmitted from the frontal bone through the roof of the skull to the vertebral column.[67]

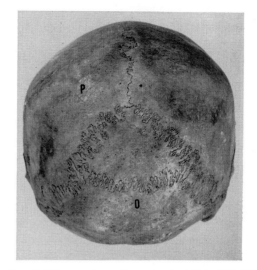

Fig. 8-33. Inca bone in an extinct New World Mongolian from the Aleutian Islands. **P,** parietal; **O,** supraoccipital. (Courtesy William S. Laughlin.[139])

Of the marginal elements of the dermatocranium, only the quadratojugal has been lost. Premaxillas sometimes unite with maxillas, however, and are then lost as independent elements (man). The jugals contribute to the zygomatic arch (Figs. 8-34 and 8-37).

Of the primary palatal elements inherited from reptiles, the **vomer,** now unpaired, lies at the base of the nasal septum and the **palatines** lie in the lateral wall of the nasopharynx (where they contribute to the orbit). The **pterygoids** are reduced to insignificant bones in the caudal wall of the nasopharynx (monotremes) or to a process of the sphenoid bone, called the **hamulus.** The **parasphenoid** has either been lost or incorporated in the vomer. Ectopterygoids have been lost.

The secondary palate consists of two parts. The "hard," or bony, palate consists of the **palatine processes** of the premaxillary, maxillary, and palatine bones (Figs. 8-22 and 8-23). Behind the bony portion is the "soft" part of the palate, so called because bone does not form in it (Fig. 12-21). Above the soft part of the secondary palate lies the nasopharynx.

Neurocranium. The neurocranium does not form above the brain in mammalian embryos (Fig. 8-4). Basioccipital, basisphenoid, and presphenoid bones ossify beneath the brain. Alongside of the brain arise exoccipital and orbitosphenoid bones. (The alisphenoid, which ossifies lateral to the brain, is a derivative of the pterygoquadrate cartilage and is a splanchnocranial element.) All sphenoid elements may unite to form a single **sphenoid bone** composed of a body (basisphenoid and presphenoid), a greater wing (alisphenoid), and a lesser wing (orbitosphenoid) (Fig. 8-6). In some mammals the relative sizes of the wings may be reversed. To the sphenoid complex is sometimes added the pterygoid bone as a hamulus. All occipital elements may also unite before birth to form a single adult **occipital bone.**

The olfactory capsules give rise to portions of the ethmoid complex. This usually consists of a cribriform plate perforated by foramina for the olfactory nerve fibers (Figs. 8-6 and 8-22), a perpendicular plate (mesethmoid) constituting much of the nasal septum (Fig. 8-22), and several **turbinal** cartilages or bones (Fig. 12-21).

The otic capsules develop a number of ossification centers that unite to become a **petrosal (periotic or petromastoid)** bone. The petrosal, at first located in the lateral wall of the neurocranium, later becomes overgrown by the expanding cerebral hemispheres so that it is relegated to a ventrolateral or even ventral position (Fig. 8-22). Where the petrosal reaches the ventral surface of the skull, it is called the mastoid portion of the temporal bone (Fig. 8-37).

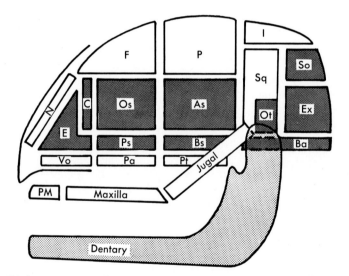

Fig. 8-34. Chief components of the neurocranium (dark) and dermatocranium (white) in a mammalian skull. The vomer, palatine, and pterygoid are components of the primary palate. The premaxilla and maxilla contribute to the secondary palate. The dentary is an element of the visceral skeleton.

As, Pleurosphenoid (alisphenoid)
Ba, Basioccipital
Bs, Basisphenoid
C, Cribriform plate of ethmoid
E, Ethmoid, perpendicular plate
Ex, Exoccipital
F, Frontal

I, Interparietal
N, Nasal
Os, Orbitosphenoid
Ot, Otic (petrous)
P, Parietal
Pa, Palatine

PM, Premaxilla
Ps, Presphenoid
Pt, Pterygoid
So, Supraoccipital
Sq, Squamosal
Vo, Vomer

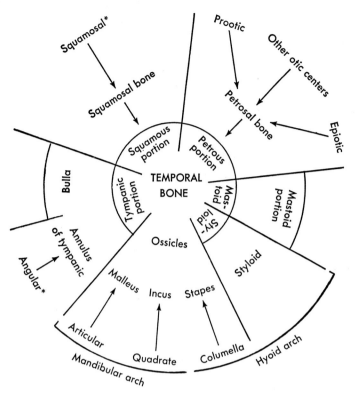

Fig. 8-35. Schematic representation of the multiple nature of the temporal bone of mammals. Note reduction in number of separate elements from the condition in reptiles (outer circle) to mammals (other circles). The two dermal elements have asterisks. The mastoid portion and tympanic bulla are mammalian innovations.

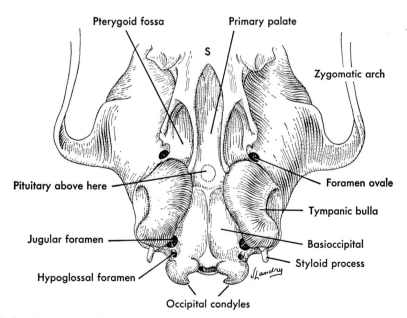

Fig. 8-36. Hamster skull, posterior part, ventral view. **S,** Secondary palate.

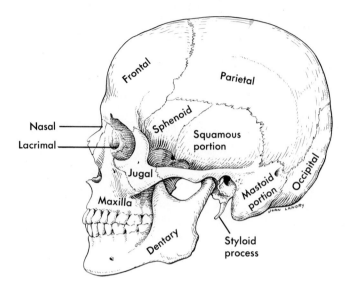

Fig. 8-37. Skull of modern man.

Temporal complex (Fig. 8-35). The temporal complex consists of petrosal, tympanic, and squamosal components of intramembranous and endochondral origin. The bony petrosal component is the ossified otic capsule, present as a separate bone in some mammals. The bony tympanic portion (new in mammals) surrounds the middle ear cavity and often swells to form a tympanic bulla (Fig. 8-36). Associated with the bulla is a bony ring, the **annulus tympanicus** (Fig. 8-12), derived (according to evidence from embryonic opossums) from the angular bone of reptiles. The tympanic membrane is attached to the annulus.

The tympanic bone, with its various parts, often unites with the petrosal bone to form a petrotympanic bone, as in rabbits, and the petrotympanic may unite with the squamosal to form an adult **temporal bone**, as in cats and man. A **styloid** process from the hyoid arch may coalesce with the temporal bone ventrally (Fig. 8-36).

Visceral skeleton (Fig. 8-15). The pterygoquadrate cartilage of the embryo gives rise posteriorly to the incus, which separates from the remainder of the cartilage to become isolated in the middle ear. The palatal portion of the pterygoquadrate cartilage remains as the **alisphenoid** bone (homologous with the **epipterygoid** of reptiles).

No significant remnant of embryonic Meckel's cartilage other than the malleus remains in adult mammals. The investing bones of the lower jaw have been reduced to a single dentary, articulating with the squamosal or with the squamous portion of the temporal bone in a mandibular fossa. No mandibular vacuities remain from reptilian precursors.

The hyomandibular element of the hyoid arch continues to give rise in mammals to a bone of the middle ear. This bone is more often called stapes than columella, but the two are partly homologous. The remainder of the visceral skeleton of mammals (hyobranchial apparatus and laryngeal cartilages) has been discussed on p. 162.

The relationship in mammals between the articular region of Meckel's cartilage (now the malleus), the quadrate region of the pterygoquadrate cartilage (now the incus), and the hyomandibula (now the stapes) is similar to the relationships in fishes exhibiting hyostyly. Only the function of the complex differs.

Chapter summary

1. The head skeleton in Chondrichthyes consists of a cartilaginous neurocranium (braincase) and a cartilaginous splanchnocranium (visceral skeleton). The neurocranium is constructed of prechordal and parachordal cartilages, notochord, and cartilaginous olfactory and otic capsules, all knit solidly together and completed by the addition of cartilaginous walls and a roof over the brain. The splanchnocranium consists of upper and lower jaws (pterygoquadrate and Meckel's cartilages), hyoid cartilages, and a series of additional gill cartilages.

2. The head skeleton in Agnatha consists of a number of loosely articulated neurocranial elements and of a branchial basket not readily homologizable with the visceral skeleton of Chondrichthyes. There are no recognizable jaws, and the notochord remains independent of the skull.

3. The bony vertebrate skull consists of a neurocranium, a splanchnocranium, and a dermatocranium. The neurocranium is cartilaginous initially but is partly or wholly replaced by endochondral bone as development progresses. Except in lower fishes, the neurocranium is generally incomplete above the brain. The splanchnocranium is cartilaginous during embryonic life but is partly replaced by bone as development progresses. The dermatocranium consists of bones that develop by intramembranous ossification. Most of these bones are derived from the dermal armor of ancient vertebrates.

4. The chief centers of endochondral ossification in the neurocranium are occipital, sphenoid, ethmoid, and otic. These contribute the floor on which the brain rests and part of the walls of the braincase. The most common replacement bones are exoccipital, basioccipital, and supraoccipital bones; presphenoid, basisphenoid, orbitosphenoid, and pleurosphenoid bones; mesethmoid and turbinal bones (amniotes); and prootic, epiotic, and opisthotic bones. These may be reduced in number by fusion in higher forms. Occipital condyles in tetra-

pods are single in ancient forms, borne jointly on the basioccipital and exoccipital bones. The condition was transmitted to modern reptiles and birds. Modern amphibians, synapsid reptiles, and mammals exhibit two condyles borne on the exoccipital bones.

5. The chief centers of endochondral ossification in the visceral skeleton are (a) **in the pterygoquadrate cartilage**—epipterygoid (alisphenoid in mammals), quadrate (incus in mammals), and metapterygoid (teleosts only); (b) **in Meckel's cartilage**—articular (malleus in mammals) and mentomeckelian (in a few lower forms); and (c) **in the hyoid and successive visceral arches**—hyomandibula (columella or stapes in tetrapods); in fishes a series of additional hyoid elements (symplectic, interhyal, epihyal, ceratohyal, hypohyal, basihyal, and urohyal) and centers in the gill arches; in tetrapods a hyobranchial apparatus and also (occasionally) laryngeal centers. The visceral skeleton remains partly cartilaginous even in bony vertebrates.

6. The early complement of dermal bones included (a) **roofing bones**—nasal, frontal, and parietal bones in all vertebrates; internasal, intertemporal, and interparietal bones chiefly in ancestral fishes; and intertemporal, posttemporal, supratemporal, squamosal, tabular, lacrimal, prefrontal, postfrontal, suborbital, postorbital, and supraorbital bones; (b) **ensheathing bones of the pterygoquadrate cartilage**—premaxillary, maxillary, jugal, and quadratojugal bones; (c) **primary palatal elements**—vomer, parasphenoid, palatine, endopterygoid, ectopterygoid, and pterygoid elements; (d) **ensheathing elements of Meckel's cartilage**—angular, surangular, splenial, coronoid, and dentary; and (e) a series of opercular elements ossifies in the operculum of fishes. No trace of these remained in the earliest tetrapods.

7. The dermatocranial elements are numerous, superficial, and scale-like in generalized fishes and smooth and more deeply situ-

ated in modern fishes and tetrapods. They have been reduced in number in tetrapods.

8. The skulls of crossopterygian fishes and those of labyrinthodonts were similar. The dermal bones were numerous and their surfaces were sculptured, resembling scales. The neurocranium was extensively ossified. The hyomandibula of crossopterygians became the columella of labyrinthodonts.

9. In modern amphibians the neurocranium exhibits small fenestras dorsally and is almost wholly cartilaginous throughout life. Fewer dermal bones develop than in any other vertebrate class. Thus, the modern amphibian skull has departed considerably from the labyrinthodont type.

10. In adult tetrapods the visceral cartilages caudal to the first have lost their function as a gill skeleton and have assumed new functions associated with life on land, chiefly as ear ossicles, hyobranchial apparatus, and laryngeal cartilages.

11. The skulls of modern reptiles have undergone reduction in number of bones from the labyrinthodont condition, but not so much as did amphibian or mammalian skulls. The number has been most reduced in lizards and snakes. The neurocranium is incomplete dorsally but is otherwise well ossified. There are many vacuities in the dermatocranium, and a partial or complete secondary (false) palate has developed. Temporal fossae result in diapsid skulls (two temporal arches) and synapsid skulls (one arch). The former were transmitted to birds and the latter to mammals.

12. The skull of birds is essentially a diapsid reptilian skull with modifications. However, the dermal bones are thin and sutures have been obliterated, except in ratites. The skull is highly domed to accommodate the expanded brain, and the jaw and facial bones are elongated to form a beak for feeding.

13. The skull of mammals is derived from the synapsid reptilian type. It is domed because of expansion of the cerebral hemispheres. The number of bones—both endochondral and dermal—has been reduced by fusion to form single units such as the occipital, sphenoid, and temporal bones. The cavum tympanum now has three ossicles (malleus, incus, and stapes), and a tympanic bulla of recent origin surrounds the cavity. Incorporation of the malleus and incus in the middle ear cavity is associated with the evolution of a new articulation of the mandible, the dentary now articulating against the squamosal region of the dermatocranium.

Chapter
9
The appendicular skeleton

The appendicular skeleton includes the cartilages and bones of the pectoral and pelvic girdles and of the fins (paired and unpaired) and limbs. The bone of the appendicular skeleton arises by endochondral ossification, except contributions of dermal origin to the pectoral girdle. Agnathans, caecilians, snakes, and some lizards have no appendicular skeleton, and it has been much reduced in certain other vertebrates.

PECTORAL GIRDLES

The pectoral girdle is embedded in the body wall at the anterior end of the trunk, where it braces the anterior appendages. In fishes it lies immediately behind the last gill arch. The basic pattern consists of two sets of elements, replacement bones and membrane bones (Fig. 9-1). The chief replacement bones are the **coracoid, scapula,** and **suprascapula.** The chief membrane bones are the **clavicle, cleithrum,** and **supracleithrum.** The membrane bones represent dermal armor, which, early in phylogeny, sank beneath the skin to invest the deeper cartilages. Early bony fish and early tetrapods exhibited a full complement of replacement and dermal bones in the pectoral girdle. Among later fishes there was a tendency to reduce the number of replacement bones, and the pectoral girdle became chiefly dermal bone (Fig. 9-2). In tetrapods, on the other hand, there has been a tendency for the dermal bones to be reduced. Urodeles and cartilaginous fishes develop no dermal bones, and so the pectoral girdle is either cartilaginous or cartilage and replacement bone (Fig. 9-3). The anterior fin or limb articulates with the pectoral girdle in a glenoid fossa of the scapula.

Pectoral girdles of the different classes

Fishes. The pectoral girdle of bony fishes (Figs. 9-2 and 9-10) exhibits a coracoid and scapula, both of which ossify from cartilages and are usually reduced in size. The embryonic coracoid and scapula may unite to form a single coracoscapula. A suprascapula may be present. Dermal bones include a ventral clavicle (usually reduced and sometimes absent), a cleithrum (enlarged and prominent), and a supracleithrum. A posttemporal bone may unite the supracleithrum with the dermatocranium, especially in primitive fishes.

The pectoral girdle of cartilaginous fishes is not typical of fishes, since dermal elements are completely lacking, and the cartilaginous elements do not ossify. The U-shaped girdle of *Squalus* (Fig. 9-3) consists of a median ventral coracoid cartilage (arising embryonically from a pair of cartilages), which articulates laterally with scapular cartilages. The latter have a gle-

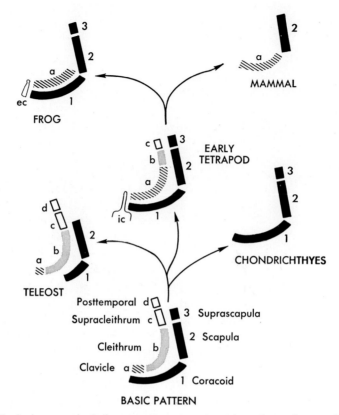

Fig. 9-1. Typical pectoral girdle components among selected vertebrates. Cartilage and replacement bones are black; other elements are of dermal origin. **ec,** Epicoracoid; **ic,** interclavicle.

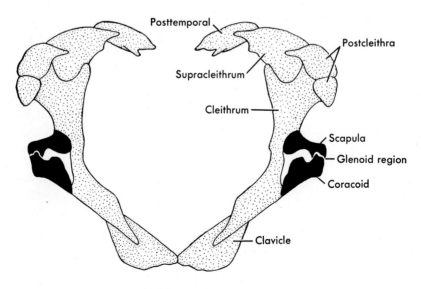

Fig. 9-2. Pectoral girdle of the ganoid fish *Polypterus*. Dermal bones are stippled; replacement bones are black.

noid fossa for articulation with the fin. A small suprascapula occurs at the dorsal tip of each scapula.

The skeleton of the girdle of fishes is interpreted as an extension into the body wall of the skeleton of the appendage. This interpretation is based partly on the fact that the cartilaginous elements of the girdle and of the fin (basals and radials, Fig. 9-9, *Squalus*) all develop from a continu-ous procartilaginous blastema extending in-to the appendage. Later, the cartilages of the girdle and fin differentiate from the blastema to become separate elements.

Tetrapods. The pectoral girdles of early tetrapods closely resembled the basic pat-tern (Fig. 9-1). A new midventral dermal element, the **interclavicle** (**episternum**), made an appearance, and the coracoid ar-ticulated with the anterior end of the sternum. The dermal elements were often sculptured like the dermal bones of the skull, thereby revealing their intimate as-sociation with the overlying skin.

In modern tetrapods (Fig. 9-4) the clei-thrum and supracleithrum have been lost. The clavicle has become prominent in all but a few specialized groups and either assists or replaces the coracoid in bracing the scapula against the sternum. The loss of the cleithrum and the expansion of the clavicle in modern tetrapods is the reverse of the condition in modern fishes, in which the cleithrum was expanded and the clav-icle was reduced. Modern generalized tetra-pods, therefore, usually exhibit two or three of the following: clavicle and inter-

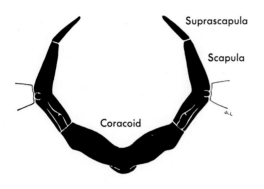

PECTORAL GIRDLE

Fig. 9-3. Pectoral girdle of the shark *Squalus*, anterior view.

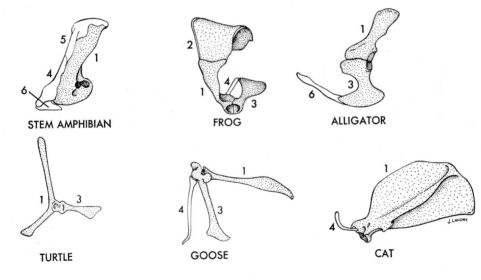

PECTORAL GIRDLES

Fig. 9-4. Pectoral girdles of representative vertebrates, left lateral view, head to left. Replacement bones are stippled. **1,** Scapula; **2,** suprascapula; **3,** coracoid; **4,** clavicle; **5,** cleithrum; **6,** interclavicle.

clavicle of dermal origin, and coracoid and scapula of cartilage origin.

The pectoral girdle of adult frogs (Fig. 7-23) consists of bony clavicles of dermal origin, and epicoracoids, coracoids, scapulas, and suprascapulas of cartilage origin. Coracoids usually ossify, epicoracoids and suprascapulas remain cartilaginous, and scapulas partly ossify. In urodeles dermal elements of the girdle fail to develop, and the cartilaginous elements remain mostly unossified. Modern amphibians have no interclavicle.

The clavicle is reduced or absent in crocodiles and legless lizards, and the entire girdle is missing in snakes. In turtles the interclavicle and clavicle are fused with the shell, but this is not surprising, since all three are dermal derivatives. The scapula of turtles (Fig. 9-4) bears a long acromial process, which assists in bracing the girdle.

In carinate birds the two clavicles unite in the midline with the interclavicle to form a **furcula** (wishbone) (Fig. 7-24). In flightless birds the clavicles do not meet ventrally and so there is no wishbone.

Monotremes have a reptilian pectoral girdle with an interclavicle, clavicle, coracoid, and scapula. Above monotremes only a clavicle and scapula remain.

The clavicle is reduced or absent in some mammals. Whales, ungulates, and some carnivores have no clavicle. In other carnivores such as cats, the clavicle has been reduced to a slender bony splinter, which fails to reach either the sternum or the scapula.

When present, a well-developed clavicle braces the shoulder against the sternum. Among mammals having a prominent, bracing type of clavicle are insectivores, bats, rodents, some marsupials, and higher primates including man.

The scapula is always present in mammals. It exhibits a broad, flat surface divided by a scapular spine into supraspinous and infraspinous fossae for the origin of muscles inserting on the humerus. The scapular spine terminates as an **acromion**

process. A **metacromion** process may also occur. A hook-like coracoid process not homologous with the coracoid bone overhangs the glenoid fossa. Monotremes exhibit a coracoid bone as well as a coracoid process.

PELVIC GIRDLES

Pelvic girdles, like pectoral girdles, are embedded in the body wall and serve to brace the appendages. Pelvic girdles have no dermal elements.

Fishes. The pelvic girdle in most fishes consists of two cartilaginous or bony **pelvic (ischiopubic) plates** articulating with the pelvic fins (Fig. 9-5, herring). The plates usually meet medially in a pubic symphysis. In sharks and lungfishes two embryonic cartilaginous plates unite to form a single adult ischiopubic plate (Figs. 9-5 and 9-6, dogfish). In some teleosts the trunk is foreshortened, and the pelvic girdle lies immediately behind the pectoral girdle and often attached to the latter.

Tetrapods. Tetrapod embryos, like fishes, form a ventral cartilaginous pelvic plate on each side. Each plate ossifies at two centers to form a **pubic bone** (**pubis**) anteriorly and an **ischium** posteriorly (Fig. 9-7). Dorsally on each side an additional cartilage becomes the **ilium**. At the junction of the three elements is the acetabulum, a socket accommodating the head of the femur.

The ilium is firmly anchored to the transverse processes of the sacral vertebrae (Fig. 7-10). The site of union is the **sacroiliac** symphysis. Bracing the pelvic girdle against the axial skeleton provides a firm base against which the femur, now bearing much of the body weight, may push. The two pubic bones typically unite ventrally in a pubic symphysis, and the ischia usually form an ischial symphysis.

Because of the pubic and ischial symphyses ventrally, and because the ilia are united with the vertebral column dorsally, the caudal end of the coelomic cavity is surrounded by a bony ring, the **pelvis.** The pelvis contains the caudal end of the diges-

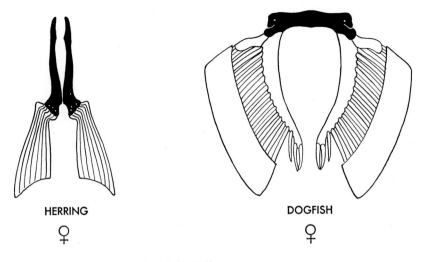

HERRING ♀ DOGFISH ♀

PELVIC GIRDLES

Fig. 9-5. Pelvic girdles of bony and cartilaginous fish, dorsal views. The girdles consist of one or two pelvic plates (black).

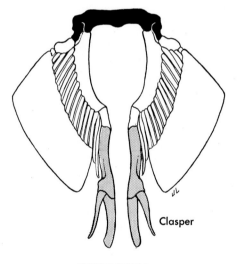

Clasper

PELVIC GIRDLE

Fig. 9-6. Pelvic girdle (black) and fin of a male shark, showing basal cartilages (gray) modified as claspers. Compare female dogfish, Fig. 9-5.

tive and urinogenital systems. The inferior border of the ring is the pelvic outlet.

In frogs (Fig. 7-10) the ilium is greatly elongated and extends from the single sacral vertebra to the end of the urostyle, where it meets the ischium and pubis and where the acetabulum is located. The ilium

is thus adapted to absorb shock when a frog lands after a jump. The pubic area does not ossify in modern amphibians, and most of the girdle remains cartilaginous in urodeles. A median Y-shaped (**ypsiloid**) cartilage develops just anterior to the pubic area in some amphibians.

In reptiles the ilium is fused with **two** sacral vertebrae, and it tends to become broader for the attachment of a greater mass of hind limb muscles. These modifications are related to the tendency of reptiles to carry the trunk elevated well above the ground, and are most pronounced in bipedal reptiles (Fig. 9-7, dinosaur). In reptiles that have lost the limbs the pelvic girdle has been reduced (legless lizards) or lost (most snakes).

In birds (Fig. 7-15) the ilium is greatly expanded to accommodate the musculature necessary for bipedalism, and it is braced against lumbar as well as the sacral vertebrae (Fig. 7-16). The pubic bones, on the contrary, are reduced to long splinters projecting caudad. They do not meet to form a pubic symphysis. Loss of the weight of the pubic bones certainly was no disadvantage to the bird on takeoff or in flight, and the absence of a pubic symphysis assured

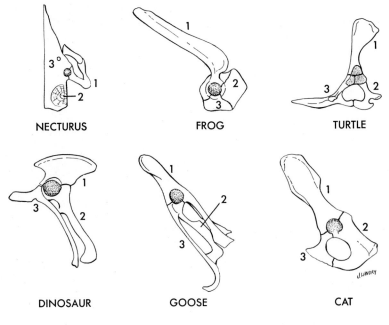

PELVIC GIRDLES

Fig. 9-7. Pelvic girdles of representative vertebrates. Left lateral views except of *Necturus*, which is a ventral view of the right girdle. **1,** Ilium; **2,** ischium; **3,** pubis. In *Necturus* the ischium appears as an ossification center at the caudal angle of the cartilaginous pubic plate. The acetabulum is stippled.

a larger pelvic outlet for laying the massive eggs.

In mammals the ilium, ischium, and pubis unite early in postnatal life to form an **innominate bone** on each side. The two innominates comprise the pelvic girdle. Occasionally an **acetabular (cotyloid) bone** ossifies in the wall of the acetabulum. Marsupials and monotremes exhibit two small **epipubic (marsupial)** bones articulating with the pubic bones and extending forward in the abdominal wall. These support the marsupial pouch in which the young of marsupials are transported. The pelvic girdle is absent in cetaceans (Fig. 9-22, porpoise).

In mammals the ligaments uniting the pubic and ischial bones ventrally are softened during pregnancy by the hormones relaxin and estrogen. This facilitates expansion of the pelvic outlet during delivery of the fetuses. In mice six days pregnant, the gap between the two pubic bones at the pubic symphysis has been shown by x-ray examination to be only 0.25 mm. Thirteen days later, on the day of birth, the gap has widened to 5.6 mm.

FINS

Most fishes (except agnathans and teleosts such as eels) have pectoral and pelvic fins. The endoskeleton of the pectoral fin is usually somewhat more specialized. Fins, whether paired or unpaired, usually serve primarily as steering devices and rudders rather than for locomotion. However, lungfishes use their fins as if they were limbs for awkward locomotion when they are temporarily out of the water. They have been induced to ascend inclined planes in this manner.

In some male Chondrichthyes the pelvic fin has been modified by the addition of skeletal elements to become a clasper (Fig. 9-6). The clasper is inserted into the cloaca of the female and guides the sperm to

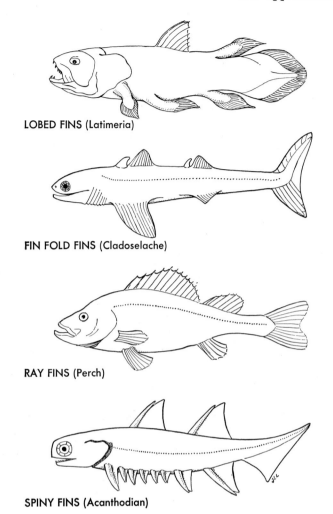

LOBED FINS (Latimeria)

FIN FOLD FINS (Cladoselache)

RAY FINS (Perch)

SPINY FINS (Acanthodian)

Fig. 9-8. Fins of fishes. *Latimeria* is a living crossopterygian. *Cladoselache* is an extinct shark. The acanthodian is an extinct placoderm.

the mouth of the uterus during copulation.

There is a wide diversity in the endoskeletal structure of fins. Three predominant skeletal variants of paired fins will be described—**biserial fins, ray fins, fin fold fins.**

Biserial fins. Characteristically, the biserial fin exhibits a central axis composed of a series of jointed cartilages or bones articulating with the girdle proximally and tapering distally (Fig. 9-9, *A*). An anterior (preaxial) and posterior (postaxial) series of radials account for the name biserial. Dermal fin rays project from the radials to

the anterior and posterior edges of the fin. The biserial fin may represent an early stage in the evolution of fins.

The biserial fin has a pronounced basal lobe consisting of muscle (Fig. 9-8, *Latimeria*). The fin is relatively narrow where it emerges from the basal lobe, but it usually has a paddle-like expansion before tapering distally. This type of fin is very ancient. It dates back to the Devonian period and is associated with the Sarcopterygii, which are often called lobe-finned fishes.

Upper Devonian crossopterygians, considered close to the ancestral line from

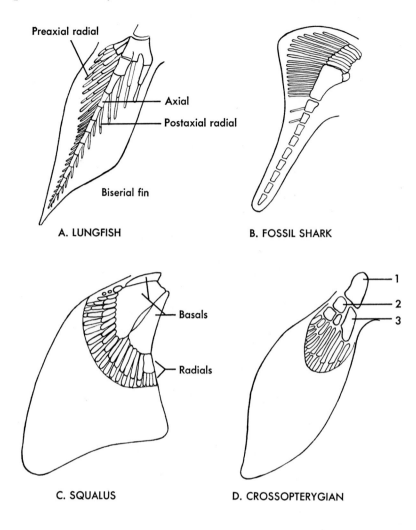

ANCIENT FIN SKELETON

Fig. 9-9. Ancient fin skeletons, reoriented. **A,** Biserial pelvic fin of *Neoceratodus*, a living lungfish. **B** to **D,** Pectoral fins of *Cladodus*, dogfish, and fossil crossopterygian, respectively. **1** to **3,** Equivalent by position to humerus, ulna, and radius of tetrapod limb, respectively.

which tetrapods emerged, show a modification of the skeleton of the biserial fin. The result has been a fin skeleton strikingly similar to the skeleton of the earliest tetrapod limb (Fig. 9-9, *D*). In modern lungfishes there is no longer any ossification in the fin skeleton.

Ray fins. The ray fin of teleosts is a more recent development. Actinopterygii (*actin* = ray) exhibit a diversity of modifications of this type of fin. Some of the variants are intermediate in morphology between the primitive biserial fin and the fin of modern teleosts.

In teleosts (Figs. 9-8, perch, and 9-10) there are no basal lobe and no skeletal elements within the fin other than radials. Even the radials may be lost, whereupon the dermal fin rays articulate directly with the girdle.

Fin fold fins (Fig. 9-8). Fin fold fins have an extended base of attachment. They

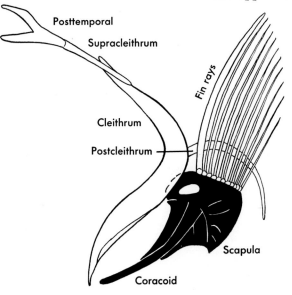

Posttemporal

Supracleithrum

Fin rays

Cleithrum

Postcleithrum

Scapula

Coracoid

MODERN FIN SKELETON

Fig. 9-10. Pectoral girdle and ray fin of a teleost (ribbon fish). Note absence of basal and radial elements. Replacement bones are shown in black. (Redrawn from James.[63])

were common in ancient sharks. Living Chondrichthyes still exhibit a fin with a relatively broad, fleshy base, but the internal skeletal elements have been modified. The fin skeleton of living Chondrichthyes (Fig. 9-9, *C*) consists of a proximal row of one to five **basal cartilages** and several distal rows of **radial cartilages**. Dermal fin rays stiffen the fin distally.

Median fins. In addition to paired fins, most fishes have unpaired (median) fins, which help to prevent torque and yaw during swimming. In some larval fishes the median fins form continuous dorsal and ventral folds uninterrupted around the tip of the tail, which suggests that this may represent a primitive state. Adults typically have one or two **dorsal fins**, a ventral **anal fin** behind the anus or vent, and a **caudal fin.** Dorsal fins may be reduced or absent, or there may be a series. The anal fin may be lost, especially in bottom dwellers. The anal fin in teleosts is sometimes modified as an intromittent organ or ovipositor.

Median fins have a skeleton composed of slender radials (or of basals and radials) embedded in the connective tissue septum dorsal to the neural spines or ventral to the hemal spines. Dermal fin rays extend from the radials to the edge of the fin.

Theories of the origin of paired fins. According to the fin fold theory, a continuous fold of the lateral body wall may have been the precursor of paired fins. If the fold became interrupted in the middle of the trunk, this would have left later fish with a pair of pectoral and a pair of pelvic fins (*Cladoselache*, Fig. 9-8). The theory is no longer widely accepted.

Another theory derives paired fins from fin spines similar to those of acanthodian fishes (Fig. 9-8). In some acanthodians pectoral and pelvic appendages were members of a series of spiny appendages that extended the length of the trunk. Associated with each spine was a fleshy web. It is possible that the first fins consisted of spines with fleshy webs that were metamerically arranged. Subsequently, all spines and webs may have been lost except one anterior and one posterior pair. These may then have evolved into pectoral and pelvic fins.[137]

Interesting as a historical note is the

gill arch theory proposed by Gegenbaur in the nineteenth century. According to this theory, which is no longer tenable, the fin skeleton is a modification of the skeleton of terminal gill arches. The location of the pectoral girdle immediately behind the last gill arch and its superficial resemblance to a gill arch prompted this hypothesis. It does not explain the primitive location of pelvic fins immediately in front of the cloaca. The first chordates lacked paired appendages, and clues to the origin of paired appendages are probably hidden forever in the obscurity of time.

TETRAPOD LIMBS

The tetrapod limb, both anterior and posterior, consists of five segments (Table 9-1 and Fig. 9-11). The skeleton within homologous segments is remarkably similar despite differences such as arms, legs, wings, and paddles. Adaptive modifications are chiefly in the nature of a reduction (rarely of an increase) in the number of bones of the hand or foot.

A few tetrapods have completely lost one or both pairs of appendages. Lacking limbs altogether are caecilians, most snakes, and snake-like lizards. Having forelimbs only are sirens (family Sirenidae, a tailed amphibian), the lizard *Chirotes,* and manatees and dugongs. Having hind limbs only are pythons, boa constrictors, a few lizards, and a few ratites. Cetaceans have lost all external manifestations of hind limbs, but vestiges sometimes remain embedded within the body wall. In many instances in which an adult limb is absent, an embryonic limb bud appears transitorily but fails to develop.

The limbs of early tetrapods were short and stout and directed outward at right angles to the body. This posture persists among many lower tetrapods, but in most reptiles and in mammals there has been a rotation of the appendages toward the body, so that the long axis of the humerus and femur more nearly parallel the vertebral column. To a marked degree the knee is directed cephalad and the elbow

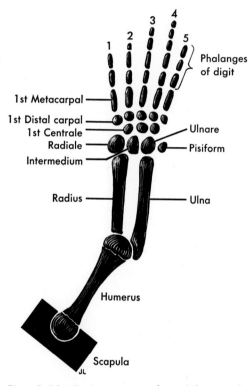

Fig. 9-11. Basic pattern of a right anterior limb, viewed from above, palm down. **1** to **5,** First to fifth digits.

Table 9-1. Homologous segments in anterior and posterior limbs of tetrapods

Anterior limb			Posterior limb		
Name of segment		Skeleton	Name of segment		Skeleton
Hand (manus)	Upper arm (brachium)	Humerus	Foot (pes)	Thigh (femur)	Femur
	Forearm (antebrachium)	Radius and ulna		Shank (crus)	Tibia and fibula
	Wrist (carpus)	Carpals		Ankle (tarsus)	Tarsals
	Palm (metacarpus)	Metacarpals		Instep (metatarsus)	Metatarsals
	Digits (digiti)	Phalanges		Digits (digiti)	**Phalanges**

is directed caudad (Fig. 9-22, A). Limbs oriented in this fashion are much better shock absorbers. They also permit much greater leverage between the axial skeleton and the appendage, as a result of which speed and agility are greatly increased. Such reorientation was an essential step toward bipedalism in reptiles.

Skeleton of the anterior limbs

Humerus. The humerus is the bone of the upper arm. The head of the humerus articulates with the scapula in the glenoid fossa. The distal end articulates with the ulna in a depression that in higher forms becomes the **semilunar notch** of the ulna. The humerus also articulates with the radius.

The similarity of the humeri of all tetrapods is more striking than their differences (Fig. 9-12). Variations in length, diameter, and shape are adaptive modifications. The humerus of the mole, for example (Fig. 9-13), flares greatly for the insertion of the massive muscles used for digging.

Humeri exhibit numerous tuberosities, ridges, condyles, and other prominences for articulation or for the attachment of appendicular muscles. One or two supracondyloid foramina occur in some species near the distal end for the transmission of blood vessels and nerves en route to the hand. All bones exhibit nutritive foramina, which transmit vessels and nerves to the interior of the bone.

Radius and ulna. The radius and ulna are bones of the forearm. The radius is a preaxial (anterior) element articulating proximally with the humerus and ulna and distally with wrist bones on the thumb

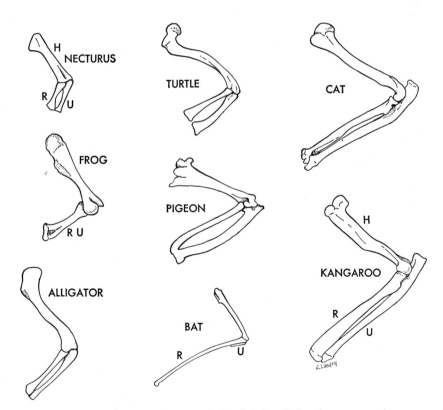

Fig. 9-12. Humerus, radius, and ulna of the left forelimb of representative tetrapods, lateral views. **H**, Humerus; **R**, radius; **U**, ulna. In the frog the radius and ulna have united to form a radioulna, **RU**. In the bat the ulna is vestigial.

side of the hand. The radius bears most of the body weight. The ulna is the longer, postaxial element articulating proximally with the humerus and radius and distally with wrist bones on the side opposite the thumb. The shaft of the ulna may fuse with the radius during embryonic development (ungulates), or it may fail to develop (bat).

Extending between the shaft of the radius and ulna is an interosseous ligament that sometimes ossifies. In man the ligament is incomplete and flexible. Place your

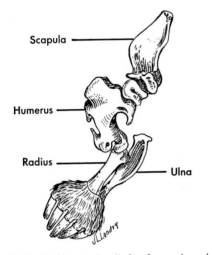

Fig. 9-13. Right anterior limb of a mole, which has been modified for digging. This is a medial view! The palms of the mole turn outward from the body.

forearm and hand on a flat surface and turn your palm down (pronation); then turn your hand over (supination). Few animals can accomplish this movement because of the interosseous ligament. In frogs the ligament is so ossified that adults appear to exhibit only one bone—the radioulna.

Manus. Wrist, palm, and digits constitute a functional unit—the hand, or manus. The manus is surprisingly uniform throughout the tetrapod series.

The skeleton of the wrist, or **carpus**, is the most stable of the regions of the manus. In generalized forms the carpus consists of three rows of carpal bones (Fig. 9-11). (1) The proximal row of carpals exhibits a **radiale** at the base of the radius, an **ulnare** at the base of the ulna, and an **intermedium** between the radiale and ulnare. At the lateral (or posterior) end of the proximal row in most reptiles and mammals is the **pisiform,** a sesamoid bone. (2) The middle row of carpals consists of one to three **centralia** (sing., **centrale**). (3) The distal row is composed of five **distal carpals** numbered 1 to 5 commencing on the thumb (medial or anterior) side. In Table 9-2 are listed the names of the carpal bones.

The metacarpals constitute the skeleton of the palm. Primitively, there were probably as many distal carpals and metacarpals as there were digits.

Each digit consists of a linear series of

In the figure: Scapula, Humerus, Radius, Ulna

Table 9-2. Synonymy of carpal bones

Terms preferred by comparative anatomists	Nomina Anatomica*	Anglicized names and synonyms
Radiale	Os scaphoideum	Scaphoid, navicular
Intermedium	Os lunatum	Lunate, lunar, semilunar
Ulnare	Os triquetrum	Triquetral, cuneiform
Pisiform	Os pisiforme	Pisiform, ulnar sesamoid
Centralia (0 to 3)	Os centrale	Centrale
Distal carpal 1	Os trapezium	Trapezium, greater multangular
Distal carpal 2	Os trapezoideum	Trapezoid, lesser multangular
Distal carpal 3	Os capitatum	Capitate, magnum
Distal carpal 4 } Distal carpal 5 }	Os hamatum	Hamate, unciform, uncinate

*Terms approved by the Eighth International Congress of Anatomists at Wiesbaden in 1965.[160]

phalanges. The primitive phalangeal formula commencing with the thumb is usually given as 2-3-4-5-3, the formula in primitive reptiles.

Modifications of the manus with few exceptions involve **reduction of elements by evolutionary loss or fusion.** Among the first elements so affected were the centralia. These elements have been reduced in number or lost altogether in most modern tetrapods. Fusion of distal carpals 4 and 5 is common and results in a **hamate** bone. Digits may be reduced in number, whereupon the corresponding metacarpals become rudimentary or lost. A less common modification is the **disproportionate lengthening or shortening of some of the elements.** Least common is an **increase in the number of phalanges.**

In modern amphibians one digit at least and one metacarpal have been reduced or lost (Fig. 9-14). However, the rudiment of the missing finger or its corresponding metacarpal remains in several species of frogs (*Rana catesbeiana* and *R. temporaria*).

A few amphibians have three fingers (*Amphiuma means tridactylum*) or two (*A. means means*). Several carpals have also been lost by fusion or deletion in amphibians. In frogs and *Necturus*, for example, the embryonic intermedium and ulnare fuse, and there are only one centrale and three distal carpals.

Since *Necturus*, along with other modern amphibians, has only four fingers, it is usually assumed that the first digit, or thumb, is missing. In accordance with this interpretation, the three distal carpals have been considered to be, commencing on the radial side (Fig. 9-15): carpal 2, carpal 3, and hamate (fused carpals 4 + 5). However, the fifth finger rather than the thumb may be missing (Fig. 9-14).[60] The three distal carpals might then represent a prepollex (an extra bone that sometimes occurs near the thumb, or pollex), carpals 1 + 2, and carpals 3 + 4. (Carpal 5 would be missing.) The theory is based on the fact that the bone labeled *P* in Fig. 9-14 has no muscle connecting it with a finger, and

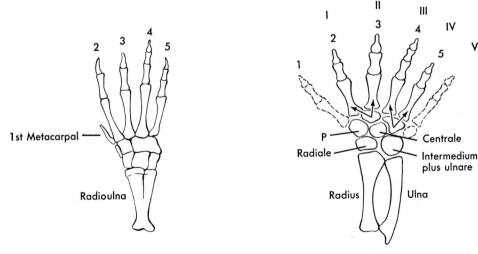

FROG   NECTURUS

Fig. 9-14. Hands of *Rana catesbeiana* and *Necturus*, dorsal views. Which finger is missing in *Necturus*? The Arabic numerals suggest that the thumb, **1,** is missing and the little finger, **5,** is present, as in many mammals. The Roman numerals suggest that the thumb, **I,** is present and the little finger, **V,** and last distal carpal and metacarpal are missing. Arrows indicate existing muscle attachments. Broken lines suggest missing elements. **P,** Prepollex or second distal carpal, depending on interpretation.

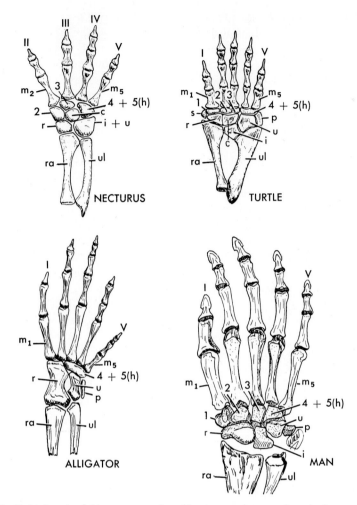

Fig. 9-15. Right hand of *Necturus*, turtle, alligator, and man, dorsal views. **c,** Centrale; **h,** hamate; **i,** intermedium; **m₁, m₂,** and **m₅,** metacarpals; **p,** pisiform; **r,** radiale; **ra,** radius; **s,** radial sesamoid; **u,** ulnare; **ul,** ulna; **1** to **5,** distal carpals; **I** to **V,** digits. Compare these examples with the basic pattern given in Fig. 9-11.

the muscle from the first and second digits attaches to the second of the three distal carpals. (The second carpal often shows double ossification centers.)

Modern reptiles, insectivores, and primates tend to remain pentadactyl and hence usually have five metacarpals and a full complement of carpals, except for centralia (Fig. 9-15). The carpal complex in alligators, however, has been reduced to four bones. Retention of an adult centrale is characteristic of numerous mammals, including beavers, hyraxes, and most monkeys. It may lie in the distal row of carpals, or, as in cats, it may fuse with the radiale and intermedium to form a bone of triple origin (scapholunar). Occasionally, human fetuses may have centralia.

Major adaptive modifications of the manus. Among modifications of the manus are those for flight, for life in the ocean, for swift-footedness, and for grasping objects. These will be discussed briefly.

Adaptations for flight. Adaptations associated with flight in birds have resulted in loss of certain bones of the manus and

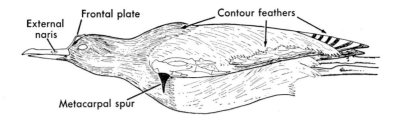

Fig. 9-16. Mexican *Jaçana,* showing the skeleton of the manus within the wing.

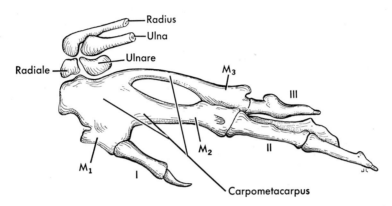

Fig. 9-17. Left manus of a bird. **I** to **III**, Digits; M_1 to M_3, metacarpals fused at their bases with three carpals to form a carpometacarpus. For position of the manus in the wing, see Fig. 9-16.

fusion of others. Pectoral muscles inserting chiefly on the humerus provide the power for flight, and feathers, often embedded in the ulna, provide the necessary air foil. The hand, therefore, plays little role in flight (Fig. 9-16). Despite this, most of the basic components of a tetrapod limb are identifiable, especially in embryos. Two carpals form in the proximal row (Fig. 9-17) and three in the distal row. As development progresses, the three distal carpals typically unite with the three metacarpals to form a rigid **carpometacarpus.** What relationship this structure bears to flight, if any, is not known. Three digits (four, in embryonic terns), inherited from reptilian precursors, are usually present and often have claws. The number of phalanges in each digit has been reduced.

Flight in extinct flying pterosaurs was associated with a different modification of the manus from that in birds. The first three digits were small hooks, perhaps for clinging to ledges. The fourth digit became tremendously elongated—as long as the length of the body—and consisted of four long phalanges, which provided the principal support for the wing membrane (**petagium**). These reptiles were evidently capable of sustained flight but were very awkward on the ground.[149]

A patagial wing also evolved in bats. The metacarpals and phalanges of the last four digits all became greatly elongated (Fig. 9-18) and incorporated in the wing membrane. The short thumb remained free and is used to suspend the body upside down from ledges. The three proximal carpal bones are fused. In flying lemurs (which do not fly, but soar) the patagium is less well developed and, although the hand is webbed, the fingers are not elongated. The formation of a patagium in such unrelated species as pterosaurs and bats is an example of evolutionary convergence.

Adaptations for life in the ocean. The

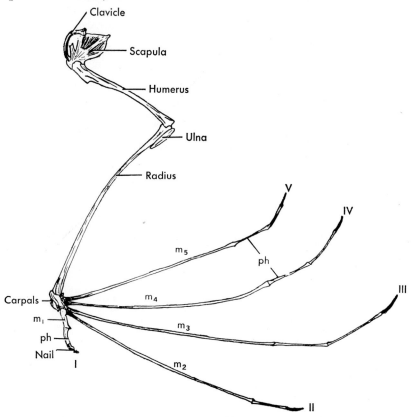

Fig. 9-18. Right pectoral girdle and anterior limb of a bat. **m₁** to **m₅**, Metacarpals; **ph**, phalanges; **I** to **V**, digits.

Ichthyosaurus

Fig. 9-19. Jurassic and Cretaceous ocean-dwelling reptile, ranging up to 10 feet or more in length. (From Colbert: Evolution of the vertebrates, New York, 1955, John Wiley & Sons, Inc.)

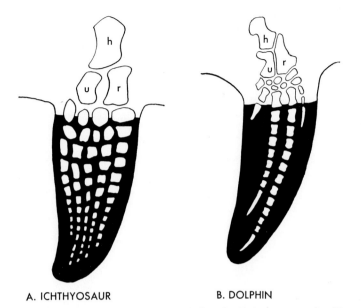

A. ICHTHYOSAUR   B. DOLPHIN

Fig. 9-20. Convergent evolution in anterior limbs. **A,** Extinct, water-dwelling reptile. **B,** Water-dwelling mammal. **h,** Humerus; **r,** radius; **u,** ulna.

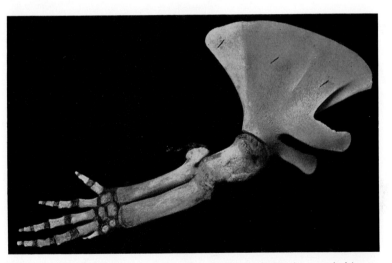

Fig. 9-21. Forelimb and pectoral girdle of a beaked whale. A remarkable resemblance to basic pattern remains, despite the fact that the limb has been modified to become a flipper. (Courtesy American Museum of Natural History, New York, N. Y.)

hands of ichthyosaurs, plesiosaurs, some sea turtles, penguins, cetaceans, sirenians, seals, and sea lions exhibit convergence as a result of adaptive modifications for life in the sea. Their anterior appendages tend to become modified as paddles. The appendage becomes flattened, short, and stout, and in several species the number of phalanges is greatly increased. In the ancient aquatic *Ichthyosaurus* (Figs. 9-19 and 9-20), which was the big reptile of the ocean, there were as many as twenty-six phalanges per digit, or over a hundred in a single "hand." Dolphins exhibit the same

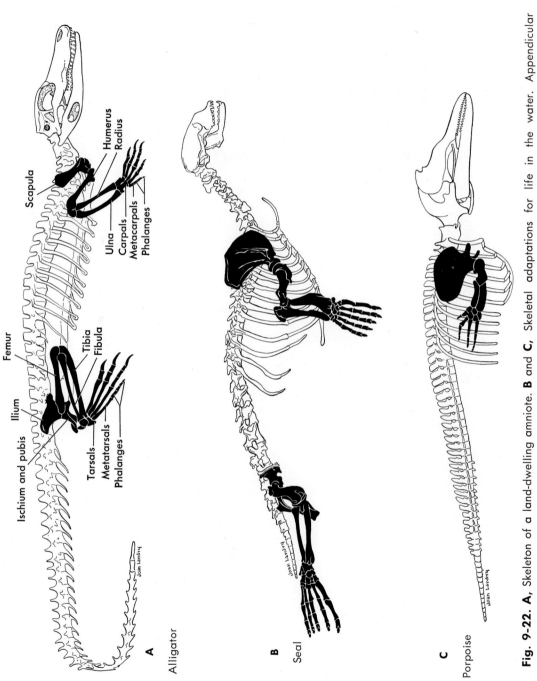

Fig. 9-22. A, Skeleton of a land-dwelling amniote. **B** and **C,** Skeletal adaptations for life in the water. Appendicular skeleton is shown in black.

kind of modification. Within the flipper of most of the other swimmers, however, the skeletal elements conform closely to the primitive tetrapod pattern (beaked whale, Fig. 9-21). Some aquatic mammals have lost all traces of hind limbs (Fig. 9-22).

Adaptations for swift-footedness. Mammals with pentadactyl hands and feet usually walk with a plantigrade (flat-footed) gait in which the palm, wrist, and digits of the manus (and the metatarsals and ankle of the foot) all tend to rest more or less flat on the ground (Fig. 9-23, monkey). This stance is a primitive characteristic among mammals. Insectivores, monkeys, apes, man, bears, and some other animals walk and run in this manner.

Mammals in which the thumb has been reduced or lost (rabbits, rodents, many carnivores, and so forth) tend to exhibit a digitigrade gait and bear their weight on the ends of their metacarpals (and metatarsals). Their digits are more or less flat on the ground and the wrist (and ankle) is well elevated (Fig. 9-23, dog). Animals exhibiting a digitigrade gait have more "spring" in their walk and usually run faster than plantigrade species. They also walk much more silently and balance well.

The extreme modification of reducing the number of digits and of walking on the remaining ones is seen in the ungulates (hoofed mammals). Ungulates walk on the very tips of their fingers (and toes) (Fig.

9-23, deer), which typically number three, two, or even one, although a few retain four digits. The claws at the ends of the digits are thickened and hardened to form tough hoofs, which bear the weight of the body. The metacarpals corresponding specifically to the missing digits are likewise reduced in size or lost along with the corresponding digits, and those metacarpals that remain are elongated and often united for added strength (Fig. 9-25, deer, camel, and horse). Hoofed animals, as a group, are the most fleet-footed of all mammals. Their digits are highly specialized for running, but the specialization also has made their fingers and toes practically useless for anything else.

Sequential evolutionary steps leading from the primitive plantigrade to the most specialized unguligrade gait may be illustrated by placing the fingers and palm flat (prone) on a table top, with the forearm perpendicular to the surface. This represents roughly the plantigrade position. Raising the palm off the table while keeping the fingers flat on the table illustrates

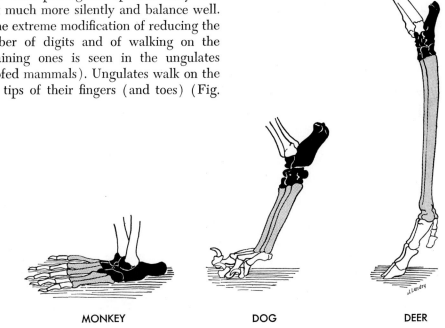

MONKEY   DOG   DEER

Fig. 9-23. Plantigrade, digitigrade, and unguligrade feet, from left to right. Ankle bones are black and metatarsals are gray.

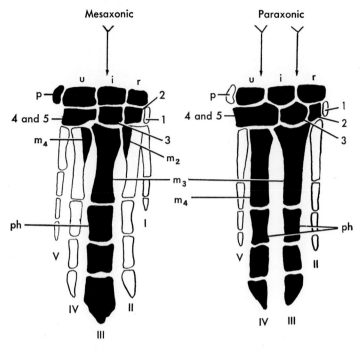

Fig. 9-24. Mesaxonic and paraxonic manus of representative ungulates, showing elements lost (white) and retained (black) and indicating distribution of body weight through wrist and digits. The horse and camel are used as specific examples, and the number of bony elements are correct for these animals. **i,** Intermedium; **m₂** to **m₄**, second, third, and fourth metacarpals; **p,** pisiform; **ph,** first phalanx; **r,** radiale; **u,** ulnare; **1** to **5,** distal carpals; **I** to **V,** digits.

roughly the digitigrade position. Unguligrade conditions may be illustrated by placing only the fingertips on the table and then raising the thumb, the little finger, the second finger, and finally the fourth finger, leaving only the third finger to bear the body weight, as in modern horses. Those fingers that fail to reach the table represent digits that have been successively reduced or lost in ungulates.

The horse underwent these successive changes commencing with the early *Eohippus,* which had four digits on the manus, and culminating in the modern *Equus,* which has a single digit. Despite extreme specialization of the manus of the modern horse, the proximal row of carpals (Fig. 9-24) is intact, and the distal row lacks only the first carpal. With the loss of

digits I, II, IV, and V, metacarpals 1 and 5 have been lost, and 2 and 4 have been reduced to splinters. Metacarpal 3, associated with digit III, is elongated.

Evolution among the ungulates seems to have progressed along two independent lines. In the line leading to artiodactyls, the weight of the body tended to be distributed equally between digits III and IV. Thus arose the "cloven" hoof (Fig. 9-26). Such a foot is said to be **paraxonic** because the body weight is borne on two parallel axes (Fig. 9-24). Artiodactyls of today have an even number of digits. In the evolutionary line leading to perissodactyls, the body weight increasingly tended to be borne on digit III, the middle digit. This is a **mesaxonic** (*mes* = middle) foot. Most perissodactyls have an odd number of

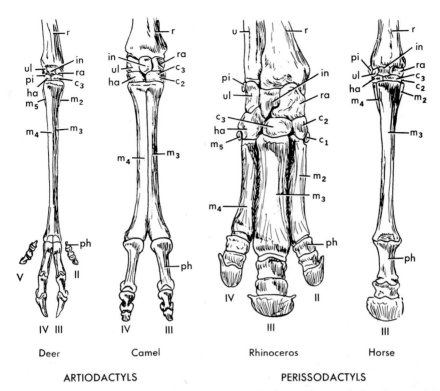

Fig. 9-25. Manus in several ungulates. c_1 to c_3, Distal carpals 1 to 3; **ha**, hamate (distal carpal 4 + 5); **in**, intermedium; m_2 to m_5, metacarpals 2 to 5; **ph**, first phalanx; **pi**, pisiform; **r**, radius; **ra**, radiale; **u**, ulna; **ul**, ulnare; **I** to **V**, digits.

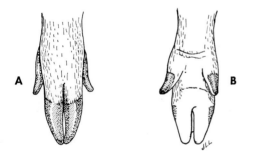

Fig. 9-26. Manus of fetal pig. There is a cleft in the foot between the digits. Thus, the hoof is "cloven." **A,** As seen from in front; **B,** as seen from behind.

digits, although some have four. It is the mesaxonic foot and not the number of digits that defines the perissodactyls.

Proboscideans give the impression of having a plantigrade gait, but they actually walk on the tips of their digits, like un-gulates. Elephants have five toes with four greatly thickened nails that are not quite hoofs. Behind the digits on the palms and soles are large pads of elastic tissue and fat. These pads assist the toes in bearing the body weight. Because of the peculiar specialization of the hand and foot, probos-cideans are referred to as subungulates. *Hyrax*, the coney, is also a subungulate.

Adaptations for grasping (Fig. 9-27). Many mammals are able to flex the hand at the joint between the palm and the base of the fingers. Rodents, for example, sit on their haunches and nibble on food held be-tween **two paws**, which are flexed in this manner and **which face one another.** A further specialization is the ability to wrap the fingers (but not the thumb) around an object such as a pencil, so that it is held securely in a **single hand.** This is accom-plished by flexing the fingers at each inter-

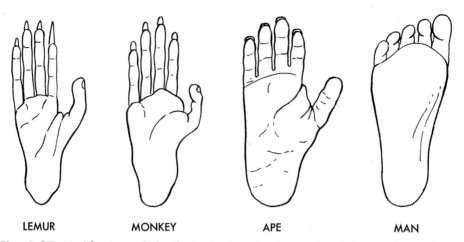

| LEMUR | MONKEY | APE | MAN |

Fig. 9-27. Modifications of the foot of selected primates. The big toe of man is not opposable.

phalangeal joint, and it is seen in primates alone. The final stage thus far in the evolution of the mammalian hand was development of an opposable thumb—one that can be made to touch the tips of each of the other digits. This was accomplished by formation of a saddle joint at the base of the thumb where it meets the palm, by setting the thumb at increasingly wider and wider angles to the index finger, and by the evolution of strong adductor pollicis (thumb) muscles. True opposability appears for the first time among living primates in Old World monkeys. Even in these forms the hand does not have the full range of functional capability that has evolved in man. Neither New World monkeys nor anthropoid apes have a perfectly opposable thumb. In evolving such a hand, man's immediate forebears and, later, man were able to fashion ever increasingly sophisticated instruments, commencing with rocks that were chipped by design and continuing to the electronic computer. Of course, his excellent brain was an essential part of his evolution; still it seems impossible that any species lacking a prehensile hand, even if it had the brain, could have evolved so sophisticated an existence. As John Napier[142] has said, "The implements of early man were as good (or as bad) as the hands that made them."

Skeleton of the posterior limbs

The bones of the hind limbs of tetrapods are comparable, segment by segment, to those of the forelimbs, although the ankle lacks an equivalent of the pisiform bone (Tables 9-1 and 9-3). A sesamoid bone, the **patella** (kneecap), protects the knee joint of birds and mammals.

Variations among femurs, tibias, and fibulas of different species are of the same nature as variations of equivalent bones of the forelimb. The fibula may unite partially or completely with the tibia (mole and frog, Fig. 9-28); it may be reduced to a splinter (parrot, Fig. 9-28), or it may be absent, as in ungulates (deer, Fig. 9-28). In birds the tibia fuses with the proximal row of tarsals to form a **tibiotarsus**, and the metatarsals fuse with the distal row of tarsals to form a **tarsometatarsus.** An ankle joint between the tibiotarsus and tarsometatarsus permits flexion at this joint. Reptiles have a similar intratarsal joint.

Digits of the foot may be more or less numerous than digits of the manus in the same animal. In most amphibians there are four fingers but five toes. Tapirs, on the contrary, have four fingers and three toes. The feet of birds have four, three, or two functional toes. Aquatic birds usually have webbed toes, as do many aquatic mammals. The great toe (hallux) is opposable in

Table 9-3. Comparison of skeletal elements of manus and pes

Manus*	Pes with synonyms	
Radiale	Tibiale	Talus or astragalus†
Intermedium	Intermedium	
Ulnare	Fibulare	Calcaneus
Pisiform		
Centralia (0 to 3)	Centralia (0 to 3)	Navicular
Distal carpal 1	Distal tarsal 1	Entocuneiform
Distal carpal 2	Distal tarsal 2	Mesocuneiform
Distal carpal 3	Distal tarsal 3	Ectocuneiform
Distal carpal 4 ⎫ Hamate	Distal tarsal 4 ⎫	Cuboid
Distal carpal 5 ⎭	Distal tarsal 5 ⎭	
Metacarpals (1 to 5)	Metatarsals (1 to 5)	
Digits (I to V)	Digits (I to V)	

*For synonyms, see Table 9-2.
†Often incorporates the intermedium.

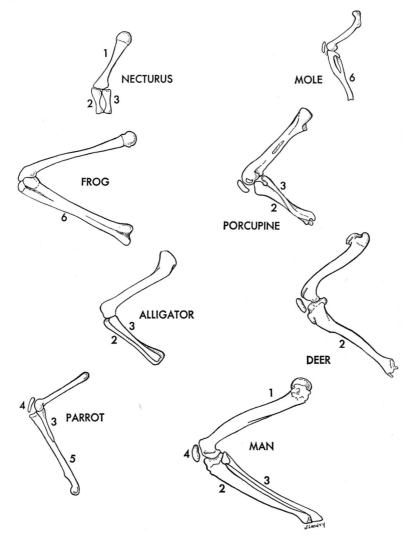

Fig. 9-28. Left thigh and shank bones of representative tetrapods, lateral views. **1,** Femur; **2,** tibia; **3,** fibula; **4,** patella; **5,** tibiotarsus; **6,** tibiofibula.

many primates, but not in man (Fig. 9-27). The primitive phalangeal formula for the foot is usually considered to be 2-3-4-5-4.

Origin of tetrapod limbs

Although the problem of the origin of paired fins may never be satisfactorily resolved, this is not true of the origin of tetrapod limbs. All evidence points to the origin of tetrapods from fishes, and so the paired fins of some Devonian fish must have been precursors of tetrapod limbs. The question then arises: Did the fin of any known Devonian fish evince the potentiality of evolving into a limb? For an answer we turn to the lobe-finned crossopterygian fishes, which resembled the first tetrapods in many respects.

The skeleton in the basal lobe of a crossopterygian fin bears a striking resemblance to that in a tetrapod limb (Fig. 9-29). In crossopterygians a single bone (we will call it a "humerus") articulates proximally with the scapula and distally with a pair of bones (we will call them "radius" and "ulna"). Loss of the dermal fin rays and relatively minor modifications of the fin distal to the radius and ulna could have produced the basic tetrapod manus.

It is possible that crossopterygian fins were employed during the Devonian period for propping the fish in a resting state on the bottom of the pond, without bearing any appreciable weight.[136] Minor modifications would have permitted locomotion on the fins along the floor of the river or pond, including close to shore. Finally, during the Carboniferous period[26] some descendants of these fishes must have ventured onto land—for short intervals at first and then for longer and longer visits. Because of the open duct to the air bladder, they had long been competent to gulp air. Several hundred species of modern fishes "walk" on their fins along the muddy or sandy bottoms. The Australian lungfish is one of these.[135] Some living fishes move several feet inland nightly.[155] Some climb inclined planes with little fins that have a remarkable resemblance to hands.[138]

The pressures that drove vertebrates onto the land must, of necessity, be conjec-

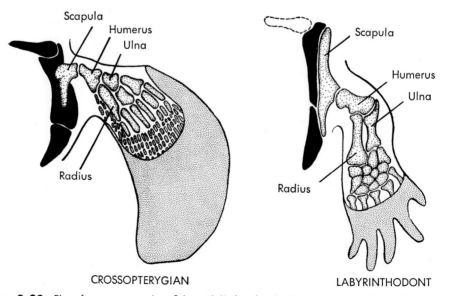

CROSSOPTERYGIAN   LABYRINTHODONT

Fig. 9-29. Fin of crossopterygian fish and limb of primitive tetrapod, oriented to show similarity of skeleton. Dermal bones are black and replacement bones or cartilage are stippled. Dotted line represents missing element. The ulna and radius of the crossopterygian are part of the radial series.

tural. It may be that there were fewer predators on land, or that there was less competition for food, or simply that food was abundant on land. It may have been the drive to abandon a pond or river that was receding or in which oxygen became low. Or perhaps it was simply a manifestation of the tendency of organisms to invade a contiguous environment whenever nothing prevents the invasion. Whatever the explanation for the early forays onto land, it seems evident that tetrapods evolved from crossopterygian-like fishes whose appendicular skeleton clearly had the potential to evolve into tetrapod appendages.

Chapter summary

1. The appendicular skeleton includes the cartilages and bones of the pectoral and pelvic girdles and of the paired and unpaired fins and limbs.

2. The pectoral girdles of all vertebrates are built in accordance with a single architectural pattern. Certain components belong to the primary endoskeleton and consist of cartilage or replacement bone. Other components arise by intramembranous ossification and are considered phylogenetic derivatives of dermal armor.

3. Coracoids, scapulas, and suprascapulas arise as cartilages, which may be partly or wholly replaced by bone. Suprascapulas are confined to fishes and amphibians. Coracoids are absent above monotremes. A scapula is almost universally present.

4. Clavicles, cleithra, and supracleithra arise by intramembranous ossification and were probably derived from dermal armor. They do not develop in cartilaginous fish. Cleithra and supracleithra are confined to fishes; they have disappeared in early tetrapods. Clavicles occur in most vertebrate classes. They are best developed in tetrapods. A midventral interclavicle (episternum) of membranous origin is essentially reptilian but was transmitted to birds, as part of the furcula, and to monotremes.

5. The pelvic girdle of fishes consists of two pelvic plates, which meet ventrally in a symphysis and articulate laterally with a pelvic fin. The plates unite to form a median ischiopubic bar in sharks and lungfishes. In tetrapods two ossification centers arise in each pelvic plate and form ischial and pubic bones. A third element, the ilium, develops on each side in tetrapods and braces the girdle against the vertebral column. There has been no contribution of dermal armor to the pelvic girdle.

6. Paired fins are of three general types: (a) biserial (lobed) fins (Sarcopterygii) with a central jointed axis and a series of preaxial and postaxial radials; (b) fin fold fins (Chondrichthyes) with a proximal row of basals and distal rows of radials; and (c) ray fins (Actinopterygii), reduced to a few short radials embedded in the body wall (or none at all) and composed chiefly of dermal fin rays. Median fins (dorsal, caudal, and anal) have a skeleton of radials and rays.

7. All tetrapod limbs are built in accordance with a pattern seen in early labyrinthodonts. Tetrapod limbs exhibit a basal segment with a humerus or femur, an intermediate segment with a radius and ulna or tibia and fibula, and a distal segment, the manus (carpals, metacarpals, and phalanges) or pes (tarsals, metatarsals, and phalanges).

8. Adaptive modifications of tetrapod limbs involve chiefly reduction in the number of bones of the hand or foot. Digits have been reduced to four (typically) in modern amphibians, three in birds, and as few as one in some ungulates. Reduction of digits is usually accompanied by reduction of carpals and metacarpals, tarsals and metatarsals. The most striking modifications are found in flying tetrapods (pterosaurs, birds, and bats), water-adapted reptiles and mammals (ichthyosaurs, cetaceans, and sirenians), and in hoofed mammals (ungulates). A few species lack anterior appendages, posterior appendages, or both.

9. Excepting sesamoids, the bones of the fins and limbs (distal to the girdles) arise by endoskeletal ossification.

10. Reliable clues to the origin of paired fins have been obscured by time. They may have evolved from a series of fin spines such as those seen in fossil acanthodians. Once established, a biserial fin may have given rise to other fin types. It is likely that paired limbs of tetrapods arose from lobed fins resembling those of crossopterygians.

Chapter

10

Muscles

Muscles may be classified in accordance with any number of criteria—location, structure, function, embryonic origin, shape, or any other useful criterion. The criterion employed at any time is dictated by its appropriateness for the moment.

Skeletal muscles attach to the skeleton. They are either axial (arising and inserting on the skull and vertebral column), appendicular (inserting on the appendicular skeleton), or branchiomeric (attaching to the visceral skeleton). Muscles not attached to the skeleton must be referred to as **nonskeletal**, since there is no other antonym.

When examined histologically, muscles may be either striated or unstriated (smooth). **Striated muscles** are usually definite organs with long, multinucleate fibers that exhibit cross striations resulting from the patterned arrangement of muscle proteins. **Smooth muscles** usually occur in sheets as part of another organ (Fig. 10-1). The cells are spindle-shaped, uninucleate, and lack striations.

Muscles may be either voluntary or involuntary according to whether or not they are subject to volitional control. **Voluntary muscles** contract at will if they are not fatigued. This does not preclude reflex operation of voluntary muscles if the body is endangered, as when the skin is brought in contact with a painful stimulus. **Involuntary muscles** cannot usually be contracted at will.

Muscles may be classified as either somatic or visceral. **Somatic muscles** enable the animal to orient itself in the environment and thus to move toward an environment conducive to its welfare (shelter, food, or a mate) and to move away from an inimical environment. Somatic muscles are locomotor, or derived from locomotor muscles. Somatic muscles arise from the myotomes of the mesodermal somites either directly during embryonic development or phylogenetically (that is, historically) and are therefore also called **myotomal** muscles. They are striated, skeletal, and voluntary.

Visceral muscles assist in regulating the internal environment. Among the visceral muscles are those of the blood vessels, tubes, and ducts, those elevating the hairs for temperature control, and those of the alimentary canal including the branchiomeric muscles attached to the visceral skeleton. Most visceral muscles are smooth, nonskeletal, and involuntary, but branchiomeric muscles are an exception.

Muscle cells function by shortening when stimulated. The shortening is a result of a chemical change in a muscle protein complex known as **actomyosin**, which makes up about 80% of the muscle cell. Shortening of a sufficient number of muscle cells will cause a corresponding shortening of the muscle. If the muscle surrounds a lumen, the lumen will be com-

213

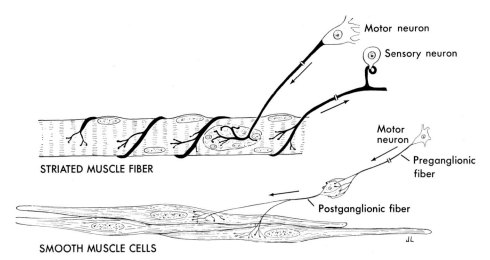

Fig. 10-1. Muscle fibers and their innervation. The striated fiber is from a muscle spindle and has a sensory innervation for proprioception. Arrows indicate direction of nerve impulse.

pressed. If the muscle extends between two structures, one (the site of **insertion**) will be drawn toward the other (the **origin**). A muscle may, however, move the part on which it originates. For example, the geniohyoid muscle attached to the hyoid bone and to the mandible can either pull the lower jaw down or draw the hyoid forward, depending on whether the jaw or the hyoid is more firmly fixed in position at the time.

It is exceptional for a single skeletal muscle to act independently of other muscles. Instead, muscles nearly always cooperate in functional groups. While one functional group is contracting, another group must relax simultaneously, or a stalemate would result. In order that muscles may act cooperatively (synergistically), they must be reflexly directed by the nervous system, just as the musicians in a symphony orchestra must be directed by a conductor. The reflex control comes chiefly from the cerebellum, which sends motor impulses to the appropriate muscles after receiving incoming information. The information is in the form of feedback over sensory pathways from the muscles being directed, from the tendons of the muscles, and from the bursas of affected joints. The

sensory phenomenon is known as proprioception.

A connective tissue sheath, the muscle **fascia,** surrounds each muscle. The fascia around the muscle merges with connective tissue from within the muscle to become the **tendon** of origin or insertion. Broad, flat muscles usually have broad tendons. Such a tendon is an **aponeurosis.** The **lumbodorsal aponeurosis** consists of several layers of such tendons.

In this chapter we will be concerned chiefly with those muscles that attach to the skeleton, namely (1) those that operate the head, trunk, tail, and limbs and thereby provide for locomotion and orientation of the animal in the environment (somatic or myotomal muscles) and (2) those that operate the visceral skeleton and thereby regulate feeding and respiratory movements (branchiomeric muscles). These two groups of muscles show many adaptive modifications. A short discussion of **integumentary muscles** will also be included.

SOMATIC (MYOTOMAL) MUSCLES

Somatic muscles are historically metameric. The metamerism is most evident in lower vertebrates, and although it tends to

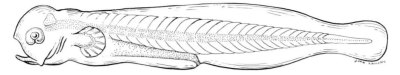

Fig. 10-2. Body wall muscle of larval teleost. Position of notochord is indicated by stipple commencing just behind the eye and extending caudad into the tail.

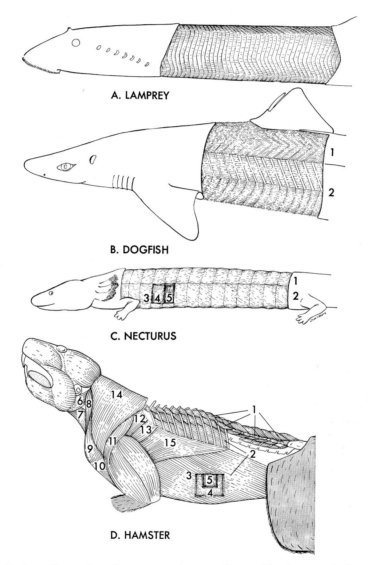

A. LAMPREY

B. DOGFISH

C. NECTURUS

D. HAMSTER

Fig. 10-3. Body wall muscles of representative vertebrates. The lamprey is *Petromyzon*. **1,** Epaxials; **2,** hypaxials; **3,** external oblique; **4,** internal oblique; **5,** transversus. The epaxial muscles shown in the hamster are intervertebral muscles. Additional muscles of the hamster are **6,** digastric; **7,** sternohyoid; **8,** clavotrapezius; **9,** levator scapulae ventralis; **10,** acromiodeltoid; **11,** spinodeltoid; **12,** infraspinatus; **13,** teres major; **14,** acromiotrapezius; **15,** latissimus dorsi.

be obscured in higher tetrapods, it is never obliterated. The metamerism is directly attributable to the fact that the somatic muscles are derived ontogenetically (in fishes) or phylogenetically (in tetrapods) from segmental mesodermal somites (Figs. 4-7 and 10-6). Mesenchyme from the myotomes of successive somites migrates into the lateral body wall to contribute to the blastema from which the muscle will develop (Fig. 4-9). Somatic muscles include muscles of the body wall, hypobranchial and tongue muscles, extrinsic eyeball muscles, and muscles of the appendages.

Body wall muscles of fishes

Metamerism is clearly evident in the muscles of the trunk and tail of fishes and urodeles (Figs. 10-2 and 10-3). The body wall musculature consists of a series of **myomeres** (muscle segments) separated by connective tissue **myosepta.** The myosepta extend from the parietal peritoneum to the skin, and the muscle fibers of each myomere typically arise on one myoseptum or on the rib within the myoseptum and insert on the next. The myosepta are often bent into zigzag shape, thereby providing additional tendinous areas for attachment of the muscle fibers.

Except in agnathans, the musculature is further divided into dorsal and ventral masses by a **horizontal septum.** This septum extends outward from the transverse processes of the vertebrae to the skin and forms a fibrous membrane, above which are the **epaxial** muscles and below which are the **hypaxial** muscles. Vertical **middorsal** and **midventral septa** separate the muscles of the two sides of the body. At the base of the midventral septum is a longitudinal ligament, the **linea alba.**

The metamerism of the hypaxial muscles is interrupted where the pectoral and pelvic girdles and fins are built into the body wall. It is also interrupted by the pharynx, since myotomal muscles do not invade the visceral arches. However, hypaxial muscles form above and below the gill chambers (**epibranchial** and **hypobranchial** muscles).

Although the bulk of the hypaxial muscle in fishes is directed horizontally from one myoseptum to the next, there is some reorientation and stratification of body wall muscles. A thin sheet of oblique fibers lies superficial to the main hypaxial mass ventrolaterally. The fibers extend ventrocaudad. Farther ventrad, on either side of the linea alba, a narrow ribbon of still more superficial fibers extends cephalocaudad. Teleosts have more of the oblique fibers than sharks, and in these fishes the cephalocaudad fibers may form a definite bundle, the **rectus abdominis.**

In agnathans there are no epaxial and hypaxial subdivisions (Fig. 10-3, *A*). This may be related to the absence of vertebrae and hence of transverse processes. Hagfishes have oblique and rectus masses, but the oblique muscles are unsegmented. In *Lampetra* all body wall muscle is unsegmented, and there are no obliques. It is not known whether the condition in *Lampetra* is primitive.

Each myomere in fishes arises from mesenchyme that invades the body wall from the somite of that body segment. The corresponding spinal nerve grows into the myomere. Thus, body wall muscles of fishes are myotomal by ontogenetic origin. In higher vertebrates some of the body wall musculature arises in situ rather than as somitic contributions. However, these muscles are clearly myotomal by phylogenetic derivation, as evidenced by their tendency to metamerism and their metameric innervation by successive spinal nerves. In all vertebrates epaxial muscle is typically innervated by the dorsal ramus of the spinal nerve supplying that body segment, and hypaxial muscle is innervated by the ventral ramus (Fig. 1-2).

The chief function of the body wall muscles of fishes is locomotion. Waves of contraction sweep along the body wall from the anterior end of the trunk to the tip of the tail and result in sinuous swimming movements (Fig. 1-3).

Body wall muscles of tetrapods

Tetrapods, like fish, exhibit epaxial and hypaxial masses, and all retain considerable muscle metamerism. Certain adaptive modifications have developed in the body wall musculature in association with life on land as follows:

1. The stratification of the hypaxial musculature into sheets is generally more pronounced in tetrapods. Typically, there are external oblique, internal oblique, transverse, and rectus masses.

2. The hypaxial muscles tend to lose their metamerism and to form broad sheets. This tendency results from fusion of successive myomeres with partial or extensive loss of myosepta. The tendency is least pronounced in tailed amphibians and most pronounced in birds and mammals. The resulting wide muscle sheet is innervated by a series of successive spinal nerves, which reveal its phylogenetic metamerism.

3. The epaxial myomeres **tend to coalesce** to form elongated bundles extending through many body segments, and they **tend to become buried** under the expanding extrinsic appendicular muscles required to operate the limbs. The deepest epaxial musculature still retains its primitive metamerism, however.

Epaxial muscles of tetrapods. Epaxial muscles lie along the vertebral column dorsal to the transverse processes and lateral to the neural arches. They extend from the base of the skull to the tip of the tail. In urodeles (Fig. 10-3, *C*) and in primitive lizards (Fig. 10-4) these epaxial muscles (**dorsalis trunci**) are obviously metameric.

In other tetrapods the superficial bundles form long muscles occupying several or many body segments, whereas the deepest bundles remain segmental. The longest bundles are the longissimus, iliocostalis, and transversospinalis groups. The shortest bundles are the intervertebral muscles connecting one vertebra with the next.

The **longissimus group** (Fig. 10-5) lies upon the transverse processes of the vertebrae and is so named because it includes the longest epaxial bundles. (The suffix *issimus* is a Latin ending meaning "the most.") Regional- subdivisions include the longissimus dorsi in the region of the trunk (dorsal) vertebrae, longissimus cervicis in the neck, and longissimus capitis inserting on the skull. The longissimus muscles are continued into the tail as the extensor caudae lateralis.

The **iliocostalis group**, located lateral to the longissimus, arises on the ilium, and the bundles pass forward to insert on the ribs near their dorsal ends or on the uncinate processes when present. Anterior extensions occur as far forward as the neck region. The muscles lie just above where the horizontal septum would be were it present in amniotes.

The **transversospinalis group** (spinalis dorsi, spinalis cervicis, spinalis capitis, and semispinales) includes the long or medium-length bundles lying close to the vertebrae and passing forward from neural spines and arches to insert several or many segments anteriorly or on the skull. These bundles are continued into the tail as the extensor caudae medialis.

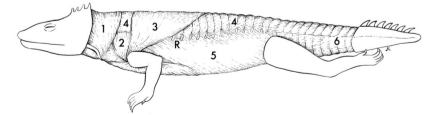

Fig. 10-4. Selected superficial muscles of *Sphenodon*. **1,** Trapezius; **2,** dorsalis scapulae; **3,** latissimus dorsi; **4,** dorsalis trunci (epaxial); **5,** external oblique of the abdomen (hypaxial); **6,** hypaxial muscle of tail; **R,** rib. (Modified from Byerly.[55])

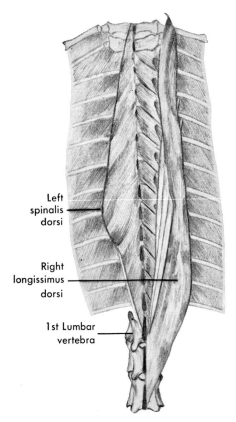

Left
spinalis
dorsi

Right
longissimus
dorsi

1st Lumbar
vertebra

Fig. 10-5. Two long epaxial bundles of a hamster, dorsal view. The longissimus dorsi has been removed on the left and the spinalis dorsi on the right. To reveal the segmental epaxial muscles shown in Fig. 10-3, D, the long bundles must be removed.

The **intervertebral muscles** (Fig. 10-3, D) have retained the primitive segmental condition of fishes and tailed amphibians. They extend chiefly from the transverse process of one vertebra to the transverse process of the next anterior vertebra (intertransversarii), beteween successive neural spines (interspinales), between successive neural arches (interarcuales), and between successive zygapophyses (interarticulares).

The foregoing epaxial muscles and many others occur in amniotes. Precise homologies between similarly named bundles in different classes have not been established in all cases. Most of the epaxial muscles are hidden by the overlying extrinsic ap-pendicular muscles of the forelimb or by the lumbodorsal aponeurosis.

Epaxial muscles in tetrapods perform the same primary function as in fishes—side-to-side and dorsoventral flexion of the vertebral column. The extent to which they are able to function depends on the flexibility of the column. In turtles and birds the vertebral column of the trunk is rigid and the associated epaxial muscles are poorly developed.* In the neck and tail they function just as in other vertebrates. Because the anteriormost longitudinal bundles attach to the skull, epaxial muscles participate in movements of the head. In a chick one of these muscles, the complexus, constitutes 2% of the body weight just before hatching. It is the chief muscle used in moving the head back and forth while trying to crack the eggshell. For this reason it is sometimes called the "hatching" muscle. It is found also in some mammals.

Hypaxial muscles of tetrapods. The hypaxial muscles of tetrapods are typically disposed in layers: an external oblique, an internal oblique, and a transverse sheet (Fig. 10-3, C and D). In addition, on either side of the linea alba is a narrow rectus muscle in which the fibers are directed longitudinally. One or more of these sheets may be split into subsidiary sheets or, on the contrary, may be greatly reduced or missing. The following examples will illustrate. In some urodeles the external oblique sheet is represented by a superficial and deep portion; in *Sphenodon*, crocodiles, and some lizards all three layers may consist of two sheets each. In anurans the internal oblique may be missing. In birds both oblique sheets are greatly reduced, and the transverse sheet may be missing. In mammals the transverse

---

*There is no scientific evidence that the muscles fail to develop because they are not needed. That they were not being used made it possible for the muscles to be poorly developed without disadvantage to the species, **whatever the cause.** Failure to develop is a mutation affecting the necessary embryonic inductors.

sheet is often incomplete in the thoracic region. In turtles, body wall muscles are almost completely absent, a condition related to the presence of the shell.*

The hypaxial muscles of urodeles and some reptiles retain myosepta the length of the body and hence are obviously metameric. But in most amniotes adjacent myomeres have tended to unite into broad sheets of muscle, except where there are long ribs. Thus the external and internal obliques of the abdomen are broad sheets without myosepta, whereas the external and internal intercostals of the thorax are metamerically disposed between ribs. The transverse sheet, on the other hand, is represented in the abdomen by the transversus abdominis and in the thorax by the transversus thoracis, both of which are broad sheets since the thoracic portion lies internal to the ribs. Precise homologies between the various sheets in different vertebrate groups have not been unequivocally determined.

Numerous muscle bundles other than intercostals become associated with ribs, especially in reptiles and mammals. These include the scalenes, the serratus dorsalis, the levatores costarum, and the transversus costarum. Whether the scalene muscles of reptiles are homologous with muscles of the same name in mammals and birds is not known; if they are, all scalene muscles may be epaxial derivatives, as is probably also the serratus dorsalis. The other muscles attaching to ribs are probably hypaxial derivatives. The innervation of muscles is one clue to their

phylogeny but probably not an infallible one.

The **mammalian diaphragm** is a muscular organ resulting from invasion of the septum transversum by hypaxial muscles that have their embryonic origin from myotomes at the level of cervical spinal nerves C4, C5, and C6. The ventral rami of these nerves unite to form the phrenic nerve, which passes down the neck and beyond the heart to reach the diaphragm. The muscle is attached to the ribs, sternum, and vertebrae and inserts on a central tendon.

The **cremaster muscle** (Fig. 14-18), wrapped around the spermatic cord of mammals, is hypaxial muscle. As the cord penetrates the internal inguinal ring, it gains an investment of oblique muscle slips, which continue for some distance within the inguinal canal. The muscle is used for retracting the testes in mammals retaining an open inguinal canal (rodents, rabbits, bats and so forth).

A **pyramidalis muscle** found in association with the marsupial pouch represents a slip of the rectus abdominis. Some higher mammals exhibit a reduced, presumably vestigial pyramidalis muscle. A small percentage of human beings exhibit, as an anomaly, a muscle that may or may not be homologous.

A column of longitudinal hypaxial muscles, known as **hyposkeletal**, or **subvertebral**, muscles, lies immediately underneath and against the transverse processes of the vertebrae. The body cavity must be opened to demonstrate these muscles. In the lumbar region these include the psoas, iliacus, and iliopsoas and in the neck, the longus colli and others. In the thorax the hyposkeletal muscles are usually reduced. They assist the epaxial muscles in movements of the vertebral column. Hyposkeletal muscles are sold commercially as tenderloin and filet mignon.

The hypaxial muscles of aquatic urodeles are used for swimming, just as in fishes, and even on land many predominantly aquatic urodeles use these muscles for wig-

---

*When dissecting adult mammals, students notice that the body wall muscles vary in thickness in the two sexes. This is especially evident when the transversus abdominis is compared in a tomcat and in a female. In females the transversus is sometimes so thin as to present considerable difficulty in separating it from the internal oblique muscle without tearing both. Testosterone, the predominating gonadal hormone in males, is anabolic on protein and thus causes amino acids to be linked together into polypeptides and proteins. Since muscle is 80% protein, testesterone will result in larger muscles in adult males.

gling along, since the limbs are scarcely suitable for locomotion. Limbless tetrapods (Apoda, snakes, and legless lizards) also use these muscles for locomotion. However, in the majority of tetrapods the muscles of the body wall are used chiefly to compress the viscera and to operate the ribs for respiration.

Hypobranchial and tongue muscles

Myomotal muscle does not invade the **lateral** pharyngeal wall. However, mesenchyme from the hypaxial region immediately behind the pharynx streams forward **beneath** the pharynx to organize somatic muscles as far cephalad as the mandibular symphysis (Fig. 10-6). These are referred to as hypobranchial muscles.

In fishes the hypobranchial muscles extend forward from the coracoid bar to insert on the mandible, on the ventral hyoid elements, and on the remaining visceral skeleton in the pharyngeal floor. The muscles are referred to collectively as the **coracoarcual** muscles, and bear names such as coracomandibular (geniocoracoid), coracohyoid, and coracobranchial muscles. They strengthen the floor of the pharynx and pericardial cavity and aid the branchiomeric muscles in elevating the floor of the mouth, lowering the jaw, and expanding the gill pouches.

In tetrapods the hypobranchial muscles extend forward from the sternum and first ribs. They retain their primitive attachments to the mandible and visceral skeleton, but their function mirrors the altered roles of the branchial skeleton, particularly the hyoid apparatus and larynx. They include the sternohyoid, sternothyroid, thyrohyoid, omohyoid, and geniohyoid muscles.

The tongue of amniotes is essentially a mucosal sac stuffed with hypobranchial muscle. The premuscle mesenchyme migrates into the developing tongue from the blastema of the hypobranchial muscles. This explains the fact that in bats some of the tongue muscles extend all the way from the sternum. The chief tongue muscles in mammals are the hyoglossus, styloglossus, genioglossus, and lingualis. The lingualis is an intrinsic mass of crisscrossing fibers. The tongue of vertebrates below amniotes has no intrinsic musculature.

Because of their origin, ontogenetically or phylogenetically, from anterior trunk somites, the hypobranchial muscles are supplied by anterior spinal nerves. These nerves in fishes arch around the caudal end of the pharynx and proceed forward under the pharynx to their destination. Their course is the pathway of migration of the myotomal mesenchyme from which the muscles arise. The nerve to the tongue muscles also arches downward and forward (Fig. 15-22). The nerve fibers supplying the hypobranchial and tongue muscles have their central nervous system origin from the same somatic motor column that gives rise to fibers supplying the muscles of the trunk and tail (Fig. 10-6).

Table 10-1. Muscles derived phylogenetically from head somites and their innervation

Head somite	Cranial nerve supply	Eyeball muscle derivative	Other derivatives
I	III (oculomotor)	Superior rectus Inferior rectus* Medial (internal) rectus Inferior oblique	Levator palpebrae superioris
II	IV (trochlear)	Superior oblique	
III (with contributions from somite II)	VI (abducens)	External (lateral) rectus	Retractor bulbi Quadratus of eye Pyramidalis of eye

*In lampreys the inferior rectus is operated by cranial nerve VI. In vertebrates with a degenerate eyeball, the muscles are correspondingly imperfect.

Extrinsic eyeball muscles

The muscles that operate the eyeball of vertebrates are striated, skeletal, and voluntary. In elasmobranch fishes they arise during embryonic development from three preotic somites of the head (Fig. 10-6). Head somite I lies at the site where cranial nerve III emerges from the brain.

This somite splits longitudinally to give rise to the four eyeball muscles operated by nerve III (Table 10-1). Head somite II lies at the site where cranial nerve IV emerges from the brain. This somite gives rise to the superior oblique eyeball muscle innervated by cranial nerve IV. Head somite III (with contributions from somite

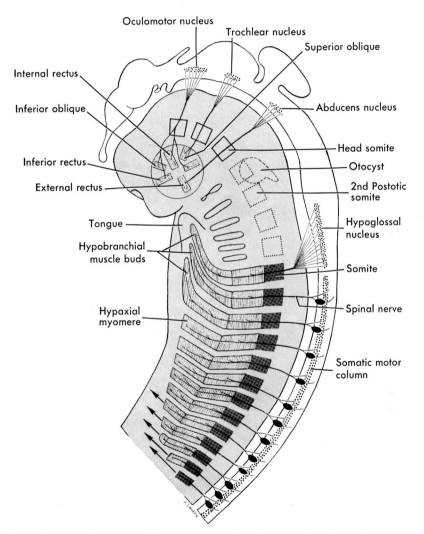

Fig. 10-6. Myotomal muscle origins and innervation in a shark embryo, diagrammatical. Muscles include eyeball muscles from three preotic head somites, segmental muscle of the trunk from trunk somites (supplied by corresponding segmental nerves), and hypobranchial musculature migrating forward into floor of pharynx and accompanied by nerve supply. The motor fibers innervating myotomal muscle have their cell bodies in the somatic motor column of the cord and brain (black dots). The four postotic somites make no contribution to the musculature of the body, and so they are shown in dotted outline. The central nervous system has been projected above the embryo for clarity.

II) lies at the site where cranial nerve VI emerges. The somite gives rise to the external (lateral) rectus muscle supplied by nerve VI. In tetrapods the head somites are evanescent, and the eyeball muscles form from mesenchyme that cannot be traced directly to the somites. Nevertheless, tetrapod eyeball muscles are obviously homologous with those of elasmobranchs. The eyeball of hagfishes is vestigial and eyeball muscles do not form.

Further evidence of the myotomal nature of the eyeball muscles is the fact that the cell bodies of the motor fibers innervating these muscles are in the somatic motor column of the central nervous system (Fig. 10-6). This column also supplies fibers to myotomal muscles elsewhere in the body—the epaxial and hypaxial muscles of the trunk and tail, the appendicular muscles, the hypobranchial muscles, and the tongue muscles. Therefore, it can be said with confidence that the eyeball muscles of all vertebrates are myotomal either by embryonic origin or by phylogenetic derivation.

In addition to two oblique and four rectus muscles, most tetrapods have a retractor bulbi, which draws the eyeball deep into the orbit, and many amniotes have muscles inserting on the lids and nictitating membrane (pyramidalis of reptiles and birds, quadratus of birds, and levator palpebrae superioris of reptiles and mammals). Innervation of these muscles by somatic nerves (Table 10-1) indicates that these, too, are myotomal derivatives.

Appendicular muscles

Appendicular muscles attach to the girdles and fins or limbs. **Extrinsic** appendicular muscles arise on the axial skeleton or on fascia of the trunk and insert on the girdles and limbs. **Intrinsic** appendicular muscles arise on the girdle or on proximal skeletal elements of the appendage and insert on more distal elements. All extrinsic muscles must be completely severed in order to permit removal of an appendage from the body; in the process, intrinsic

muscles remain intact. Appendicular muscles are striated, skeletal, voluntary, and myotomal. In fishes they usually contribute little to locomotion.

Morphogenesis. Paired fins and limbs first appear in vertebrate embryos as elevated folds or buds protruding from the lateral body wall (Fig. 10-7). Into the fin folds of fishes migrate buds of hypaxial muscle from several body segments. These invading buds differentiate to become the appendicular muscles. Other hypaxial muscles attach to the girdles.

In most tetrapods the appendicular musculature does not arise by invasion of the appendage by muscle buds from the body wall. Instead, the muscles arise from blastemas that organize intrinsically within the developing limb bud. These blastemas give rise not only to the intrinsic musculature of the limb, but also to some of the extrinsic muscles when buds from within the appendage spread trunkward to achieve an attachment on the axial skeleton. Muscles that arise from blastemas within the limb bud are sometimes called **primary appendicular muscles**. The latissimus dorsi is such a muscle.

Other extrinsic muscles of the tetrapod appendage (**secondary appendicular muscles**) develop from blastemas in the body wall. The pectoral muscles are hypaxial muscles that arose in this way. The rhomboideus and cucullaris (trapezius) also arose this way. The former is thought to represent epaxial musculature (although the innervation does not confirm this), and the latter is probably a contribution from the gill arches.

Invading the fin or limb bud are branches of the ventral rami of spinal nerves. In fishes these nerves belong to the same body segments that contribute the muscle buds. It seems probable that the number of spinal nerves entering each limb in tetrapods is indicative of the number of mesodermal somites that contributed the musculature phylogenetically.

Appendicular muscles of fishes. Hypaxial muscle bundles from several successive

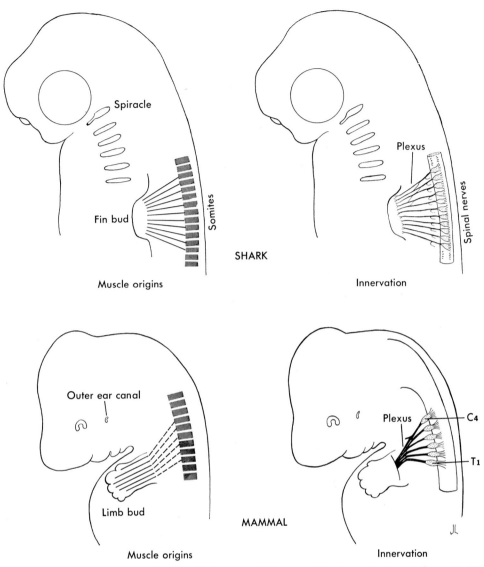

Fig. 10-7. Direct origin of appendicular muscle from somites in a shark and theoretical phylogenetic origin of appendicular muscle of mammal from specific somites based on innervation. Solid lines in shark indicate direct developmental relationship. Dotted lines in mammal indicate probable phylogenetic relationship. C_4 and T_1 are the fourth cervical and first thoracic spinal nerves of the brachial plexus of the mammal.

myomeres attach to the girdles and to the skeleton at the base of the fins of fishes. Those at the base of the fin form two opposing muscle masses, dorsal **extensors** and ventral **flexors** (Fig. 10-8, A). Specialized bundles other than occasional rotators are uncommon and muscles located entirely **within** the fin (intrinsic muscles) are almost nonexistent. The movements of the

fins are as uncomplicated as the muscles that move them.

A few fishes, such as sunfishes, swim entirely with paired fins, and the pelvic fins are attached just behind the pharynx. In these species the appendicular muscles are well developed and trunk and tail muscles are almost nonexistent.

Appendicular muscles of tetrapods. The

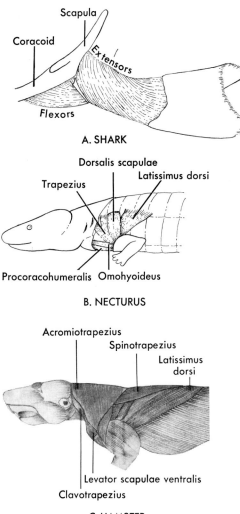

A. SHARK

B. NECTURUS

C. HAMSTER

Fig. 10-8. Increased specialization of appendicular muscles above fishes. Extensors in fish are also called levators. Flexors are adductors. Compare *Sphenodon*, Fig. 10-4.

appendicular muscles of even the lowest tetrapods are far more complicated than those of fishes. The greater leverage required for locomotion on land necessitates extension of the extrinsic muscles over a greater area of the trunk. Also, the jointed appendage requires a much larger number of intrinsic muscles for independent operation of the segments.

The extrinsic musculature of the anterior limb of tetrapods is disposed in dorsal and ventral groups. The dorsal group of the

forelimbs attaches to the posterior part of the skull and to the vertebral column or fascia of the trunk to a point well caudal to the scapula. The fan-shaped muscles converge to insert chiefly on the scapula or proximal end of the humerus (Fig. 10-8, *Necturus* and hamster). The trapezius, sternomastoid, and cleidomastoid are represented in fishes by the cucullaris. Unlike other appendicular muscles, the cucullaris is thought to be derived from branchomeric muscles of posterior gill arches.

The ventral extrinsic muscles of the anterior limb attach to ventral skeletal elements (sternum, coracoid, and gastralia) and converge to insert primarily on the humerus. Chief among ventral extrinsic muscles are the pectorals.

The extrinsic muscles of the posterior limbs play a lesser role in locomotion than do counterparts in the shoulder because the pelvic girdle of tetrapods is usually immovably ankylosed to the vertebral column. Most of the locomotor muscles of the hind limb are intrinsic. A recitation of the posterior appendicular muscles would yield little additional insight into vertebrate phylogeny.

The intrinsic appendicular musculature of tetrapods operates the various segments of the appendages. In the generalized plan the intrinsic appendicular muscles are divisible into dorsal and ventral masses, but with rotation of the appendages and subdivision of the bundles, these primitive relationships became obscured. In Table 10-2 are listed the chief intrinsic muscles of the anterior limbs. Precise homologies between the intrinsic muscles of lower tetrapods and those of mammals have not been clarified.

The extrinsic appendicular muscles of amphibians are relatively simple; they converge on the limbs as fan-shaped muscles dorsally and ventrally. The dorsal muscles of the anterior limb attach to fascia on the dorsolateral wall of the trunk, thus failing to reach the neural spines, and they extend a relatively short distance posteriorly. The dorsal group of the posterior limbs is

Table 10-2. Chief intrinsic muscles of the forelimbs of tetrapods

Muscles of girdle
Girdle to humerus, proximally
Deltoideus (dorsalis scapulae)
Subcoracoscapularis
Subscapularis (derived from the subcoracoscapularis)
Procoracohumeralis
Teres minor (of amniotes only)
Supracoracoideus
Infraspinatus (possible derivative of procoracohumeralis)
Supraspinatus
Coracobrachialis
Teres major (splits from latissimus dorsi for the first time in reptiles)

Muscles of upper arm
Girdle or humerus to proximal end of radius or ulna
Triceps brachii (anconeus)
Humeroantebrachialis (biceps of amphibians)
Biceps brachii
Brachialis

*Muscles of forearm**
Distal end of humerus and proximal end of radius and ulna to manus
Extensors and **flexors** of wrist, hand, and digits
Supinators and **pronators** of hand

Muscles from bones of hand to individual digits†
Flexors, extensors, abductors, and **adductors** of fingers

*All these muscles are best developed in mammals but are greatly simplified in digitigrade and unguligrade forms. They have long tendons of insertion.
†These muscles are reduced in species in which the digits are reduced.

poorly developed. As a result, the movements of the limbs are much more restricted than in amniotes. The intrinsic musculature is subdivided into a relatively small number of independent bundles.

The appendicular muscles have become more numerous, more diversified, and more powerful in reptiles. A teres major connecting the scapula with the humerus separates from the latissimus dorsi. The cucullaris is subdivided into the trapezius and sternocleidomastoideus. The extrinsic muscles spread over a greater span of the body wall than in urodeles, the deeper ones become more numerous, and a rhomboideus group is contributed, probably by the epaxial myomeres. The intrinsic muscles become more specialized functionally and thus provide greater support for the weight of the body. They also permit increased mobility of the distal segments of the appendages. Reptilian appendicular muscles presage the development of powerful pectoral muscles for flight in birds and of muscles for digging, flying, swimming, climbing, and general agility in mammals.

The largest and most powerful muscles of birds are those that operate the wings. These, the pectoral and supracoracoid muscles, constitute the white meat of the breast and are attached to the sternum. The intrinsic musculature of the wings is vestigial. In contrast, the intrinsic musculature of the hind limbs is well developed in support of bipedalism. A specialization for perching is seen in the very long tendons of the muscles of the shank. These tendons pass behind the "heel" to insert on the digits. Because of the extreme length of the tendons, the muscles do not have to shorten so much; this reduces the energy expenditure of the perching bird.[150]

The appendicular muscles of mammals are not greatly different from those of reptiles, although they often bear different names. Some further subdivision of bundles occurs and some bundles are lost. The latissimus dorsi extends far caudad to attach, via the lumbodorsal aponeurosis, to the neural spines of the lumbar vertebrae. The pectoral muscles are divided into major and minor and superficial and deep bundles and are assisted by additional slips (xiphihumeralis and pectoantebrachialis). The rhomboideus is also assisted by an extra slip, the rhomboideus capitis. Further subdivision of bundles or addition of bundles from the trunk musculature occurs. In ungulates the muscles associated with the reduced skeleton of the feet become correspondingly reduced. A few representative somatic muscles of the head, trunk, and forelimb of mammals are listed in Table 10-3.

Table 10-3. Representative myotomal muscles of head, trunk, and forelimbs in mammals*

Head	
Eyeball	*Tongue*
Superior oblique	Genioglossus
Inferior oblique	Hyoglossus
Medial rectus	Styloglossus
Lateral rectus	Lingualis
Superior rectus	
Inferior rectus	

Trunk	
Epaxial muscles	*Hypaxial muscles*
Longissimus	Hyposkeletal muscles
Iliocostales	Longus colli
Tranversospinales	Psoas
Intervertebrales	Iliacus
Extensors and adductors of tail	Quadratus lumborum
	Oblique group
	Internal and external intercostals
	Internal and external obliques of abdomen
	Diaphragm
	Cremaster
	Pyramidal
	Transverse group
	Transversus thoracis
	Transversus abdominis
	Rectus abdominis
	Hypobranchial muscles
	Geniohyoideus
	Sternohyoideus
	Sternothyroideus
	Thyrohyoideus
	Omohyoideus

Forelimb	
Extrinsic muscles	*Intrinsic muscles*
Levator scapulae	See Table 10-2
Latissimus dorsi	
Rhomboideus	
Serratus ventralis	
Pectorales	

*The student may wish to add other muscles encountered in dissection.

BRANCHIOMERIC MUSCLES

Associated with the pharyngeal arches of vertebrates (Fig. 10-9) is a series of striated, skeletal, voluntary, branchiomeric muscles. The basic architectural pattern of these muscles is illustrated in fishes, in which they operate the jaws and successive gill arches (Fig. 10-10, A). In tetrapods they continue to operate the jaws. However, with loss of gills, the more posterior branchiomeric muscles have assumed new functions.

That branchiomeric muscles must be classified as visceral is evident from the following facts:

1. Embryonically, they arise from splanchnic mesoderm, which, elsewhere in the alimentary canal, gives rise to smooth, involuntary muscles.

2. Their position in the wall of the pharynx, an organ of the alimentary canal, identifies them as visceral.

3. The motor nerve fibers innervating them arise in motor columns located lateral to the somatic columns, just as in the case of fibers innervating smooth muscles and glands. Their innervation is therefore by visceral efferent fibers.

4. Functionally, they are associated with two important visceral processes: nutrition and respiration. The foregoing criteria set these apart from myotomal muscle.

Muscles of the first pharyngeal arch. In *Squalus* (Fig. 10-10, A) the branchio-meric muscles of the first pharyngeal arch operate the jaws. A large adductor mandibulae inserts on Meckel's cartilage and, together with the intermandibularis, raises the lower jaw. A levator maxillae superioris arises on the postorbital process and otic capsule and inserts on the pterygoquadrate cartilage and raises the upper jaw. These muscles are innervated by the mandibular branch of the fifth cranial nerve.

Above fishes, the muscles of the first arch continue to operate the jaws (Fig. 10-10, B and C). Adductors of the mandible now include internal and external pterygoideus muscles. The adductor mandibulae is usually divided into a masseter and temporalis. The intermandibularis, now called the mylohyoideus, may give rise to a separate digastricus. Levators of the upper jaw become relatively useless when the upper jaw fuses with the rest of the skull. But in the streptostylic snakes, lizards, and birds, levators continue to function. In mammals only, a tensor tympani muscle inserts on the malleus (posterior tip of Meckel's cartilage) and puts tension on the eardrum. Other muscles derived from the first arch occur in the orbit of reptiles and birds as protractors or levators of the eyeball and as a depressor of the lower lid (depressor palpebrae inferioris). All the foregoing muscles are innervated by motor branches of the fifth cranial nerve.

Muscles of the second pharyngeal arch. The hyoid arch in *Squalus* is functionally associated with the jaws and also bears a demibranch. Constrictors include the epihyoideus dorsally (otic capsule to hyomandibular and ceratohyal cartilages) and the interhyoideus ventrally (midventral raphe of the pharynx to the hyoid cartilages). Other constrictors and levators of the second arch extend forward to insert on the lower jaw. In bony fishes a constrictor muscle of the hyoid arch operates the bony operculum.

In tetrapods the functional association of some of the muscles of the second arch with the lower jaw has been retained. A

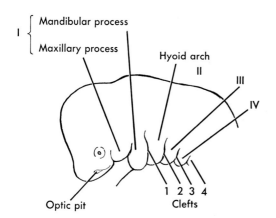

Fig. 10-9. Pharyngeal arches of a tetrapod.

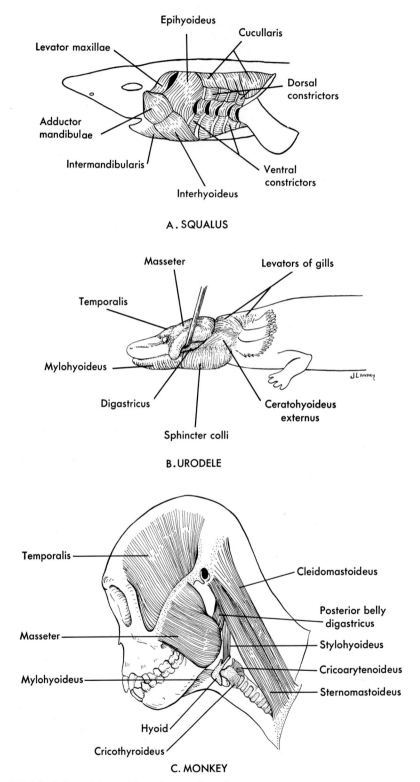

Fig. 10-10. Selected branchiomeric muscles of representative vertebrates. The sphincter colli derivatives have been omitted in the monkey.

large depressor mandibulae from the second arch attaches to the lower jaw, except in mammals. (Replacing it functionally in mammals is the posterior belly of the digastricus.) The anterior portion of the mylohyoideus of amphibians represents arch I, whereas the posterior portion represents arch II.

The muscles of facial expression (mimetic muscles) are derived from the sphincter colli of arch II (Fig. 10-11). In reptiles and birds the sphincter colli spreads upward around the rear of the skull to insert on the skin of the head as the platysma. In mammals it spreads forward over the facial area and becomes subdivided into numerous muscles used for displaying the canine tooth as in snarling, in wiggling the ears and moving the nostrils, and in other ways. In primates

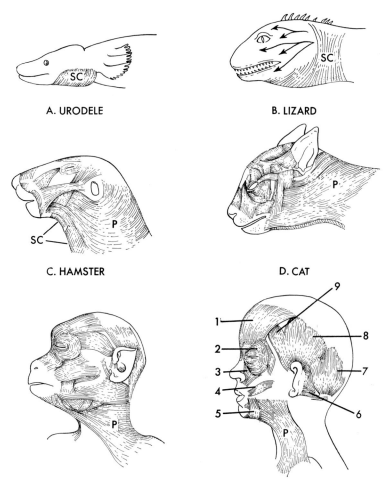

Fig. 10-11. Evolution of mimetic muscles from hyoid arch muscles of lower tetrapods. The sphincter colli, **SC**, spreads onto the neck, where it is called the platysma, **P**, and then forward onto the head and face, as indicated by arrows in **B**. Compare extent of differentiation of mimetics in the four mammals shown. In man are shown **1**, frontalis; **2**, orbicularis oculi; **3**, quadratus labii superioris; **4**, risorius; **5**, triangularis; **6**, posterior auricular; **7**, occipital; **8**, superior auricular; **9**, anterior auricular. (**C** after Priddy and Brodie[66]; **E** and **F** after Huber.[62])

they become further developed and are used in smiling, frowning, sneering (displaying the canine tooth), puckering the lips, furrowing the brow, and in other facial expressions. They thus contribute to the integumentary musculature.

In mammals the stylohyoideus and stapedius are of second arch origin. The former inserts on the anterior horn of the hyoid and is used in swallowing. The latter inserts on the stapes. These and all the other muscles of the second arch are innervated, as in the shark, by motor branches of the seventh cranial nerve. It

is called the facial nerve because its terminal branches are widely distributed just under the skin of the face.

Muscles of the third and successive pharyngeal arches. In *Squalus* the branchiomeric muscles of the third through the sixth pharyngeal arches are represented by a series of constrictors located above and below the gill chambers and by other serially arranged bundles that assist in compressing or expanding the gill pouches. The constrictors of the third arch are innervated by the ninth nerve, and those of the remaining arches are innervated by the

Table 10-4. Homologous branchiomeric muscles and their innervation in Squalus and in tetrapods

| Pharyngeal arch | Pharyngeal skeleton in Squalus | Chief branchiomeric muscles | | Cranial nerve innervation |
		Squalus	Tetrapods	
I (Mandibular arch)	Meckel's cartilage	Intermandibularis	Intermandibularis or its derivatives: Mylohyoideus Anterior belly digastricus	V
		Adductor mandibulae	Adductor mandibulae or its derivatives: Masseter Temporalis Pterygoidei Tensor tympani	
	Pterygoquadrate cartilage	Levator maxillae superioris	Protractors and levators of eyeball	
II (Hyoid arch)	Hyomandibula Ceratohyal Basihyal	Epihyoideus	Stapedius Stylohyoideus	VII
		Interhyoideus	Interhyoideus (may contribute to mylohyoideus) Depressor mandibulae Posterior belly digastricus Sphincter colli (gularis) or its derivatives: Platysma Mimetics	
III	Gill cartilages	Constrictors	Stylopharyngeus	IX
IV to VI	Gill cartilages	Constrictors	Striated pharyngeal muscles Thyroarytenoideus Cricoarytenoideus Cricothyroideus	X
VII	Seventh visceral cartilages	Cucullaris (probably derived also from constrictors III to VI)	Trapezius Sternomastoideus Cleidomastoideus	Occipitospinal nerves in anamniotes XI In amniotes

tenth. In bony fishes the branchiomeric musculature caudal to the hyoid arch has become reduced since the operculum, to a large extent, controls the respiratory stream.

In amphibians with gills the muscles of the gill arches may be well developed. In fully metamorphosed amphibians and in amniotes, the absence of gills is correlated with reduction of the branchiomeric muscles of the third and successive visceral arches. In amniotes the chief branchiomeric muscle derived from the third arch is the stylopharyngeus innervated by the ninth nerve. Derivatives of the fourth and remaining visceral arches (thyroarytenoideus, cricoarytenoideus, and cricothyroideus) are associated with the larynx and are innervated by branches of the tenth cranial nerve. The cucullaris muscle (represented by the trapezius and sternocleidomastoid in amniotes) is partly derived from the branchiomeric series. The innervation of the branchiomeric muscles is given in Table 10-4.

INTEGUMENTARY MUSCLES

The skin of fishes and tailed amphibians is firmly attached to underlying tissues, and so there is no independent movement of the skin. In anurans the skin is firmly attached in some places, as around the external nares, whereas in other places it is separated from the underlying muscle by large lymph sinuses. In either case the skin is scarcely movable. One bundle, the cutaneous pectoris derived from slips of the pectoral muscles, inserts on the skin of the pectoral region of anurans.

Integumentary muscles capable of twitching the skin or drawing it in one direction or another occur first in reptiles and have been exploited by mammals. The sphincter colli, of branchiomeric origin, has spread over the head and neck to form a platysma inserting on the skin (Fig. 10-11). Mammals have inherited the platysma, which has spread forward over the facial area to form numerous mimetic muscles of the scalp and face. These are best developed in primates. Man exhibits the largest number of separate mimetic muscles—about thirty (Fig. 10-11).

Slips of hypaxial myotomal musculature insert on the ventral scutes of snakes. These costocutaneous muscles erect the scutes and thereby provide friction for locomotion. The pectoral muscles of birds give rise to petagial muscles inserting on the skin of the wing membrane.

Integumentary muscles reach their peak of evolution in the lower mammals. The trunk region of many mammals is almost completely wrapped in a muscle sheet, the panniculus carnosus (Fig. 10-12), which is thought to be derived from the latissimus dorsi and pectoral muscles. In armadillos it helps roll the body into a ball when the animal is endangered. In marsupials a sphincter portion of this muscle surrounds the entrance to the marsupial pouch. Horses and cows use the muscle to shake the skin to displace flies or other irritants. In other mammals the panniculus carnosus (also referred to as the cutaneous maximus) may be more restricted. It is poorly developed or absent in primates.

Muscles intrinsic to the skin occur only in birds and mammals. They attach to the base of the feather follicle (arrectores plu-

CAT   MONKEY

Fig. 10-12. Panniculus carnosus of a cat and primate, for comparison.

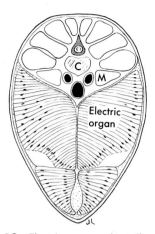

Fig. 10-13. Electric organs in tail of electric eel. Each nucleated horizontal plate (electroplax) is a modified hypaxial muscle cell. **C,** Centrum; **M,** epaxial myomere.

marum) or to the hair follicle (arrectores pilorum). These erector muscles permit fluffing the feathers or elevating the hairs for insulation or as an emotional response to danger. They are smooth muscles operated by visceral motor fibers.

ELECTRIC ORGANS

In a few elasmobranchs and a wide variety of marine and freshwater teleosts, certain muscle masses are highly modified to produce, store, and discharge electricity. In *Torpedo*, the electric ray, an electric organ lies in each pectoral fin near the gills. This muscle mass is probably of branchiomeric origin, since it is supplied by the seventh and ninth cranial nerves. In *Raia* (another ray) and *Electrophorus* (electric eel) electric organs lie in the tail and are modified hypaxial musculature (Fig. 10-13). The total charge produced by these organs in eels amounts to 500 volts. Other fishes exhibit electric organs of lower voltage, which are thought to serve as locating mechanisms. Electric organs consist of a large number of electric discs (up to 20,000 in the tail of one ray) piled in vertical (ray) or horizontal (eel) columns.

Each disc (electroplax) is a modified multinucleate muscle cell embedded in a jelly-like extracellular gelatinous material and surrounded by connective tissue. Nerve endings terminating on each disc cause the discharge. Capillaries develop in the jelly layer. Electric organs seem to have no systematic distribution among fishes, and the various types probably result from convergent evolution. The electric organ of one teleost (a catfish native to Africa) is said to be a derivative of a skin gland embedded in the dermis, rather than of musculature.

Chapter summary

1. The body wall and tail muscles of fishes are arranged as a metameric series of myomeres separated by myosepta. Myomeres are divided into an epaxial mass located dorsal to the horizontal septum and a hypaxial mass located ventral to the septum. Metamerism is interrupted at the level of the girdles and at the base of each appendage. The body wall muscle does not invade the lateral pharyngeal wall but extends forward under the pharynx as hyprobranchial muscle.

2. Lower tetrapods exhibit a similar metamerism of the body wall muscles. In higher forms the metamerism tends to be obscured by obliteration of myosepta and fusion of successive myomeres to form long muscle bundles (especially epaxially) or continuous broad muscle sheets (especially hypaxially).

3. The metamerism in all vertebrates is an expression of the serial arrangement of the embryonic mesodermal somites. The mesenchyme of each myotome in lower vertebrates grows downward into the somatopleure to give rise to a muscle segment innervated by the spinal nerve of the same body segment. Epaxial and hypaxial masses

are innervated by dorsal and ventral rami, respectively.

4. Muscles derived from mesodermal somites are referred to as myotomal (somatic) muscles. They are striated, skeletal, and voluntary. Included in this category, in addition to the muscles of the body wall and tail, are extensions of the hypobranchial muscles into the head, epibranchial muscles, eyeball muscles and certain accessory muscles of the lids, appendicular muscles (except the cucullaris), and certain integumentary muscles.

5. Epaxial muscles in anamniotes tend to retain much metamerism. In amniotes they tend to form elongated bundles extending over several or many segments, but the deepest masses (intervertebral muscles) retain their primitive vertebra-to-vertebra segmentation. In animals with ankylosed vertebrae (turtles and birds), the epaxials tend to be reduced.

6. Hypaxial muscles are segmentally arranged in fishes and tailed amphibians. In the remaining tetrapods the contributions from successive myotomes tend to coalesce and form large muscle sheets or long bundles innervated by a series of spinal nerves. Where long ribs develop, the metamerism is reinforced.

7. Hypaxial muscles are disposed in strata represented by horizontal, oblique, and rectus sheets, which are better differentiated in tetrapods. The strata may further split or become reduced. Extensive reduction of body wall muscles of turtles is associated with the presence of bone in the skin.

8. The mammalian diaphragm and cremaster muscles are hypaxial muscles that have migrated from their origin, accompanied by the spinal nerves that innervate them.

9. The hypaxial muscles located directly ventral to the transverse processes of the vertebrae are disposed in longitudinal bundles and constitute the hyposkeletal muscles.

10. The hypaxial muscle extends forward under the pharynx as far as the mandible and constitutes the hypobranchial musculature. These muscles contribute the musculature of the amniote tongue. All hypobran-

chial muscles are innervated by the most anterior spinal nerves or (in amniotes) the last cranial nerves.

11. Eyeball muscles arise from three head somites in elasmobranchs but develop "in situ" in higher vertebrates. They are innervated by somatic motor fibers of the same cranial nerves (III, IV, and VI) in all vertebrates and must be considered myotomal derivatives in all instances.

12. The appendicular muscles of fishes migrate into the fins as buds of the hypaxial muscles. In tetrapods primary appendicular muscles arise from blastemas located within the developing appendages, and some of these spread out of the appendage to assume axial attachments. Secondary appendicular muscles originate from body wall blastemas.

13. Branchiomeric muscles are striated, skeletal, voluntary muscles, which arise in the pharyngeal arches and are innervated by visceral motor fibers of cranial nerves V, VII, IX, X, and IX. They operate the visceral skeleton.

14. The branchiomeric muscles of the first pharyngeal arch operate the jaws. In mammals a derivative of the first arch musculature operates the malleus. The muscles of the first arch are innervated by the fifth cranial nerve regardless of location or function.

15. The muscles of the second arch are attached to the hyoid cartilages and their derivatives. Strong bundles extend forward to the lower jaw and backward onto the operculum. The sphincter colli spreads over the head to give rise to an integumentary muscle, the platysma. In mammals the platysma spreads and subdivides to become the muscles of facial expression. A derivative of the second arch musculature operates the stapes. The muscles of the second visceral arch are innervated by the seventh cranial nerve regardless of location or function.

16. Muscles of the third and successive arches operate gill cartilages in fishes and in amphibians with gills. In higher vertebrates they perform new functions but remain at-

tached to their original skeletal elements. Certain muscles of the pharynx are derivatives of the third arch, and the intrinsic muscles of the larynx are derivatives of more posterior arches. The muscles of the third arch are innervated by the ninth cranial nerve, and those of successive arches by the tenth cranial nerve.

17. The cucullaris is a branchiomeric muscle, which in tetrapods has become an extrinsic appendicular muscle. In amniotes it becomes subdivided into trapezius, sternomastoideus, and cleidomastoideus muscles. It is innervated by occipitospinal nerves in fishes and by the spinal accessory nerve in amniotes.

18. Extrinsic integumentary muscles insert on and move the skin. They are restricted to amniotes. They reach peak development as the panniculus carnosus (cutaneous maximus) of mammals and the mimetics of primates. The integumentary muscles of the trunk are myotomal muscles innervated by spinal nerves, and those of the face and scalp are branchiomeric muscles innervated by the seventh cranial nerve.

19. Intrinsic integumentary muscles are smooth muscles innervated by visceral nerve fibers.

20. Electric organs are columns of modified muscle cells (electroplaxes) capable of producing, storing, and discharging electric potential. A few appear to be modified integumentary glands. They are widespread among fishes.

Digestive system

The digestive tract is a tube, seldom straight and often tortuously coiled, commencing at the mouth and terminating at the vent or anus (Fig. 11-1). It functions in ingestion, digestion, and absorption of foodstuffs and in elimination of undigested wastes. Major subdivisions of the tract are the oral cavity, pharynx, esophagus, stomach, and small and large intestines. Associated with the tract are accessory organs, such as the tongue, teeth, oral glands, pancreas, liver, and gallbladder. The digestive tract and accessory organs constitute the digestive system.

Cilia are often present in the digestive tract. The entire tract is ciliated in the adult amphioxus and in many larval vertebrates including ammocoetes. Cilia occur in the stomach of many teleosts, in the oral cavity, pharynx, esophagus, and stomach of some adult amphibians, in the ceca of some birds, and in numerous other locations in various species. They occur transitorily in the stomach of the human fetus.

Embryonic tract. The most anterior part of the oral cavity arises from the **stomodeum,** a midventral invagination of the ectoderm of the head. A similar ectodermal invagination, the **proctodeum,** becomes the terminal segment of the digestive tract just beyond the endodermal part of the cloaca. The stomodeum and proctodeum have been discussed elsewhere. The remainder of the digestive tract is lined by

endoderm, as are any evaginations of it. Surrounding the endoderm is the splanchnic mesoderm, from which differentiate the muscular and connective tissue coats and the visceral peritoneum.

The entire embryonic digestive tract consists of three regions. The part containing the yolk or to which the yolk sac is attached is the **midgut.** The part anterior to it is the **foregut,** and the part caudal to the midgut is the **hindgut.** The foregut elongates to become the endodermal part of the oral cavity, pharynx, esophagus, stomach, and most of the small intestine. The hindgut becomes the large intestine and cloaca.

The embryonic gut from the stomach to the cloaca is attached to the dorsal body wall of the trunk by a continuous dorsal mesentery and to the ventral body wall by a ventral mesentery (Fig. 11-15). Much of the dorsal mesentery remains throughout life, but the ventral mesentery disappears, except at two sites. The ventral mesentery remains in all vertebrates at the level of the liver, where it is called the falciform ligament, and in amniotes it remains as the ventral mesentery of the urinary bladder.

MOUTH

In a restricted sense the term "mouth" refers only to the anterior **opening** into the digestive tract. In a broader sense the term

235

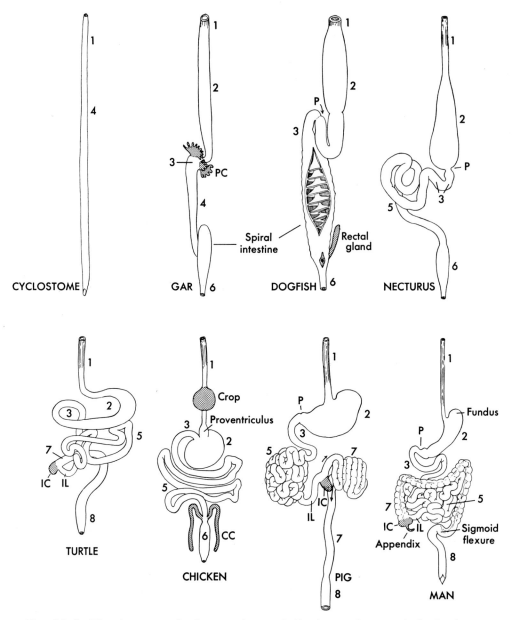

Fig. 11-1. Digestive tracts of a few vertebrates. **1,** Esophagus; **2,** stomach; **3,** duodenum; **4,** intestine; **5,** small intestine; **6,** large intestine; **7,** colon; **8,** rectum; **CC,** paired ceca of bird; **IC,** ileocolic cecum; **IL,** ileum; **P,** pyloric sphincter; **PC,** pyloric ceca. All ceca are shown in gray.

is used as a synonym for oral cavity. The mouth (sensu strictu) is usually terminal in position, although in some fishes, especially elasmobranchs and sturgeons, it is located ventrally and often well back from the cephalic tip of the head. In cyclostomes the opening is at the vortex of the buccal funnel and is always open, since there are no hinged jaws or other mechanisms for closing it. In the amphioxus it is in the velum. In lower chordates it is sometimes encircled by papillae, as in amphioxus, larval cyclostomes, and catfish. The papillae usually have chemoreceptors akin to taste buds. The mouth of the larval tadpole is a small oval suctorial organ surrounded by horny lips and rasping papillae consisting of rows of horny teeth. These are adaptations for feeding on plants. At metamorphosis the horny parts are shed, the mouth changes to a wide slit, and the frog becomes carnivorous. Fleshy muscular lips surround the mouth in most mammals. Muscular lips and cheeks are adaptations for suckling. In most other vertebrates the mouth is surrounded by an unmodified or heavily cornified epidermis.

ORAL CAVITY

The oral cavity commences at the mouth and ends at the pharynx. In adult fishes the oral cavity is very short and almost nonexistent. It is longer in tetrapods and culminates in mammals in a sucking and masticatory organ bounded laterally by muscular cheeks. The muscular floor of the oral cavity of all tetrapods assists in swallowing, and is used in breathing in the absence of a diaphragm (amphibians and reptiles). A primary or secondary tongue occupies part of the floor of the oral cavity.

The roof of the cavity is the palate. If it is a primitive (primary) palate, the nasal passageways, when present, empty into the oral cavity anteriorly. With formation of a secondary palate in amniotes, the internal nares open farther to the rear of the oral cavity or into the pharynx. The bony secondary palate of mammals is continued caudad as a membranous soft palate, from which a uvula sometimes dangles into the laryngeal pharynx. Teeth usually occur on the jaws and, in lower vertebrates, on the palate as well.

In mammals a trench, the **oral vestibule,** separates the gums from the cheeks and lips. In many rodents large membranous **cheek pouches** extend from the vestibule backward under the skin and muscle of the side of the head. These pouches are used for transporting grain and other foods to the nest. Cheek pouches also occur in other species including duckbills and some primates.

Since the oral cavity is lined by stomodeal ectoderm anteriorly and by gut endoderm posteriorly, the mucous membrane is supplied anteriorly by the same nerves that innervate the ectoderm of the head and posteriorly by those nerves innervating the endoderm of the foregut.

Teeth

Teeth composed primarily of a core of dentine surmounted by a crown of enamel are characteristic of vertebrates. They are seen in a relatively generalized and perhaps primitive condition in sharks, in which the placoid scales of the head exhibit a gradual transition to teeth as they approach the cutting edges of the jaws (Fig. 5-7). In the development of placoid scales and also of teeth, a dermal papilla produces a **root** composed of dentine, and part of this root erupts through the overlying epidermis (Fig. 11-2). An enamel organ, an ingrowth of the germinal layer of the epidermis, usually secretes enamel on that part of the tooth which erupts. The enamel organ may be functionless in sharks, and is clearly functionless in armadillos and a few other fishes and mammals. In such cases any hard material on the surface of the tooth is a product of the dermis. It is plausible that vertebrate teeth are specialized remnants of an ancient dermal armor. Teeth vary among vertebrates in number, distribution within the oral cavity, degree of permanence, mode of attachment, and shape.

A few toothless species are found in every vertebrate class. Agnathans, sturgeons, some toads, sirens (urodeles), turtles, and modern birds are toothless. However, one species of terns develops an embryonic set that fails to erupt, and at least one genus of turtles develops an enamel organ. Toothless mammals develop a set that either do not erupt or that, after erupting, are soon lost.

Number and distribution. Teeth are very numerous and widely distributed in the oral cavity and pharynx of fishes. They develop in relation to the bones of the jaws,

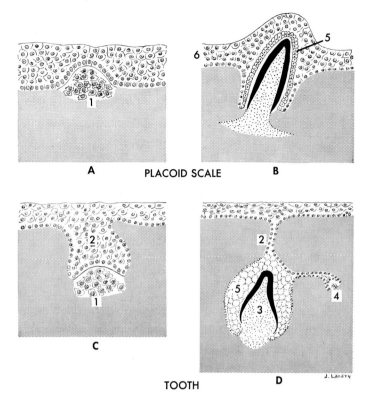

Fig. 11-2. Development of a placoid scale, **A** and **B**, and of a mammalian tooth, **C** and **D**, from a dermal papilla. Dermis, except papilla, is shown in gray; enamel of tooth and scale is shown in black. **1,** Dermal papilla; **2,** dental ridge; **3,** deciduous (milk) tooth; **4,** permanent tooth; **5,** enamel organ; **6,** epidermis. The scale and tooth form from the dermis, but the epidermis contributes the enamel of the mammalian tooth, at least.

palate, and even the pharyngeal skeleton. When teeth are lacking on some of these areas in fishes smaller tooth-like **stomodeal denticles** may appear.

In early tetrapods, as in fishes, teeth were widely distributed on the palate, and even today most amphibians and many reptiles have teeth on the vomer, palatine, and pterygoid bones and occasionally on the parasphenoid. In crocodilians, toothed birds, and mammals, teeth are confined to the jaws and are least numerous among mammals. The teeth of vertebrates have therefore tended toward reduced numbers and restricted distribution. Only in mammals is there a precise number of teeth in a given species.

Degree of permanence. Most vertebrates through reptiles have a succession of teeth that develop from new dermal papillae, and the number of replacements is indefinite but numerous (**polyphyodont dentition**). Mammals for the most part develop two sets of teeth (**diphyodont dentition**)—deciduous (milk) teeth and permanent teeth (Fig. 11-2). Milk teeth usually erupt after birth, but in guinea pigs and a number of other mammals milk teeth develop and are shed before birth. A few mammals (platypus, sirenians, and toothless whales) develop only a first set of teeth (**monophyodont dentition**), and these may not erupt; if they do, they are usually shed shortly afterward. In the platypus they are replaced by horny teeth.

Mode of attachment (Fig. 11-3). Teeth

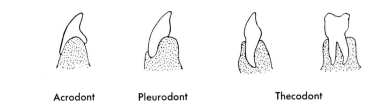

| Acrodont | Pleurodont | Thecodont |

Fig. 11-3. Types of dentition based on attachment of tooth to jaw. The acrodont tooth is attached to the summit and outer surface of the jawbone. The pleurodont tooth is attached to the inner surface of the jawbone. (From Atwood: Comparative anatomy, ed. 2, St. Louis, 1955, The C. V. Mosby Co.)

are intimately attached to the underlying skeleton by fibrous connective tissue. Jaw teeth may be attached to the outer surface or to the summit of the jawbone, as in frogs (**acrodont dentition**), or to the inner side of the jawbone, as in *Necturus* and lizards (**pleurodont dentition**). Or they may be rooted in individual bony sockets called **alveoli** (**thecodont dentition**). Socketed teeth occur in crocodilians, in toothed birds and mammals, and in many fishes.

Morphological variants. Some of the teeth of a shark bear spines for tearing flesh; others exhibit flattened or rounded surfaces for sawing or crushing (Fig. 5-7). The fangs of some poisonous snakes are specialized teeth borne on the maxillary bones (Fig. 11-7). These teeth may be permanently erect, or the jaw may move so that when the mouth is opened, the teeth are brought into a striking position (Fig. 8-31). The poison from the gland pours into a groove or tube built into the tooth. For the most part, in any single individual below mammals all teeth are alike except for size (**homodont dentition**). In mammals alone the teeth of each individual exhibit morphological varieties—incisors, canines, premolars, and molars—each in a specific number (**heterodont dentition**). Extinct reptiles in the mammalian line (synapsids) were the first to exhibit heterodont dentition (Fig. 8-27).

Incisors, located anteriorly, are usually specialized for cutting (cropping). The incisors of rodents and lagomorphs continue to grow throughout most of life. Enamel is deposited only on the anterior face of the

incisor teeth of these orders, and since enamel wears down more slowly than dentine, sharp chisel-like edges result. Incisors may be totally absent (sloth) or lacking on the upper jaw (Fig. 11-4, ox). Elephant tusks are modified incisor teeth (Fig. 11-4, mastodon).

Canines lie immediately behind the incisors. In generalized mammals, incisors and canines scarcely differ in appearance (Fig. 11-4, shrew). In carnivores the canines are spear-like and are used for piercing the prey and tearing flesh (Fig. 11-4, dog). They are the tusks of the walrus (Fig. 11-4). Canine teeth are always absent in rodents and lagomorphs and thus leave a toothless interval, the **diastema,** between the last incisor and the first cheek tooth (Fig. 11-4, rabbit).

Cheek teeth (**premolars** and **molars**) are used for macerating food (Fig. 11-5). Molars are cheek teeth that are not replaced by a second set. In herbivorous mammals (Fig. 11-4, ox) the cheek teeth tend to be numerous, and there is little differentiation between premolars and molars. Cusps (conical elevations) are not pronounced, soon wearing down to form broad, flat grinding surfaces. In carnivores the number of cheek teeth is often reduced, and cusps are prominent. At least one pair on each jaw may have very sharp cusps for cracking bones and shearing tendons (**carnassial teeth,** Fig. 11-4, dog). Grinding is deemphasized and the food is swallowed with little chewing. The cheek teeth of other mammals are more or less intermediate in shape between the ex-

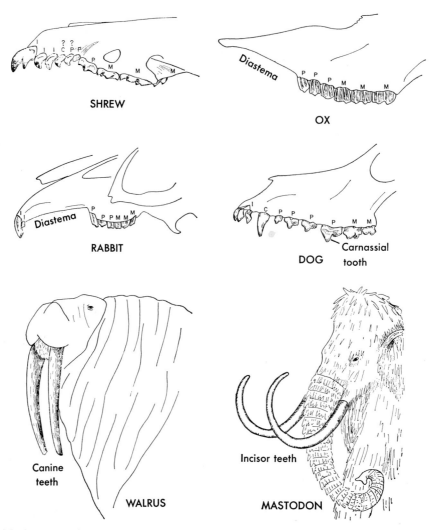

Fig. 11-4. Mammalian upper teeth showing generalized pattern (shrew), and specializations in a herbivore (ox), in a gnawing animal (rabbit), and in a carnivore (dog). Extreme specialization of canines is seen in the walrus and of incisors in the mastodon. **I,** Incisor; **C,** canine; **P,** premolar; **M,** molar.

tremes exhibited by herbivores and carnivores. Eruption of the last cheek tooth (wisdom tooth) is delayed in higher primates, and this tooth is sometimes imperfectly formed or absent in man.

Primitive placental mammals exhibited 3 incisors, 1 canine, 4 premolars, and 3 molars on each side of each jaw, a total of 44 teeth. This condition may be expressed by the formula $\frac{3\text{-}1\text{-}4\text{-}3}{3\text{-}1\text{-}4\text{-}3}$. The same formula applies to modern horses, although the first

premolar may be missing. Dental formulas of other representative adult mammals are cat, $\frac{3\text{-}1\text{-}3\text{-}1}{3\text{-}1\text{-}2\text{-}1}$; man, $\frac{2\text{-}1\text{-}2\text{-}3}{2\text{-}1\text{-}2\text{-}3}$; and rabbit, $\frac{2\text{-}0\text{-}3\text{-}3}{1\text{-}0\text{-}2\text{-}3}$. The formulas for the shrew, ox, and dog can be derived from Fig. 11-4. These formulas illustrate some of the adaptive modifications displayed among mammals.

Epidermal (horny) teeth. Agnathans lack true teeth, but the buccal funnel and

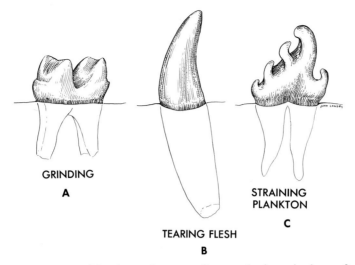

GRINDING

A

TEARING FLESH

B

STRAINING PLANKTON

C

Fig. 11-5. Adaptive modifications of mammalian teeth, lateral views. **A,** Lower left molar from a tapir. **B,** Lower left canine from a jaguar. **C,** Lower right molar from a crabeater seal.

"tongue" are provided with horny epidermal teeth for rasping flesh. Horny teeth arise from the stratum corneum. The lips of the suctorial mouth of some species of frog tadpoles develop horny teeth. In turtles, *Sphenodon*, crocodiles, birds, and monotremes a horny "egg tooth" for cracking the egg develops on the upper jaw at the tip of the beak prior to hatching. (In lizard and snake embryos that develop within an eggshell, the egg tooth is a transitory genuine tooth of dentine.) The adult duck-billed platypus has only horny teeth. Horny beaks in some turtles and birds have serrations that simulate teeth.

Tongue

In agnathans the tongue is a thick, fleshy, rasping organ in the floor of the buccal cavity, armed with horny teeth and supported by an elongated lingual cartilage. When protracted by its extensive myotomal musculature, it extends into the buccal funnel. It does not appear to be homologous to the tongue of higher vertebrates.

The tongue of gnathostome fishes consists of a crescentic or angular, relatively immobile elevation in the floor of the pharynx that marks the site of the under-lying hyoid cartilages. The lean hyoid tongue of fishes is a **primary tongue.** *Necturus* and other perennibranchs likewise have only a primary tongue (Fig. 11-6).

The tongue of most amphibians consists of the primary tongue and, in addition, of a contribution (the **glandular field**) from the pharyngeal floor anterior to the hyoid arch (Fig. 11-6). The tongue of frogs and toads is attached in the glandular field area and the posterior end is free (Fig. 11-8). Protraction is effected chiefly by rapid injection of lymph into a sublingual lymph sac. A few frogs lack tongues.

The tongue of reptiles and mammals has a primary tongue contribution from arch II. A small median elevation in the floor of arch I, the **tuberculum impar,** may be homologous with the glandular field of lower vertebrates. However, this median field makes little or no contribution to the mammalian tongue. Instead, a pair of **lateral lingual swellings** of the first arch give rise to the anterior two thirds of the tongue. In addition, in mammals the third arch contributes by growing forward over the second arch and excluding it from the surface. As a result, the adult mammalian tongue is covered with first arch endoderm in the anterior two thirds, and with third

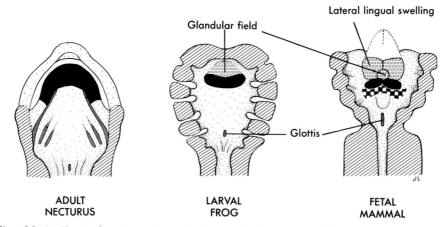

Fig. 11-6. Floor of oral cavity and pharynx depicting probable stages in evolution of the tongue. Black represents the primary tongue, a contribution of the second arch. Circles in frog and mammal represent contributions of the first arch. Black and white squares in mammal represent the contribution of the third arch, and dotted outline shows final extent of lateral lingual swelling. The glandular field in mammals is also known as the tuberculum impar.

arch endoderm in the posterior one third. This accounts for the fact that the anterior two thirds of the tongue is supplied by the fifth nerve (mandibular arch nerve) for general sensation, whereas the posterior third is supplied by the ninth nerve (third arch nerve) for general sensation.

In birds the lateral lingual swellings are suppressed and intrinsic musculature is lacking. Instead, an entoglossal bone of the hyobranchial apparatus extends into the tongue. An entoglossal bone also develops in lizards.

In turtles, crocodilians, some birds, and whales most of the tongue is immobilized in the floor of the oral cavity and cannot be extended. At the other extreme, in snakes, some lizards, and some birds the tongue is very long and darts in and out of the mouth at remarkable speed. This is also true of insectivorous mammals and pollen- and nectar-eating bats in which the tongue may be as long as the rest of the body. In most mammals the tongue is attached to the floor of the oral cavity by a ligament, the **frenulum.** In man, if the frenulum extends unusually far forward, the individual is "tongue-tied."

The surface of the tongue of most am-

niotes bears hair-like, scale-like, knobby, or spiny papillae. Taste buds are associated with fungiform or circumvallate papillae. The spiny fungiform papillae on the tongue of a cat assist in grooming the fur.

Oral glands

Few multicellular glands are found in the oral cavity of fishes, aquatic amphibians, aquatic turtles, birds that obtain their food from the water, and whales. Their absence is no handicap, since any secretion would be constantly washed away and is unnecessary for lubrication or digestion. In fact, most oral glands below mammals contain no digestive enzymes. Goblet cells, however, are common. The males of some species of catfish carry the fertilized eggs of the species in their mouths, and during the breeding season the oral epithelium becomes extensively folded and exhibits numerous crypts supplied with large goblet cells. These brood pouches shelter the eggs and atrophy after the breeding season.

In land-dwelling vertebrates unicellular and multicellular oral glands, chiefly mucous, are scattered about in the roof, walls, and floor of the oral cavity, but the secretion does not become copious below mammals. The largest oral glands are the poison glands of snakes and the compound salivary glands of mammals (Fig. 11-7).

Oral glands are often named according to their location. **Labial** glands open at the base of the lips and **palatal** glands open onto the palate. **Intermaxillary (internasal)** glands lie between the premaxillary bones. In frogs they consist of up to twenty-five small glands, each emptying a sticky secretion by its own duct into the oral cavity. **Sublingual** glands lie under the tongue. In *Heloderma*, the only poisonous lizard, they secrete the toxin. In mammals sublingual glands are usually classified as salivary glands, although they do not always secrete ptyalin. The **parotid** (*par-otic*, near the otic region) is the largest salivary gland. Its duct crosses the masseter muscle and opens into the vestibule opposite one of the upper molars. **Submandibular** salivary glands open on papillae behind the lower incisors. Rabbits have a fourth salivary gland, the **infraorbital** in the caudal floor of the orbit, and cats have a fifth, the **molar** gland under the skin of the lower lip just anterior to the masseter muscle (Fig. 11-7). It opens via several ducts onto the mucosa of the cheek. Monotremes apparently do not have salivary glands. Saliva comprises a number of components, including mucin (a mixture of water-soluble glycoproteins), a watery fluid, and ptyalin for the digestion of starch, but not all salivary glands secrete all these components. Saliva is chiefly a mammalian product, but ptyalin-secreting glands occur in some frogs and birds, at least.

PHARYNX

The vertebrate pharynx is that part of the digestive tract exhibiting in the embryo a series of paired lateral diverticula called pharyngeal pouches. These pouches may perforate to the exterior and give rise to transitory or permanent pharyngeal slits associated with respiration by gills. In adult vertebrates in which the slits have closed, the pharynx may be identified as that part of the foregut immediately anterior to the esophagus. The pharynx as a respiratory organ is discussed in Chapter 12. The taste buds and endocrine derivatives of the pharynx are discussed in Chapters 16 and 17.

An interesting modification of the pharynx apparently not associated with respiration is seen in some teleosts. An elongated tubular muscular canal evaginates from the posterior roof of the pharynx on each side of the entrance to the esophagus, and extends cephalad above the membranous roof of the pharynx, behind the skull. The canals then turn caudad and terminate as blind sacs. The structures are known as **suprabranchial (pharyngeal) organs.** Elongated gill rakers from the last two gill arches form funnel-shaped baskets that extend directly into the entrances of the organs. Each organ is enveloped by a cartilaginous capsule, which provides attachment for the striated muscle in its walls. The epithelium of the terminal blind

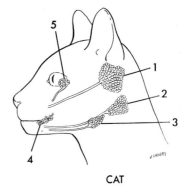

SNAKE   CAT

Fig. 11-7. Oral glands in a reptile and mammal. Birds have no large oral glands. **1** to **5,** Parotid, submandibular, sublingual, molar, and infraorbital salivary glands, respectively; **6,** poison gland of rattlesnake; **7,** tooth with groove for transfer of toxin; **8,** tongue.

sacs exhibits many goblet cells, and the sacs contain quantities of plankton, sometimes compressed into a bolus. The function of these organs is not yet known. One suggestion is that they entrap plankton from the incoming water as in filter feeders and that they concentrate the plankton into mucified masses, which are then expressed and swallowed.[159]

Most constant features of the tetrapod pharynx are the **glottis** (a slit leading into the larynx), the openings of the eustachian tubes, and the opening into the esophagus (Fig. 11-8, *A*). Commencing in reptiles, a cartilaginous flap, the **epiglottis**, overlies the glottis. In swallowing, the larynx is drawn forward against the epiglottis, which blocks the glottis and prevents foreign substances from entering the lower respiratory tract.

The secondary palate of mammals (Fig. 11-8, *B*) divides the adult pharynx into an **oral pharynx** ventral to the soft palate, a **nasal pharynx** dorsal to the soft palate receiving the openings of the eustachian tubes, and a **laryngeal pharynx** surrounding the glottis. The airstream from the nasal passageways crosses the food pathway in the laryngeal pharynx. The crossing is sometimes referred to as the pharyngeal chiasma.

Guarding the entrance to the laryngeal pharynx is a ring of lymphoidal masses— palatine tonsils in the oral pharynx, adenoids in the nasal pharynx, and lingual tonsils at the root of the tongue. Associated with the palatine tonsils are crypts, which are remnants of the first pharyngeal pouches of the mammalian embryo.

ESOPHAGUS

The esophagus is a distensible muscular tube (shortest in neckless vertebrates) connecting the pharynx with the stomach. In a few vertebrates the esophagus is lined with finger-like papillae (fleshy in elasmobranchs and horny in one marine turtle), but more often the lining has longitudinal folds or is smooth. The striated muscle at the cephalic end of a long esophagus is gradually replaced farther down by smooth muscle; however, striated muscle fibers may continue onto the stomach wall, especially in cud-chewing mammals that regurgitate their food for leisurely chewing.

Only one notable specialization occurs in the esophagus—the **crop** (or **crop sac**) of some birds (Fig. 11-9). The crop is a

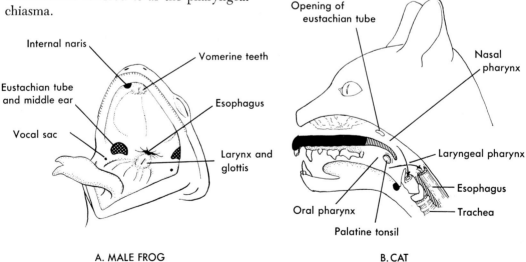

A. MALE FROG   B. CAT

Fig. 11-8. Oral cavity and pharynx of an amphibian and mammal. Only the mammal has a secondary palate. The bony part of the secondary palate is shown in black and the soft part is cross hatched. Crossed arrows indicate the pathways for food and air. (**A** courtesy J. H. Roberts, Baton Rouge, La.)

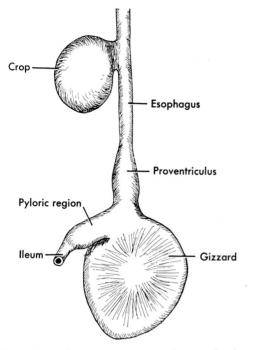

Fig. 11-9. Crop, esophagus, and stomach of a grain-eating bird.

paired or unpaired membranous sac-like diverticulum of the esophagus occurring primarily in grain-eaters and used for initial storage of food. An enzyme for preliminary digestion may also be secreted. In male and female pigeons a part of the epithelial lining of the crop undergoes fatty degeneration under the stimulation of prolactin. The cells are then shed and regurgitated along with partially digested food as "pigeon's milk," which is fed to nestlings.

STOMACH

The stomach is a muscular chamber or a series of chambers immediately preceding the small intestine. Among other functions it serves as a storage and macerating site for ingested solids. It secretes certain digestive juices and renders solid foods semiliquid prior to injection into the small intestine, which is the chief absorbing area of the digestive tract.

No stomach is present in protochordates, and the stomach is poorly delineated in cyclostomes. A boundary between the esophagus and stomach is indefinite or lacking in vertebrates below birds. Even in birds and mammals the mucosa of part or all of the stomach may resemble that of the esophagus (Fig. 11-10). The stomach terminates at the pyloric sphincter.

The stomach is straight when it first develops in the embryo (Fig. 11-15) and may remain so throughout life in lower vertebrates. More often, flexures develop producing a J-shaped or U-shaped stomach (Figs. 11-1 and 11-11). As a result, the stomach may exhibit a concave border (**lesser curvature**) and a convex border (**greater curvature**). The stomach also undergoes torsion in higher forms, so that it may finally lie across the long axis of the trunk. As flexion and torsion become pronounced during development of mammalian stomachs, the dorsal mesentery of the stomach (**mesogaster**) becomes twisted and finally suspended from the greater curvature, which was originally the dorsal border of the stomach. The part of the dorsal mesentery attached to the greater curvature is then called the **greater omentum.** Because it is derived from mesentery, it is a membranous sac, and because the mesentery became twisted, it encloses a **lesser peritoneal cavity** continuous with the main peritoneal cavity via an **epiploic foramen.**

The stomach of some vertebrates, especially of fishes, exhibits one or more blind pouches, or **ceca** (Fig. 11-11). The stomach of some marine teleosts is lined with a horny membrane that may exhibit spiny projections.

In many birds the stomach is divided into two regions: **proventriculus** and **gizzard** (Fig. 11-9). The proventriculus (glandular stomach) secretes a digestive enzyme. The gizzard is a thick muscular bulb strengthened by fibrous connective tissue, lined with a horny membrane, and often containing pebbles. The gizzard macerates the food, after which the partially digested mash enters the small intestine. The proventriculus and gizzard are best developed in birds that eat seeds and grain

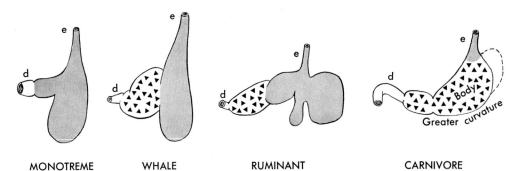

MONOTREME   WHALE   RUMINANT   CARNIVORE

Fig. 11-10. Distribution of typical gastric glands (triangles) and of esophageal-like epithelium (gray) in the stomachs of selected mammals. **d,** Duodenum; **e,** esophagus.

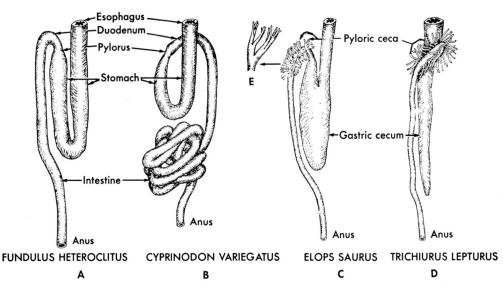
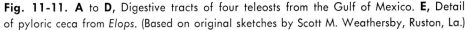

FUNDULUS HETEROCLITUS   CYPRINODON VARIEGATUS   ELOPS SAURUS   TRICHIURUS LEPTURUS
A   **B**   **C**   **D**

Fig. 11-11. A to **D,** Digestive tracts of four teleosts from the Gulf of Mexico. **E,** Detail of pyloric ceca from *Elops.* (Based on original sketches by Scott M. Weathersby, Ruston, La.)

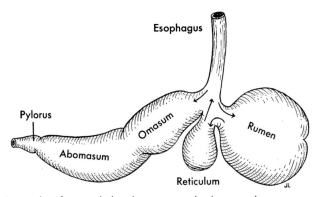

Fig. 11-12. Stomach of a cud-chewing mammal. Arrows show course of food.

and least developed in carnivorous birds.

The mammalian stomach is sometimes divided into several chambers. This is especially true in ruminants (Fig. 11-12). Grasses or grain is chewed briefly and then swallowed, after which it passes into the **rumen,** where preliminary digestion, especially by bacterial action, occurs. From the rumen food moves into the **reticulum,** the lining of which is honeycombed (reticulated) by ridges and deep pits. In the reticulum, food is formed into a cud, which is regurgitated at will to be leisurely rechewed. After thorough chewing of the cud, the food is again swallowed. This time it passes into the **omasum,** where salivary enzymatic action continues. Finally it enters the **abomasum.** This segment exhibits the usual varieties of gastric glands, and the lining (along with that of the omasum) exhibits longitudinal ridges (**rugae**) found also in other vertebrate stomachs. The rumen, reticulum, and perhaps the omasum should possibly be considered specialized parts of the esophagus analogous to the crop of birds (Fig. 11-10, ruminant).

The region of the mammalian stomach adjacent to the esophagus is the **cardiac portion,** and that preceding the pyloric sphincter is the **pyloric portion.** The remainder of the stomach, especially when it is not multichambered, is the **body.** Gastric glands of a specific histological nature usually characterize each of these areas—**cardiac glands** near the esophagus, **fundic glands** somewhere in the body, and **pyloric glands** near the pylorus. However, the precise distribution of these varieties differs among the species, and no general rule can be employed to predict where any one histological type will be found. The mucosa of the entire stomach of monotremes exhibits no typical gastric glands, and only the last segment of the stomach of whales and ruminants exhibits them (Fig. 11-10).

INTESTINE

The intestine is that part of the digestive tract between the stomach and the cloaca or anus. In fishes and caecilians the intestine is usually relatively short and straight, whereas in tetrapods it is tortuous and therefore proportionately longer. Compensating for the short absorptive area in the intestinal tract of lampreys, elasmobranchs, some of the older bony fishes, and an occasional teleost, a **spiral valve** or **typhlosole** (Fig. 11-1, dogfish) traverses much of the lumen. The part occupied by the spiral valve (**spiral intestine**) is equivalent to the small intestine of higher vertebrates. The **duodenum** is the first part of the small intestine. It is short, has characteristic glands, and receives the bile duct and one or two pancreatic ducts. In mammals the small intestine beyond the duodenum is divided into **jejunum** and **ileum** on the basis of differences in the shape of the villi and the nature of the glands and walls. The small intestine is the chief site of digestion and absorption.

The entire small intestine contains villi in most higher vertebrates. These are broad, leaf-like, and numerous at the upper end of the intestine and are longer and more finger-like farther along. Like the spiral valve of some fishes, villi increase the absorptive area. By contracting and relaxing they facilitate the emptying of the lacteals that are within them.

In tetrapods the junction between the small and large intestine is marked by an **ileocolic sphincter** (Fig. 11-13). In many fishes it is not possible to identify small and large intestines, since the size may not change abruptly and there is no ileocolic valve.

The large intestine is usually straight and short in fishes and amphibians. In reptiles, birds, and mammals it is divided into a colon and rectum. The colon begins at the ileocolic sphincter, and part of it (**transverse colon**) usually lies transversely in the body cavity. The terminal part of the colon (**descending colon**) passes caudad. The colon may have many coils, as in some carnivores and pigs (Fig. 11-1). In man alone the descending colon terminates in a **sigmoid flexure** (*sigma* = S-shaped). The

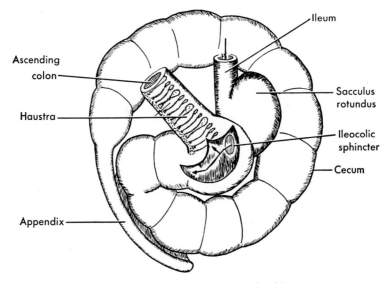

Ascending colon

Haustra

Appendix

Ileum

Sacculus rotundus

Ileocolic sphincter

Cecum

Fig. 11-13. Ileocolic junction, cecum, and appendix of rabbit.

rectum (*rectus* = straight) passes from the descending colon straight back to the cloaca or, when an adult cloaca is lacking, to the anus. The rectum of mammals is derived from partitioning of the cloaca. Therefore, the rectum of mammals is not homologous with the rectum of reptiles and birds.

Ceca. In addition to spiral valves and villi, **ceca** increase the surface area of the intestine. Although spiral valves are usually lacking in modern fishes, pyloric and duodenal ceca are common and may be numerous (Fig. 11-11). Up to 200 have been described in a mackerel. In some species adjacent ceca expand and unite to form a large sac.

Ceca beyond the duodenum are not common in fishes or amphibians, but they are the rule in reptiles, birds, and mammals. On the large intestine of amniotes, often at the ileocolic junction, is found an ileocolic cecum, usually two in birds (Fig. 11-1, turtle, chicken, fetal pig, and man). These ceca are sometimes long enough to equal in capacity the rest of the large intestine. They are sometimes coiled in animals feeding heavily on cellulose (Fig. 11-13), and may house cellulose-digesting bacteria as commensal parasites. The cecum

may be absent or small in turtles, crocodiles, woodpeckers, pigeons, a few other birds, and mammals with an insectivorous or carnivorous diet. Man and other primates also have a relatively short ileocolic cecum.

The ileocolic cecum of some mammals is abruptly reduced in diameter near the blind end and terminates in a **vermiform appendix** with a greatly restricted or semi-occluded lumen. Such is the case in monkeys, apes, man, rodents, rabbits, and numerous other mammals.

Ceca farther along on the colon are not rare by any means. *Hyrax*, for instance, has a large bicornuate cecum on the descending colon (Fig. 11-14).

The liver and pancreas arise in the embryo as though they were going to be ceca, and in the amphioxus the liver is, in fact, a cecum throughout life. The **rectal gland** of elasmobranchs is a cecum that secretes sodium chloride.

LIVER

The liver and common bile duct of vertebrates arises early as a hollow diverticulum, the **liver bud,** from the ventral wall of the foregut just caudal to the developing stomach (Fig. 11-15). The bud, occa-

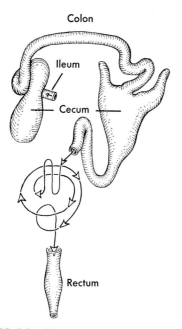

Fig. 11-14. Digestive tract of a mammal (*Hyrax*) with ceca on large intestine. Arrows indicate direction taken by missing segment of large intestine.

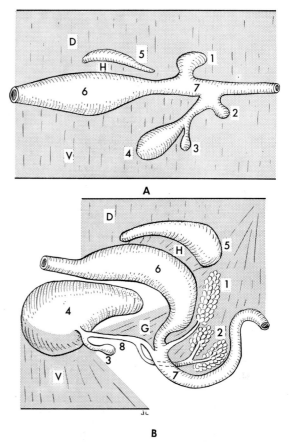

Fig. 11-15. Development of liver, pancreas, spleen, stomach, and associated mesenteries. **A,** Representing an earlier stage. **1,** Dorsal pancreatic bud; **2,** ventral pancreatic bud; **3,** gallbladder; **4,** liver bud in **A** and liver in **B**; **5,** spleen; **6,** stomach; **7,** duodenum; **8,** common bile duct. **D,** Dorsal mesentery; **G,** lesser omentum; **H,** gastrosplenic ligament; **V,** ventral mesentery, which remains as the falciform ligament in **B**.

sionally multiple, invades the ventral mesentery of the intestine and then grows cephalad into the ventral mesentery of the stomach. The distal tip of the bud gives rise to numerous sprouts that will become the lobes of the expanding liver. One sprout becomes the gallbladder. The anterior pole of the liver finally becomes anchored to the septum transversum by a coronary ligament.

Although most of the embryonic ventral mesentery disappears during subsequent development, the mesentery ventral to the duodenum and stomach that was invaded by the liver bud remains as the **hepatoduodenal ligament** connecting the duodenum with the liver, and as the **gastrohepatic ligament** connecting the pyloric end of the stomach with the liver. These two ligaments are continuous with one another and constitute, collectively, the **lesser omentum**. The omentum serves as a bridge transmitting the common bile duct, the hepatic artery, and the hepatic portal vein.

The shape of the adult liver conforms to the space available in the coelom. In forms with an elongated trunk the liver is elongated. In animals with short trunks it is broad and flattened. It is variously subdivided in different species but usually exhibits median and lateral lobes.

The liver is a digestive organ and manufactures bile, which aids in fat digestion and absorption. Other roles, associated with homeostasis—maintenance of a suitable internal milieu, are more vital. It helps regulate blood sugar levels by glycogenesis, glycogenolysis, and glyconeogenesis; it deaminates amino acids taken in as food and hence is a source of urea; and it manufactures several blood proteins including certain clotting substances. The fetal liver is an important source of red blood cells, and the adult liver excretes the breakdown products of hemoglobin as bile pigments. This list of functions is not exhaustive.

GALLBLADDER

One sprout of the liver bud expands to become the **gallbladder** and **cystic duct.** A gallbladder develops in most vertebrates including hagfishes; however, no gallbladder develops in lampreys, many birds, rats, perissodactyls, or whales. Since the gallbladder serves primarily to store bile for emulsifying ingested fats, and since animals lacking a gallbladder have little fat in their diet, loss of the genetic factors necessary to induce differentiation of a gallbladder was no disaster. Human beings live many years after surgical removal of the gallbladder, but they must avoid fats in the diet. Because of its embryonic origin from liver bud, the gallbladder is embedded in, or suspended from, one of the lobes of the liver and empties into the common bile duct. The terminal segment of the common bile duct is embedded in the wall of the duodenum for a short distance and is called the **ampulla of Vater.**

PANCREAS

The pancreas consists of two histologically distinct and functionally independent components—an exocrine portion secreting pancreatic juice into pancreatic ducts, and an endocrine (ductless) portion (the pancreatic islands) secreting the hormones insulin and glucagon into the bloodstream. Although the endocrine masses are almost always interspersed as islands among the exocrine cells, in a few vertebrates (agnathans, some teleosts) the endocrine masses are entirely apart from the exocrine pancreas. The endocrine pancreas is discussed in Chapter 17.

Generally, in fishes and lower tetrapods the pancreatic tissue is distributed more or less diffusely in the mesenteries, but in elasmobranchs and higher tetrapods the pancreas consists of a discrete ventral lobe, or **body,** and a dorsal lobe, or **tail.** In addition to pancreatic strands in the mesenteries, exocrine pancreatic tissue develops in lampreys and some teleosts among the cells of the intestinal epithelium. In the amphioxus no pancreatic tissue has been identified.

Typically, the pancreas arises as one or two ventral outgrowths from the liver bud, and as one dorsal outgrowth from the gut near the liver bud. The ventral anlagen invade the mesentery and form a ventral lobe (Fig. 11-15, *B, 2*). The dorsal anlage becomes the dorsal lobe. (In sharks the entire pancreas develops from a dorsal bud. In mammals it develops from one ventral and one dorsal bud.) It is likely that the development of three pancreatic buds is the more primitive condition.

Vertebrates may have as many pancreatic ducts as there are embryonic pancreatic buds. More often, one or more of the ducts lose their connection with the gut or bile duct, and all adult pancreatic tissue is drained by the remaining duct or ducts. Mammals illustrate the variety of conditions that exist. In sheep and man the duct of the dorsal pancreas (**duct of Santorini**) loses its connection, and the pancreas drains into the common bile duct. In other mammals such as pigs and oxen the ventral pancreas (**duct of Wirsung**) loses its connection, and the entire pancreas drains directly into the duodenum. In still other mammals such as horses, dogs, and cats both ducts remain. One of the two ducts is usually the larger, and the other may be referred to as the **accessory pancreatic duct.** As an anomaly in any individual, both ducts may remain functional.

CLOACA

The cloaca (L., *sewer*) is a chamber at the end of the digestive tract of most vertebrates. It receives the large intestine and the ducts of the urinary and genital tracts. It opens to the exterior via the **vent**. In cyclostomes and teleosts the cloaca is shallow during embryonic development and practically nonexistent in adults. In most placental mammals the embryonic cloaca becomes subdivided into dorsal and ventral passageways. The dorsal passageway becomes the rectum and the ventral passageway becomes the urogenital sinus (Fig. 14-35). The cloaca is discussed more extensively elsewhere.

Chapter summary

1. Major subdivisions of the digestive tract are the oral cavity, pharynx, esophagus, stomach, and intestine. Chief accessory digestive organs are the tongue, teeth, oral glands, pancreas, liver, and gallbladder.
2. The mouth is an aperture leading into the oral cavity.
3. The oral cavity is shallow in fishes but becomes increasingly prominent in tetrapods. Nasal passageways open into the oral cavity in lung-breathing vertebrates lacking a secondary palate.
4. Teeth typically arise from mesodermal papillae that secrete dentin and from ectodermal enamel organs that secrete enamel. Teeth are built on the same plan as placoid scales and may represent vestiges of an early dermal armor. In lower vertebrates, teeth are more numerous, more widespread in the oral cavity, more readily replaceable, less intimately associated with the jawbones, and more alike in all parts of the mouth. A few vertebrates in every class are toothless.
5. The primary tongue of fishes and tailed amphibians is a nonmuscular crescentic elevation in the floor of the pharynx overlying the basal hyoid skeleton. In higher amphibians a glandular field (tuberculum impar) contributes to the tongue, and in amniotes paired lateral lingual swellings also contribute. Tongue muscle is of myotomal origin.
6. Oral glands are scarce in water-dwelling vertebrates. They are numerous in most land-dwelling forms, except birds, and are specialized as poison glands in some reptiles and as salivary glands in mammals.
7. The pharynx is that portion of the foregut exhibiting visceral pouches in the embryo. The fate of the pouches depends on the adult mode of respiration. They become gill slits in gill-bearing forms but otherwise usually disappear. In mammals the pharynx is divided into nasal, oral, and laryngeal pharynges by the hard and soft secondary palate.
8. The esophagus connects the pharynx with the stomach, is short in neckless forms, and is not readily delimited from the stomach in most lower vertebrates. The avian crop sac is a diverticulum of the esophagus.
9. The stomach is a muscular enlargement at the base of the esophagus. It is compartmentalized in birds (proventriculus and gizzard) and ungulates (rumen, reticulum, omasum, and abomasum). It terminates at the pylorus.
10. The intestine is the chief site of absorption of digested foodstuffs. It is relatively straight in lower forms and tortuous in higher ones. The first part is the duodenum. Practically all vertebrates exhibit small and large intestines, the former with villi.
11. The liver, gallbladder, and pancreas arise as evaginations of the future duodenum. There are usually a single liver bud (occasionally two) and two or three pancreatic buds. One or more of the embryonic pancreatic ducts tends to disappear.
12. Numerous ceca occur along the digestive tract, chief among which are gastric and duodenal ceca in fishes, colic ceca in most vertebrates, and an esophageal cecum (crop) in birds. Some mammals exhibit a vermiform appendix at the termination of the colic cecum.
13. The cloaca is characteristic of most adult vertebrates, is practically nonexistent in modern fishes, and is absent in adult placental mammals. The mammalian rectum is a derivative of the embryonic cloaca.

Respiratory system

The respiratory system functions chiefly in the exchange of oxygen and carbon dioxide between the organism and the environment. In early vertebrate embryos respiration occurs directly between the individual blastomeres and their surroundings. Later, highly vascularized membranes may serve in some species for respiration during embryonic life. Larval fishes and larval amphibians use external gills, which usually project from the surface of the pharynx, and internal gills located in chambers in the pharyngeal wall. Adult fishes rely chiefly on internal gills; amphibians rely on external gills, lungs, skin, and the buccopharyngeal membranes. The remaining tetrapods rely chiefly on lungs for respiration. Accessory respiratory organs are present in some species. In this chapter gills, swim bladders, lungs, and accessory respiratory organs will be discussed in that order.

ADULT GILLS

Gills (branchiae) are respiratory organs typically associated with the pharyngeal wall of fishes and amphibians. No embryonic or adult amniote uses gills at any time. Since the gills of the dogfish are generalized in structure and relationships, they will be considered first.

Squalus. The pharyngeal wall of the dogfish *Squalus acanthias* is specialized for respiration. A series of five external gill slits are visible on the surface of the pharynx (Fig. 12-1). The formation of gill slits from pharyngeal pouches and ectodermal grooves has been described on p. 7. The slits in *Squalus* are not covered by an operculum, and so they are **naked gill slits.** If any of these slits is probed, the instrument will enter a gill chamber. The anterior and posterior walls of the first four gill chambers exhibit a gill surface or **demibranch.** The last (fifth) gill chamber lacks a demibranch in the posterior wall. The hyoid cartilages lie in the anterior wall of the first gill chamber. The relationships of the remaining demibranchs to associated structures are diagrammed in Fig. 12-1. The demibranch in the anterior wall of a gill chamber is a **pretrematic demibranch** (*pre* = before, *trema* = hole). The demibranch in the posterior wall of a gill chamber is a **posttrematic demibranch.** Separating the two demibranchs of a gill is an interbranchial septum. The septum is strengthened by delicate cartilaginous gill rays radiating from the gill cartilage. Gill rakers protrude from the cartilage into the pharynx. The two demibranchs of one gill, together with the interbranchial septum, the cartilage, the associated blood vessels, and the branchiomeric muscles, nerves, and connective tissues, constitute a **holobranch.** *Squalus* has four holobranchs. An additional demibranch occurs on the hyoid arch. Other fishes may have more or fewer demibranchs.

Water enters the pharynx via the mouth

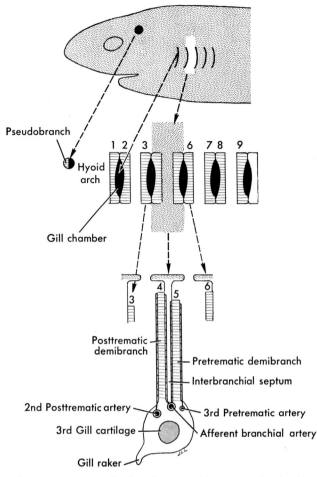

Fig. 12-1. Respiratory structures in *Squalus acanthias*, an animal with a spiracle, five naked gill slits, nine demibranchs (**1** to **9**), and four holobranchs. The second holobranch (fourth visceral arch) has been excised and demonstrated in cross section.

or spiracle and passes into the gill chambers, where it comes in contact with the demibranchs. The latter consist of many gill filaments rich in capillary beds. The capillaries are supplied by afferent branchial arterioles and drained by efferent branchial arterioles. Water is forced from the gill chambers by branchiomeric muscles.

Anterior to the first gill slit is a spiracle. In the embryo the spiracle is the same size as the gill slits, but it fails to keep pace in growth with the other slits. What appears to be a vestigial demibranch grows in its anterior wall. It is called a pseudobranch.

Other chondrichthyes. Most elasmo-branchs are of the pentanchid type; that is, they have five gill slits. *Hexanchus*, another shark, has six gill slits and a spiracle, and *Heptanchus* has seven gill slits and a spiracle, the largest number of slits in any jawed vertebrate. The five gill slits in adult rays are on the underside of the flattened body (Fig. 12-20), but the spiracle is located dorsally, behind the eyes, and is used for water intake. This development is an adaptation to the niche that rays occupy. Rays are bottom-dwellers and quantities of mud and sand would be taken into the ventral mouth if respiratory water were to enter by that route. In embryonic rays, however, the gill slits and spiracle lie in series on the side of the pharynx, as in sharks. In some adult elasmobranchs, the spiracle is closed· by a membrane.

The branchial region of *Chimaera* is further modified. The spiracle is closed, there are only four adult gill pouches, the interbranchial septa are short and do not reach the skin, and a fleshy operculum extends backward from the hyoid arch and hides the gills. In several of these respects *Chimaera* resembles bony fishes.

Osteichthyes (Fig. 12-2). The gill apparatus of bony fishes is basically similar to that in *Squalus*. Five gill slits is the rule, but there are exceptions. The chief differences between elasmobranchs and bony fishes will be listed.

PRESENCE OF BONY OPERCULUM. An operculum, a bony flap arising from the hyoid arch, projects backward over the gill chambers. The result is an opercular cavity that opens by a crescentic cleft (a small aperture in eels) just anterior to the pectoral girdle. In a few fishes the left and right opercular chambers open by a common aperture in the midventral line. Opercular movements assist in the expulsion of water from the gill chambers. The operculum of many bony fishes (sturgeon, spoonbill, *Polypterus*, *Lepidosteus*, and many teleosts) bears on its inner surface either a functional gill or a pseudobranch.

REDUCTION OF INTERBRANCHIAL SEPTA. The interbranchial septa in bony fishes do not reach the skin, and in teleosts the interbranchial septa are shorter than the demibranchs.

ELIMINATION OF SPIRACLE. A spiracle occurs in the relatively ancient chondrosteans,

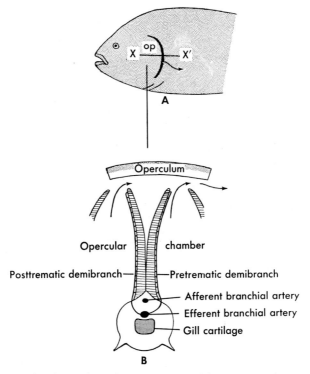

Fig. 12-2. A, Gill chambers of a teleost are covered by an operculum, **op,** which grows caudad over the gills from the hyoid arch. **B,** Cross section of one holobranch in the plane **X** to **X'**. Note the absence of an interbranchial septum in the teleost holobranch. Arrows indicate direction of efferent water flow.

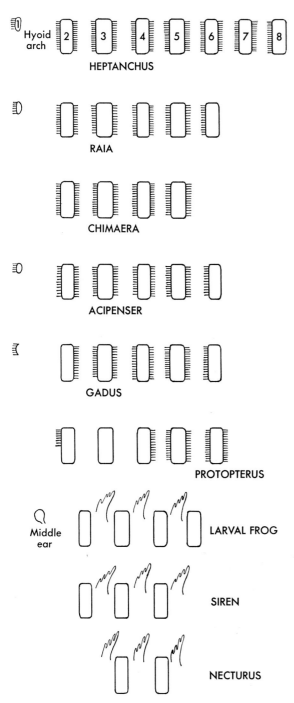

Fig. 12-3. Open visceral slits in selected aquatic vertebrates and distribution of gill surfaces (horizontal lines) in fish. *Heptanchus* closely resembles the basic pattern. *Gadus* (the cod) has a pseudobranch on the operculum and the spiracle is closed. **1** to **8**, Visceral slits. External gills are indicated in the amphibians.

but it closes during embryonic life in modern fishes and in lungfishes.

REDUCTION IN NUMBER OF DEMIBRANCHS (Fig. 12-3). Most teleosts lose the demibranch on the hyoid arch. It is retained in the more ancient Chondrostei and Holostei. Additional demibranchs are lost in some lungfishes. Since lungfishes breathe partly by lungs, the loss was no disadvantage.

Agnathans. Living agnathans have six to fifteen pairs of gill pouches. *Myxine glutinosa* usually has six pairs of gill pouches but occasionally there are five or seven (Fig. 12-4). The various species of *Bdellostoma* have from five to fifteen pairs, and within the species *B. stouti* the number varies between ten and fifteen, although twelve pairs are common. Among the lampreys, *Petromyzon* has eight embryonic and seven adult pairs.

The gill pouches in agnathans are connected to the pharynx by afferent branchial ducts and to the exterior by either separate or common efferent branchial ducts. Lampreys and *Bdellostoma* have separate ducts that open by separate apertures. In *Myxine* and its relatives the efferent ducts unite to open via a common external aperture on each side.

The pathway of flow of the respiratory stream differs in lampreys and hagfishes. In the former the water can be taken in through the external gill slits and ejected by the same route. This is essential when the lamprey is attached by its buccal funnel to a host fish, since the nasal duct does not extend to the pharynx. In hagfishes the water enters via the naris and passes along the nasopharyngeal duct into the pharynx.

The flow of respiratory water in hagfishes is maintained by the pumping action of a **velar chamber** at the anterior end of the pharynx, into which the nasopharyngeal duct empties. The walls of the velar chamber are composed partly of constrictor muscles and are strengthened by posterior projections of the first branchial cartilages. The walls pulsate 50 to 100 times per minute in alert animals and 25 to 30 times per

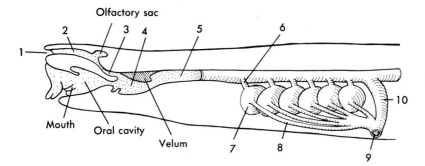

Fig. 12-4. Respiratory pathway in the hagfish *Myxine glutinosa*, lateral view. **1**, Naris; **2**, nasal duct; **3**, nasopharyngeal duct; **4**, velar chamber; **5**, pharynx; **6**, afferent branchial duct; **7**, gill pouch; **8**, efferent branchial duct; **9**, common external gill aperture; **10**, pharyngocutaneous duct. The gill pouches and ducts are paired, but only the left side is shown.

minute in sleeping animals. The action directs and pumps water into the pharynx and thereby creates a vacuum, which is filled by additional water drawn in through the nostrils and nasopharyngeal duct.

In hagfishes, behind the gills on the left side only, is a pharyngocutaneous duct connecting the pharynx with the last efferent branchial duct or with the exterior. Embryonically, the pharyngocutaneous duct forms just like the gill pouches and their ducts, and it is probable that it is a modified gill pouch. Periodically, the pharynx accumulates debris or particles too large to enter the afferent branchial ducts. These particles are forcefully ejected through the pharyngocutaneous duct to the exterior. An interesting account of the respiratory mechanisms of hagfishes will be found in a book by Brodal and Fänge.[4] References to recent literature will also be found in that volume.

During metamorphosis in lampreys the pharynx becomes subdivided into an esophagus dorsally and pharynx ventrally so that in adults the pharynx terminates blindly (Fig. 12-5).

Excretory function of gills. Chloride-secreting glands occur on the gills of many marine fish. These secrete excess chlorides into the excurrent water streams. Gills also serve in marine fishes for excretion of nitrogenous wastes. The excretory func-

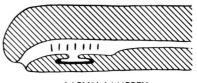

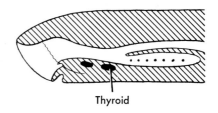

Fig. 12-5. Location of gill slits in larval and adult lampreys. A subpharyngeal gland, shown in black, is open to the pharynx in the ammocoete and isolated as a ductless thyroid gland in the adult lamprey.

tion of the gills supplements excretion by other organs, including rectal glands and kidneys.

LARVAL GILLS

External gills arising from the exposed outer surfaces of the branchial arches are typically larval and hence usually temporary organs. They occur in lampreys, a few larval fishes including *Polypterus* (an an-

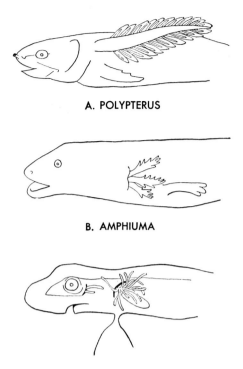

A. POLYPTERUS

B. AMPHIUMA

C. SCYLLIUM

Fig. 12-6. Larval external gills of a bony fish, **A**, an amphibian, **B**, and an elasmobranch, **C**. In *Amphiuma* the operculum is growing caudad over the gills.

cient chondrostean, Fig. 12-6, *A*), lung-fishes, some larval teleosts, and in all larval amphibians including caecilians.

In tadpoles, before the gill slits perforate, external gills develop as finger-like elevations on the outer surfaces of visceral arches III to V. A rudimentary gill may develop on arch VI. Later, the mouth perforates and pharyngeal pouches II to V rupture to the exterior to form four gill slits. The first pouch typically becomes the middle ear cavity. When the gill slits have developed, their walls become folded to form a second set of gills, and from the hyoid arch a fleshy operculum grows backward over the entire gill apparatus and encloses it in an opercular cavity, as in bony fishes. This second set of gills is referred to as internal gills because they lie within the opercular chamber. After vascularization of the internal gills, the first set ceases

to function and atrophies. The internal gills serve throughout most of the tadpole stage, which is several months in frogs and years in bullfrogs.

The operculum grows caudad beyond where the anterior limbs will later sprout, and the imprisoned limbs must therefore push through the operculum. Histolytic substances given off by disintegration of the internal gills at metamorphosis erode the operculum so that the anterior limbs may break through. The left and right opercular cavities become confluent ventrally, and the right opercular cleft usually closes, whereas the left remains as a spiracle (*spiraculum* = air hole), not homologous with the spiracle of fishes.

In larval urodeles and caecilians a vestigial operculum makes an appearance, but it becomes only a small fold on the hyoid arch. At metamorphosis the gill slits usually close and the gills are resorbed. However, two families of urodeles (Proteidae and Sirenidae) retain gills throughout life, thereby exhibiting incomplete metamorphosis. The external gills of the adult *Necturus* are found on visceral arches III to V. The slits represent embryonic pouches III and IV. In two other families of urodeles (Table 3-1, p. 43), external gills disappear but a slit remains.

In elasmobranchs with prolonged periods of development within the uterus or eggshell, larval gills form and project into the uterine fluid or under the eggshell. They serve not only for respiration but also for the absorption of nutrients.

PHYLOGENY OF GILLS

There has been a trend toward reduction in the number of gill slits in modern vertebrates, as indicated by the following observations:

1. The spiracle, when present, arises during embryonic life in series with the gill slits and sometimes bears a pseudobranch. This strongly suggests that the spiracle is a vestigial gill slit.

2. The spiracle remains open in phylogenetically ancient fishes but in modern

forms the first pharyngeal pouch either fails to perforate or closes during later development.

3. The last pharyngeal pouch fails to achieve an opening to the exterior in some relatively recent fishes.

4. Demibranchs fail to develop in some gill chambers of certain fishes.

5. Larval amphibians develop a larger number of pharyngeal pouches than rupture to form gill slits.

6. The number of open slits in amphibians is reduced when compared with the number in fishes.

7. Amniote embryos develop a series of four to six pharyngeal pouches that may rupture to form temporary slits in lower amniotes but tend not to rupture in higher amniotes.

From the foregoing facts we may deduce that modern fishes and amphibians are descendants of vertebrates that had a larger number of functional gill slits and that amniotes may be descendants of gill-breathing vertebrates.

How much reduction may have occurred in the number of gill slits? If the number of gill chambers in generalized fishes such as elasmobranchs is a reasonable criterion, it may be concluded that the number of gill pouches in jawed vertebrates may have been not much larger than seven. In ostracoderms, the most ancient known fishes, the largest number yet discovered is ten.

Did pharyngeal pouches and slits originally serve for respiration? Or did they have a still more primitive function in the earliest chordates? The suggestion has been advanced that their original use may have been the ensnaring of minute foodstuffs from an incurrent water stream in the anterior end of the digestive tract (filter feeding). In the lowest living chordates (ammocoetes, amphioxuses, and urochordates) food particles collected by the ciliated branchial apparatus represent the entire diet. Such a hypothesis is attractive because it furnishes an explanation for the presence of the respiratory system in the anterior wall of the digestive tract. The hypothesis is, of necessity, based on speculation, but without hypotheses, science could make little progress.

SWIM BLADDERS AND THE ORIGIN OF LUNGS

During embryonic life nearly every vertebrate from the most ancient to the most recent develops an unpaired evagination from the pharynx or esophagus that terminates as one or a pair of sacs filled with gases derived directly or indirectly from the atmosphere. These pneumatic sacs are called either lungs or swim bladders (Fig. 12-7), depending on their chief role. The only adult vertebrates that do **not** have such pneumatic sacs are agnathans, cartilaginous fishes, a few marine teleosts, some bottom-dwellers such as flounders, and a few tailed amphibians that have not fully metamorphosed. Furthermore, some of the elasmobranchs and teleosts that lack these organs as adults develop them in the embryo. Except for the agnathans, about whose ancestry we know nothing, it can be said with considerable confidence that vertebrates that fail to develop pneumatic sacs have lost the genetic factors necessary to induce them. After evaginating, the pneumatic organs may retain their unpaired connection with the foregut, in which case the connection may serve as an air duct, or the embryonic connection may be lost, whereupon the adult sac is ductless.

The sacs are known as lungs when they function as respiratory organs and take in atmospheric air via the duct, remove oxygen, add carbon dioxide, and return the gas to the atmosphere. This is one role of these organs in lungfishes, probably in some ancient ray fins (*Polypterus*), and in tetrapods. Lungs connect to the foregut ventrally via an unpaired air duct, but distally they are typically paired.

The sacs are known as swim bladders when their function is primarily hydrostatic, as in fishes with gills. By hydrostatic is meant that the contained gas in the sac

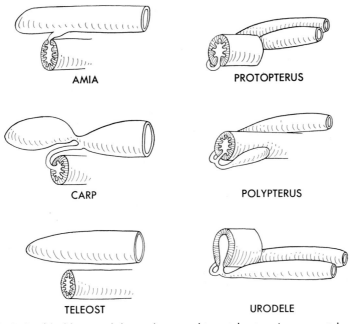

Fig. 12-7. Swim bladders and lungs in aquatic vertebrates. In many teleosts the embryonic pneumatic duct later closes, and the swim bladder thereafter has no connection with the gut.

is reflexly regulated as to volume and thereby affects the specific gravity of the animal. If the fish is experimentally placed under increased pressure such as would occur at greater depths, the gas in the swim bladder increases in volume. The fish thereby maintains an appropriate bouyancy. When serving as a swim bladder, the pneumatic organ may lose its duct.

The gas in the hydrostatic swim bladder comes from the blood. It is actively transported into the bladder from a remarkable network of small arteries and veins (rete mirabile) in the lining of the bladder, which, because of its appearance, is called the **red body** (**red gland**). The arterial supply to the gland is from the celiac artery, and the veins empty into the hepatic portal vein. The gas is reabsorbed in an area of modified epithelium near the caudal end of the bladder, or in physostome (open-duct) fishes it may be bubbled through the mouth. If a pneumatic duct is present, air can also be gulped, but it is mostly provided by the rete, and the

duct is an escape route. The gases differ among fishes. Some swim bladders contain almost pure (99%) nitrogen, some contain up to 87% oxygen, and all contain at least traces of four atmospheric gases—nitrogen, oxygen, carbon dioxide, and argon. In deepwater fishes the nitrogen may be transported from the blood to the chamber against a nitrogen pressure of as high as 10 atmospheres.

In ray-finned fishes the swim bladder is an unpaired sac lying in the roof of the coelom ventral to the kidneys (Fig. 14-14). It may bulge into the coelom and may be short or long. There may be one or more diverticula, including extensions into the head. The walls contain elastic tissue and smooth muscle and may be strengthened by vertebral processes. The internal lining is relatively smooth. The pneumatic duct usually connects the swim bladder with the esophagus, rarely with the stomach or pharynx. Except in lower ray fins and primitive teleosts, the duct is confined to the embryo. When present, the duct usually enters the esophagus dorsally, but it may enter on the right or left (Fig. 12-7).

Swim bladders in some species perform other functions in addition to their hydrostatic role. In one suborder of teleosts (Cypriniformes), a series of small bones, the weberian ossicles, connects the anterior end of the swim bladder and the sinus impar, a projection of the perilymph cavity (Fig. 16-6). Low-frequency vibrations of the gas within the swim bladder, evoked by waves of similar amplitude in the water, are transmitted by the ossicles to the membranous labyrinth. Therefore these fish can hear.

In certain herring-like teleosts a diverticulum of the swim bladder comes into direct contact with the membranous labyrinth, since there are no interposed bones. It has not been clearly established that this modification is a hearing device. There is a possibility that the specialization may reflexly regulate gas pressure in the swim bladder or perhaps be related to depth perception.

In a few fishes, contractions of extrinsic striated muscles (supplied by spinal nerves) cause the swim bladder to emit thumping sounds or may force air back and forth between one chamber and another through muscular sphincters, as in croakers and grunters. The urinary bladder of one species of marine teleost from Indian waters passes through the air bladder. The advantage, if any, is not known.

The striking similarity between swim bladders and lungs suggests a phylogenetic relationship between the two. Exhibition in the placoderm Bothriolepis of a terminal pair of ventral pharyngeal pouches, which seem to have served as pneumatic organs, could be an instance of convergence or may provide a clue to the phylogeny of swim bladders and lungs. Since placoderms are very ancient vertebrates, pneumatic evaginations of the foregut would have been early acquisitions indeed. It is possible that swim bladders and lungs both arose phylogenetically from a pair of expanded posterior pharyngeal pouches in series with the gill chambers. No vertebrate exhibits both a swim bladder and a lung. Supporting the view that lungs (at least) may be derivatives of pharyngeal pouches is the fact that the musculature in the proximal portion (larynx) of tetrapod air ducts is branchiomeric, as are the supporting skeletal elements. Since air ducts are always unpaired at their origin from the foregut, it must be assumed that the originally paired pharyngeal diverticula fused in the midventral line or else that one of the pouches became lost. The latter actually happens during embryonic development in the lungfish Neoceratodus. The migration of the pneumatic duct of fishes to a lateral or dorsal aspect would have been a later innovation. Present evidence permits only speculation on this matter. Complete closure of the duct in some species seems to have been a relatively recent mutation.

It seems likely to many authorities who have studied the problem that the original function of pneumatic evaginations from the foregut may have been to serve as accessory respiratory organs. It has been pointed out that the hydrostatic swim bladder may serve as an accessory respiratory organ in cases of need.[12] Certainly, the respiratory function of these sacs is ancient, and lungs were present in crossopterygians long before their descendants became permanent land-dwellers.

LUNGS AND THEIR DUCTS

The lungs of the chondrostean fish Polypterus are asymmetrically bilobed (Fig. 12-7), and the duct opens into the pharynx ventrally. The lining is not smooth as is that of swim bladders, and there are at least a few furrows that increase the surface in contact with the air.

The lungs of dipnoans may be bilobed, as in African and South American lungfishes (Fig. 12-8), or there may be a single sac, as in Australian lungfishes. The duct in dipnoans opens into the esophagus slightly to the right of the midventral line, but the sacs grow into the dorsal mesentery of the gut, where they lie in the adult.

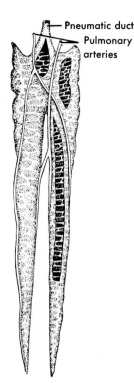

Pneumatic duct
Pulmonary
arteries

Fig. 12-8. Lungs of the African lungfish.

There is some subdivision of the lining into pockets. In both *Polypterus* and lungfishes the lungs are supplied by pulmonary arteries off the sixth embryonic aortic arch, but in *Polypterus* the venous return is to the hepatic veins, and in dipnoans it is to the left atrium, as in tetrapods. There are authorities who refer to any pneumatic sac in fishes as a swim bladder even though it performs a respiratory role, but this is a matter of semantics.

Tetrapod lungs have an unpaired trachea that enters the pharynx ventrally. They are supplied by pulmonary arteries arising from the embryonic sixth aortic arch and are drained by veins emptying into the left atrium. The lining is septate or pocketed. The lungs and air ducts of successively higher tetrapods exhibit increasing specializations. The lungs of amphibians are simplest and resemble those of lungfishes.

Embryogenesis of lungs. Lungs in tetrapods arise as midventral evaginations from the caudal floor of the pharynx (Fig. 1-5). The lung bud in amphibians commences as

a solid outgrowth but soon develops a lumen. In higher vertebrates the bud is hollow from the beginning. The median opening in the pharyngeal floor becomes the glottis. The unpaired lung bud elongates only slightly before bifurcating to form the primordia of the two lungs (Fig. 12-16). The primordia push caudad underneath the foregut to finally bulge into the embryonic coelom lateral to the heart. As they push into the coelom, the primordia carry along an investment of peritoneum, which becomes the visceral pleura. The lungs of amphibians and of many reptiles cease development after two relatively simple sacs have been formed within the coelomic cavity, but those of higher reptiles, birds, and mammals become more complex.

Amphibians

The lungs of amphibians are two simple sacs (Fig. 12-9), elongate in tailed amphibians and bulbous in anurans since the latter have a foreshortened trunk. They occupy the pleuroperitoneal cavity along with other viscera. The left lung is usually the longer except in caecilians, in which it is rudimentary. The internal lining of the amphibian lung may be smooth throughout, there may be simple sacculations in the proximal portion, or the entire lining may be pocketed.

The trachea is usually very short. Its wall is reinforced by small cartilaginous plates or open rings. The trachea usually bifurcates to form two short bronchi lined with cilia, but in some lower urodeles bronchi are absent, and the two lungs are confluent with the trachea.

The larynx is simplest in some urodeles such as *Necturus*, in which a pair of lateral cartilages surrounds the glottis just behind the hyoid apparatus. In most amphibians there are arytenoid and cricoid cartilages (Fig. 12-10). Stretched across the laryngeal chamber of frogs and toads is a pair of vocal cords. Air forced back and forth between the oral cavity and lungs causes the vocal cords to vibrate and produce various calls. The larynx is therefore truly a voice

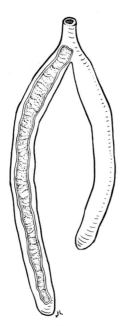

Fig. 12-9. Lungs of *Necturus*.

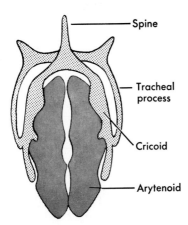

Fig. 12-10. Laryngeal skeleton of a frog. The glottis lies between the two arytenoid cartilages.

box in such forms. Males of many species of both frogs and toads possess paired (occasionally median) sac-like evaginations of the oral cavity (vocal sacs, Fig. 11-8), which lie under the skin and act as resonating chambers that amplify the sounds. Most female anurans and urodeles of both sexes emit weak sounds or none that are audible. However, the female *Rana pipiens* is quite a conversationalist!

The lungs of perennibranchiate urodeles

apparently function primarily as hydrostatic organs except when the dissolved oxygen in the water is insufficient to permit branchial respiration. A few urodeles (plethodonts) develop no lung bud and lack both gills and lungs throughout life. An especially vascularized region of the pharyngeal and esophageal lining is used as a respiratory membrane, along with the skin. Salamanders inhabiting swift mountain streams may have only vestigial lungs a few millimeters in length. Perhaps reduction of the lungs made it possible for these species to inhabit this particular niche in nature, since buoyancy would be a disadvantage in swiftly flowing currents,

Amphibians force air into their lungs by a pumping action of the floor of the oral cavity in association with the action of valves that surround the external nares. In inhaling, the external nares are opened and the floor of the oral cavity is depressed by contraction of the mylohyoid and other muscles. Atmospheric pressure forces the air into the oral cavity. The nares are then closed by smooth muscle action and the floor of the mouth is elevated. This action forces the air in the only direction available, through the glottis and into the lungs. (The esophagus is tightly constricted, and the eustachian tubes lead only to the middle ear cavity.) In male anurans, opening the entrance to the vocal sacs would also force air to enter those chambers. In exhaling, the body wall muscles compress the contents of the pleuroperitoneal cavity and thus force the air up into the oral cavity. It may then be pumped back into the lungs again or released through open nostrils. Urodeles gulp air as well as inhale via the external nares, but the floor of the mouth must still force the air into the lungs. In exhaling, some urodeles with open gill slits (*Siren*, for example) eject the air forcefully through the slits.

Reptiles

In *Sphenodon* (Fig. 12-11) and snakes the lungs remain simple sacs. The posterior

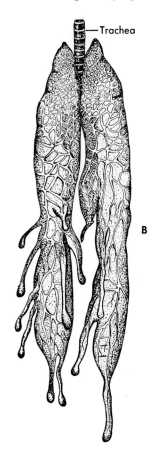

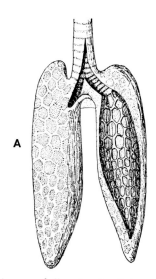

Fig. 12-11. Lungs of lizards. **A,** *Sphenodon.*
B, Chameleon.

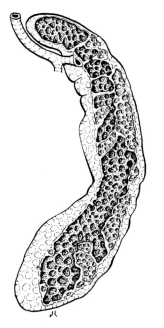

Fig. 12-12. Left lung of the lizard *Heloderma.*
This is a more specialized lung than that of
Sphenodon illustrated in Fig. 12-11.

third of the lining of the snake lung is septate and is filled with stored (residual) air.
In higher lizards, crocodilians, and turtles
the septa are so constructed that there are
numerous large chambers, each with a mul-
titude of individual subchambers (Fig.
12-12). The trachea bifurcates to form two
bronchi, and from each bronchus arise
numerous bronchioles, which lead to the air
chambers. These lungs are spongy because
of the numerous pockets of trapped air.

The left lung in legless lizards and in
snakes is rudimentary or absent altogether,
except in occasional forms such as black-
snakes. An enormous diverticulum of the
left lung extends into the neck region of
the puffing adder. Inflation of the diverticulum causes the neck to spread characteristically, and superinflation of the
lungs causes the body to swell. In the
spotted king snake the lung and its bronchus extend fully two thirds the length of
the body.

Whereas in amphibians and *Sphenodon* the bronchus perforates the anterior pole of the lung, in amniotes it usually enters at a hilus on the median aspect.

The reptilian trachea is long, since it must traverse the neck. Tracheal cartilages form complete rings in the cephalic portion and are generally incomplete dorsally in the remainder of the trachea except in crocodilians, in which the rings are complete and bony. Irregular cartilages support the walls of the bronchi. In some turtles and crocodiles the trachea is convoluted. In turtles this convoluted trachea is probably related to the fact that the neck may be drawn into the shell or extended.

The larynx of most reptiles shows little advance over that of amphibians, since it exhibits a pair of arytenoid cartilages and a cricoid ring. Crocodilians add a thyroid cartilage, found otherwise only in mammals. An epiglottis-like fold of mucous membrane lies anterior to the glottis in some reptiles and may be the precursor of the mammalian epiglottis. Some lizards have vocal cords that produce guttural sounds, but most reptiles lack these and are silent except for hissing sounds or for the bellowing sounds emitted by male alligators during the breeding season. The anterior wall of the larynx of some male lizards (*Anolis* and others) exhibits a gular pouch, which, when inflated with air, causes a ballooning of the ventral surface of the neck. The beautiful bright crimson color, especially when viewed in sunlight, has sex appeal for females of the species. Elongated processes of the hyobranchial apparatus assist in supporting the gular pouch.

Although many reptiles utilize the floor of the oral cavity as an air pump, others take advantage of atmospheric pressure to help fill the lungs. The long bony ribs are elevated and this action enlarges the body cavity. The pressure in the coelom around the lungs falls, and just as quickly atmospheric pressure inflates the lungs until they occupy the additional available space. By this method the lungs are passively rather than forcibly inflated. Turtles cannot use the ribs for respiration, since they are fused with the shell, but muscular peritoneal membranes increase the cavity (for inhalation), and others compress the lungs (for exhalation).

Oblique septum and pleural cavities of reptiles and birds. Most reptilian lungs occupy the pleuroperitoneal cavity along with the other viscera, as in anamniotes. A few higher reptiles—crocodilians, snakes, and some lizards—and all birds develop membranous folds of the dorsal and lateral parietal peritoneum, which grow into the embryonic pleuroperitoneal cavity and enclose each lung in an individual pleural cavity (*pleura* = rib). In birds the resulting partition is called the oblique septum because of its orientation. As a result, these specific reptiles and birds have three subdivisions of the coelom: (1) the pericardial cavity found in all vertebrates, (2) the paired pleural cavities, and (3) the abdominal cavity containing the other coelomic viscera. Mammals achieve the same subdivisions by developing the muscular diaphragm. The septum transversum at the caudal end of the heart participates in the formation of the mammalian diaphragm and the oblique septum of birds.

Birds

The lungs and their ducts in birds are modified in contrast with those of generalized reptiles in several respects: (1) formation of extensive diverticula (air sacs) of the lungs that invade most parts of the body, (2) the anastomosing nature of the air ducts within the lungs so that no passage terminates blindly within the lung, (3) development of a syrinx superceding the larynx as a vocal organ, and (4) isolation of the lungs in pleural cavities. Air sacs and pleural cavities are not unique in birds, since they occur also in a few of the more specialized reptiles.

Blind, thin-walled, distensible diverticula of the lungs, known as air sacs, extend into all major regions of the body. They lie

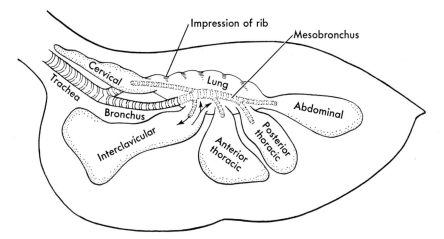

Fig. 12-13. Air sacs of a bird. Axillary sacs are less common and are not illustrated. Air leaves the sacs via recurrent bronchi, which lead to the respiratory surfaces of the lung. The position of one recurrent bronchus is indicated by the double arrow.

between layers of pectoral muscle, project among the viscera, and even penetrate the bone marrow cavity (Fig. 12-13). Air sacs may be thought of as the distended blind ends of certain bronchi that project from the surface of the lung. Most birds have five or six pairs of air sacs: (1) cervical sacs at the base of the neck, (2) inter-clavicular sacs dorsal to the furcula and sometimes united across the midline, (3) anterior thoracic sacs lateral to the heart, (4) posterior thoracic sacs within the oblique septum, (5) abdominal sacs extending caudad among the abdominal viscera, and (6) axillary sacs (less common) lying between two layers of pectoral muscle. Air sacs are found in reptiles but not commonly. In chameleons they extend among the viscera, often as far caudad as the pelvis (Fig. 12-11, *B*). In dinosaurs they extended into the vertebrae.

From the air sacs long, slender diverticula penetrate the marrow cavity of many of the bones via pneumatic foramina. Pneumatic bones are generally well developed in flying birds, but this is not universally true. Gulls have none, and in many small fliers pneumatic bones are poorly developed. Most ratities lack them, as did the ancient *Archaeopteryx*.

Air sacs have a poor vascular supply and contain no respiratory epithelia; hence, they play no direct role in gaseous exchange. They do play an important accessory role in respiration. During flight they are constantly compressed by the fluctuating pressures of adjacent muscle masses and by rhythmical movements of the appendages and other body parts. They thus serve as a bellows and maintain a constant flow of air over the respiratory epithelia. At other times, body wall muscles inserted on the ribs and oblique septum apparently suffice to operate the bellows. A thermoregulatory role has also been ascribed to to the air sacs, since they may dispel excess heat. Much remains to be learned about respiratory mechanisms in birds.

Air ducts and air flow. The air duct system within the avian lung is unique (Figs. 12-13 and 12-14). A bronchus enters each lung and continues toward the caudal pole. Within the lung the bronchus, called a **mesobronchus** in this location, gives off ducts, which emerge from the surface of the lung to enter each air sac of the neck and thorax. The mesobronchus finally emerges at the caudal pole to enter the abdominal air sac. Within the substance of the lungs, each mesobronchus receives **secondary bronchi.** The latter are interconnected by many small ducts of uniform

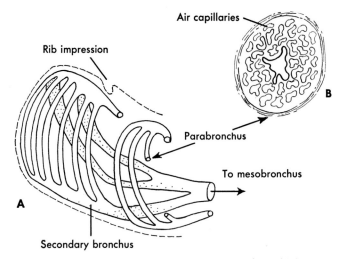

Fig. 12-14. A, Small section of bird's lung, showing parabronchi interconnecting larger channels. B, Parabronchus has been transected to display the air capillaries. The section illustrated is on the airflow path between an air sac and the mesobronchus. (A based on Locy and Larsell.[77])

diameter, the **parabronchi**. From each parabronchus tiny air capillaries sprout as diverticula, which loop back into the lumen of the parabronchus. These air capillaries contain the respiratory epithelium. Each air sac is connected with the secondary bronchi within the lungs by **recurrent bronchi**, which return the air from the sacs to the duct system within the lung.

During embryonic life the trachea, bronchi, secondary bronchi, and parabronchi arise as sprouts off the unpaired lung bud, which is a midventral pharyngeal diverticulum. The recurrent bronchi, however, sprout from the air sacs and invade the lung to establish connections with secondary bronchi.

Inhaled air, under atmospheric pressure, passes nonstop through the mesobronchi and into the air sacs, which are thereby inflated. The next compression of the sacs forces the air, via the recurrent bronchi, into the secondary bronchi and parabronchi within the lung. It is prevented from reflux directly into the mesobronchus by valves. The air flows gently and steadily over the respiratory surfaces of the air capillaries of the parabronchi. The air then returns to the mesobronchus, bronchus, trachea, pharynx, nasal passageway, and external nares, where it escapes. Lacking blind endings as it does, the anatomosing duct system of the lungs makes possible a relatively free and steady flow of air and yet provides an ample respiratory surface. Since the air in the air capillaries is constantly and completely replaced as a result of the bellows action of the air sacs, bird lungs contain only completely fresh air. This is in contrast to the condition in the lungs of other vertebrates, in which there is always residual unexpired air partly depleted of its oxygen content. The respiratory system of birds is therefore highly efficient and well adapted to satisfy the oxygen demand resulting from sustained flight. At rest the oxygen demand is lower, and the flow is presumably less swift.

Trachea, syrinx, and larynx. The length of the trachea in birds is generally proportional to that of the neck, but it may be longer and may exhibit convolutions. Tracheal rings are bony and are often complete dorsally. At the bronchial bifurcation is a small or large syrinx, a special voice box found only in birds (Fig. 12-15).

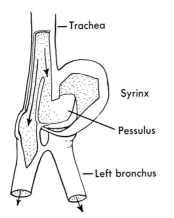

Fig. 12-15. Asymmetrical bronchotracheal syrinx of a canvasback duck. Arrows indicate path of inhaled air.

Syringes are of three types: bronchotracheal, tracheal, and bronchial.

In the **bronchotracheal** type of syrinx the last several tracheal rings and the first bronchial half rings support the walls of an expanded resonating chamber (tympanum), into which project membranous folds of the lining of the syrinx. A bony pessulus bearing a semilunar membrane may be present within the chamber. When air is expelled from the lungs and the striated syringeal muscles are contracted, the membranes become taut, and bird calls and songs are produced. The bronchotracheal syrinx may be median or asymmetrical.

The other two types of syringes are simpler. In the **tracheal syrinx** the lateral portions of the last several tracheal rings are absent, and the resulting membranous wall vibrates and thereby produces the sound. In the **bronchial syrinx** the membrane between two bronchial cartilages becomes folded into the lumen of the bronchus when the cartilages are drawn together. Vibration of this simple vocal cord produces the sound.

The larynx of birds lacks vocal cords, and is functionless as a vocal organ. **Arytenoid** cartilages or bones lie at the margins of the glottis, and a **cricoid** cartilage follows. The cricoid often exhibits a separate dorsal

segment, the **procricoid.** There is no thyroid cartilage.

Mammals

The lungs of mammals are multichambered and usually divided into lobes, with more lobes on the right. Man has two left and three right lobes (Fig. 12-16); rabbits have three lobes on each side, but the right posterior lobe is subdivided; cats have three left lobes, four right ones, and several are subdivided. The lungs of whales, sirenians, elephants, perissodactyls, and *Hyrax* lack lobes, and in monotremes and rats, among others, only the right lung is lobed.

The trachea, supported by incomplete cartilaginous rings united dorsally by smooth muscle, bifurcates to become two primary bronchi. Each bronchus penetrates a lung and divides into secondary and tertiary branches, which give rise to many bronchioles. The latter redivide multitudinously. The walls of the bronchi and the larger bronchioles are strengthened by irregular cartilaginous plates, which finally disappear in the smaller branches. Terminal bronchioles lead into delicate thin-walled **alveolar ducts,** the walls of which are evaginated to form clusters of **alveoli,** or respiratory pockets, estimated at 400,-000,000 in man. Each alveolar duct is similar to, but much smaller than, the simple lungs of lower tetrapods.

The skeleton of the larynx of mammals consists of paired **arytenoid cartilages** in the dorsal rim of the glottis, a ring-like **cricoid cartilage** anterior to the first tracheal ring, and an unpaired **thyroid** (shield-shaped) **cartilage** (Fig. 12-17). The latter develops from paired lateral anlagen and is found below mammals only in crocodilians. A **procricoid cartilage** may occur as a dorsal segment of the cricoid, as in birds. A cartilaginous **epiglottis** develops in the pharyngeal wall anterior to the glottis. Other small cartilages (**cuneiforms, corniculates,** etc.) develop in association with the arytenoid in occasional species (man, rabbit, etc.). The posterior horn of the

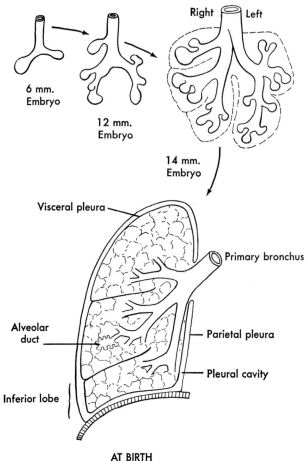

Fig. 12-16. Development of mammalian lung. Embryo lengths are applicable approximately to both the fetal pig and to man.

hyoid bone usually articulates with the thyroid cartilage, and the thyroid and hyoid are also connected by a thyrohyoid ligament and by muscle. In monotremes the adult laryngeal cartilages are all bilaterally represented (Fig. 8-14), as are the visceral cartilages from which they were derived.

A pair of fleshy vocal cords are stretched across the cephalic part of the laryngeal chamber. These are internal folds of the laryngeal wall composed of elastic connective tissue covered by the mucous membrane of the larynx. Changes in the tension of the folds brought about by contraction of intrinsic laryngeal muscles are responsi-

ble for vocal sounds made by mammals—the most vociferous of all vertebrates. A more delicate pair of false vocal cords is located anterior to the true folds in some mammals. In cats, vibration of the false cords results in purring. True vocal cords are absent in a few mammals such as the hippopotamus.

The larynx exhibits several interesting modifications among mammals. That of the howler monkey (*Alouatta*) is massive (Fig. 12-18). The thyroid bones are enormous plates, causing the larynx to bulge in the neck like a goiter. The hyoid bone is also enlarged. Between the true and false vocal cords on either side is a sac-like re-

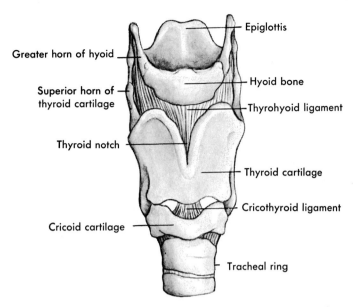

Fig. 12-17. Human larynx, frontal view. (From Francis: Introduction to human anatomy, ed. 5, St. Louis, 1968, The C. V. Mosby Co.)

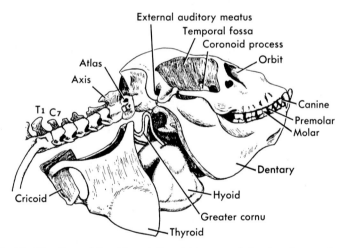

Fig. 12-13. Modification of hyoid and thyroid bones of the howler monkey.

cess, the laryngeal ventricle, or sinus of Morgagni. This is a resonating chamber that makes the weird howl of this monkey carry far into the jungle. Similar sinuses occur in some of the apes, and a vestigial recess is found in most mammals. In whales the arytenoid cartilages are elongated and project into the nasopharynx (Fig. 12-19), permitting rapid inspiration of air while there is much water in the oral cavity. A

similar modification in baby marsupials prevents milk, pumped by the mother's teat into the baby's esophagus, from being drawn into the lungs when the baby inhales. The baby marsupial can breathe while suckling. Mammals lacking this adaptation must stop breathing to swallow.

When foreign particles accidentally reach the larynx, they are engulfed by mucus and swept back into the pharynx by cilia.

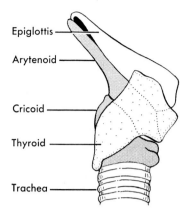

Epiglottis

Arytenoid

Cricoid

Thyroid

Trachea

Fig. 12-19. Adaptation of laryngeal skeleton of a whale to life in the ocean. The epiglottis and arytenoid extend into the nasopharynx.

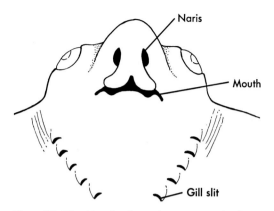

Naris

Mouth

Gill slit

Fig. 12-20. Head of a skate, ventral view, showing oronasal groove connecting naris with mouth.

Cilia continue down the trachea and into the smaller bronchioles.

Morphogenesis of the mammalian diaphragm and the pleurae. Mammals develop a tendinous and muscular dome-shaped diaphragm subdividing the coelom into thoracic and abdominal cavities. The dome bulges into the thorax. The diaphragm, assisting the ribs, plays a primary role in respiration in mammals. Contraction of the diaphragmatic muscles flattens the diaphragm and increases the size of the thoracic cavity. As a result of the combined action of the diaphragm and of the ribs, the pressure around the lungs is reduced. Atmospheric pressure then forces air into the lungs, which instantly inflate. Exhaling is partly a passive return of the rib muscles and diaphragm to a relaxed state. Forced expiration requires muscular effort.

The diaphragm develops when pleuroperitoneal membranes grow into the coelomic cavity and, in association with the septum transversum at the caudal end of the heart, form a partition at the level of the last rib. The membranous part of the partition becomes invaded by body wall muscle derived from several cervical myotomes. The septum transversum contributes the central tendon of the diaphragm. Completion of the diaphragm establishes a separate thoracic cavity.

As the embryonic lungs invade the de-veloping thoracic cavity, they carry peritoneum along as a visceral investment, and they assume a position on either side of the median septum (**mediastinum**) occupied by the heart. Each lung therefore occupies a separate pleural cavity. The peritoneum of each pleural cavity lines the inner surface of the thoracic wall as the **parietal pleura** (Fig. 12-16) and covers the cephalic face of the diaphragm as the **diaphragmatic pleura.** It is reflected over the **root** of the lung (where the bronchus and pulmonary vessels enter and leave) to completely cover the lung as the **visceral pleura.** The space enclosed by these pleurae is the pleural cavity.

NARES AND NASAL CANALS

The external nares of most fishes lead to blind olfactory sacs containing the olfactory epithelium. They often have an incurrent and excurrent opening separated by a partition. The naris in agnathans is a dorsomedial opening. In lampreys it leads into the nasal canal, which terminates blindly a short distance beyond the olfactory sac. In hagfishes (Fig. 12-4) it continues as the nasopharyngeal duct to the velar chamber of the pharynx and permits a flow of respiratory water into the pharynx while the buccal funnel containing the mouth is otherwise occupied.

In Sarcopterygii, also known as Cho-

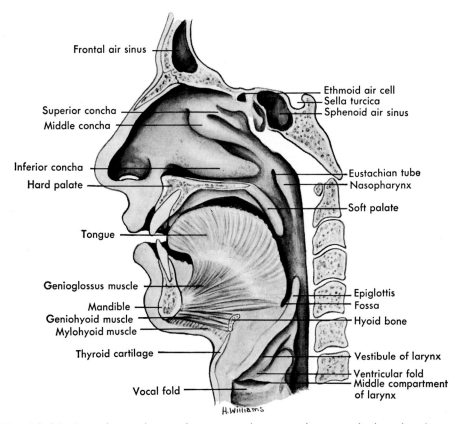

Fig. 12-21. Secondary palate and upper respiratory pathway, sagittal section, in man. The concha constitute the turbinal elements. (From Francis: Introduction to human anatomy, ed. 5, St. Louis, 1968, The C. V. Mosby Co.)

anichthyes (*choana* = funnel-like opening), the nasal canals continue beyond the olfactory sacs and open into the oral cavity as **internal nares** (Fig. 8-20). But in these lung-breathing fishes air is gulped through the mouth and the nasal canal is not used. *Latimeria*, the living crossopterygian, lacks internal nares. Only in tetrapods do the nasal canals serve as passageways for air.

An oronasal groove frequently connects each naris with the angle of the mouth in elasmobranchs (Fig. 12-20). In rays the groove becomes almost tubular as a result of folding together of the lateral walls. By the same process, an oronasal groove is converted into nasal canals in sarcopterygians, amphibians, and amniotes.

With formation of a secondary palate, the nasal canals are extended caudad. The more complete the secondary palate, the

farther caudad are the internal nares. Therefore, in sarcopterygians and amphibians, which lack a secondary palate, the internal nares are forward in the oral cavity and are laterally situated; in lower reptiles and in birds they are farther caudad and nearer the midline; and in crocodilians the internal nares are far to the rear of the oral cavity. In mammals the nasal canal opens into the nasopharynx above the soft palate.

The external nares of tetrapods fulfill their primitive role of admitting the surrounding medium (air, instead of water) to the olfactory epithelium. In higher vertebrates the olfactory epithelium is found in the upper chambers of the nasal canals, and the ventral part of the canals has a ciliated, glandular epithelium like that of the trachea and bronchi. Within the nasal passages of mammals the air is warmed by venous plexuses lying under the epithelium of the turbinal elements (concha, Fig. 12-21). Hairs within the external nares serve as filters which trap coarse particles and insects. Certain bones of the skull of mammals contain air **sinuses** that open into the nasal canals. These increase the resonance of vocal sounds. A fleshy, partially cartilaginous proboscis (nose) may develop in mammals and carry the external nares to characteristic positions (compare the location of nostrils in cats, man, and elephants). In whales, on the other hand, there is no proboscis, and external nares or blowholes are dorsally situated, although in fetal whales so far studied the external nares are farther forward. In some whales the two nares become a median blowhole during later development.

ACCESSORY RESPIRATORY ORGANS

Adult fishes rely chiefly on pharyngeal gills for respiration. However, other devices also occur, such as the bushy gills attached to the pectoral fin of the male *Lepidosiren*. In aquatic amphibians the lining of the cloaca, rectum, oral cavity, or pharynx may assist the gills or take their place. Respiration via the skin is common in amphibians. African hairy frogs have filamentous hair-like projections of the skin of the posterior trunk region and hind limbs, which serve as gills. Obligatory terrestrial vertebrates are unable to utilize the skin effectively since it is cornified.

Practically all embryonic vertebrates use the yolk sac and its vitelline circulation for respiration, at least until other respiratory devices have developed. Ovoviviparous vertebrates may use the yolk sac in contact with the uterine wall as a respiratory device, as do marsupials. The pericardium in embryonic viviparous fishes may be enlarged to serve as a respiratory membrane, and embryonic anal gill filaments also occur in these fishes. Reptiles, birds, and mammals except marsupials soon supplant the yolk sac and its vitelline vessels with the allantois and its allantoic (umbilical) vessels. The allantois in association with the chorion lies against the porous eggshell or against the uterine wall of the mother and is a respiratory organ.

Chapter summary

1. Pharyngeal gills are respiratory structures that develop on certain of the visceral arches of fishes and amphibians. They are supported by the visceral skeleton and are moved by branchiomeric muscles. They are of two types—external and internal.

2. Internal gills develop in the walls of gill chambers or opercular chambers and are characteristic of fish throughout life. Certain cells of the gill epithelium are excretory in function, particularly in marine forms.

3. External gills are outgrowths of the external surfaces of the branchial arches of larval fishes and larval amphibians. Perennibranchiate amphibians retain external gills throughout life.

4. Frog tadpoles exhibit an initial set of external gills. Later a set develops within the opercular chamber. Although the latter arise from the outer surfaces of branchial arches, they are often considered internal gills because of their location.

5. Elasmobranchs have naked gill slits. An operculum covers the gill chambers in chimaeras, bony fishes, and tadpoles. The operculum is bony in Osteichthyes.

6. The number of gill chambers in primitive vertebrates may not have been much higher than that in living elasmobranchs. One known ostracoderm had ten chambers; cyclostomes have six to fifteen; elasmobranchs have five to seven; modern bony fishes usually have five. Neotenous amphibians may retain three to no gill slits. Most amniote embryos exhibit four to six pharyngeal pouches, thought to be transient vestiges of earlier gill chambers.

7. The spiracle of fishes is derived from the

first pharyngeal pouch and is considered a vestigial gill slit. It is open in most elasmobranchs and chondrosteans but closed in *Chimaera* and higher bony fishes. The first pharyngeal pouch in tetrapods becomes the eustachian tube and contributes to the middle ear cavity.

8. Associated with reduction in number of gill chambers is reduction in number of demibranchs. The hyoid demibranch tends to disappear in higher fishes, and the number of demibranchs is reduced still further in lungfishes.

9. Pharyngeal chambers and visceral slits originally may have been food-ensnaring devices, the walls of which subsequently became highly vascularized respiratory membranes.

10. Almost all bony fishes exhibit a hydrostatic swim bladder arising embryonically as an unpaired dorsal or lateral evagination of the foregut caudal to the last gill chamber. A few embryonic elasmobranchs exhibit a rudimentary swim bladder, but it is absent in cyclostomes.

11. The striking similarity between swim bladders and tetrapod lungs suggests a phylogenetic relationship between the two. The common precursor may have been a terminal pair of pharyngeal pouches, which became, through mutations, swim bladders in most fishes and lungs in other fishes and tetrapods.

12. Tetrapod lungs arise, embryonically, as a midventral evagination of the foregut. Lungs are usually paired and asymmetrical. They are usually absent on one side in caecilians and in legless reptiles.

13. Lungs occupy the pleuroperitoneal cavity in amphibians and lower reptiles and pleural cavities in higher reptiles, birds, and mammals. In reptiles and birds an oblique septum separates the pleural from the peritoneal cavity. In mammals the diaphragm forms the partition.

14. The lungs of amphibians and reptiles are simple sacs. The lining is relatively smooth in lower forms and septate in others. In higher lizards, crocodilians, and turtles, sep-

ta result in formation of many chambers and produce a spongy lung.

15. The lungs of birds are unique because of the parabronchi and the direction of airflow.

16. Air sacs, which are diverticula of lungs, occur in some reptiles and reach maximum development in birds, in which they often penetrate the bones.

17. The lungs of mammals are multichambered. Instead of septa initiating the formation of chambers, the developing bronchi sprout innumerable branches that result in formation of a large number of minute terminal sacs, each of which is similar to, but much smaller than, the simple lungs of lower forms.

18. The glottis, larynx, trachea, and primary bronchi are almost universal in tetrapods. Branching bronchi appear first in reptiles.

19. A cartilaginous epiglottis is characteristic of mammals. An epiglottis-like fold in the reptilian pharynx may be a precursor.

20. The larynx is supported by lateral cartilages in lower urodeles and by arytenoid and cricoid cartilages in higher tetrapods. A thyroid cartilage is added in crocodilians and mammals. Other small cartilages may develop. Laryngeal cartilages are derivatives of the last several visceral arches. The larynx is reduced in birds. It is usually supplanted as a vocal organ by the syrinx.

21. External nares in fishes lead typically to the olfactory sacs only. In hagfishes, sarcopterygians, (Choanichthyes), and tetrapods, nasal canals connect the external nares with the oral cavity or pharynx. Internal nares are displaced caudad in animals with a secondary palate.

22. In lower vertebrates the pharyngeal gills and lungs are often supplemented by the skin and less often by the lining of the oral cavity, pharynx, cloaca, rectum, and gill-like projections of the trunk and limbs. Embryonic respiratory devices include the pericardium and anal gill filaments in occasional fishes, yolk sac in most vertebrates, and allantois in most amniotes.

23. External respiration (exchange of oxygen

and carbon dioxide between the organism and the environment) is promoted by skeletal muscle action. Branchiomeric muscles operate branchial arches. In amphibians and some reptiles the floor of the mouth acts as a pump, body wall muscles assist in most tetrapods, ribs are depressed and elevated in practically all amniotes, and the diaphragm is operated in mammals. In birds the air sacs, serving as bellows, are compressed and expanded by skeletal muscle action, which forces air through the air ducts.

The circulatory system of vertebrates consists of the heart, arteries, veins, capillaries, and blood (the **blood vascular system**) and of lymph channels and lymph (**lymphatic system**). The blood carries oxygen collected in respiratory organs, nutrients from extraembryonic membranes of embryos and from the adult digestive tract, hormones from endocrine tissues, substances associated with maintaining homeostasis and immunity to disease, and waste products of metabolism to be removed in suitable excretory organs. Lymph channels collect interstitial tissue fluids not taken up by the bloodstream and emulsified fats absorbed in the small intestine and terminate in one or more of the large venous channels of the blood vascular system.

BLOOD VASCULAR SYSTEM

Arteries carry blood away from the heart. They have muscular, elastic, and fibrous walls capable of distention with each intrusion of blood (feel your pulse!) and of active constriction and dilatation in response to nerve impulses (Fig. 13-1). Arteries thereby assist in regulating blood pressure. They terminate in capillary beds. **Veins** have proportionately less muscle and elastic tissue and more fibrous tissue and are therefore capable of less distention or constriction. They carry blood toward the heart from capillary beds. Although all

venous streams ultimately terminate in the heart, portal veins have a capillary bed interposed en route. Capillaries consist of endothelium only, with a lumen just large enough to accommodate red blood cells in single file. In fact, the red cells must "squeeze through" and, in so doing, become deformed. Connecting small arteries with capillaries are arterioles, and venules connect capillaries with small veins. Arteries have branches, but veins, except portal veins, are more correctly spoken of as having tributaries, since small veins flow into larger ones. The **heart** is a modified blood vessel with very muscular walls. Valves preventing backflow of blood occur in the veins and heart.

In vertebrates respiring by gills, blood is pumped from the heart to the gills, where external respiration takes place. From the gills it typically flows via arteries to capillaries throughout the body. The blood finally returns to the heart via venous channels. Venous blood from the capillaries of the tail typically passes via a **renal portal system** to capillaries of the kidney before returning to the heart. Blood from the digestive tract, pancreas, and spleen must pass via a **hepatic portal system** to the capillaries of the liver before continuing to the heart. Except, perhaps, in Osteichthyes blood from the hypothalamus must pass via a **hypophyseal portal system** to the pars distalis of the pituitary gland before con-

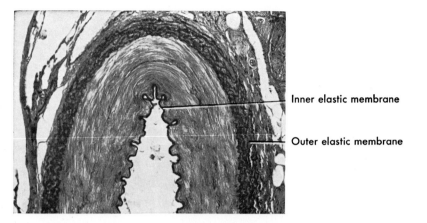

Fig. 13-1. Section of a medium-sized artery. Between the two elastic membranes lies a thick layer of smooth muscle. (From Bevelander: Essentials of histology, ed. 5, St. Louis, 1965, The C. V. Mosby Co.)

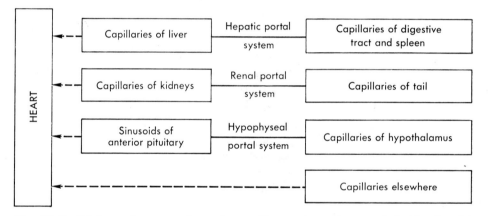

Fig. 13-2. Chief portal systems of vertebrates. The renal portal system is lacking in adult Theria, and the hypophyseal portal system has not been demonstrated in most Osteichthyes.

tinuing to the heart. A portal system is a system of veins commencing in the capillaries of one or more organs and terminating in the capillaries of another organ (Fig. 13-2).

In typical tetrapods blood is pumped from the heart to the lungs via pulmonary arteries. Pulmonary veins return oxygenated blood to the heart. In adult tetrapods well adapted to terrestrial life, particularly birds and mammals, the blood then passes through systemic arteries to all parts of the body other than the lungs and returns ultimately to the heart via systemic veins in-

cluding portal systems. Adult mammals lack the renal portal system.

Heart

The heart is a modified blood vessel exhibiting the three layers characteristic of arteries: an inner lining (**endocardium**) of endothelium and elastic tissue; a muscular layer (**myocardium**) that is very thick, especially in the ventricular region; and an outer fibrous tunic (**epicardium**), on the surface of which is the visceral pericardium. The heart pulsates as a result of the response of the muscle cells to electro-

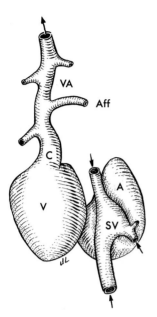

Fig. 13-3. Heart of the cyclostome *Myxine glutinosa*, ventral view. **A,** Atrium; **Aff,** afferent branchial artery; **C,** conus arteriosus; **SV,** sinus venosus; **V,** ventricle; **VA,** ventral aorta. Arrows indicate direction of bloodflow.

lytes that infuse it. The rhythmicity of the pulsations is regulated reflexly by the autonomic nervous system, except in hagfishes, in which no nerve fibers supply the muscle. The heart occupies the pericardial cavity, a subdivision of the coelom anterior to the septum transversum.

The heart of all vertebrates is built in accordance with a basic architectural pattern. It is demonstrated in its simplest form in hagfishes (Fig. 13-3) where it exhibits a series of four chambers: **sinus venosus, atrium, ventricle,** and **conus arteriosus,** through which blood flows in that sequence.

Fishes. The heart of the dogfish is typical of most fishes. The sinus venosus is thin walled; it has little muscle and much fibrous tissue in its walls. It receives blood from large veins returning from all parts of the body. Although the sinus shows some contractility, it serves chiefly as a reservoir. Its blood gushes through the sinoatrial aperture into the atrium as soon as

the latter begins to relax after having emptied its contents. The caudal wall of the sinus venosus is anchored to the anterior face of the septum transversum.

The atrium is a large, thin-walled muscular chamber dorsal to the ventricle and anterior to the sinus venosus. Blood from the atrium pours into the ventricle through an atrioventricular aperture guarded by valves.

The ventricle has very thick muscular walls. The anterior end is prolonged as a muscular tube of small diameter, the conus arteriosus, which passes forward to the cephalic end of the pericardial cavity, where it is continuous with the ventral aorta. The lining of the conus in elasmobranchs is projected into a series of semilunar valves that prevent backflow of blood. Because of its contractility and elasticity, the conus helps to maintain a steady arterial pressure into and through the gills. In bony fishes the conus is foreshortened and its function is assumed by the elastic **bulbus arteriosus,** an enlargement of the ventral aorta. The bulbus inflates like a balloon when the ventricle contracts.

Lungfishes and amphibians. The heart in lungfishes and amphibians differs in only two notable features from that just described. These differences are associated with the development of lungs and enable oxygenated blood returning from the lungs to be separated from deoxygenated blood returning from elsewhere. One modification is the establishment of a partial or complete partition within the atrium, so that there is a right and left atrium (Fig. 13-4, Dipnoi, urodele, anuran). The partition is complete in anurans and some urodeles. The pulmonary veins empty into the left atrium so that the blood in this chamber is oxygen rich. The sinus venosus empties into the right atrium; hence, the blood in this chamber requires oxygenation. A partial partition also forms in the ventricle in lungfishes. The other modification associated with the advent of lungs involves the valves of the conus arteriosus. One of the valves may become a **spiral valve,**

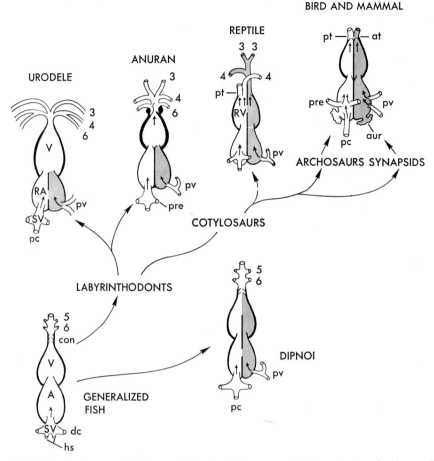

Fig. 13-4. Successive modifications of the heart for economical respiration by lungs. The parts of the heart shown are **A,** atrium; **RA,** right atrium; **V,** ventricle; **RV,** right ventricle; **SV,** sinus venosus; **con,** conus arteriosus; **aur,** auricle of mammalian heart. **3** to **6,** Third through sixth aortic arches. Other vessels are **at,** aortic trunk; **dc,** common cardinal vein; **hs,** hepatic sinus; **pc,** postcava; **pre,** precava (common cardinal vein); **pv,** pulmonary veins; **pt,** pulmonary trunk. Gray chambers contain chiefly, or only, oxygenated blood.

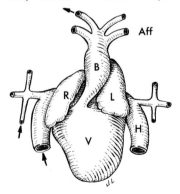

Fig. 13-5. Heart of *Necturus,* ventral view. **Aff,** Afferent branchial artery; **B,** bulbus arteriosus, a swelling on the ventral aorta; **H,** hepatic sinus; **L,** left atrium; **R,** right atrium. The sinus venosus is dorsal to the ventricle, **V,** and hence is not seen in this view. Arrows indicate direction of bloodflow.

which shunts deoxygenated blood into the pulmonary vessels and oxygenated blood into other channels. Because the two atria do not empty simultaneously into the ventricle, there is little mixing of blood in the single ventricle as long as the spiral valve is functioning. The conus (or truncus) arteriosus is prominent in anurans. In urodeles it is distinguishable only internally by the semilunar valves guarding the exit from the ventricle. However, urodeles have a bulbus arteriosus, which replaces the conus functionally (Fig. 13-5). The bulbus is a modification of the ventral aorta.

Amniotes. Amniotes exhibit two atria and two ventricles. The two ventricles are completely separate only in birds and mammals. A sinus venosus is also present in adult reptiles. It may be large, as in turtles,

or reduced in size but distinct internally, as in crocodilians. Birds and mammals exhibit a sinus venosus during early development, but it fails to keep pace with the growth of the right atrium into which it empties and finally becomes incorporated into the wall of that chamber. Thereafter, the vessels that emptied into the sinus venosus empty directly into the right atrium. Its earlier location is marked by the sinoatrial node of neuromuscular tissue, which plays a role in regulation of the heartbeat.

In adult amniotes the right and left atria are completely separated by an interatrial septum. Nevertheless, they are confluent during embryonic development via an **interatrial foramen (foramen ovale)**, which closes at the time of hatching or birth. The site of the obliterated foramen ovale is

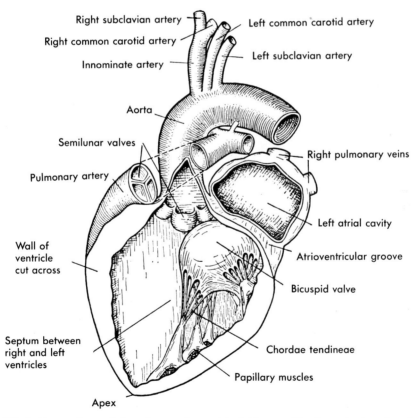

Fig. 13-6. Human heart, left atrium and ventricle laid open. The innominate artery is also named brachiocephalic. (From Tuttle and Schottelius: Textbook of physiology, ed. 15, St. Louis, 1965, The C. V. Mosby Co.)

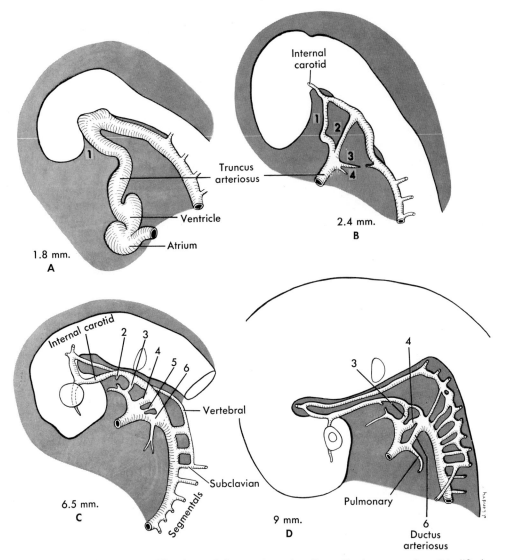

Fig. 13-7. Embryonic modifications of the aortic arches **(1** to **6)** of a porcupine. (Modified from Struthers.[90])

marked in the adult heart by a depression (**fossa ovalis**) in the medial wall of the right atrium. The right atrium receives the sinus venosus (reptiles) or the venous blood that previously emptied into the sinus venosus (birds and mammals). In addition, it receives blood from the coronary veins. The left atrium receives blood from the pulmonary veins.

In mammals each atrium has an ear-like flap, the **auricle,** containing a blind, sac-like chamber. The term "auricle" is often applied to the atrium in lower vertebrates but perhaps should be reserved for this special appendage of the mammalian heart. Any functional advantage of the mammalian auricle has yet to be demonstrated.

The ventricles of amniotes are paired. The condition results when an interventricular septum develops within the embryonic ventricle and divides it into right and left chambers. In reptiles other than crocodilians the interventricular septum is incomplete. The internal walls of the ventricles exhibit anastomosing muscular columns and ridges (**columnae carneae** and **trabeculae carneae**).

Valves guard the passage from the atria into the ventricles (Fig. 13-6). These valves are fibrous flaps (muscular flaps on the right in crocodiles and birds) connected by tendinous cords (**chordae tendineae**) to papillary muscles projecting from the ventricular walls. During relaxation of the ventricle (diastole), blood from the atria falls freely past the valves into the ventricles. During the contraction phase (systole), ventricular pressure rises, and the flap-like valves are forced upward into the atrioventricular passageway, thereby preventing reflux of blood into the atria. The valves have one or two flaps (cusps) in reptiles and birds. In most mammals the left valve has two flaps (**bicuspid** or **mitral valve**); the right has three (**tricuspid valve**).

Guarding the exit of the great arteries from the ventricles are **semilunar valves**, which prevent backflow into the ventricles. The semilunar valves may be vestiges of similar valves in the conus arteriosus of fishes.

Morphogenesis of the heart. In lower fishes and in amphibians the heart commences as a midventral longitudinal blood vessel. In further development the vessel becomes bent into a U shape so that the atrial region, previously at the caudal end, is carried to a dorsal position. In teleosts and amniotes the heart likewise passes through a single-tube stage (Fig. 13-7, A). Prior to this, however, a pair of longitudinal vessels in the splanchnopleure unite in the midline beneath the pharynx.[44, 45] The initial paired condition in teleosts, reptiles, and birds appears to be correlated with the presence of a large amount of yolk; and, although there is no yolk in mammals, the embryo develops as if there were. After fusion of the paired vessels, the usual U-shaped twisting of the primitive heart tube occurs. In amniotes the developing atria are carried so far cephalad that they finally lie anterior to the ventricles.

In view of its embryonic development, it is probable that the heart evolved from a primitive midventral longitudinal blood vessel with muscular, pulsating walls. The ventral aorta of the amphioxus, which lacks a heart, is such a vessel.

Arterial channels

Arterial channels supply most organs with oxygenated blood, but they carry deoxygenated blood to the respiratory organs. In the basic architectural plan, as seen in all vertebrate embryos (Fig. 13-8), the major arterial channels consist of (1), a **ventral aorta** emerging from the heart and passing forward beneath the pharynx; (2) a **dorsal aorta,** paired anteriorly, passing caudad above the pharynx and digestive tract; and (3) typically six pairs of aortic arches connecting the ventral aorta with the dorsal aorta. Branches of these major arterial channels supply all parts of the vertebrate body. Modifications of the embryonic pattern, which occur during later ontogeny are of such a nature that they either adapt the aortic arches for respiration by gills or adapt them for respiration by lungs.

Ventral aorta and aortic arches. These vessels will be discussed in fishes and in tetrapods as a group. Then their modifications will be described in tailed amphibians, anurans, reptiles, birds, and mammals.

Fishes. Adaptive modifications of the ventral aorta and aortic arches for respiration by gills may be illustrated in developing sharks (Fig. 13-9). The ventral aorta in *Squalus* extends forward under the pharynx and connects with the developing aortic arches. The aortic arch in the mandibular arch is the first to develop. Shortly thereafter the other five aortic arches appear. Before the sixth aortic arch has completely developed, the first loses its connection with the ventral aorta, and the dorsal part remains as the efferent pseudobranchial artery. From the second aortic arch sprouts a bud destined to become the first pretrematic artery. Other buds sprouting from the third, fourth, fifth, and sixth aortic

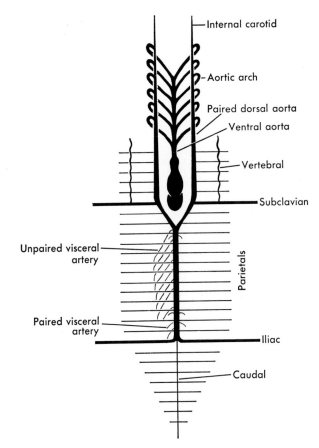

Fig. 13-8. Basic pattern of the chief arterial channels of vertebrates.

arches become posttrematic arteries. The embryonic posttrematic arteries sprout cross trunks, which grow caudad and form the last four pretrematic arteries. In the meantime, capillary beds are developing within the nine demibranchs. Afferent branchial arterioles (not shown in Fig. 13-9) sprout from the aortic arches into each demibranch and lead to the capillaries of the gill filaments. Aortic arches II to VI soon become occluded at one site (broken lines in Fig. 13-9, A). The segments ventral to the sites of occlusion become afferent branchial arteries. The dorsal segments become efferent branchial arteries. A cross trunk (Fig. 13-9, As), sprouting from the first pretrematic artery, becomes the afferent pseudobranchial artery. Efferent branchial arterioles connect the capillaries with the pretrematic and posttrematic

arteries. The paired dorsal aortas extend forward as internal carotid arteries.

As a result of formation of this system of pharyngeal arteries in *Squalus,* coupled with occlusion of each aortic arch at one point, blood entering the aortic arch from the ventral aorta is forced to detour into the gill capillaries before proceeding to the dorsal aorta. The aortic arches have thus been modified to serve the respiratory apparatus, and all blood emerging from the heart must be aerated in the gills before being distributed to the body (Fig. 13-24, fish).

The same general changes convert aortic arches of other fishes into arteries supplying and draining the gills. The specific number of afferent and efferent arteries in adult fishes depends on the number of functional gills. In most teleosts the first

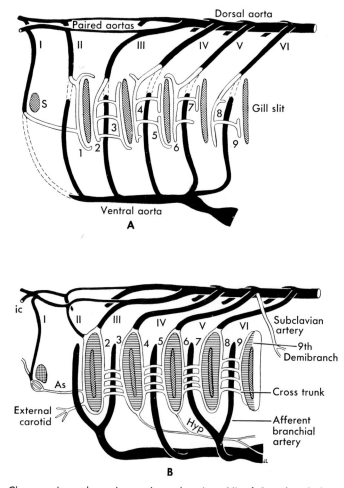

Fig. 13-9. Changes in embryonic aortic arches I to VI of *Squalus* during development, lateral view. In **A** buds (in white) off the aortic arches are establishing pretrematic and posttrematic arteries and cross trunks. Broken lines indicate sections of the aortic arches that become occluded, forcing blood into afferent branchial arterioles (not shown). In **B** the first arch, **I**, has become the efferent pseudobranchial artery; **II** has become the hyoidean efferent; **III** to **VI** have become the remaining efferent branchial arteries. **1, 3, 5, 7**, and **9**, Pretrematic arteries; **2, 4, 6**, and **8**, posttrematic arteries. **As**, Afferent pseudobranchial; **Hyp**, hypobranchial; **ic**, internal carotid; **S**, spiracle.

and second aortic arches tend to disappear.

In Dipnoi and in *Polypterus* a pulmonary artery develops off the efferent part of the sixth arch on each side and grows to the developing lung. In *Protopterus* the third and fourth embryonic aortic arches are uninterrupted by gill capillaries (Fig. 13-10, *B*).

Tetrapods as a group. Like fishes, tetra-

pods construct a ventral aorta and six embryonic aortic arches (Fig. 13-11). The first and second aortic arches are transitory, often disappearing before the sixth arch has differentiated. A pulmonary artery sprouts off the sixth aortic arch, as in lungfishes. With the exception of a few tailed amphibians, tetrapods lose the fifth aortic arch during embryonic life. The remaining arches (III, IV, and VI) are seldom com-

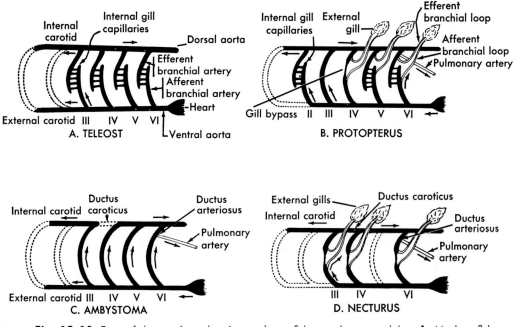

Fig. 13-10. Fate of the aortic arches in two bony fishes and two urodeles. **A,** Modern fish. **B,** Lungfish. **C,** Terrestrial urodele. **D,** Perennibranchiate urodele. **II** to **VI,** Aortic arches; broken lines indicate vessels confined to embryo.

pletely interrupted by gill capillaries even in gill-breathing amphibian larvae. A knowledge of the ventral aorta and aortic arches in amphibians will serve as a background for understanding the changes occurring in these vessels in amniotes.

Tailed amphibians. Immediately upon emerging from the heart, the ventral aorta becomes modified to form a swollen, muscular **bulbus arteriosus** leading to the aortic arches (Fig. 13-5). Four aortic arches (III to VI) are present in some families of urodeles (*Ambystomatidae* [Fig. 13-10], *Salamandridae*, and *Plethodontidae*), but in several members of these families (*Ambystoma, Notophthalmus,* and *Plethodon*) aortic arch V unites partly or wholly with IV, the degree of union varying among individuals of the same species. In *Necturus* (Fig. 13-10, *D*), *Siren,* and *Amphiuma* the fifth aortic arch never completely forms, or it fuses with IV, or it disappears during development. These forms, therefore, typically have only three adult aortic arches (III, IV, and VI). A pulmonary artery

sprouts from arch VI. Tailed amphibians illustrate the transition from four to three aortic arches.

In urodeles that retain external gills throughout life, the aortic arch of each gill arch gives off an afferent branchial arteriole, which passes into the gill (Fig. 13-10, *D*). From the gill capillaries an efferent branchial arteriole returns blood to the aortic arch. A short section of the aortic arch therefore serves as a gill bypass. Some blood apparently uses these bypasses at all times. When the gills in *Necturus* are highly active, their vessels are dilated; the gills are large, bushy, and crimson in color; and they stand out from the pharyngeal wall and wave gently back and forth in the water. If, however, the water is depleted of oxygen, the gill vessels constrict and the gills shrink. More blood then utilizes the bypasses. When the gills of *Siren* are caused to be absorbed experimentally under the influence of thyroid extract, the bypasses carry all the blood entering the arches.

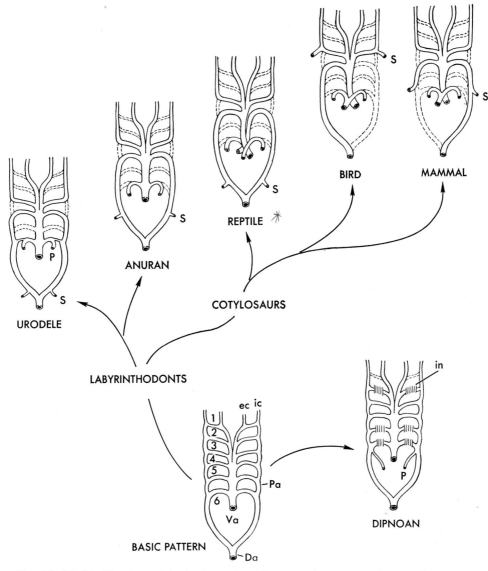

Fig. 13-11. Modifications of the basic aortic arch pattern in representative vertebrates. **Da,** Dorsal aorta; **ec,** external carotid; **ic,** internal carotid; **in,** aortic arch interrupted by gill capillaries; **P,** pulmonary artery; **Pa,** paired (dorsal) aorta; **S,** subclavian artery; **Va,** ventral aorta.

Even when gills and lungs are both present, some blood goes from the ventral to the dorsal aorta without being aerated. The only completely unmixed, oxygenated blood in *Necturus* is therefore in the efferent branchial arterioles or in the pulmonary veins and in superficial veins from the skin. The lack of economy of circulation in perennibranchiate amphibians is attributable to the fact that these animals are in a state analogous to suspended transition from gill to lung respiration. A blood cell may circulate through the body of *Necturus* precisely as it would through that of the dogfish, or it may circulate exactly as it would through the body of an amniote (Fig. 13-24). Figuratively speaking, these animals do not seem to be able to make up their minds whether they want to stay in the water and breathe through gills, like fishes, or to live on land and breathe with lungs, like tetrapods.* Because of this transitional state, the anomalous, somewhat uneconomical circulatory pattern persists.

Anurans. The larval frog is in all respects a tailed amphibian. From the ventral aorta arise six aortic arches. After aortic arches I and II disappear, the paired dorsal aortas anterior to aortic arch III are named internal carotid arteries. Aortic arch VI sprouts a pulmonary artery to the developing lung. Aortic arches III to V give rise to afferent and efferent branchial loops, which pass into the three gills. Gill bypasses are prominent at first, but may later become greatly reduced or may even temporarily disappear. When the first set of external gills is discarded, new loops grow from the original aortic arch into each developing opercular (internal) gill.

At metamorphosis, with loss of gills three notable changes occur in the aortic arches (Fig. 13-11, anuran). (1) Aortic arch V drops out altogether. (2) The dorsal aorta between aortic arches III and IV (**ductus caroticus**) disappears. As a result, blood entering aortic arch III (carotid arch) **must** continue to the head via the carotid arteries. (3) The dorsal segment of aortic arch VI (**ductus arteriosus**) from the origin of the pulmonary artery to the dorsal aorta disappears. Blood entering arch VI now **must** continue to the lungs (or skin). Aortic arch IV (**systemic arch**) on each side continues to the dorsal aorta to distribute blood to all parts of the body except the head and lungs (Fig. 13-12).

A longitudinal septum-like spiral valve, probably a modified semilunar valve, develops in the truncus arteriosus. It shunts deoxygenated blood to the pulmonary arch

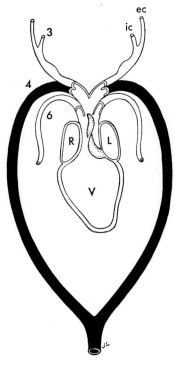

FROG

Fig. 13-12. Systemic arch (black), carotid arch **(3)**, and pulmonary arch **(6)** of a frog. **ec,** External carotid; **ic,** internal carotid; **R** and **L,** right and left atria; **V,** ventricle. The spiral valve is stippled.

and oxygen-rich blood to the carotid and systemic arches.[86]

The preceding modifications in frogs, resulting from complete loss of gills, tend to make circulation more economical than in tailed amphibians. Deoxygenated blood from the systemic veins is shunted to the lungs (or skin) for aeration. Elimination of the ductus arteriosus prevents this blood from returning prematurely to the systemic circulation. Oxygen-rich blood, upon returning to the left atrium and to the ventricle, is shunted to the head, trunk, and limbs. The frog has thus moved a step closer to a return to an economical circulation that was disrupted by imposing lungs on a gill-bearing animal.

Reptiles. Modern reptiles exhibit three adult aortic arches—III, IV, and VI. The

---

*This is unfair to perennibranchiate amphibians. Actually, they lack the genetic factors necessary for complete metamorphosis.

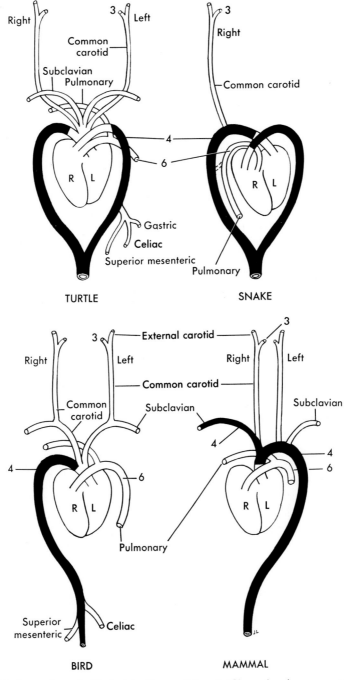

Fig. 13-13. Systemic arch **(4,** in black), carotid arch **(3),** and pulmonary arch **(6)** of amniotes. **R** and **L,** Right and left ventricles.

ductus caroticus and ductus arteriosus usually drop out during embryonic development. However, both remain in *Sphenodon,* and in a few other reptiles one or the other may persist.

An innovation has been introduced in the ventral aorta. Instead of developing a spiral valve to shunt fresh and deoxygenated blood into the proper arches, reptiles underwent a series of mutations that split the ventral aorta into three separate passages—two aortic trunks and a pulmonary trunk.* The effects of these changes were as follows (Fig. 13-13, turtle and snake): (1) The pulmonary trunk emerges from the right ventricle and connects with the **sixth aortic arches.** Deoxygenated blood from all parts of the body is therefore sent to the lungs to be oxygenated. (2) One aortic trunk leads out of the right ventricle and connects with the **left fourth aortic arch.** Some deoxygenated blood is thereby carried to the dorsal aorta. (3) The other aortic trunk leads out of the left ventricle and carries oxygenated blood to the **right fourth aortic arch** and to the carotid arches. This is advantageous, since oxygen-rich blood is sent into the head and dorsal aorta by this route. Since the dorsal aorta receives deoxygenated blood from the right ventricle, the circulation is not as economical as would be the case if the aortic trunk from the right ventricle were eliminated. This is precisely what happened in birds and mammals!

Birds. In birds, for the first time since the introduction of lungs and the formation of two atria, a circulatory route has been reestablished in which there is no mixing of oxygenated and unoxygenated blood.

*It is an attractive hypothesis that in early reptiles the ventral aorta was divided into two trunks that corresponded to the two ventricles. A pulmonary trunk from the right ventricle would have led to the sixth aortic arches, and an aortic trunk from the left ventricle would have led to the fourth and third aortic arches. This simple condition would then have been altered in three directions: toward the three trunks of modern reptiles, toward the condition in modern birds, and toward the condition in modern mammals.

This has been achieved by closing the interventricular foramen and by dividing the ventral aorta into two trunks. The pulmonary trunk (Fig. 13-13, bird) emerges from the right ventricle and leads only to the sixth aortic arches and lungs. The aortic trunk emerges from the left ventricle and leads to the third and right fourth aortic arches. The **left fourth aortic arch disappears** (Fig. 13-11, bird). As a result of these modifications, all blood returning to the right side of the heart passes next to the lungs. From there it returns to the left side of the heart to be recirculated.

In birds, therefore, six aortic arches are developed in the embryo, and the first, second, fifth, and the left fourth aortic arches, the ductus caroticus, and the ductus arteriosus are eliminated. (The ductus ateriosus remains until hatching to shunt unoxygenated blood into the dorsal aorta, which leads partly to the respiratory channels of the allantois.)

Mammals. Mammals have achieved the same economical circulation as birds and by the same modifications, with one difference. In mammals most of the **right fourth aortic arch** is lost instead of the left. The only part of the right fourth aortic arch that remains is represented by the proximal part of the right subclavian artery (Figs. 13-11 and 13-13, mammal). The left fourth aortic arch is referred to as "the" aortic arch by anatomists.

The common carotid and external carotid arteries are the embryonic ventral aortas. The internal carotid arteries represent the third aortic arches and the paired dorsal aortas (Fig. 13-7, *C*). A few individual and species differences in the vessels arising from "the" aortic arch in mammals are illustrated in Fig. 13-14.

The circulatory route in adult birds and mammals is as follows: systemic venous channels, right atrium, right ventricle, pulmonary trunk, pulmonary artery, capillaries of the lung, pulmonary vein, left atrium, left ventricle, aortic trunk (ascending aorta), and then to the fourth aortic arch (left in mammals and right in birds) or to

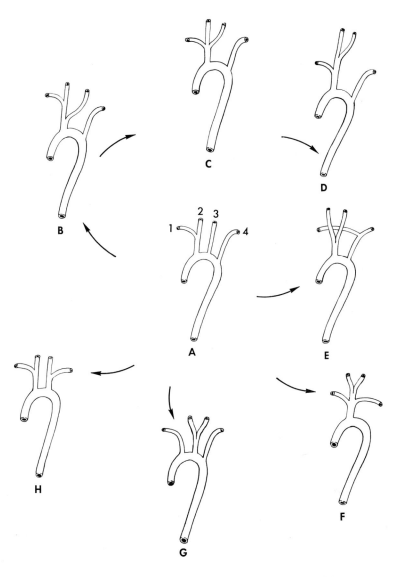

Fig. 13-14. Individual and species differences in the relationships of the common carotid and subclavian arteries in selected mammals. **A,** Basic pattern, seen also in man,* porcupine,* rabbit,* pig; **B,** cat,* dog,* monkey,* pig,* man, and rabbit; **C** and **D,** domestic cat, **D** occurring with lower frequency; **E,** anomalous right subclavian artery in cat,[92] man, and rat; **F,** many perissodactyls; **G,** walrus; **H,** man, porcupine, and rabbit. An asterisk indicates this to be a predominant condition in the populations examined. If there is no asterisk, the condition is common but not predominant. **1,** Right subclavian; **2,** right common carotid; **3,** left common carotid; **4,** left subclavian.

The anomalous condition shown in **E** has been induced experimentally in rats by irradiation of the fetus and also by a single injection of trypan blue into the mother immediately after mating.[83] Because all the conditions illustrated are derived ontogenetically from the same pattern, any of the variants could, and probably do, occur with measurable frequency in every mammalian species.

the carotid circulation. The circulatory pattern of the fetal mammal and the changes that occur at birth are described later.

Interpretation. The development of six aortic arches in all vertebrate embryos and the systematic elimination of first one vessel and then another illustrate clearly that all vertebrates are built in accordance with a basic architectural pattern. The successive changes in the arches in higher forms also illustrate the biogenetic law that lower and higher vertebrates during embryonic development pass through the same fundamental changes and in the same sequence. Development of six aortic arches in all vertebrates supports the theory that lower and higher vertebrates are genetically related (theory of organic evolution) and that structures exhibited by more recent forms were inherited from earlier, more generalized forms.

Dorsal aorta. The dorsal aorta in the head and pharyngeal region is represented bilaterally. Portions are known as **paired aortas** in sharks and as **radices** (singular, **radix**) **aortae** in amphibians. The internal carotid arteries are extensions of the dorsal aortas anterior to the third aortic arches. The ductus caroticus is the segment of each paired aorta between the third and fourth aortic arches (Fig. 13-10, *C*).

The dorsal aorta of the trunk is an unpaired vessel commencing at the union of the paired portions or at the end of the arch of the aorta (in birds and mammals). It passes caudad ventral to the vertebral column. It gives off a segmental series of paired somatic branches to the body wall and appendages, a series of unpaired visceral branches to the digestive organs and spleen, and paired visceral branches to the kidneys, gonads, and adrenal glands. In mammals the aorta is subdivided by the diaphragm into thoracic and abdominal portions. The aorta continues into the tail as the caudal artery.

Somatic branches. The subclavian arteries are an anterior pair of enlarged segmental arteries (Fig. 13-7, *C*). Primitively, the subclavian arteries arose from the paired aortas and they continue to do so in many vertebrates. They usually arise from the embryonic third aortic arches in birds and from the fourth in mammals (Figs. 13-11, 13-13, and 13-14).

In tetrapods the subclavian artery traverses the axilla as the axillary artery, parallels the humerus as the brachial artery, and divides in the forearm into ulnar and radial arteries. The subclavian artery gives off branches to the body wall before entering the appendage. One branch on each side (**vertebral artery,** Fig. 13-7, *C*) passes cephalad (in the vertebrarterial canal when cervical vertebrae are present) to join the circle of Willis (Fig. 13-15). Branches of the vertebral artery in the neck take the place of segmental arteries.

A series of segmental arteries arise from the aorta along the length of the trunk. These give off short vertebromuscular branches to the epaxial muscle, skin, and vertebral column, and long parietal branches, which encircle the body wall. The long branches traverse intercostal spaces as intercostal arteries. In the lumbar and sacral regions the segmentals are known as lumbar and sacral arteries. Caudally directed branches of the subclavian artery (anterior epigastric, long thoracic, and so forth) anastomose in the body wall with anteriorly directed branches (posterior epigastric, superficial epigastric, and so forth) of more caudal segmental arteries (iliolumbar, iliac, and so forth). Phrenic arteries to the mammalian diaphragm arise directly or indirectly from the aorta.

The iliacs are segmental arteries supplying the fins or hind limbs. In tetrapods the iliac becomes the femoral where it follows the femur, popliteal in the knee, and tibial in the shank.

Unpaired visceral branches. Primitively, an extended series of unpaired arteries pass via dorsal mesenteries to the stomach, liver, gallbladder, spleen, pancreas, and small and large intestine. Such a primitive series occurs in *Necturus*. In most living vertebrates the number of unpaired visceral branches is reduced to about three. The

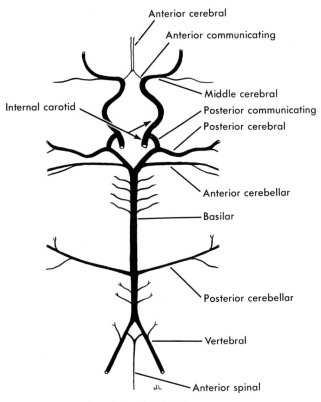

Anterior cerebral

Anterior communicating

Middle cerebral

Posterior communicating

Posterior cerebral

Internal carotid

Anterior cerebellar

Basilar

Posterior cerebellar

Vertebral

Anterior spinal

Fig. 13-15. Blood supply to the circle of Willis in a cat.

arteries supplying the stomach, liver, gall-bladder, and spleen may arise from a single celiac trunk, which gives off gastric, hepatic, and splenic branches. The arteries supplying the small intestine are usually reduced to a single anterior (superior) mesenteric. Those arteries supplying the large intestine are reduced to a posterior (inferior) mesenteric. Branches of the anterior mesenteric anastomose with branches of the celiac and posterior mesenteric so that each part of the digestive tract is supplied by more than one major vessel.

Among anastomosing channels on the digestive organs of mammals are a superior pancreaticoduodenal branch of the celiac, which anastomoses with an inferior pancreaticoduodenal branch of the superior mesenteric; a middle colic branch of the superior mesenteric, which anastomoses with a left colic branch of the inferior

mesenteric; and a superior rectal branch of the inferior mesenteric, which anastomoses with a middle rectal branch of the internal iliac. Anastomoses are also common on the greater curvature of the stomach between the short gastric arteries (branches of the splenic) and the right gastroepiploic (a branch of the gastroduodenal branch of the hepatic), and on the lesser curvature they are common between the left and right gastric arteries. Other anastomoses occur.

Paired visceral branches. Paired visceral branches of the aorta include large internal iliacs to the organs in the pelvic cavity (rectum, urinary bladder, and lower reproductive tract), genitals (ovarian or spermatic arteries) to the gonads, renals to the kidneys, and adrenals to the adrenal glands. The internal iliac arteries may arise as a bifurcation of the caudal end of the dorsal aorta or from a common iliac. Genital and renal arteries are a series in lower vertebrates, several pairs in reptiles and birds, and usually a single pair in mammals. In the thorax the aorta gives off a series of esophageal and bronchial arteries.

The umbilical (allantoic) arteries are embryonic branches of the dorsal aorta that carry unoxygenated blood to the allantois and to the chorioallantoic placenta (Fig. 13-23). The arteries commence during embryonic development as branches off the dorsal aorta near its caudal end. These branches vascularize the allantois. Later, the umbilical arteries sprout internal iliacs to the differentiating pelvic viscera (Fig. 13-23, 7).

Coronary arteries. The walls of all arteries and veins except the smallest ones are supplied, like all other tissue, with blood vessels. The vessels are called **vasa vasorum** (vessels of the vessels). The heart is no exception, and here the vasa vasorum are the **coronary arteries.** In elasmobranchs the coronary arteries arise from hypobranchial arteries. These receive aerated blood from several efferent arterial loops around the gill chambers (Fig. 13-9). Definitive coronary arteries do not occur in urodeles, and the heart has many small vasa vasorum. In frogs the coronary artery arises off the carotid arch. In reptiles and birds, coronary arteries arise from the aortic trunk leading to the right fourth arch or from the brachiocephalic. In mammals they lead from sinus-like dilations at the base of the ascending aorta just beyond the semilunar valves. The coronary arteries of reptiles, birds, and mammals, in essence, arise from a derivative of the ventral aorta.

Venous channels

During early embryonic life the venous channels of all vertebrates are relatively simple and are constructed in accordance with a basic pattern. As embryonic development progresses, the arrangement of the basic channels is slowly modified by deletion of some vessels and addition of others.

Modifications are few in elasmobranchs and more numerous in higher forms.

The major basic venous channels and their tributaries comprise several streams, which may be grouped for convenience as follows: **cardinal** (anterior, posterior, and common cardinal veins), **renal portal, lateral abdominal, vitelline** (hepatic portal and hepatic sinuses), and **coronary** streams. Two additional streams are characteristic of lungfishes and tetrapods—the **pulmonary** stream from the lungs and the **postcava** from the kidneys. The foregoing streams and their tributaries drain the entire body —head, trunk, tail, and appendages. Certain minor streams, such as the inferior jugular veins draining the lower jaw of sharks, are not included because they have no apparent evolutionary significance and cannot be considered basic to the architectural pattern.

Basic venous channels and their modification in sharks. The embryonic shark exhibits the basic venous channels (Fig. 13-16, basic pattern). As development progresses, these channels become modified (Fig. 13-16, adult dogfish). However, the modifications are relatively few, and the adult shark is an almost ideal living, swimming blueprint of the basic architectural pattern. A knowledge of the venous channels of sharks is therefore an excellent starting point for understanding the channels of higher vertebrates.

Common cardinal stream. The sinus venosus receives all the venous blood returning to the heart of fishes. Most of this blood, except that from the digestive system, enters the sinus by a pair of common cardinal veins that use the transverse septum as a bridge to the heart from the lateral body walls. The common cardinal veins therefore are important venous channels. These vessels appear in the embryo and remain with the same relationships in adults.

Anterior cardinal stream. Blood from all parts of the head except the lower jaw is collected by a large anterior cardinal vein (or sinus) lying dorsal to the gills on each

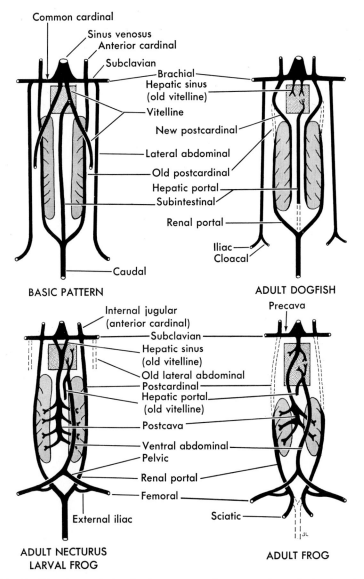

Fig. 13-16. Modifications of the basic pattern of venous channels in the dogfish and amphibians. Broken lines represent embryonic vessels that have disappeared. The pelvic veins are the paired caudal ends of the embryonic lateral abdominal veins. Vessels that are obviously the same have not been labeled on every diagram.

side. These vessels pass caudad and empty into the common cardinal veins at the level of the transverse septum. They remain essentially unchanged throughout life.

Posterior cardinal stream. The earliest embryonic posterior cardinal (also called postcardinal) veins are continuous with the caudal vein. They pass cephalad **lateral** to

the kidneys and empty into the common cardinal veins. At their anterior ends they expand to form posterior cardinal sinuses. Their tributaries include a series of renal veins draining the kidneys.

While these embryonic posterior cardinal veins are functioning, a network of **sub-cardinal** channels is forming **between** the kidneys close to the dorsal aorta. These new channels soon become confluent with the posterior cardinal veins at the anterior end of the mesonephroi. As more and more blood flows from the kidneys into the subcardinal channels, the older posterior cardinal veins become interrupted at the anterior end of the kidney (Fig. 13-16, adult dogfish). Thereafter, the name posterior cardinal is applied to the new posterior cardinals medial to the kidneys.

The chief tributaries of the adult posterior cardinal stream in sharks are a series of renal veins from the kidneys, a series of parietal veins from the dorsal body wall, and a series of veins from the gonads.

Renal portal stream. At an early stage in morphogenesis the blood from the caudal vein continues forward beneath the digestive tract as a subintestinal vein. Later, the caudal vein achieves a connection with the posterior cardinal stream, and the connection with the subintestinal vein is lost. All blood from the tail must thereafter utilize the posterior cardinal stream. Afferent renal veins from the caudal ends of the old posterior cardinals invade the kidneys and contribute blood to the capillaries surrounding the mesonephric tubules (but never to the glomeruli). Thereafter, more and more blood from the tail enters the kidneys. Finally, the old posterior cardinals become interrupted near the cephalic ends of the kidneys and all blood from the tail passes into the kidneys. The old posterior cardinals anterior to the caudal vein are then called renal portal veins.

In many bony fishes the renal portal system exhibits unexpected variations, including connections with the liver. Cyclostomes have no renal portal system.

Lateral abdominal stream. Commencing at the base of the pelvic fin from which it receives a femoral vein, and passing forward in the lateral body wall on each side is a lateral abdominal vein. At the level of the pectoral fin it receives a brachial vein, after which the abdominal vein turns abruptly toward the heart to enter the common cardinal vein. That part of the abdominal stream between the brachial and the common cardinal veins is the subclavian vein. In addition to collecting blood from the anterior and posterior appendages, the abdominal stream also receives a cloacal vein and a metameric series of parietal veins from the lateral body wall. Other minor tributaries contribute to the abdominal stream in the subclavian region. Bony fishes, except lungfishes, lack abdominal veins.

Hepatic portal system and hepatic sinuses; vitelline and subintestinal veins. The first venous channels to differentiate in the vertebrate embryo are paired vitelline (omphalomesenteric) veins (Fig. 13-16, basic pattern) passing from the yolk sac (or midgut region when a yolk sac is lacking), up the yolk stalk, then cephalad to enter the developing sinus venosus. One of the vitelline veins, usually the left, is joined by a subintestinal vein, which commences in the tail as a caudal vein and passes forward beneath the straight embryonic digestive tract. As the liver develops, it encompasses the vitelline veins, causing them to be broken into many sinusoidal channels. The left and right vitelline veins later unite to form a single channel, the hepatic portal vein, in the lesser omentum. The connection of the subintestinal with the caudal vein is soon lost and the subintestinal and vitelline veins caudal to the liver become a hepatic portal system draining the yolk sac, stomach, and intestine. Between the liver and the sinus venosus the vitelline veins remain paired in adult sharks and are known as hepatic sinuses.

In addition to draining the digestive tract, the hepatic portal stream receives veins from the pancreas, gallbladder, and rectal gland, which are derivatives of the tract. Veins from the spleen also find their way into the system.

Venous channels of tetrapods. The embryonic venous channels of tetrapods are essentially identical with those of embryonic sharks. We have just seen how, by a series of relatively minor modifications, the pattern of the shark embryo is converted into a pattern characteristic of the adult shark. We will now see how the same embryonic pattern is converted into channels characteristic of adult tetrapods.

Cardinal veins and the precavae. Embryonic tetrapods exhibit anterior, posterior, and common cardinal veins emptying into the sinus venosus. In the development of *Necturus* and anuran tadpoles these embryonic cardinal vessels are scarcely altered. The postcardinal veins retain their primitive connection with the caudal vein posteriorly and with the common cardinal anteriorly (Fig. 13-16, adult *Necturus*). In this respect, *Necturus* is less modified than the shark, which interrupts the primitive postcardinal vein during development.

In anurans (Fig. 13-16, adult frog) the postcardinal veins anterior to the kidneys disappear at metamorphosis after having drained the tail in the tadpole. As a result, the common cardinal veins in adult frogs drain chiefly the anterior limbs and the head.

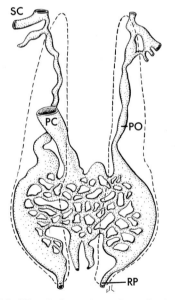

Fig. 13-17. Embryonic subcardinal venous plexus and origin of the postcava in the turtle *Chrysemys*, ventral view. Location of the mesonephros is indicated in outline. **PC,** Postcava; **PO,** postcardinal; **RP,** renal portal vein; **SC,** subclavian vein. (Modified from De Ryke.[81])

In tetrapods the common cardinal veins are better known as **precavae,** and the anterior cardinals are called **internal jugular** veins. Precaval veins therefore flow into the sinus venosus, or they flow into the right atrium when the sinus venosus is absorbed into the right atrial wall (Fig. 13-22).

Although most mammals retain both the left and right precaval veins (Fig. 13-22), some lose the left precava during embryonic life (Fig. 13-25). A new vessel (left brachiocephalic) connects the left and right anterior cardinal streams. Thereafter, the blood from the left side enters the heart via the right precava. In these mammals **brachiocephalic** (formerly called innominate) is the name given to the veins resulting from the confluence of the subclavian, external, and internal jugular veins.

The left precava does not disappear completely. A portion proximal to the heart remains as the **coronary sinus.** This drains the coronary veins and empties into the right atrium, since the sinus venosus is absorbed by the right atrial wall during development.

The fate of the postcardinal veins in tetrapods is affected by the development of a new venous channel, the postcava, to be discussed next.

The postcava. The postcardinal veins in lungfishes and tetrapods are supplemented or replaced functionally by a postcaval vein (inferior vena cava). The postcava is a large, median vessel arising during embryonic life between the kidneys in the subcardinal venous plexus (Fig. 13-17). The plexus receives renal veins from the kidneys. One subcardinal channel predominates, usually the right, establishes a channel in the ventral mesentery in which the liver is developing, and becomes confluent with the vitelline veins (hepatic sinuses). The enlarging liver finally envelops the postcava, but the latter is not interrupted by capillaries. Thus, the postcava becomes a large expressway passing directly through the substance of the liver. The

Right   Left

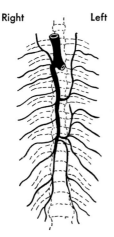

Fig. 13-18. The azygos (on the animal's right) and the hemiazygos (on the animal's left) in a rhesus monkey, ventral view. This condition is one of many variants in this species of monkey. The variants are essentially the same in man. (Redrawn from Seib.[89])

hepatic sinus is thereafter considered part of the postcava, and the hepatic veins become tributaries of the latter. In alligators, birds, and mammals (Figs. 13-19, 13-21, and 13-22), veins from the hind limbs establish direct connections with the postcava near the caudal ends of the kidneys, and the postcava finally becomes the chief drainage channel of the hind limbs as well as of the kidneys and most other organs of the trunk.

With establishment of a postcava, blood that would otherwise flow from the kidneys to the heart via the postcardinal veins now uses the postcava. Postcardinal veins in tetrapods therefore are reduced in size and importance (*Necturus*) or disappear (frogs, most reptiles, and birds). Remnants of postcardinal veins, when present, are variable, and parts of them masquerade un-

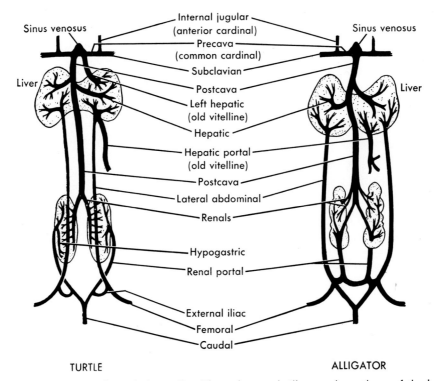

Fig. 13-19. Venous channels in reptiles. The only vessels illustrated are those of the basic plan shown in Fig. 13-16. When necessary, channels have been diagrammatically oriented to emphasize homologous vessels. A strong branch of the renal portal vein in crocodilians continues directly to the postcava. The pelvic vein occurs in turtles but has been omitted to facilitate identification of the basic pattern.

der new names such as azygos and hemi-azygos veins (Fig. 13-18). They chiefly drain the intercostal spaces and vertebral column.

Abdominal veins and limb drainage. In fishes the abdominal veins drain the pelvic fins, pass cephalad in the body wall, receive brachial veins from the pectoral fins, and then turn mediad as subclavian veins to enter the common cardinal veins. In tetrapods the abdominal veins undergo a series of embryonic modifications, which (1) causes them to terminate in the liver and thus dissociates them from the anterior limb drainage; (2) dissociates them from the hind limb drainage in birds and mammals; and (3) restricts them to the embryo in mammals.

Dissociation of the abdominal stream from the anterior limb drainage is seen in amphibians (Fig. 13-16). The embryonic abdominal veins of *Necturus* and frogs are at first paired and lie close together on either side of the midventral line. As development progresses, the paired abdominal veins caudal to the level of the liver fuse in the midventral line to form a median ventral abdominal vein. Blood in this median vessel finds its way into channels in the falciform ligament. Soon the blood in the abdominal stream has opened a new channel across the falciform ligament to the capillaries of the liver. The abandoned abdominal veins anterior to the liver disappear. Thus, the abdominal stream no longer drains the anterior limbs.

Formation of a ventral abdominal vein also occurs in some lungfishes *(Neoceratodus),* but in this species the adult ventral abdominal bypasses the liver and ends in the sinus venosus.

Dissociation of the abdominal stream from the hind limb drainage may have been initiated with formation of a communicating vessel (external iliac [Figs. 13-16, and 13-19]) between the hind limbs and the renal portal system. This new vessel provides blood from the hind limbs with an alternate route to the heart. The blood may now pass from the hind limbs,

into the renal portal system, deserting the abdominal stream.

The abdominal veins of reptiles remain paired throughout life but, as in amphibians, they terminate in the liver and are connected with the renal portal system by an external iliac vein.

The abdominal stream of birds and mammals is confined to the embryo. When the abdominal veins first appear, they commence in the posterior limb bud region, pass forward close together in the ventral body wall, and end in the common cardinal veins, just as in the embryo of sharks. As the allantois grows, its vascular channels become tributaries of the abdominal stream. Soon the abdominal stream loses its connection with the limb bud region and drains only the allantois. During further development the abdominal veins, now called allantoic veins in birds and umbilical veins in mammals, undergo the same changes as in amphibians. They unite over much of their route in the ventral body wall and acquire the tetrapod detour across the falciform ligament. Their function is to carry blood from the allantois (or from the chorioallantoic membrane of the placenta of mammals) into the embryo (Fig. 13-23, 2). Thus, a very ancient vessel, the abdominal vein, which at one time drained the posterior limbs, emerges in the highest vertebrates as a vessel draining the allantois and placenta. The allantoic (or umbilical) veins carry a considerable volume of blood, and they finally cut a broad channel, the **ductus venosus** (Fig. 13-23), straight through the upper end of the liver and into the postcava.

When the chicken hatches or the mammal is born, the extraembryonic membranes are discarded. This severs the allantoic or umbilical circulation. Thereafter, in mammals no blood flows through the stump of the vein that remains between the umbilicus and the liver. The blood that is left in the vessel clots, and the vessel becomes the **round ligament of the liver.** It is seen in the dissecting room as a fibrous cord in the free border of the falciform ligament connecting the umbilicus and liver. In birds the stump within the body may remain as the epigastric vein. The ductus venosus at birth becomes a **ligamentum venosum** embedded in the liver.

Blood from the posterior limbs of birds and mammals is no longer associated with the abdominal stream because of the changes just discussed. Instead, this blood enters the postcava.

Renal portal system. In *Necturus* the renal portal vein is directly continuous with the postcardinal veins, as in the basic pattern (Fig. 13-16). Afferent renal veins from the renal portal penetrate the kidney and

supply the capillaries surrounding the mesonephric tubules. Any blood entering the kidney capillaries may return to the postcardinal veins, or it may enter the postcava. In anurans, however, all the blood in the renal portal must enter the kidneys since the postcardinal veins are missing anteriorly (Fig. 13-16).

In amphibians an external iliac vein connects the abdominal and renal portal streams (Fig. 13-16), and blood from the hind limbs may be shunted to the renal portal system. Loss of the caudal vein at metamorphosis in tailless amphibians reduces the renal portal stream to draining the hind limbs and nothing else.

In reptiles the renal portal system is essentially the same as in amphibians. It drains the tail and the hind limbs. The blood that passes to the kidney via the renal portal in alligators (Fig. 13-19) may pass nonstop through the kidneys via channels leading directly into the postcava. Since snakes have no limbs, their renal portal system drains only the tail (Fig. 13-20).

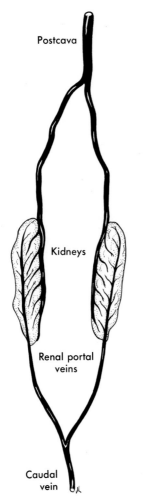

Fig. 13-20. Renal portal veins and venous drainage of the kidneys in a snake.

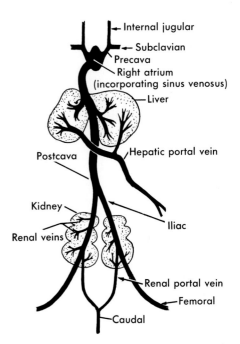

Fig. 13-21. Basic venous channels of a bird, ventral view.

Birds have inherited from reptiles the tendency for the blood in the renal portal stream to bypass the kidneys (Fig. 13-21). As a result, very little venous blood actually enters the kidney capillaries, and most of the blood from the caudal vein and hind limbs continues directly to the postcava.

Except for monotremes, adult mammals lack a renal portal system and blood from the caudal vein and hind limbs goes directly into the postcava (Fig. 13-22). The system appears transitorily in all embryonic mammals, however.

Hepatic portal system. The hepatic portal system is essentially similar in all verte-brates; it chiefly drains the digestive tract from the stomach caudad, its derivatives, and the spleen. It terminates in the capillaries of the liver. Its morphogenesis has been discussed earlier. The abdominal stream (allantoic in birds and umbilical in mammals) becomes a tributary of the hepatic portal system, commencing with lung-fishes. In bony fishes, veins from the swim bladders are usually tributaries of the system, as is the posterior coronary vein of frogs.

Pulmonary and coronary veins of verte-brates. Pulmonary veins drain the lungs and terminate in the left atrium. In lung-

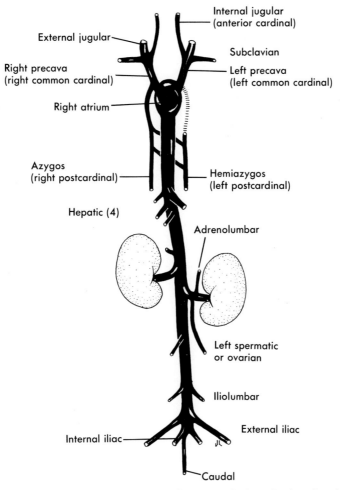

Fig. 13-22. Basic venous channels of a rabbit, ventral view. Broken line indicates obliterated segment of left posterior cardinal vein.

fishes and amphibians the left and right pulmonary veins unite to form a single vessel before entering the chamber. When one lung is absent, the corresponding pulmonary vein fails to develop.

Coronary veins in fishes enter the sinus venosus. Many amphibians lack a definitive venous coronary system. In frogs an anterior coronary vein enters the left brachiocephalic and a posterior coronary vein empties into the abdominal vein. In reptiles, birds, and mammals the coronary veins empty into the right atrium via a coronary sinus, which, in mammals, is a remnant of the left common cardinal of the embryo.

Circulation in the fetus and changes at birth

In the avian or mammalian fetus blood passes from the caudal end of the dorsal aorta into the umbilical (allantoic) arteries (Fig. 13-23). These extend out the umbilical cord to the allantois or placenta, where exchange of respiratory gases takes place and oxygen is acquired. From the allantois or placenta the blood returns to the fetus via umbilical (allantoic) veins, which traverse the falciform ligament to enter the liver. Some of this blood enters the hepatic portal system, but most of it continues nonstop via the ductus venosus, into the postcava, and finally into the right

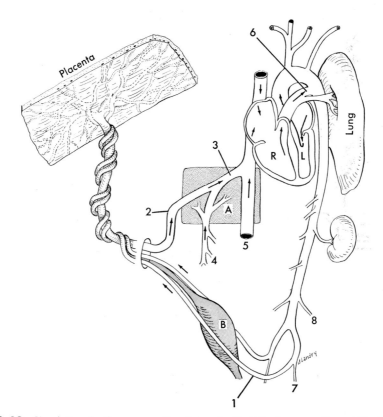

Fig. 13-23. Circulation in the mammalian fetus. **1,** Umbilical artery; **2,** umbilical vein; **3,** ductus venosus; **4,** hepatic portal vein; **5,** inferior vena cava; **6,** ductus arteriosus; **7,** internal iliac; **8,** external iliac growing into hind limb bud; **A,** liver; **B,** base of the allantois, which is developing into a urinary bladder; **L,** left ventricle; **R,** right ventricle. Much of the blood returning to the right atrium from the inferior vena cava passes through a foramen ovale (not illustrated) leading into the left atrium.

atrium. From the right atrium most of it passes via an **interatrial foramen** (**foramen ovale** of the heart) into the left atrium. The rest of the blood, along with that returning to the right atrium from the head, enters the right ventricle and is pumped into the pulmonary trunk. Because the ductus arteriosus is patent (Fig. 13-23), much of the blood in the pulmonary trunk is shunted into the dorsal aorta. This is an advantage, since this blood is mostly unaerated, and some of it will pass down the dorsal aorta to enter the umbilical arteries leading to the fetal respiratory membranes.

Blood coming from the lungs, which is unaerated and in small quantities, enters the left atrium and, along with the semioxygenated blood coming into the left atrium via the interatrial foramen, passes into the left ventricle. This blood is then pumped into the ascending aorta to be distributed to the head, trunk, and limbs. From the account just given, it can be seen that the blood in the fetus is either venous or mixed, except in the umbilical (allantoic) veins.

At birth or at hatching, major circulatory changes adapt the organism for pulmonary respiration:

1. The ductus arteriosus becomes occluded as a result of nerve impulses passing to its muscular wall. These impulses are initiated reflexly when the lungs are first filled with air. In birds this is usually the day before hatching, when the imprisoned chick pecks a hole in its extraembryonic membranes and starts breathing the air entrapped between these membranes and the shell. (When the chick inside the shell starts to peep, it already has air in its lungs!) In mammals the air enters the lungs with the first gasp after delivery. In either case, the ductus arteriosus constricts and blood ceases to flow through it. Thereafter, all blood entering the pulmonary trunk goes to the lungs. The ductus arteriosus soon becomes converted into the **arterial ligament** (**ligamentum arteriosum**).

2. A valve-like flap is pressed against the interatrial foramen by the increased pressure in the left atrium, the result of the greatly increased volume of blood entering from the lungs. This prevents the unoxygenated blood in the right atrium from entering the left atrium, which now contains only oxygenated blood from the lungs. Within a few days the foramen ovale is permanently sealed and only a scar, the fossa ovalis, remains.

3. At birth or at hatching, the umbilical arteries and vein are severed at the body wall when the extraembryonic membranes are shed. Thereafter, no blood passes through the umbilical arteries beyond the distal tip of the urinary bladder, which the arteries continue to supply. From the tip of the bladder to the navel the umbilical arteries become converted into **lateral umbilical ligaments**. These lie in the free border of the ventral mesentery of the bladder.

4. Blood no longer flows through the umbilical vein, and this vessel becomes converted into the **round ligament of the liver**. At the same time, the ductus venosus is converted into the **ligamentum venosum**. (This occurs half way through gestation in whales.[30]) As a result of these changes, the fetal bird or mammal is changed from a water-dwelling, allantoic-respiring organism to one capable of living on land and breathing air.

Failure of the interatrial foramen to close or of the ductus arteriosus to fully constrict results in cyanosis (blueness) of the skin of the newborn, since blood continues to be shunted away from the lungs and hence has the bluish color of venous blood.

Circulation in the different classes: an overview

Fishes (Figs. 13-3, 13-4, 13-9 to 13-11, 13-16, and 13-24). The pericardial cavity and heart lie ventral to the pharynx. The atrium (incompletely paired in Dipnoi) is dorsal to the ventricle. The heart contains only venous blood, which passes from the sinus venosus to the atrium, to the ventricle, and finally to the conus arteriosus (reduced in teleosts). The conus arteriosus

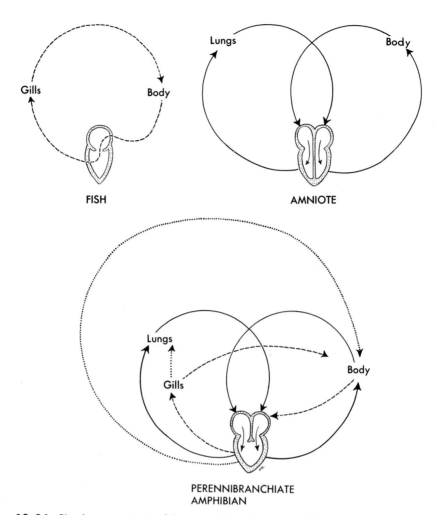

Fig. 13-24. Circulatory routes in fish, perennibranchiate amphibians, and amniotes. Perennibranchiates exhibit the fish type of circulation (broken lines) and the amniote type (solid lines). Dotted lines in lower figure indicate routes theoretically available in perennibranchiates.

exhibits three to eight transverse rows of valves (elasmobranchs) or a single row (teleosts). From the conus arteriosus venous blood passes into a ventral aorta and thence to afferent branchial arteries, derived from embryonic aortic arches and corresponding in number with the gill-bearing arches (five to seven in elasmobranchs and seldom more than four in teleosts and ganoids). After branchial aeration, blood is collected in efferent branchial arteries derived from the dorsal parts of certain of the embryonic aortic arches and

passes to the head (via carotids and other vessels), to the trunk and paired fins (via branches of the dorsal aorta), and to the tail (via the caudal artery). Venous blood is collected from the stomach, intestine, pancreas, spleen, and (elasmobranchs) rectal gland and is carried to the capillaries of the liver via a hepatic portal system. Hepatic veins empty into the sinus venosus via hepatic sinuses. Blood from the tail passes via a renal portal system (caudal vein and renal portal veins) to the capillaries of the kidneys. Anterior cardinal veins drain most of the head. Posterior cardinal veins drain the kidneys, body wall, gonads, and (in most bony fishes) pelvic fins. Anterior and posterior cardinals along with subclavian veins from the pectoral fins empty into common cardinal veins. In elasmobranchs but not in teleosts lateral abdominal veins drain the body wall and posterior appendages and empty into subclavian veins. Valves are common in elasmobranch veins. There is no mixed blood in circulation.

Lungfishes exhibit the following modifications, some of which are encountered also among amphibians. There is an incomplete atrial septum correlated with development of lungs. The right atrium receives blood from the sinus venosus; the left atrium receives the pulmonary vein. The conus arteriosus is spirally twisted, and the ventral aorta shows the beginning of a division separating arterial from venous blood. Aortic arches III and IV (*Protopterus*) are uninterrupted by gill capillaries. A pulmonary artery arises off arch VI. A postcaval vein drains the kidneys. Only the left postcardinal remains. In *Neoceratodus* the abdominal veins fuse to form a single ventral abdominal that terminates in the sinus venosus. The abdominal stream develops a connection with the renal portal system.

Amphibians (Figs. 13-4, 13-5, 13-10 to 13-12, 13-16, and 13-24). The heart has moved caudal to the pharynx. It exhibits a sinus venosus, a single ventricle, and two atria (interatrial septum is fenestrated in urodeles, complete in anurans, and lacking in lungless forms). The atria are anterior to the ventricles. The right atrium receives the sinus venosus; the left atrium receives the pulmonary veins (absent in lungless forms). Aortic arches III (carotid), IV (systemic), and VI (pulmonary) are always present, and V occurs in some urodeles. Blood entering external gills, when present, does so via detours off the aortic arches. The ventral aorta forms a swollen bulbus arteriosus, subdivided in anurans and many urodeles by a spiral valve shunting freshly oxygenated blood to arches III and IV (to the head and dorsal aorta) and deoxygenated blood to arch VI. In anurans the dorsal aorta between arches III and IV (ductus caroticus) disappears, as does the segment of arch VI (ductus arteriosus) above the pulmonary artery. These persist in many urodeles.

The venous channels in tailed amphibians are essentially similar to those of fishes. Anterior cardinal (internal jugular veins), postcardinal, and subclavian veins empty into common cardinal veins (precaval veins). A postcava is present, and a connection is established between the abdominal stream from the hind limbs and the renal portal vein from the tail. The ventral abdominal vein in amphibians terminates in the capillaries of the liver.

The venous channels of adult anurans have been further modified. Postcardinal veins are mostly lacking and hence the postcava assumes added importance in draining the kidneys and hind limbs. The renal portal system drains only the hind limbs following loss of the tail at metamorphosis. The ventral abdominal vein functions as in urodeles.

Most of the blood vessels in amphibians contain blood not wholly oxygenated. This is attributable to the introduction of lungs without concomitant changes in the ventricle and to the fact that aortic arches are retained as gill bypasses when external gills are present.

Reptiles (Figs. 13-4, 13-11, 13-13, 13-17, and 13-19). The reptilian heart has two completely separated atria, two ventricles incompletely separated except in crocodilians, and a sinus venosus that is often indisinguishable externally but is distinct internally. The right atrium receives the sinus venosus and coronary veins; the left atrium receives the pulmonary veins. Aortic arches III, IV, and the ventral part of VI remain in adults. The ductus caroticus and ductus arteriosus disappear, except in primitive species such as *Sphenodon*. Instead of exhibiting a spiral valve for distributing deoxygenated and oxygenated blood, the ventral aorta becomes split into three channels. A pulmonary trunk passes from the right ventricle to the sixth aortic arches and supplies the lungs. An aortic trunk passes from the **left ventricle** to the **right fourth aortic arch** and to the **third aortic arches** and supplies the head, anterior limbs, and dorsal aorta. Another aortic trunk passes from the **right ventricle** to the **left fourth aortic arch** and supplies the dorsal aorta.

The incomplete ventricular septum plays an interesting role in aquatic turtles that are capable of remaining under water for prolonged periods. When the turtle is ventilating its lungs a valve-like cartilage in the right ventricle partially blocks the passage of blood between ventricles. A small amount of blood is shunted from left to right, which increases the amount of oxygenated blood going to the stomach and intestines but does not affect blood going to the head (Fig. 13-13, turtle). During prolonged periods under water the oxygen tension of the blood falls, the interventricular cartilaginous valve is displaced, and the two ventricles become confluent. Blood is then shunted from right to left, hence away from the lungs. Under these conditions the turtle derives energy by glycolysis, an anaerobic process.[84] Just how a change in oxygen tension is translated into displacement of the valve is not known. It may result from constriction of the pulmonary arteries, which would raise the pressure within the right ventricle and cause blood to be shunted to the left.

Although crocodilians have a complete ventricular septum, a foramen of Panizza connects the two aortic trunks, providing an interaortic shunt. The role of this shunt in crocodilians has not been studied experimentally.

Snakes and legless lizards lose their embryonic left sixth aortic arch when the left lung fails to develop. In snakes the left third aortic arch also disappears. This leaves the adult snake with a right pulmonary artery and a right common carotid only (Fig. 13-13). The latter has a bilateral distribution.

Precaval (common cardinal) veins receive internal jugular (anterior cardinal) veins from the head, subclavian veins from the anterior limbs, and vertebral veins (replacing the postcardinal veins which are apparently lost in most adult reptiles). The two ventral abdominal veins receive blood from the hind limbs and terminate in the capillaries of the liver as tributaries of the hepatic portal system. Renal portal veins drain the tail and hind limbs and empty into the kidneys or (crocodilians) partly bypass the kidneys to terminate in the postcava. The postcava is the sole drainage of the kidneys, carries some blood from the tail, receives hepatic veins from the liver, and terminates in the sinus venosus. The major modification in the venous channels of reptiles is the bypassing of the kidneys by some of the blood in the renal portal system.

Birds (Figs. 13-4, 13-11, 13-13, and 13-21). The two atria and two ventricles are completely separated in birds. The sinus venosus becomes incorporated into the wall of the right atrium during embryonic development and is absent in adults. The right atrium therefore receives two precaval veins, the postcava, and the coronary veins. The left atrium receives pulmonary veins. Emerging from the right ventricle is the pulmonary trunk leading to the sixth aortic arches. The ductus arteriosus disappears. Emerging from the left ventricle is a single aortic trunk leading to aortic arches III and IV. Arch IV is bilaterally represented in embryos, but as development progresses, the left fourth aortic arch disappears. The only adult aortic arch leading to the dorsal aorta is therefore the right fourth, commonly referred to as "the" aortic arch. The subclavian artery develops as an enlarged segmental artery associated with the carotid arch (III).

Venous channels are essentially crocodilian in nature. Precaval veins approach the heart from a cephalic direction and receive internal jugular and subclavian veins.

The blood from the hind limbs that earlier in phylogeny went, via the external iliac, to the renal portal system and kidneys now bypasses the kidneys and goes directly into the postcava via iliac veins. The iliac veins receive renal veins en route. The renal portal is greatly reduced by loss of blood from the hind limbs and by the decreased size of the tail. Since postcardinal veins are absent, the postcava is the chief vessel draining most of the body, except for the head, neck, and wing. Coursing in the falciform ligament and associated mesenteries is a small epigastric vein, peculiar to birds, draining the greater omentum. This vessel may be a remnant of the abdominal vein of lower vertebrates.

No mixing of oxygenated and deoxygenated blood occurs in the heart or arches. Deoxygenated blood passing through the right side of the heart goes to the lungs. Oxygenated blood in the left side of the heart passes next to the various organs of the body other than the lungs. There is thus established a double circulation, with pulmonary and systemic streams completely separated.

Mammals (Figs. 13-4, 13-6, 13-7, 13-11, 13-13 to 13-15, 13-18, and 13-22 to 13-25). The two atria and two ventricles are separated by an interatrial and interventricular septum. A foramen ovale connects the two atria until birth and then closes. The atria exhibit unique, ear-like appendages—the auricles. The sinus venosus is absorbed into the wall of the right atrium. The left fourth aortic arch remains as the sole connection between the heart and descending aorta, a condition opposite to that in birds. The brachiocephalic artery is a remnant of the right fourth arch, as is the proximal part of the right subclavian. The subclavian arteries arise as segmental branches from the fourth arch. The left ductus arteriosus remains patent until birth and shunts blood out of the pulmonary trunk, away from the lungs, and into the dorsal aorta. The terminal branches of the latter in the embryo lead to the fetal respiratory organ (placenta).

In many mammals two precavals (common cardinal veins) empty into the right atrium after receiving internal jugular and subclavian veins in accordance with the basic plan. In some mammals (whales, carnivores, primates, and so forth) the left precava disappears (except the part that becomes the coronary sinus) and blood is shunted from the left to the right precava (Fig. 13-25). Azygos and hemiazygos veins draining intercostal spaces and emptying into the precava represent, in part, postcardinal veins.

The renal portal system appears transitorily during embryonic development and, except in monotremes, disappears along with the embryonic mesonephric kidney. The abdominal (umbilical) stream is likewise confined to the fetus. The postcava becomes the sole drainage for the hind

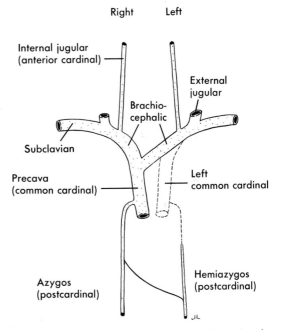

Fig. 13-25. Basic anterior venous channels of cat and man, ventral view. Broken lines indicate obliterated vessels. These sometimes remain as anomalies in adult mammals, including cats, pigs,[82] and man.[85] Compare channels with those of a rabbit given in Fig. 13-22.

limbs and tail. It terminates in the right atrium at a site marking the location of the embryonic sinus venosus.

Physiological adaptations of the circulatory system take place when whales dive to great depths. Extensive thick, spongy masses of twisted generous-sized arteries, the **retia mirabilia,** are located in protected positions along the vertebral column under the transverse processes, within the bony neural canal, and within the thoracic cavity. The retia are all confluent, are supplied by branches of the thoracic aorta, and are drained by efferent arteries. When a whale dives, the abdominal wall is compressed dorsad against the vertebral column by external pressure, the abdominal viscera are forced into the thorax, and the air is expressed from the lungs into the trachea. There is a general constriction of all arteries, except those of the brain (which are protected by the skull). Blood forced out of the various organs collects in great quantities in the retia, which permit adaptation to the two chief effects of diving— increased pressure on the body and reduced oxygen supply.[30] (In some species the dive may last up to two hours.) The retia accommodate the blood expressed from the soft organs, and the pools thus formed constitute protected reservoirs of oxygenated blood that is spared for use by the brain. Other physiological adaptations to diving also take place. Similar retia occur in most aquatic mammals. Retia serving other functions in mammals are found in ungulates, edentates, and carnivores. Retia are not confined to mammals, however.

LYMPHATIC SYSTEM

The lymphatic system resembles the blood vascular system in that it consists of vessels, fluids in transit, and associated organs (Fig. 13-26). A major difference is that lymph flows in only one direction— toward the heart.

The lymphatic system of all vertebrates consists of thin-walled **lymphatics** and, in birds and mammals, of **lymph nodes** interposed along the course of the lymphatics.

The lymphatics penetrate nearly all the soft tissue of the body and commence as blind-end **lymph capillaries** that collect interstitial fluids. Once inside the lymph capillaries, the fluid is called **lymph,** a colorless or pale-yellow fluid containing metabolites and secretions, which constantly collect in the intercellular spaces. The lymph from any area mirrors the metabolic activities of that area from moment to moment. Lymph from different areas passes into successively larger tributaries. The largest vessels empty into one or more veins of the blood vascular system, often, but not universally, in the vicinity of the heart. The walls of the larger lymphatics are strengthened by smooth muscle, and the largest resemble veins, although they lack uniform diameter. The lymphatics in the villi of the small intestine collect globules of fat absorbed from the intestine after a meal. If the meal has been particularly fatty, the lymph in these vessels is milky. For this reason, the lymphatics of the intestinal villi are called **lacteals,** and the lymph therein is called **chyle** (juice). Some of the lymphatics in cyclostomes, cartilaginous fishes, and even in man contain red blood cells as well as lymph. The fluid in these vessels is called **hemolymph.**[4, 7, 91]

Lymph nodes are masses of hemopoietic tissue interposed along the course of lymph channels, especially at the site of confluence of several channels. They may be no larger than a pinhead, or they may measure several centimeters in length or diameter. These are the "swollen glands" that can be palpated in the neck, axilla, and groin of man when there is inflammation in the areas drained. Lymph nodes consist of a connective tissue reticulum enmeshing large numbers of lymphocytes, many of which are formed therein, and of sinusoidal passageways lined by phagocytic cells. Lymph enters a node via several afferent lymphatics, filters through the sinusoidal spaces, and leaves via a single large efferent lymphatic. Lymph nodes constitute one line of defense against disease, since they contain cells that can ingest and destroy bacteria as well as other foreign particles. In this respect they serve as filtering beds for interstitial fluid before the latter is returned to the general circulation. In addition, the nodes contain large numbers of special cells that are produced in the thymus gland and that, when stimulated by antigens, produce antibodies.

Lymph channels in birds and mammals are provided with fibrous valves that assist in preventing backflow. Valves may also guard the exits from the major lymph ducts into the veins. In fishes, amphibians, and reptiles muscular pulsating **lymph hearts** are situated at strategic locations along the lymphatics. Frogs have four pairs of lymph hearts, two near the thigh and two under the scapula. Urodeles have as many as sixteen pairs and caecilians as many as one hundred pairs. Amphibians have a much more active seepage from their vascular channels than other vertebrates, and so their lymph hearts move a large volume of fluid hourly.

In addition to the pulsations of lymph hearts, the flow of lymph results from numerous other factors. These include the pressure of the incoming fluid, contraction of the muscular walls of the larger ducts, contraction of surrounding skeletal muscles, movements of the viscera that squeeze the lymphatics and "milk" the fluids along the

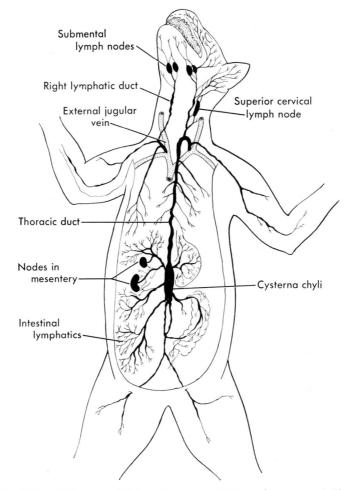

Fig. 13-26. A few of the superficial and deep lymphatics of a mammal. The large veins of the neck approaching the heart are stippled.

vessels, and respiratory movements in birds and mammals, which have alterations in intrathoracic pressure because of the muscular diaphragm or oblique septum.

Sinus-like enlargements of lymph channels occur in some locations. Such are the **subcutaneous lymph sinuses** under the skin of frogs, the **sublingual lymph sac** of frogs that, when suddenly filled with lymph, causes the tongue to dart out, and the **cisterna chyli** in the abdominal cavity of many higher vertebrates that collects chyle from lacteals. From these and other lymph sinuses emerge large lymphatics that terminate in an appropriate vein.

Lymph channels in lower vertebrates empty into one or more major venous channels, including caudal, iliac, subclavian, and postcardinal veins, depending on the species. Most mammals have two major lymph channels—a **thoracic duct** commencing in the cisterna chyli and draining the abdomen, left side of the head, neck, thorax, and the left arm and shoulder and

a **right thoracic duct** (sometimes several) draining the right side of the head, neck, thorax, and the right arm and shoulder (Fig. 13-26). Each empties into the subclavian vein on its own side or into an adjacent vein such as the brachiocephalic or external jugular. Birds and a few mammals have two thoracic ducts arising from a single cisterna chyli.

Although lymph nodes have not been described in fishes, tailed amphibians, or reptiles, lymphoid masses are scattered throughout the body of all vertebrates. Lymph does not flow **into** these, but they are drained by lymph capillaries. Lymphoid masses are numerous in the coelomic mesenteries. The spleen is the largest of such masses. It is absent in cyclostomes. Other masses include the tonsils at the upper end of the digestive tube in mammals, the thymus (a derivative of the pharynx), Peyer's patches in the mucosa of the small intestine of amniotes, and the bursa Fabricii of birds.

Chapter summary

1. The circulatory system includes the blood vascular and lymphatic systems. The blood vascular system consists of the heart, arteries, capillaries, and veins. The lymphatic system consists of lymphatics (including lacteals), lymph capillaries, lymph sinuses, lymph nodes in higher forms, and lymph hearts (absent in birds and mammals). Lymph is transported from tissue spaces, and chyle is transported from the intestinal villi to certain major venous channels.

2. A sinus venosus occurs in fishes, amphibians, and reptiles. It is absent in adult birds and mammals, having been incorporated in the right atrial wall during embryonic development.

3. A single atrium that receives blood from the sinus venosus occurs in most fishes. In lung-breathing vertebrates the atrium is partitioned into two chambers by a septum, which is incomplete in lungfishes and some urodeles but complete in other tetrapods.

The right atrium receives the sinus venosus or the largest systemic veins and, in amniotes, the coronary veins. The left atrium receives pulmonary veins.

4. A single ventricle occurs in fishes and amphibians. Lungfishes have an incomplete ventricular septum partially dividing the ventricle in two. Amniotes exhibit two ventricles separated by an incomplete septum in most reptiles, and a complete septum in crocodilians, birds, and mammals.

5. A conus arteriosus provided with valves emerges from the ventricle in most fishes and amphibians and constitutes part of the heart. It is absent in adult amniotes. A spiral valve within the conus shunts oxygenated and deoxygenated blood to appropriate aortic arches.

6. A ventral aorta leads cephalad from the heart to all aortic arches in fishes and amphibians. In urodeles the ventral aorta exhibits a swelling, the bulbus arteriosus. In

reptiles the ventral aorta is split longitudinally into an aortic trunk leading from the left ventricle to the third and right fourth arches, a second aortic trunk from the right ventricle leading to the left fourth arch, and a pulmonary trunk from the right ventricle leading to arch VI. In birds and mammals two trunks instead of three emerge from the heart: an aortic trunk (ascending aorta) leads to the carotid and systemic arches, and a pulmonary trunk leads to the pulmonary arch. Formation of the separate trunks eliminates the need for a spiral valve.

7. An aortic arch is a blood vessel connecting the ventral and dorsal aortas and located, in the embryo at least, in a visceral arch. Typically, six pairs of aortic arches develop in each vertebrate embryo. During ontogeny the aortic arches are reduced in number, the highest vertebrates retaining the fewest arches. In most fishes the aortic arches become interrupted by gill capillaries. In gill-bearing amphibians, detours from the aortic arches carry blood into and out of the external gills. In lung-bearing vertebrates the sixth aortic arch sprouts pulmonary arteries.

8. Aortic arch I (mandibular) usually disappears, in part at least, in all vertebrates. Elasmobranchs retain parts of aortic arches II to VI inclusive, and some species develop additional arches caudal to the sixth. Modern fishes and many tailed amphibians retain arches III to VI inclusive. Other tailed amphibians and all higher vertebrates retain arches III, IV, and VI only. The dorsal aorta between III and IV (ductus caroticus) disappears in some urodeles, some reptiles, and all higher vertebrates. The dorsal segment of aortic arch VI (ductus arteriosus) disappears in anurans and in amniotes. Birds lose, in addition, the left fourth aortic arch, and mammals lose much of the right fourth. As a result of loss of the ductus caroticus, blood entering arch III (carotid) must continue to the head (as in frogs) instead of passing caudad into the dorsal aorta (as in *Necturus*). As a result of loss of the ductus arteriosus, blood entering arch VI must pass to the lungs (as in frogs) instead of having the alternative of continuing into the dorsal aorta (as in *Necturus*). Birds and mammals therefore lose arches I, II, V, the left or right side of IV, the dorsal segment of VI on both sides, and the ductus caroticus.

9. As a result of successive modifications in the heart and ventral aorta and deletions in the aortic arches, a system of vessels appropriate for respiration via gills is translated phylogenetically and ontogenetically into one suitable for lung respiration. There is no mixing of oxygenated and deoxygenated blood in fishes. Considerable mixing occurs in tailed amphibians, and less occurs in frogs and reptiles. Mixing occurs in fetal birds and mammals but not in adults. In adults the pulmonary channels are completely separate from the systemic channels, and the heart is completely divided into right (oxygen-poor) and left (oxygen-rich) sides.

10. Major venous channels in the basic pattern of vertebrate circulation are anterior cardinal (internal jugular) veins from the head, postcardinal veins from the trunk and kidneys, and subclavian veins from the anterior appendages—all flowing into common cardinal veins; abdominal veins from the hind limbs; renal portal system from the tail; and hepatic portal system from the chief digestive organs. Hepatic sinuses drain the liver, coronary veins drain the musculature of the heart, and pulmonary and postcaval veins are added in lung-breathing forms.

11. Postcardinal veins are absent in frogs but are partially present in reptiles, birds, and mammals under new names (azygos, hemiazygos, and so forth, with homologies not entirely clarified). Postcardinal veins are superceded functionally by postcaval veins in lungfishes and tetrapods.

12. Abdominal veins unite with veins from the anterior fins in most fishes (absent in teleosts) and become tributaries of the hepatic portal system in amphibians and reptiles. In birds and mammals they drain the allantois or placenta.

13. The renal portal system drains only the tail

in fishes. It acquires a connection (external iliac) with the hind limb drainage in amphibians and reptiles. In crocodilians and birds the connection may bypass the kidneys and go directly to the postcava. The renal portal system is absent in adult mammals, except monotremes, and the hind limbs and tail are drained solely by the postcava.

14. The postcava becomes increasingly prominent in higher vertebrates. Commencing in lower forms as an alternate outlet for blood from the kidneys, it finally assumes drainage of the hind limbs and tail. It utilizes the hepatic sinus from the liver to the heart.

15. Remnants of embryonic vascular channels in adult mammals include the round ligament of the liver (remnant of the left umbilical vein between the umbilicus and liver), ligamentum venosum of the liver (remnant of the ductus venosus), ligamentum arteriosum (remnant of the left ductus arteriosus connecting the pulmonary and systemic arches), lateral umbilical ligaments (remnants of paired umbilical arteries from the urinary bladder to the umbilicus via the ventral mesentery of the bladder), and fossa ovalis of the median wall of the right atrium (site of occluded foramen ovale connecting the left and right atria of the fetus). The circulatory changes at hatching or birth are described on p. 300.

Chapter

14

Urinogenital system

Although the function of the kidneys is different from that of the gonads, the ducts of the two organs are so intimately related developmentally and functionally that neither the urinary nor the genital system can be discussed without constant reference to the other. For this reason, it is convenient to discuss both systems in a single chapter. Although the more formal term "urinogenital" has been used in the chapter title to describe the combined systems, the more popular term "urogenital" has been used in the text.

VERTEBRATE KIDNEYS AND THEIR DUCTS

Paleontological evidence has been interpreted as indicating that the earliest vertebrates lived in fresh water, and that the early stages of evolution of fishes took place in that medium. Animals that are submerged in fresh water inevitably acquire excess water either by absorbing it through the skin or by swallowing it with the food. Therefore, a mechanism for elimination of excess water is necessary. On the other hand, salt is scarce in fresh water, the only source being food. Freshwater organisms therefore must prevent any wasting of salt from the body.

The elimination of water and the reclamation of salts may have been the earliest functions of vertebrate kidneys. Tufts of blood vessels (**glomeruli**) filtered water

out of the bloodstream into the body cavity, and **convoluted tubules** with openings into the coelom collected the filtrate, retrieved any salts from it, and emptied the final filtrate into a **longitudinal duct** that passed to the cloaca.

When ancestral fishes were later adapting to salt water, they faced a different water-salt problem. Instead of accumulating too much water in their tissues, they were in danger of accumulating too much salt. The problem of survival then became one of conserving water and excreting salts. Structural modifications of the kidneys helped to solve this problem. One modification was the shortening or loss of distal segments of the kidney tubules, these being the segments responsible for reabsorption of salts and participating in water excretion. Another modification was loss of the glomeruli in some marine teleosts (toadfish, sea horse, pipefish, and others), and in many others the glomeruli are poorly vascularized or cystic. Shortening of the tubules and loss of glomeruli permitted increased salt excretion and water retention in marine fishes.

There are a few freshwater teleosts with aglomerular kidneys and no distal segments of the tubules. Water excretion in these species is principally tubular. These teleosts are thought to represent former saltwater species that assumed a freshwater habitat.

The concept that vertebrates arose in

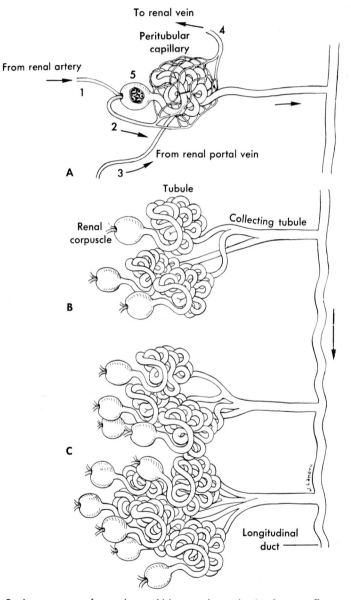

Fig. 14-1. Basic structure of vertebrate kidney, schematic. In **A,** an afferent glomerular arteriole, **1,** penetrates a Bowman capsule, **5** (part of the wall removed), to form a glomerulus. The glomerulus plus the capsule constitute a renal corpuscle. Emerging from the capsule is an efferent glomerular arteriole, **2,** that terminates in a peritubular capillary bed. Also contributing to the peritubular capillaries is a venule, **3,** from the renal portal system. Emerging from the peritubular capillaries is a renal venule, **4. B,** Replication of renal corpuscles in a single body segment. **C,** The increased number of tubules per segment disrupts the metamerism of the nephrogenic mesoderm.

fresh water is not universally accepted. There are some who think vertebrates arose in a marine environment and that the kidneys were used originally to eliminate aqueous solutions of inorganic salts from animals whose blood was isotonic with the sea, and that later, when fresh water was encountered, the same mechanism was useful in eliminating excess water.

Thus far nothing has been said about excretion of the nitrogenous wastes of metabolism because in no fish is the kidney of primary importance in excretion of nitrogenous wastes.[24] These wastes are eliminated via the gills and other extrarenal tissues. The excretion of metabolic wastes is probably a more recent function of the kidney. The kidneys of reptiles, birds, and mammals serve to dispose of nitrogenous wastes as well as to regulate water and electrolyte levels.

Kidney structure and the archinephros

From the preceding discussion it is evident that vertebrate kidneys (**nephroi**) are all built in accordance with a basic pattern incorporating three components: (1) **glomeruli**, (2) **tubules**, and (3) **a pair of longitudinal ducts** (Fig. 14-1). Variations in the structure of the adult kidneys of vertebrates from fish to man are primarily in the nature of alterations in number, complexity, and arrangement of the glomeruli and tubules.

Glomeruli are tufts of blood capillaries, which, assisted by blood pressure, filter water, salts, and certain other substances out of the bloodstream. In some species the glomeruli are large enough to be seen with the naked eye or a hand lens. In others they are microscopic. The glomeruli arise as localized modifications of blood vessels, which can be traced from segmental branches of the dorsal aorta. Supplying a glomerulus is an afferent glomerular arteriole and emerging from the glomerulus is an efferent glomerular arteriole. The latter leads to capillary beds that surround the kidney tubules. From the peri-tubular capillary beds emerge venules that lead to renal veins.[*]

The most anterior embryonic or larval glomeruli may be suspended in the coelomic cavity (Fig. 14-2, A). They are sometimes called "external" glomeruli to differentiate them from "internal" glomeruli, which are encapsulated by the kidney tubule.

Kidney tubules (Fig. 14-1) are microscopic convoluted urinary vessels that collect glomerular filtrate and transport it to the longitudinal duct. Certain substances are reabsorbed from the filtrate during passage through the tubules, and certain others are secreted into the filtrate. Each tubule typically commences as a **Bowman capsule** (Fig. 14-2, B). This is a blind end of the tubule that surrounds an internal glomerulus and receives glomerular filtrate. The capsule plus the encapsulated glomerulus constitute a renal corpuscle.

The more anterior tubules may exhibit a ciliated, funnel-shaped **nephrostome**, which is an opening into the coelom (Fig. 14-2, B). Nephrostomes are usually confined to embryos and larvae. If the embryonic tubule that exhibits a nephrostome is not lost altogether, the nephrostome may close at a later stage of development. Nephrostomes are thought to be vestiges of a hypothetical primitive kidney (archinephros) in which all glomeruli were external. If the hypothesis is correct, the more intimate relationship of a glomerulus and tubule via a Bowman capsule is a later development.

Kidney tubules arise from the intermediate mesoderm. This is a ribbon of nephrogenic tissue extending uninterruptedly from the level of the heart to the cloaca. It lies just lateral to the segmental (dorsal) mesoderm. Almost the entire ribbon of intermediate mesoderm produces kidney tubules. Commencing at the anterior end,

---

[*]It has been postulated that efferent glomerular vessels at one time led directly to renal venules, as in living myxinoids, and that the peritubular capillaries belonged solely to the renal portal system.

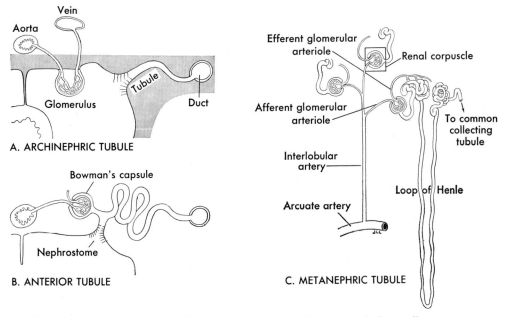

Fig. 14-2. Increasingly specialized kidney tubules and associated glomeruli.

new tubules are added more and more caudad as differentiation progresses. The anteriormost tubules are always metameric, since one tubule develops at the level of each mesodermal somite. Farther back numerous tubules develop in each segment and the metamerism is lost. The segmental tubules usually disappear later in development, with the result that the adult kidney does not extend as far forward as did the embryonic kidney. Neither does it contain evidence of its earlier metamerism. Nevertheless, the blood supply to the kidney is from segmental arteries.

The **longitudinal ducts** of the basic pattern appear first at the anterior end of the nephrogenic mesoderm as caudally directed extensions of the first tubules (Fig. 14-3). Each duct grows caudad until it achieves an opening into the cloaca. At this time it is known as the **pronephric duct.** The pronephric duct participates in the induction of additional tubules farther and farther back. The later tubules achieve an opening into the pronephric duct unless a separate duct is destined to drain them.

Archinephros (Fig. 14-4). Comparative studies in anatomy and embryology sug-

gest that the earliest vertebrate kidneys extended the entire length of the coelom, that all tubules were segmentally disposed, and that each tubule opened via a nephrostome. A single glomerulus, suspended in the coelomic cavity, is postulated for each tubule. The kidney would then have drained off the coelomic fluid, as does the excretory system of annelid worms and amphioxus. From such a hypothetical archinephros may have arisen the kidneys of later vertebrates. The kidney of myxinoid cyclostomes (p. 316) closely resembles an archinephros.

Pronephros

The first embryonic kidney tubules in all vertebrates arise from the anterior end of the intermediate mesoderm. Because they are the first of what will become a series of tubules extending the length of the coelom, they are called pronephric tubules (Figs. 4-9 and 14-3). Pronephric tubules are segmentally arranged, one opposite each of the more anterior mesodermal somites.

Each pronephric tubule arises in the intermediate mesoderm as a solid bud of

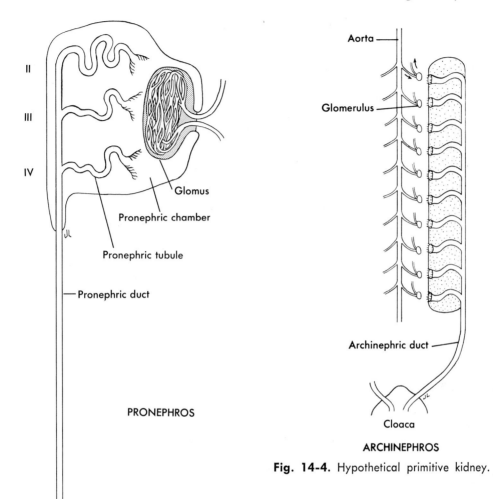

II

III

IV

Glomus

Pronephric chamber

Pronephric tubule

Pronephric duct

PRONEPHROS

Aorta

Glomerulus

Archinephric duct

Cloaca

ARCHINEPHROS

Fig. 14-4. Hypothetical primitive kidney.

Fig. 14-3. Encapsulated pronephric kidney of the 15 mm. larval frog, schematic. **II, III,** and **IV,** levels of the second, third, and fourth somites. The next tubule to form will be a mesonephric tubule at the level of somite VII. The glomus represents three fused glomeruli.

cells, which later organizes a lumen and, in most anamniotes, a nephrostome. Associated with each tubule typically is a glomerulus. The number of pronephric tubules is never large—three in frogs, seven in the human embryo at the level of somites VII to XIII, and about a dozen in chicks commencing at somite V. The tubules lengthen and become coiled. The region of the nephrogenic mesoderm exhibiting these segmentally arranged tubules

is called the pronephros. The duct of the pronephros is the pronephric duct.

The pronephros in most vertebrates is evanescent. It is functional only until such time as the tubules farther back are prepared to supercede it. This is at the end of the larval stage in amphibians or at an equivalent developmental stage in fishes. Although a pronephros invariably develops as a necessary step in nephrogenesis in amniotes, it is doubtful that it ever functions as an excretory organ in these higher vertebrates.

Occasionally, in larvae, several glomeruli unite to form a single **glomus** (Fig. 14-3). The glomus and the pronephric tubules are generally enclosed in a pronephric chamber derived from the lining of the pericardial or pleuroperitoneal cavity.

At involution the glomeruli of the pronephric region lose their vascular connection with the dorsal aorta, and the segmental tubules commence to regress. Tubules involute more slowly than glomeruli, and traces of some of them may remain in adult fishes, at least. However, the pronephric duct remains.

Only in adult cyclostomes and a few teleosts is the pronephros retained throughout life. But it ceases functioning as a kidney and involutes into a lymphoid mass (Fig. 17-8, primitive teleost). In fact, lymphoid tissue is present in the teleost pronephros from the appearance of the first tubules and gradually increases in quantity as the fish grows. The pronephros of the larval *Rana pipiens* at ninety days of development is likewise lymphoidal. The pronephros of the adult *Myxine* gives rise to white blood cells periodically, and it is likely that any remnants of the pronephros of teleosts do the same. Some of the cells in the pronephros of *Myxine* and adult teleosts give a chromaffin reaction, which suggests they may produce epinephrine or a related amine. Because it lies immediately behind the head, the pronephros is often called a "head kidney."

Mesonephros

Under the stimulus of the pronephric duct acting as an inductor, additional tubules develop sequentially in the intermediate mesoderm behind the pronephric region. The new tubules establish connections with the existing pronephric duct. For at least several segments these tubules, too, may be segmentally disposed, they exhibit the same convolutions as the ones anterior to them, and often they have open nephrostomes. In fact, there is seldom justification for drawing at any specific point a boundary between the embryonic pronephros and the rest of the kidney. Typically, there is a gradual transition from tubules characteristic of the pronephric region to those found farther back.

In the transitional area secondary and tertiary tubules develop in each segment (Fig. 14-1). As these additional tubules

enlarge and encroach on one another, the metamerism of the developing kidney is first obscured and then lost altogether. Another transitional feature, in most species at least, is the development of internal glomeruli and of tubules that lack nephrostomes. Many embryonic fishes and amphibians develop nephrostomes for a long distance back, and some fishes retain a number of them as adults. Nephrostomes are prominent in the embryos of reptiles, but in birds and in mammals only a few of the cephalic tubules ever have more than rudimentary nephrostomes. Except in cyclostomes, the portion of the kidney that persists behind the pronephric area is not metameric, the tubules have longer convolutions, and there is a more intimate relationship between glomerulus and tubule than was the case farther forward.

With the disappearance of the pronephric region, the old pronephric duct is thereafter called the mesonephric duct, and the newer kidney region that it serves is the mesonephros (Fig. 14-19). The mesonephros is the functional adult kidney of fishes and amphibians. (It is sometimes called the **opisthonephros** in these adults.) The mesonephros is also a functional embryonic kidney in reptiles, birds, and mammals.

Myxine. The kidney of the adult cyclostome *Myxine* is much more primitive than that of gnathostomes. It occupies a 10 cm. segment of the nephrogenic mesoderm commencing some distance behind the persistent (nonurinary) pronephros, and terminating anterior to the cloaca. This segment consists of thirty to thirty-five large renal corpuscles up to 1.5 mm. in diameter connected to the mesonephric duct by very short tubules. The tubules are so short that they probably play no role in altering the content of the glomerular filtrate by addition or reabsorption. However, there is a possibility that the mesonephric duct itself has assumed some of the functions of the missing segments of the tubules.[4] The corpuscles are strictly segmental. Each glomerulus is supplied by an afferent glomerular arteriole from the dorsal aorta and gives off a branch to the postcardinal vein.

Between the functional kidney and the persistent nonurinary pronephros there are a variable number of corpuscles that lack glomeruli, or that have lost their connection with the longitudinal duct. Caudal to the functional kidney are additional aglomerular corpuscles. The total number of corpuscles in *Myxine*, typical and atypical, numbers about seventy, and they are rigidly segmental. The presence of segmentally arranged corpuscles anterior and posterior to the main uriniferous mass suggests that the kidney extended farther forward and backward in ancestral forms, and supports the concept of the hypothetical archinephros. Because the entire kidney of myxinoids is segmental, it is sometimes called a **holonephros.** The kidney duct after metamorphosis no longer extends to the persistent pronephros (Fig. 14-5, agnatha).

Jawed fishes and amphibians. The adult kidneys of fishes and amphibians commence somewhere behind the pronephric region—just how far behind depends on how many embryonic tubules disappear at the region of transition with the pronephros. In dogfish sharks and caecilians, for example, the adult kidney begins far forward and extends the length of the coelom, lying against the dorsal body wall behind the peritoneum (Figs. 14-5, shark, and 14-6). In many other fishes and in most amphibians a longer series of transitional tubules disappears (Figs. 14-5, urodele, and 14-7).

In males some of the anteriormost tubules of the mesonephros are used as a pathway for sperm from the testes to the mesonephric duct. This part of the kidney, together with the part of the mesonephric duct that drains it, is the **epididymis** (Figs. 14-6, 14-7, and 14-20). In such cases the anterior end of the kidney of the male may be larger or smaller than the remainder. The corresponding part of the mesonephros of the female may or may not involute.

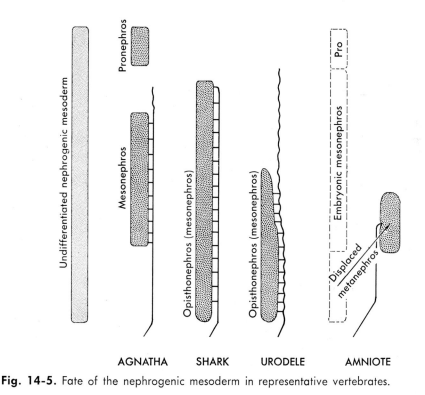

Fig. 14-5. Fate of the nephrogenic mesoderm in representative vertebrates.

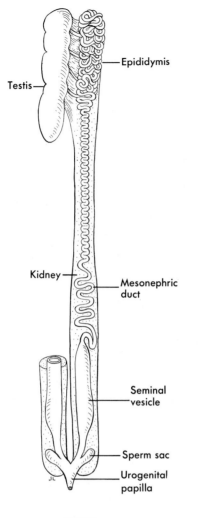

SHARK

Fig. 14-6. Urogenital system of a male shark. Accessory urinary ducts exist but are not shown.

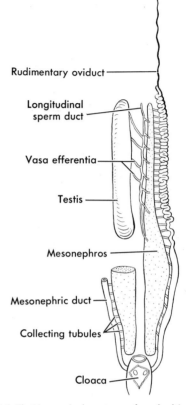

Fig. 14-7. Urogenital system of male *Necturus*. The narrow cephalic end of the kidney serves for sperm transport and is sometimes called epididymis.

Whether extending the length of the coelom or confined to the more caudal regions, and regardless of shape, the urinary regions of the kidneys of adult fishes and amphibians are basically alike, consisting of renal corpuscles and convoluted tubules, occasionally with nephrostomes in fishes, and emptying into a longitudinal duct. In the basic architectural pattern the duct is the pronephric duct of the embryo. When it ceases to serve the pronephros, it is named **mesonephric (or opisthonephric) duct.**

The mesonephric duct may lie along the lateral edge of the kidney (Fig. 14-7), or on the ventral surface (Fig. 14-6), or embedded within it. The caudal ends of the mesonephric ducts may enlarge or evaginate to form **seminal vesicles** (Fig. 14-6) for temporary storage of sperm, or urinary bladders (Fig. 14-14). In some fish (lamprey, dogfish shark, and so forth) the two ducts unite at their caudal ends before entering the cloaca to form a papilla (Fig. 14-6). Accessory urinary ducts may drain many or all the mesonephric tubules. When accessory urinary ducts are numerous, the male mesonephric duct may be used chiefly or entirely for the transport of sperm (Fig. 14-21, *B*).

Amniote embryos. The mesonephros of amniote embryos has essentially the same

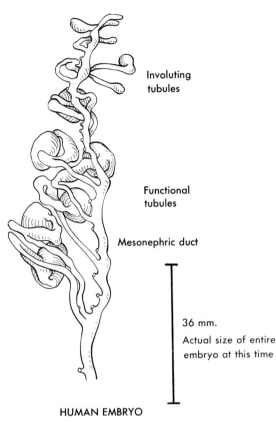

Involuting
tubules

Functional
tubules

Mesonephric duct

36 mm.

Actual size of entire
embryo at this time

HUMAN EMBRYO

Fig. 14-8. Right functional mesonephros of a 36 mm. human embryo. (Redrawn from Altschule.[93])

structure as the adult kidneys of fishes and amphibians, except that nephrostomes are rudimentary in most birds and seldom appear in mammals. In embryonic chicks the mesonephros reaches its peak of development on the eleventh day of incubation, halfway through embryonic life. In mammals it reaches its peak earlier—at nine weeks of gestation in man, approximately one-fourth through gestation.

The first mesonephric tubules in a human fetus appear after four weeks of embryonic life (twenty-somite stage). A wave of differentiation sweeps along the nephrogenic mesoderm, so that even before the last mesonephric tubules at the caudal end of the series have formed, the earliest ones at the anterior end have already involuted (Fig. 14-8). The result is that at peak development of the human meso-

nephros there are about thirty functioning renal corpuscles, although as many as eighty have formed by that time.[93] The most caudal ones are the latest to be formed; the most anterior ones are the oldest. In man all mesonephric tubules have involuted by the time the embryo reaches 40 mm. in length.

The mesonephroi of various species of mammals differ in the number of mesonephric tubules formed. Those in man, cats, and guinea pigs are relatively small compared with the mesonephroi of rabbits.

Although the mesonephros is basically an embryonic kidney in amniotes, it functions for a short time after birth in reptiles, monotremes, and marsupials—as late as the first hibernation in some lizards and the first molt in some snakes.

During the time that the mesonephros is functioning, a new kidney to be used by the amniote the rest of its life, the metanephros, is in the process of development. When the metanephros takes over the functions of a kidney, the mesonephros involutes and only remnants remain after birth.

Mesonephric remnants in adult amniotes. Remnants of the mesonephroi remain in adult amniotes of both sexes. In mammals the remnants consist of groups of blind tubules known collectively as the **paradidymis** and the **appendix of the epididymis**, both located near the epididymis (Fig. 14-19); and as the **epoophoron** and **paroophoron** near the ovary (oophoron) (Fig. 14-9). The mesonephric ducts remain as sperm ducts in male amniotes (Fig. 14-19), but they involute in females and remain only as short, blind **Gartner's ducts** in the mesentery of the oviducts (Fig. 14-9).

Metanephros

The metanephros, or adult amniote kidney, organizes from the caudal end of the nephrogenic mesoderm, which, however, is displaced anteriorad and laterad during development (Fig. 14-5). This is the same mesoderm that gives rise to the caudal

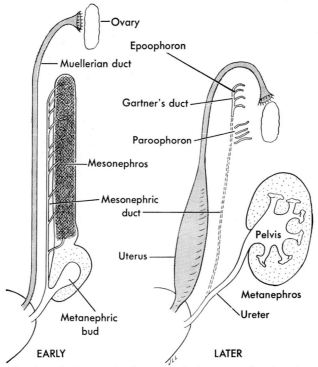

Fig. 14-9. Developmental changes in the urogenital system of a female amniote. In the early stage (left) the mesonephric kidney and duct are present, and the metanephric bud has formed. The undifferentiated muellerian duct is present. In the later stage the mesonephric kidney and duct have regressed, and the muellerian duct has differentiated to form a female reproductive tract.

part of the mesonephric kidney of fishes and amphibians. The number of tubules that forms from this caudal section in amniotes is extremely large (up to an estimated 4½ million), and the tubules are highly convoluted. The displaced section has a duct of its own, the **metanephric duct** (**ureter**). As long as the mesonephros is functioning, the metanephric duct is merely an accessory urinary duct. When the mesonephros involutes, the metanephric duct becomes the sole kidney duct.

The metanephric duct arises as an anteriorly directed **metanephric bud** off the caudal end of the embryonic mesonephric duct (Fig. 14-9). Surrounding the distal tip of the bud is the caudal end of the nephrogenic mesoderm, from which metanephric tubules will organize. The metanephric bud pushes cephalad and, in so doing, carries along the nephrogenic mesoderm as a cap. The caudal end of the nephrogenic mesoderm is thus displaced cephalad from its original site. The proximal portion of the metanephric bud becomes the metanephric duct (ureter). The dilated distal tip becomes the **pelvis** of the kidney. Many finger-like outgrowths of the pelvis invade the nephrogenic mesoderm to become **collecting tubules.** Meanwhile, the cap-like nephrogenic blastema organizes metanephric tubules. They commence as S-shaped tubules. The upper arm of each tubule grows toward, and finally opens into, a collecting tubule. The lower arm becomes invaginated by a developing glomerulus to become a Bowman capsule.

The mammalian metanephros exhibits a greater organization than that of lower amniotes. The organization is the result of

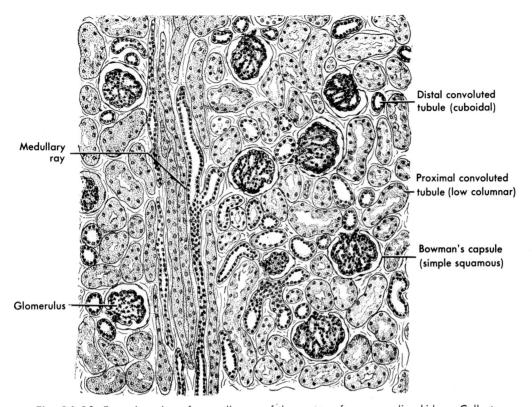

Fig. 14-10. Frontal section of a small area of the cortex of a mammalian kidney. Collecting tubules and loops of Henle extend from the cortex into the medulla as medullary rays. (From Bevelander: Essentials of histology, ed. 5, St. Louis, 1965, The C. V. Mosby Co.)

the formation of a long, thin, U-shaped **loop of Henle** (Fig. 14-2) interposed between the proximal and distal convolutions of the metanephric tubule. As the loops of Henle elongate, they grow away from the surface of the kidney and toward the renal pelvis. The kidney therefore comprises a **cortex,** in which are concentrated the renal corpuscles (Fig. 14-10), and a **medulla,** consisting of the hundreds of thousands of loops of Henle and of common collecting tubules. The loops of Henle function in the reabsorption of salt and water.

The loops and the collecting tubules give the renal medulla a striated appearance in frontal section. They are aggregated into one or several conical **pyramids** (Fig. 14-13), depending on the species. The pyramids taper to a blunt tip (**renal papilla**) projecting into the pelvis. Each collecting tubule

drains a small number of metanephric tubules and empties into the pelvis near the tip of a pyramid.

The metanephric tubules of reptiles have no loop of Henle, and those of birds have only a very short equivalent segment. The glomeruli are reduced in size and exhibit only two or three short vascular loops within Bowman's capsules. Small glomeruli result in the conservation of water.

Many metanephric kidneys are lobulated (reptiles, birds, proboscideans, cetaceans, some carnivores, and so forth), each lobe consisting of clusters of many tubules. Lobulation occurs also in human infants (Fig. 14-11) but later disappears. In snakes and legless lizards the kidneys are elongated to conform to the slender body. The kidneys of birds are flattened against the sacrum and ilium and fit snugly against

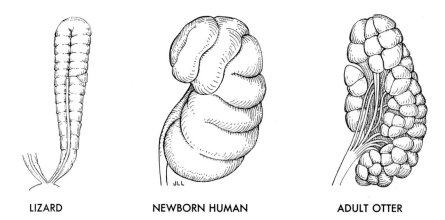

LIZARD   NEWBORN HUMAN   ADULT OTTER

Fig. 14-11. Several lobulated metanephric kidneys.

the contours of these bones. In most adult mammals the surface of the kidney is smooth, and the outline is roughly bean-shaped. At the notched medial aspect (**hilum**) the renal artery, renal vein, nerves, and ureter enter or leave.

Because of their embryonic origin as buds off the mesonephric ducts, the ureters at first terminate in the latter, but as a result of further differential growth, they finally empty independently into the cloaca (reptiles, birds, and monotremes; Figs. 14-12, and 14-25) or into the urinary bladder (placental mammals; Fig. 14-22). In a few male reptiles, however, the ureters continue to empty into the mesonephric ducts.

The arterial blood supply to the metanephros is via a series of two or more renal arteries of segmental origin in reptiles and birds. In mammals there is usually a single renal artery, but it very often bifurcates before reaching the kidney. Upon entering the kidney of mammals, the renal artery divides into numerous branches, which pass radially toward the cortex as interlobar arteries (Fig. 14-13). At the base of the cortex the interlobar arteries give off arcuate arteries which arch along the base of the cortex more or less parallel to the surface of the kidney. From the arcuate arteries arise tiny interlobular arteries, which, in turn, give off afferent glomerular arterioles terminating in glomeruli. Emerging from each glomerulus is an efferent glomerular arteriole, which passes directly to a capillary bed surrounding the tubule. The metanephros is drained by one or several renal veins.

The kidney of reptiles and, to a lesser degree, that of birds and monotremes has an afferent venous supply via a renal portal system. The portal vessels terminate in the peritubular capillary beds.

Extrarenal salt excretion in vertebrates

Marine fishes have developed extrarenal structures for excretion of salts. Large quantities are excreted by chloride-secreting glands on the surface of the gills and, in elasmobranchs, by the rectal gland. The rectal glands of bullsharks (*Carcharhinus leucas*) caught in fresh water are smaller than the glands of specimens caught in salt water, and they show regressive changes.[162] Because of extrarenal salt excretion, and because of modifications of the kidney tubules and glomeruli, which result in water conservation, the urinary output of most saltwater fishes is low.

Tetrapods likewise have evolved organs for excreting salts independently of the kidneys. This is especially true of those species that drink salt water, such as marine reptiles or marine birds. It is also true of some terrestrial lizards and snakes that, because of an arid habitat, must conserve water. Prominent salt-excreting nasal glands fulfill this function.

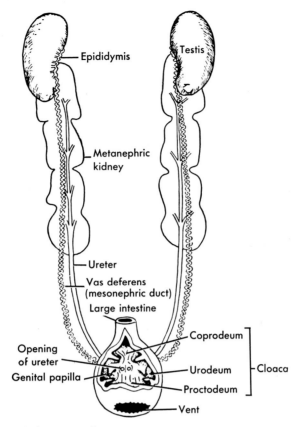

Fig. 14-12. Urogenital system of a rooster.

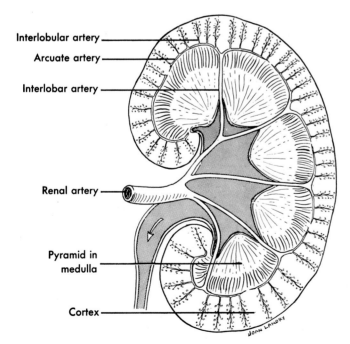

Fig. 14-13. Representative mammalian kidney, sagittal section. The renal vein has been removed. Gray area indicates the pelvis and ureter. The glomeruli are confined to the cortex. The loops of Henle extend into the pyramids of the medulla along with the common collecting tubules. The kidneys of domestic cats have only one pyramid.

The nasal gland of marine birds is a large paired gland located in a bony socket above the orbit. It is drained by a long duct that opens in association with a nostril. A groove extends from the opening to the tip of the beak. Within fifteen minutes or so after these birds have drunk water containing sodium chloride and potassium, minute drops of fluid containing these salts trickle down the groove and drip or are shaken off the tip of the beak.

In lizards the nasal glands are chiefly potassium glands located in the floor of the nasal canal and emptying into the nasal passageways via small ducts. Whitish encrustations of sodium chloride and potassium can be seen in the nasal canal or at the nostrils. In sea turtles the gland secretes chiefly sodium.

Other adaptations associated with the excretion of salt and the conservation of water in arid environments have also evolved. Some tetrapods excrete a concentrated urine, since water is reabsorbed by the lining of the cloaca. Others excrete nitrogenous wastes in the form of uric acid, which is not water soluble. Some desert snakes are said to have aglomerular kidneys that conserve water. Physiological adaptations of vertebrate kidneys to salt, brackish, and fresh water, and the nature of vertebrate urine under varying conditions of life in water and on land are discussed in textbooks of comparative physiology.[24] Electrolyte excretion is regulated chiefly by hormones.

URINARY BLADDERS

Most vertebrates have a urinary bladder. Exceptions include chiefly the cyclostomes and elasmobranchs among the fishes; snakes, crocodilians, and some lizards; and nearly every bird. The bladders of most fishes are terminal enlargements of the mesonephric ducts known as **tubal bladders** (Fig. 14-14). The bladders of dipnoans evaginate from the dorsal wall of the cloaca. Those of tetrapods are evaginations of the ventral wall (Fig. 14-15).

In amniote embryos the evagination that gives rise to the bladder is prolonged beyond the ventral body wall as an extraembryonic membrane, the allantois (Figs. 4-11 and 13-23). Only the base of the allantois—the part proximal to the cloaca—contributes to the adult bladder. After

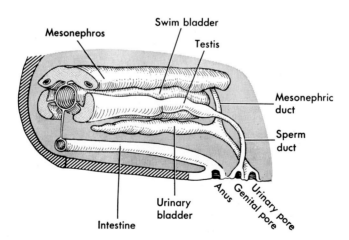

TELEOST

Fig. 14-14. Caudal end of urogenital system of a male teleost (pike), left lateral view. The urinary bladder arises as a bud off the kidney duct. Note absence of a cloaca as a result of having become increasingly shallow during embryonic life. (Redrawn from Goodrich.[12])

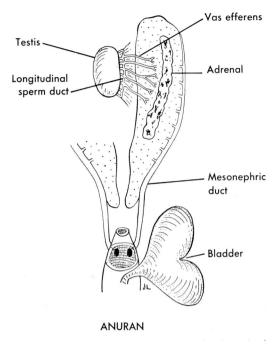

Testis

Longitudinal
sperm duct

Vas efferens

Adrenal

Mesonephric
duct

Bladder

ANURAN

Fig. 14-15. Urogenital system and adrenal of a male frog, ventral view.

birth the distal part of the allantois within the body remains in mammals as a **middle umbilical ligament (urachus)** connecting the tip of the bladder with the umbilicus. The urachus lies in the anterior border of the ventral mesentery of the bladder along with the obliterated umbilical arteries.

Turtles and most lizards have large bladders, and turtles may have two **accessory bladders** as well (Fig. 14-30). The latter are used by females for carrying water for moistening the soil when building a nest for the eggs. If they have other functions such as serving as accessory respiratory organs during prolonged submersion, these have not been demonstrated. In amphibians and reptiles the urine backs up into the bladder from the cloaca. In mammals the kidney ducts empty into the bladder, and the bladder is drained by the urethra.

It is possible that the adaptive value of the tetrapod urinary bladder was its capacity to store water that would otherwise be eliminated from the body. Antidiuretic hormone (ADH) from the pituitary gland

causes active water reabsorption from the bladder in several species, and ADH release can be evoked by hypertonic blood. The bladder would be especially useful in tetrapods living in an arid habitat, where water conservation would have survival value.

GONADS

The embryonic gonads arise as a pair of elevated **gonadal,** or **genital,** ridges. These are thickenings in the coelomic epithelium just medial to the mesonephroi (Fig. 14-16). The ridges are longer than the resulting mature gonad, which suggests that at one time the gonads extended the length of the pleuroperitoneal cavity. Although the gonadal ridges are paired, a few adult vertebrates have a single gonad because of fusion of the two ridges across the midline (most cyclostomes, perch, and some other fishes), or because one of the juvenile gonads fails to differentiate or involutes (hagfishes, some elasmobranchs, alligators, some lizards, and most birds). As the gonads approach sexual maturity, they enlarge and usually acquire a dorsal mesentery, the **mesorchium** in males and **mesovarium** in females.

The ovary in teleost fishes and amphibians is hollow, that is, saccular. The hollow condition in some teleosts results from entrapment of a small part of the coelomic cavity within the ovary (Fig. 14-17). This process occurs especially in viviparous teleosts in which the young develop within the ovary. The ovarian cavity in these species is lined by germinal (coelomic) epithelium, and the eggs are released into the cavity. In other teleosts there is no entrapment of the coelom. Instead, the cavity within the ovary results from a secondary hollowing out of the center, which occurs only at the time of ovulation. The eggs are discharged into the cavity (which is not, however, coelom and is not lined by germinal epithelium). The cavity of teleost ovaries is continuous with the lumen of the oviduct (Fig. 14-28). In amphibians the ovarian cavity results from a secondary hol-

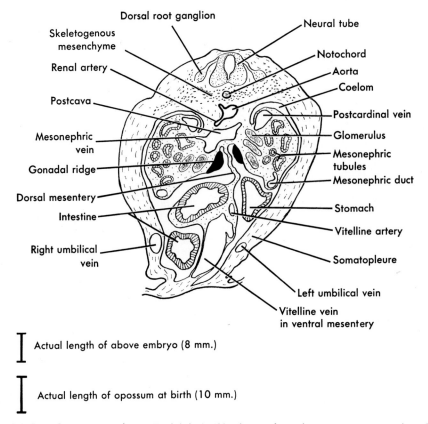

Fig. 14-16. Opossum embryo *(Didelphis)* 6½ days after cleavage, cross section. The gonadal primordia are shown in black.

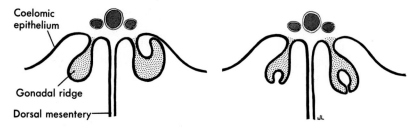

Fig. 14-17. Entrapment of coelom to form a saccular ovary in teleosts. The gonadal ridge is shown in cross section.

lowing out of the medulla, but the eggs are extruded into the coelom rather than into the ovarian cavity. After ovulation the ovaries of fishes and amphibians rapidly shrink.

The ovaries of other fishes and of amniotes are not saccular. However, those of reptiles, birds, and monotremes develop numerous irregular, fluid-filled cavities by rearrangement of the medullary tissues. The mammalian ovary is compact, and the eggs are extruded into the coelom or, in some orders, into an ovarian bursa. This is a periovarian membranous sac that entraps some of the coelom and the ostium of the oviduct.

The mature testes of vertebrates are usually smaller than the corresponding ovaries because sperm, although more numerous,

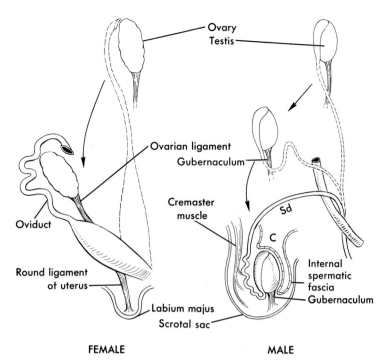

Fig. 14-18. Caudal displacement of mammalian gonads. The ovarian ligament and the round ligament of the uterus collectively are homologous with the male gubernaculum. Arrows indicate the route of translocation of the left gonads, ventral view. **C,** Scrotal recess of coelom; **Sd,** spermatic duct arching over the ureter.

are very much smaller than eggs. The testes of mammals, on the contrary, are larger than the ovaries because mammalian ova ripen a few at a time.

The embryonic testis of anurans is subdivided into an anterior portion (**Bidder's organ**), which usually disappears, and a more caudal portion, which becomes the adult testis. Bidder's organ persists in adult male toads as a small body at the cephalic pole of the testis and contains large cells resembling immature ova (Fig. 14-26). If the testes are removed experimentally, Bidder's organs will develop into functional ovaries, and the vestigial female duct system will enlarge under the influence of the increased titers of female hormones.

Instances of sex reversal occur in nature in many submammalian vertebrate groups. Hens have been known to cease laying eggs, to commence to crow, and to develop other rooster-like characteristics. This condition comes about when the single left

ovary atrophies, and the rudimentary right gonad develops into one producing male hormones. In alligators and some lizards it is the right ovary only that develops.

During early development the gonads are indistinguishable as to sex, and the anlagen for both male and female ducts appear in every embryo regardless of future sex. Under the influence of sex chromosomes and hormones the indifferent gonads develop into testes or ovaries, and the appropriate ducts (male or female) enlarge while the other pair remains rudimentary or disappears. Hermaphroditism* (production of eggs and sperm by the same individual) is common in cyclostomes and bony fishes, but it is rare among other lower vertebrates and absent among higher ones.

Translocation of ovaries and testes in

---

*Hermaphroditos was the son of Hermes and Aphrodite. When bathing in the mythical fountain of Salmacis, he became united in one body with the nymph living in the fountain.

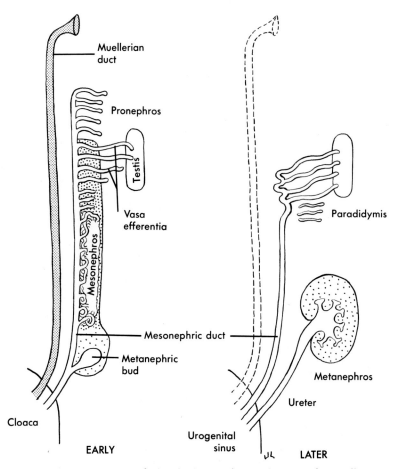

Fig. 14-19. Urogenital system of developing male amniote. In the earlier stage (left) some of the kidney tubules have invaded the testis to become vasa efferentia. In the later stage (right) the mesonephros has regressed except for remnants (paradidymis), the muellerian duct has regressed, but the mesonephric duct remains as a spermatic duct.

mammals. The caudal pole of each embryonic ovary and testis is connected to a shallow evagination in the coelomic floor on each side of the embryonic external genitalia by a fibrous ligament (Fig. 14-18). In males it is the **gubernaculum.** In females the cephalic part of the ligament is named **ovarian ligament,** and the caudal part is named **round ligament of the uterus.** The evaginations where the embryonic ovarian ligaments terminate in females remain shallow and become the two **labia majora** of the vulva. In males the evaginations are larger and become the **scrotal sacs.** The sacs are lined with parietal peritoneum, known in this location as the **in-ternal spermatic fascia.** Partly as a result of shortening of the ligaments, but more especially because elongation of the ligaments does not keep pace with elongation of the trunk, the ovaries and testes are displaced caudad toward the labia or scrotal sacs. The ovaries are not displaced as far caudad as the testes.

The testes remain retroperitoneal and descend permanently into the scrotal sacs in most mammals (Fig. 14-18). In others they are lowered into the sacs and retracted at will (rodents, rabbits, bats, and so forth). The passage between the abdominal cavity and the scrotal cavity is the **inguinal canal.** The opening of the

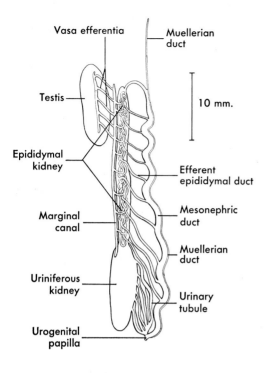

Vasa efferentia

Muellerian
duct

Testis

10 mm.

Epididymal
kidney

Efferent
epididymal duct

Marginal
canal

Mesonephric
duct

Muellerian
duct

Uriniferous
kidney

Urinary
tubule

Urogenital
papilla

AMBYSTOMA

Fig. 14-20. Urogenital system of a typical male urodele of the family Ambystomatidae. The testis is reflected to the right and the urinary tubules to the left. (Redrawn from Baker and Taylor.[94])

canal into the abdominal cavity is surrounded by a fibrous **inguinal ring** (often the site of abdominal hernia). In species that retract their testes, the canal remains broadly open. In species in which the testes are permanently confined to the scrotum, the inguinal canal is only wide enough to accommodate the spermatic cord. The cord contains the spermatic duct and spermatic arteries, veins, lymphatics, and nerves, which are dragged into the scrotum along with the testes. Scrotal sacs do not develop in monotremes, some insectivores, elephants, whales, and certain other mammals. In these the testes, like the ovaries, are permanently intra-abdominal.

MALE DUCTS

Jawed fishes and amphibians. In the basic plan of the male vertebrate the mesonephric duct transmits sperm as well as urine (Fig. 14-15). Connections between the mesonephroi and testes are established early in embryonic life (Fig. 14-19). Some of the anterior mesonephric tubules—a few to two dozen or more, depending on the species—grow across the mesorchium into a network of channels (the **rete testis**) within the testis. These modified mesonephric tubules become the **vasa efferentia,** which carry sperm to the mesonephric duct. In modifications of the basic plan the mesonephric duct may be given over solely to the transport of either sperm or urine, depending on the species, and a new duct may develop to carry the other substance (Fig. 14-21).

Variations in the extent to which, even in closely allied species, the mesonephric duct may serve to transport sperm may be illustrated by contrasting two tailed amphibians: *Necturus* and *Ambystoma.* In *Necturus* (Fig. 14-7) four vasa efferentia connect the testis to a longitudinal sperm duct, and the latter sends ductules into the epididymis (the modified anterior end of the kidney). The epididymis is composed of a single row of twenty-six modified mesonephric tubules, which are ciliated for sperm transport and are said to have no urinary function. These tubules drain into the mesonephric duct, which also collects urine from the kidney tubules farther back. In *Ambystoma* (Fig. 14-20) the mesonephric duct collects no urine from the kidney. Instead, twelve to fourteen small ducts, depending on the species, drain the urinary part of the kidney and empty into a short common urinary duct terminating in the cloaca on a small urogenital papilla.[94] A longitudinal duct carrying only sperm is a **vas deferens.** The mesonephric duct has also been freed from urine transport in Salamandridae and Plethodontidae.

In some fishes, too, the mesonephric duct may be used primarily or solely for sperm transport. In still other fishes the mesonephric duct retains its connection with the kidney and a separate sperm duct develops (Fig. 14-21, *Polypterus*). This sperm duct may terminate in the mesonephric duct or in the urogenital sinus; or it may open independently to the exterior (Fig. 14-14).

Amniotes. Although in amniotes the mesonephric kidneys disappear during embryonic development, the mesonephric ducts of males remain to serve as spermatic ducts emptying into the cloaca or into a derivative of the cloaca.

The relationships of the spermatic duct in mammals are affected by (1) separation of the embryonic cloaca into a urogenital sinus and rectum and (2) caudal migration of the testes. As a result of separation,

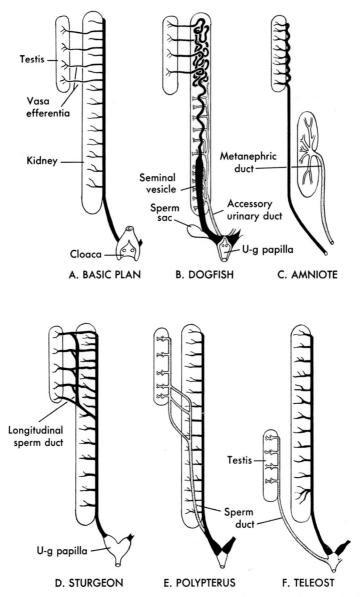

Fig. 14-21. The mesonephric duct (black) as a carrier of sperm and urine. **A,** Basic plan, carrying both sperm and urine. **B,** Carrying little urine, chiefly a spermatic duct. **C,** Carrying sperm only. **D** to **F,** Increasing tendency toward a separate sperm duct, the mesonephric duct carrying urine. **U-g,** Urogenital papilla.

the spermatic ducts finally empty into the urogenital sinus, which is the male **urethra** (Fig. 14-22). As a result of descent of the testes, the spermatic ducts become "caught" or "hung up" on the ureters in such a way that they must loop over the ureters en route to the urethra (Fig. 14-

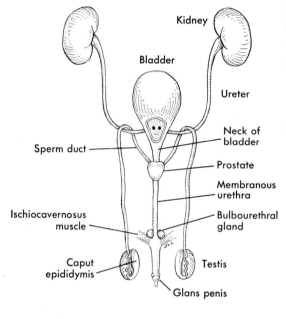

Fig. 14-22. Urogenital system of a male cat, ventral view.

18). At the junction of the spermatic ducts and the urethra in mammals are one or more accessory sex glands that produce some of the constituents of semen (Fig. 14-23).

The male urethra in mammals is sometimes divided into three regions: (1) **prostatic urethra** associated with the prostate glands, (2) **membranous urethra** from the prostate to the penis, and (3) **spongy urethra** within the penis.

Abdominal pores. Cyclostomes lack reproductive ducts, and the sperm and eggs are shed into the coelom. They then pass through a pair of **abdominal pores** in the caudal abdominal wall. In myxinoids the pores lead to an unpaired genital sinus. In lampreys the pores lead to the urogenital papilla.

Abdominal pores are not confined to cyclostomes. Similar pores, opening from the coelom to the exterior near the cloaca are found in most elasmobranchs, including the dogfish, and in a sizable number of teleosts, especially marine forms. They have also been described in turtles and alligators. Their distribution in a broad spectrum of fishes—Agnatha, Chondrichthyes, and Osteichthyes—suggests that these are probably ancient structures. One family of elasmobranchs lacks reproductive ducts and the pores serve the genital system, as in cy-

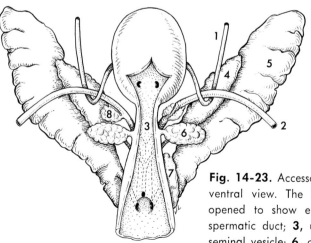

Fig. 14-23. Accessory sex organs of a male hamster, ventral view. The bladder and urethra have been opened to show entrances of ducts. **1,** Ureter; **2,** spermatic duct; **3,** urethra; **4,** coagulating gland; **5,** seminal vesicle; **6,** cranial (dorsal) prostate; **7,** caudal (ventral) prostate; **8,** ampullary gland.

clostomes. What advantage the pores may hold for the fishes having reproductive ducts is not known. In some marine teleosts the pores are open only during the breeding season, and in these forms they may still play a role in reproduction. On the contrary, they may be functionless vestiges that remain responsive to one or more of the hormones associated with reproduction.

INTROMITTENT ORGANS

When fertilization is internal, the male vertebrate usually develops intromittent (copulatory) organs for introducing sperm into the reproductive tract of the female. Intromittent organs are particularly characteristic of reptiles and mammals. They are absent in most fishes and all amphibians, since fertilization is usually external in these groups, or a spermatophore is employed. They are likewise absent in all but a few birds.

The intromittent organs of elasmobranchs are modifications of the pelvic fins and are known as **claspers.** These are grooved, digitiform appendages that are inserted into the cloaca of the female. Embedded in the fin at the base of the clasper in dogfish sharks and some related forms is a muscular siphon sac. The sac can be filled with sea water, which is used to flush sperm along the groove and into the female cloaca. In many teleosts the anal fin is modified for sperm transport and thus becomes a **gonopodium.**

Intromittent organs of amniotes are of two types—paired hemipenes and penis. Male snakes and lizards have a pair of stubby, grooved, sac-like **hemipenes** lying in pockets under the skin beside the cloaca. These can be everted for insertion into the

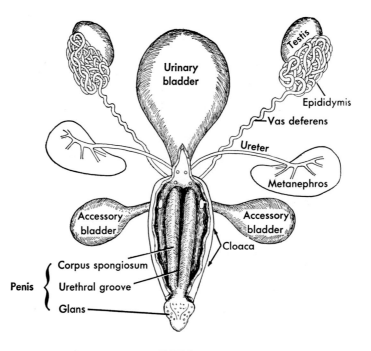

TURTLE

Fig. 14-24. Urogenital system and cloaca of a male turtle, dorsal view. The roof of the cloaca has been cut away, and the penis is in an extended position. The rectum, which enters the cloaca dorsal to the urinary bladder, has been removed. The vas deferens is the old mesonephric duct.

cloaca of the female. After sperm transfer, they are retracted by modified body wall muscles. Hemipenes, when everted, have been mistaken for legs in snakes. They are present, but much smaller, in females.

Male turtles, crocodilians, a very few birds (drakes, ganders, ostriches, and so forth), and male mammals exhibit an unpaired erectile **penis.** In its simplest form (reptiles, Figs. 14-24 and 14-25), the penis is a thickening of the floor of the cloaca, which consists chiefly of a mass of erectile tissue, the **corpus spongiosum,** containing blood sinuses. When the sinuses are distended with blood, the penis is swollen and firm. The surface of the reptilian penis bears a groove for the passage of sperm, and the distal tip protudes as a genital prominence in the floor of the cloaca. A rudimentary penis, the **clitoris,** develops in females.

In mammals the groove in the embryonic penis becomes folded into the penis as a tube continuous with the urethra. The part of the urethra within the penis is the **spongy urethra.** In monotremes the penis is carried in the cloaca. In higher mammals the cloaca disappears during development

and the penis becomes external. The glandular tip (**glans penis**) is richly supplied with sensory endings and is covered with loose skin (**prepuce**), except during erection. An os penis has been described in Chapter 6.

Male and female mammalian embryos each develop a **genital prominence** between the evaginating scrotal sacs or labia majora. The prominence in males becomes grooved, then tubular, and elongates to form the penis. In females no tube develops, and the prominence becomes a clitoris. The clitoris remains embedded in the wall of that part of the female tract (urogenital sinus or vagina) which receives the penis and, like the penis, is reflexly erectile.

FEMALE GONODUCTS

Almost every vertebrate embryo above cyclostomes, male and female, develops a pair of longitudinal **muellerian ducts** opening into the coelom anteriorly and terminating in the cloaca. The muellerian ducts remain rudimentary or disappear in males but grow larger and become the reproductive tract (sometimes referred to as the gonoduct) in females (Figs. 14-9 and 14-19). Eggs are transported in the adult tracts. Certain regions of the differentiated ducts may perform special functions such as providing the eggs with protective and nutrient coats. In most animals that bear living offspring, the lower ends of the ducts (**uteri**) provide a haven and often all necessary substances for the developing young prior to birth. The terminal segment of the ducts may be modified to receive the male intromittent organ. When fertilization is internal, sperm penetrate the eggs in the upper reaches of the ducts. The eggs are propelled along the tract by cilia or smooth muscle.

In elasmobranchs the ostia are derived from the union of three pronephric nephrostomes, and the muellerian ducts arise by longitudinal splitting of the pronephric ducts. In most other vertebrates each muellerian duct arises as a longitudinal groove,

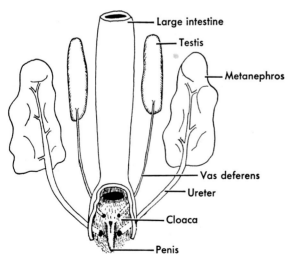

Large intestine

Testis

Metanephros

Vas deferens

Ureter

Cloaca

Penis

ALLIGATOR

Fig. 14-25. Urogenital system of a sexually immature male alligator.

which, except at the ostium, closes over to form a tube. The tube grows backward in close association with the mesonephric duct to achieve an opening into the cloaca. The ducts eventually develop a dorsal mesentery.

Although the muellerian ducts do not fully differentiate in male vertebrates, they often remain as prominent structures. A complete, although rudimentary, female tract develops in some male amphibians (Fig. 14-26). After removal of the testicular hormones by orchidectomy, the rudimentary tract develops into a functional female tract consisting of oviduct and uterus. The sperm sac at the base of each mesonephric duct of a dogfish shark is a remnant of the caudal end of the muellerian duct, and a clearly discernible vestige encircling the anterior end of the liver and ending in the falciform ligament is found in a large percent of dogfishes. In male mammals remnants include the **appendix testis**, a remnant of the anterior part of the muellerian duct, and the **prostatic sinus** (syn., vagina masculina, prostatic utriculus, and utriculus masculina), an unpaired sac near the base of the spermatic duct representing the fused caudal ends of the muellerian ducts.

Fishes. Cyclostomes do not develop muellerian ducts. The eggs leave the coelom via a pair of abdominal pores.

In female elasmobranchs the muellerian ducts give rise to oviducts with shell glands, and to paired uteri that open to the cloaca (Fig. 14-27). The two ostia of the embryonic ducts unite to form a single adult ostium in the falciform ligament. The condition is not typical of vertebrates. The uterus of viviparous species is capable of great distention.

The gonoducts in teleosts have been the subject of much investigation. They are directly continuous with the central cavity

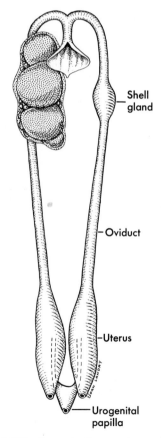

Fig. 14-27. Reproductive system of female *Squalus.* The left ovary has been removed.

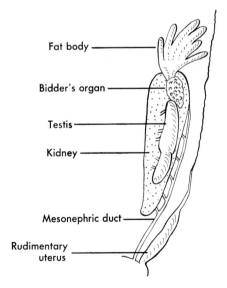

MALE TOAD

Fig. 14-26. Bidder's organ and the rudimentary female reproductive tract in a male *Bufo*, ventral view. Only the left organs are illustrated.

of the ovary (Fig. 14-28). Whether the gonoduct is a modified muellerian duct has not been settled. The two gonoducts often unite before emptying to the exterior, and the cloaca is practically nonexistent. The terminal segment of the gonoducts may lie within a papilla-like **ovipositor.** The perch has a single gonoduct leading from a single ovary.

Amphibians. Muellerian ducts in amphibians give rise to oviducts that are long and convoluted, and the caudal portions may enlarge to form uteri. Amphibian uteri, except in the few ovoviviparous urodeles, serve only as temporary storage depots for eggs about to be laid. The oviducal lining in amphibians is richly supplied with glands that secrete several jelly envelopes around each egg as it moves down the tube. The paired ostia are far forward.

Reptiles and birds (Figs. 14-29 to 14-31). Crocodilians, some lizards, and many birds exhibit a single adult female tract and ovary. The other muellerian duct remains rudimentary. The oviducts are coiled and lined with glands that, except in snakes and lizards, secrete albumin around the ovum. Caudally, the muellerian ducts become thick-walled uteri that serve only as shell glands, except in ovoviviparous reptiles. The short, muscular terminal segment of the female tract, sometimes called the vagina, empties into the cloaca.

Mammals. The muellerian ducts of mammals give rise to oviducts, uteri, and vaginas. The embryonic muellerian ducts almost always fuse at their caudal ends. As a result, the uteri are typically paired anteriorly and unpaired posteriorly, and there is a single vagina. The oviducts, or **fallopian tubes** as they are sometimes called, are relatively short, of small diameter, and convoluted. The ostium is surrounded by a fimbriated infundibulum.

UTERUS. In monotremes and many marsupials there is no fusion of the mullerian ducts. Therefore, the uteri are double

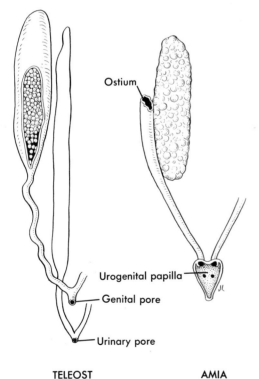

Ostium

Urogenital papilla

Genital pore

Urinary pore

TELEOST   AMIA

Fig. 14-28. Female reproductive systems of two bony fishes. In the teleost ova are shed into the cavity of the ovary.

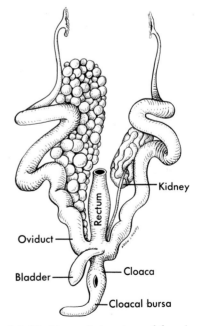

Kidney

Oviduct

Rectum

Bladder

Cloaca

Cloacal bursa

Fig. 14-29. Urogenital system of female turtle, *Trionyx euphraticus,* ventral view. The left ovary has been removed.

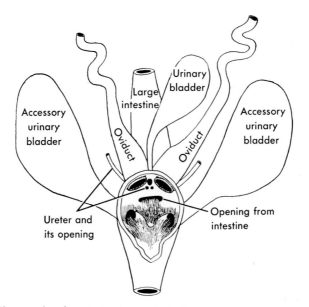

Fig. 14-30. Cloaca of a female turtle, ventral view.

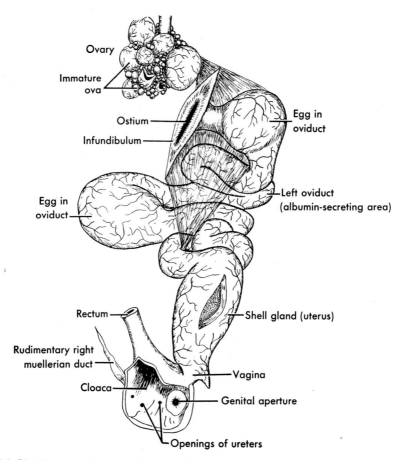

Fig. 14-31. Reproductive tract of a hen. Two eggs in the oviduct is an unusual condition.

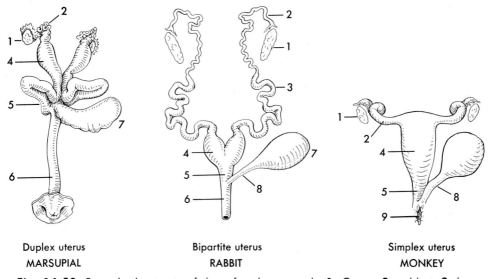

Duplex uterus
MARSUPIAL

Bipartite uterus
RABBIT

Simplex uterus
MONKEY

Fig. 14-32. Reproductive tracts of three female mammals. **1,** Ovary; **2,** oviduct; **3,** horn of uterus; **4,** body of uterus; **5,** vagina; **6,** urogenital sinus; **7,** urinary bladder; **8,** urethra; **9,** vestibule of primate. In the primate (rhesus monkey) the urethra opens independently into the vestibule just anterior to the opening of the vagina. The marsupial (redrawn from McCrady[97]) is an opossum, shown also in Fig. 14-33.

throughout (**duplex uterus,** Figs. 14-32 and 14-33). In placental mammals there are varying degrees of fusion of the caudal ends of the muellerian ducts, which results in two uterine **horns** and a single uterine **body** (Fig. 14-34). When there are two complete lumens within the body of the uterus, it is said to be a **bipartite** uterus. When there is a single lumen within the body and two horns, the uterus is said to be **bicornuate.** There are, however, many species with uteri transitional between the bipartite and bicornuate condition. When there are uterine horns, the blastocysts implant in the horns. In some mammals one horn is much larger, and the blastocysts always implant in the enlarged horn—the right in impala—even though both ovaries produce eggs.

In apes, monkeys, man, some bats, and armadillos no uterine horns develop, and the oviducts are the only remaining paired portion of the muellerian duct system (Fig. 14-34). They open directly into the body of the **simplex** uterus. Except in ectopic pregnancies—pregnancies in which

blastocysts implant in abnormal locations such as the oviduct (tubal pregnancies) or the coelom (abdominal pregnancies)—the usually single fetus or the twins, triplets, quadruplets, or quintuplets all implant in the body of the uterus.

The body of the uterus narrows to form a **cervix** (neck), the lower end of which projects into the vagina as the **lips** of the cervix. The lips surround the opening (**os uteri,** mouth of the uterus) leading from the uterus into the vagina. The cervix must dilate for the young to be delivered. After mating, sperm pass through the os uteri en route to the upper part of the oviducts where the sperm penetrate the egg. The uterine lining, or **endometrium,** becomes highly vascular under the stimulus of hormones prior to implantation of a blastocyst. The thick, muscular layer of the uterine wall, the **myometrium,** assists in ejection of the young at birth provided it, too, has been hormonally prepared for this action.

VAGINA. Typically, the vagina is the fused terminal portion of the muellerian ducts opening into the urogenital sinus (Fig. 14-34, ungulates). In many rodents and primates, however, the vagina is extended so as to open directly to the exterior (Fig. 14-34, primates). Like other muellerian duct derivatives, the vagina has a muscular wall. The lining is specialized for reception of

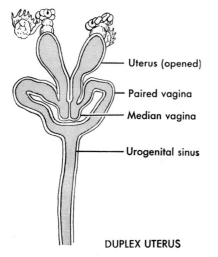

Fig. 14-33. Internal passageways in the female reproductive tract of an opossum. Compare with the external view in Fig. 14-32. The bladder has been omitted.

the male intromittent organ. In monotremes the uteri open directly into the cloaca so that there is no vagina.

The vagina in marsupials is unusual. Just beyond the uteri the two muellerian ducts meet to form a **median vagina,** which may or may not be paired internally (Fig. 14-33). Beyond the median vagina the two muellerian ducts continue to the urogenital sinus as **paired (lateral) vaginas.** The pouch-like median vagina projects caudad and lies against the urogenital sinus. At birth the fetus is forced through the partition directly into the urogenital sinus and thus bypasses the lateral vaginas. In kangaroos both routes may be used as a birth canal. The opening between the median vagina and the urogenital sinus may remain throughout life, once perforated. In opossums, however, the temporary passageway (**pseudovagina**) usually grows shut after the young are delivered. In opossums, therefore, the chief function of the lateral vaginas is to serve as a path for sperm in their journey to the egg. As an adaptation to dual vaginas, the penis of the male is forked at the tip, and one tip enters each lateral vaginal canal, where the semen is discharged.

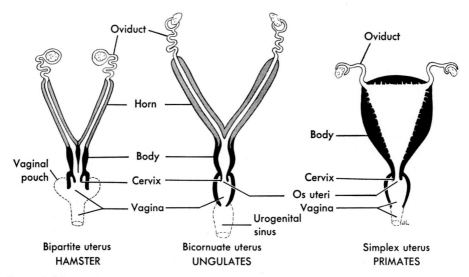

Fig. 14-34. Uterine types among mammals. See also Fig. 14-33. Blackened regions represent fused caudal ends of the muellerian ducts; broken lines represent the cloaca or a derivative thereof.

Entrance of ova into oviduct. After seeing the very large size of the shark's egg as it lies in the ovary, students in the laboratory usually inquire how such a large egg can get into the ostium and down the relatively small oviduct. We know the answer with reference to the equally large egg of the chick or the egg of the monotreme. Although apparently not observed in sharks as yet, the same process undoubtedly occurs. Surrounding the ostium is a large membranous funnel-shaped **infundibulum**, the fringes of which constitute the **fimbria** (Fig. 14-31). Under the influence of hormones at the time of ovulation, the fimbria wave about slowly in an undulating movement. When they come in contact with the egg, whether it is still in the ovary or separated from it, they clasp the egg, delicately at first and then more firmly, until they embrace it. By this process the egg is engulfed by the infundibulum. Muscular contraction of the infundibulum forces the egg slowly into the oviduct, while preventing its escape into the coelom. Peristalsis of the muscular wall of the oviduct moves the egg caudad. Although the infundibulum is usually ciliated, the cilia play a relatively unimportant role in the passage of large eggs into the ostium. In the case of the tiny eggs of mammals, however, the cilia are more important. Nevertheless, active movements of the fimbria still play a role. In most mammals the ovary is partially surrounded by the ciliated infundibulum at all times and thus the probability increases that the egg will enter the oviduct rather than become lost in the coelom. In mammals with an ovarian bursa the egg can go nowhere except into the ostium.

THE CLOACA

The role of the cloaca comes into focus whenever the digestive, urinary, or genital tracts are discussed, since this chamber is the termination of all three tracts in most vertebrates. A few adult vertebrates lack a cloaca. It is so shallow as to be practically nonexistent in agnathans, in ray-finned fishes, and in marsupials, and it is confined to the embryo in placental mammals. With these exceptions, the cloaca remains essentially unchanged throughout the vertebrate series and appears in its primitive condition even in the embryo of man.

The cloaca is usually described as representing an enlarged terminal segment of the hindgut, which acquires an opening to the exterior when the cloacal membrane, separating the hindgut from the proctodeum, ruptures. This is an adequate statement of the origin of the mammalian cloaca, which is lined entirely by embryonic endoderm. In amphibians, however, the major part of the adult cloaca is derived from the proctodeum and therefore has an extensive lining of ectoderm. Once the cloacal plate ruptures in most embryos, it is difficult to determine where the endodermal contribution stops and the proctodeal portion begins.*

In many vertebrates a partition separates the cloaca into two **partial** chambers, a **coprodeum** and a **urodeum.** The terminal portion remains undivided. In such cases, and they are numerous—many elasmobranchs, some reptiles, birds, monotremes, and marsupials—the large intestine opens into the coprodeum, whereas the urinary and genital ducts open into the urodeum. In adult placentals the partition (here called the **urorectal fold**) is continued to the cloacal membrane. The cloaca is thereby **completely** divided into two compartments, and there are at least two openings to the exterior, instead of a single vent.

Fishes and amphibians. The cloaca of fishes and amphibians receives the large intestine and the mesonephric ducts. The latter open independently or via a papilla. The female cloaca also receives the muel-

---

*For this reason, it is sometimes believed that the urinary bladder of amphibians is not homologous with the allantoic bladder of amniotes. It has not been established, however, that the amphibian bladder is not lined by endoderm. It arises from the cloaca after the cloacal membrane has ruptured.

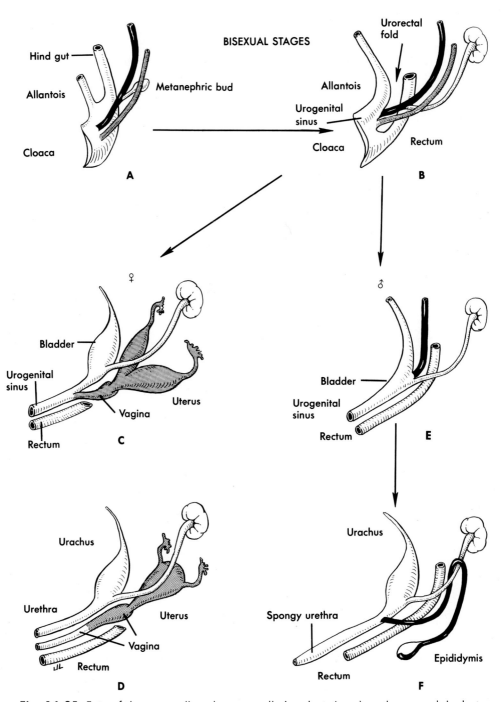

Fig. 14-35. Fate of the mammalian cloaca, muellerian ducts (gray), and mesonephric duct (black). **A** and **B,** Bisexual stages. Only the left muellerian and mesonephric ducts are shown. In **B** the cloaca is becoming subdivided by the urorectal fold into a urogenital sinus ventrally and a rectum dorsally. **C,** Typical adult female mammal. **D,** Female primate. In **C** and **D** the contributions of both the left and right muellerian ducts are shown. **E,** Developing male, showing reorientation of mesonephric and metanephric ducts. **F,** Adult male.

lerian ducts. A urinary bladder, when present, opens into the dorsal wall of the cloaca (dipnoans) or into the ventral wall (amphibians). The cloaca opens to the exterior via the vent. Cloacal glands produce a number of secretions, including scents, and the jelly-like portion of the spermatophores in male urodeles.

Reptiles, birds, and monotremes. The cloacas of these amniotes receive the same structures as in amphibians—large intestine, mesonephric ducts (but in males only and now draining only the testes), muellerian ducts in females, and urinary bladder unless absent. In addition, the ureters open into the cloaca except in a few male reptiles, in which the ureter retains its embryonic connection with the mesonephric duct. The penis or clitoris is embedded in the cloacal floor of turtles, crocodilians, a few birds, and monotremes. A lymphoid pouch, the **bursa Fabricii**, opens into the proctodeal portion of the cloaca of young birds.

Fate of the cloaca in placental mammals. Placental mammals during early embryonic development exhibit a typical cloaca with a urorectal fold in the cephalic end separating the coprodeum and urodeum. The embryonic urodeum, better known as the **urogenital sinus** in mammals (Fig. 14-35, *B*), receives the mesonephric ducts, the muellerian ducts (present at this stage in both sexes), and the allantois (future urinary bladder), precisely as in lower amniotes. Further development eliminates the cloaca. The urorectal fold extends farther and farther caudad and finally reaches the cloacal membrane, except in some marsupials. By this process the cloaca of placentals is completely divided into a rectum dorsally and a urogenital sinus ventrally. (Fig. 14-35, *C* and *E*). The cloacal membrane ruptures at two points and provides an **anus** and a **urogenital aperture.**

As development progresses in males, the muellerian ducts disappear (Fig. 14-35, *E*), and the urogenital sinus becomes continuous with the spongy urethra that has developed independently in the penis (Fig. 14-35, *F*). The ureters become reoriented to open into the bladder, whereas the mesonephric ducts (now sperm ducts) continue to empty into the urogenital sinus (urethra). Thus, the urethra, excepting the spongy part, is derived from the cloaca. Various glands contributing to the semen develop as outpocketings of the urethra (Fig. 14-23).

As development progresses in females, the mesonephric ducts disappear, and the muellerian ducts unite at their caudal ends to form the body of the uterus and the vagina (Fig. 14-35, *C*). The vagina empties dorsally into the urogenital sinus. That portion of the urogenital sinus between the allantois (future bladder) and the entrance of the vagina becomes the **urethra.** As a result of these changes, most adult female mammals exhibit a urogenital sinus that receives the urethra and vagina. The cloaca is thus converted into a urogenital sinus and urethra **ventral** to the urorectal fold and a rectum **dorsal** to the fold.

In female primates (including man) and in some rodents, the urogenital sinus becomes partitioned into dorsal and ventral passageways (Fig. 14-35, *D*). The ventral passageway is then known as the urethra all the way from the bladder to the exterior, and the dorsal passageway becomes a continuation of the vagina to the exterior. As a result of these changes, the embryonic cloaca of the female rodent and primate becomes subdivided into three passageways: (1) urethra ventrally, (2) terminal part of the vagina, and (3) rectum dorsally. Each passageway leads to the exterior via its own aperture.* The urinary tract is thereby completely separated from the genital tract for the first time. To this extent the female rodent or primate is more highly specialized than the male of the same species.

---

*The urethra and vagina in primates actually open into a shallow vestibule derived from the distal end of the urogenital sinus. In rodents the vestibule is absent.

Chapter summary

1. All vertebrate kidneys arise from a ribbon of intermediate mesoderm extending the length of the embryonic coelomic cavity. A wave of differentiation sweeps along the nephrogenic ribbon and gives rise to convoluted tubules of increasing complexity. The tubules are typically associated with glomeruli and open into a longitudinal duct usually terminating in the cloaca or a derivative thereof.

2. Glomeruli are arterial tufts from which are filtered water and other substances. The glomeruli sometimes dangle into the coelomic cavity (external glomeruli), but mostly they are surrounded by a Bowman capsule (internal glomeruli).

3. Kidney tubules are convoluted ductules that collect the glomerular filtrate, selectively reabsorb some of the substances, add others, and conduct the final filtrate to the longitudinal duct. The most primitive tubules have nephrostomes opening from the coelom. These occur especially at or near the anterior end of the kidney and chiefly in embryos and larval anamniotes. Nephrostomes in amniotes are mostly rudimentary.

4. The first tubules to form are segmental and often have open nephrostomes, except in birds and mammals. These first tubules are evanescent and constitute the pronephros. When a pronephros persists as a prominent adult organ, it is nonurinary and predominantly lymphoidal.

5. The mesonephros is that kidney which organizes behind the pronephric region. It is the functional adult kidney of fishes and amphibians. In amniotes it is a functional embryonic kidney only. For this reason, the adult kidney of fishes and amphibians is sometimes called an opisthonephros. The pronephric duct usually persists to serve the mesonephros or opisthonephros, but accessory mesonephric ducts may develop.

6. The mesonephric kidney is not metameric. Nephrostomes are generally lacking, and Bowman's capsules surround the glomeruli. Glomerulus and capsule constitute a renal corpuscle.

7. Tubules near the anterior end of the mesonephros invade the mesorchium, grow into the testes, and serve as vasa efferentia for sperm transport. Typically, these tubules empty into the mesonephric duct, which thereby becomes a reproductive duct as well as a urinary duct in males. In male amniotes, when the mesonephros involutes, the mesonephric duct remains to serve as a vas deferens.

8. The metanephros is the adult amniote kidney. It organizes from the caudal end of the nephrogenic mesoderm, which is displaced craniad and laterad. Its duct (ureter) arises as a bud off the mesonephric duct and empties into the cloaca or the urinary bladder (mammals). In mammals the renal corpuscles aggregate to form a cortex.

9. The adult kidney of amniotes differs from that of fishes and amphibians chiefly in its origin from only the caudal end of the nephrogenic ribbon and in having a new duct. The mammalian kidney exhibits additional features: (1) a loop of Henle (rudimentary in birds) between proximal and distal convolutions of the tubule; (2) the lack of an afferent venous supply from a renal portal system, except in monotremes; and (3) a discrete cortex and medulla.

10. Most vertebrates other than birds have urinary bladders. In fish other than dipnoans they are enlargements of the posterior ends of the mesonephric ducts (tubal bladders). In dipnoans they are evaginations of the dorsal cloacal wall. In amphibians and amniotes their anlagen evaginate from the ventral cloacal wall.

11. Intromittent organs accessory to internal fertilization are present in male elasmobranchs, a few teleosts, reptiles, a few birds, and mammals. Those of fishes are chiefly modifications of the pelvic or anal fins. Turtles, crocodilians, and some birds exhibit an unpaired penis derived from the cloacal floor. In mammals the penis is external, except in monotremes, and is traversed by the spongy urethra. Female amniotes develop a clitoris, which is a homologue of the penis. In

snakes and lizards paired hemipenes are present.

12. Gonads arise from a pair of genital ridges in the coelomic wall. Unpaired gonads result when paired anlagen unite across the midline or when one gonad remains undifferentiated. Gonads tend to migrate caudad in mammals. The ovaries of teleosts and amphibians are saccular. Some mammalian ovaries are surrounded by a bursa of peritoneum.

13. Cyclostomes and a few elasmobranchs lack reproductive ducts, and the eggs and sperm escape the coelom via abdominal pores. Other vertebrates of both sexes develop two sets of ducts which may function in reproduction: The mesonephric duct typically becomes the sperm duct in males and the muellerian duct becomes the gonoduct in females. Parts of the female tract become specialized for transport, nourishment, or harboring of the young and for receiving the male intromittent organ. The caudal ends of the muellerian ducts in mammals typically unite to form an unpaired body of the uterus and a vagina. Depending on the extent of fusion, uteri are duplex, bipartite, bicornuate, or simplex.

14. An adult cloaca, opening to the exterior via a vent, is characteristic of all vertebrates except agnathans, ray-finned fishes, and placental mammals. It is confined to the embryo in placentals. A horizontal partition usually partially divides the cloaca into a coprodeum receiving the large intestine and a urodeum receiving kidney ducts, reproductive ducts, and urinary bladder. The partition extends to the cloacal membrane in placentals, in which case the coprodeum is known as the rectum and the urodeum is the urogenital sinus. In female primates and some female rodents the embryonic urogenital sinus is further divided into a urethra and vagina. The part of the vagina arising from the cloaca in primates is added to the muellerian portion and thus provides the vagina with a direct exit to the exterior.

15. In early embryonic development the gonads are indistinguishable as to sex, and the duct systems for both sexes are present. In Table 14-1 are listed the chief sexually indifferent structures of mammalian embryos and the fate of these in adult males and females.

Table 14-1. Some homologous structures in male and female mammals

Indifferent structure	Mature male	Mature female
Mesonephric duct	Vas deferens	Gartner's canal*
	Epididymis	
	Seminal vesicle	
Mesonephric tubules	{Appendix of epididymis*	Epoophoron*
	{Paradidymis*	Paroophoron*
	Vasa efferentia	
Muellerian duct	{Appendix testis*	Oviduct
		Uterus
	Vagina masculina*	Vagina
Genital ridge	Testis	Ovary
Dorsal mesentery	Mesorchium	Mesovarium
	Gubernaculum	Ovarian ligament Round ligament of uterus
Coelomic pouch	Scrotal sac	Labium majus
Genital prominence	Penis	Clitoris
Urogenital sinus (from cloaca)	Urethra†	Urethra
		Urogenital sinus
		Lower vagina (rodents and primates only)

*Nonfunctional vestige.
†Sometimes called urogenital sinus in an adult male.

15

Nervous system

The vertebrate nervous system plays three vital roles. It acquaints the organism with, and enables the organism to orient itself favorably in, the surrounding environment; it makes possible the integrated control of the internal environment; and it serves, in higher vertebrates at least, as a center for learning. These functions are accomplished by the nerves, spinal cord, and brain in association with receptors (sense organs) and effectors (muscles and glands).

The organism exists in an external environment that is sometimes friendly, sometimes inimical, and seldom neutral. An environment is friendly if it contains food for nourishment, a mate for the propagation of the species, and shelter from enemies. An environment is unfriendly if it leads to the weakening of the organism or of the species.

The organism must be constantly informed of the nature of its environment in order that it may go deeper into a friendly area or withdraw from an unfriendly one. The information is supplied by afferent (sensory) nerves commencing in sense organs. The response (body movement) is initiated by nerve impulses over efferent (motor) nerves that stimulate the skeletal muscles of the body and thus cause the fish to swim or the tetrapod to crawl, run, or fly deeper into, or out of, an area. Information from the external en-

vironment is also employed in the neurosecretory regulation of internal secretions such as seasonal release of reproductive hormones (Fig. 17-3).

The organism has an internal environment that must be continually scanned and controlled. Afferent nerves from visceral receptors transmit impulses to the central nervous system, and efferent nerves carry impulses from the center to smooth and cardiac muscles and to glands. Many of the myriad functions associated with homeostasis are thereby reflexly regulated. The endocrine roles of the nervous system in homeostasis are discussed in Chapter 17.

Memory is a function of the nervous system. Without a single memory no animal can have a single thought, since there would be nothing to think about. As experiences multiply, memory accumulates. The penalty of past errors and the rewards of successes may later be recalled in new situations, and subsequent responses may be modified accordingly. The brain is the center for recall.

The nervous system is subdivided for convenience of discussion into central and peripheral nervous systems. The central nervous system consists of the brain and spinal cord. The peripheral nervous system consists of cranial, spinal, and autonomic nerves, their branches, and certain associated autonomic ganglia and plexuses. The autonomic nervous system is the part of the nervous system innervating smooth and cardiac muscles and glands.

THE NEURON

To understand the anatomy of the nervous system, one must be acquainted with the neuron—the living nerve cell. The neuron is to the nervous system what a muscle cell is to the muscle system: it performs the specific function of the system. The neuron, rather than the nerve, transmits the nerve impulse. Neurons exhibit many shapes (Fig. 15-1), but all have a **cell body** and **processes.** One process, distinguished cytologically by the absence of Nissl material, is the **axon,** or **nerve fiber.** It transmits nerve impulses to a synapse or to an effector. It may extend a long distance up or down the brain and spinal cord or into

the nerves. In fact, a **nerve** is a bundle of nerve fibers outside of the central nervous system, wrapped in a connective tissue sheath and supplied by blood vessels (vasa nervorum). Other processes exhibiting Nissl material and seldom extending far from the cell body may be present. These are dendrites. They receive synaptic endings. To observe how the neuron fits into the vertebrate nervous system, we shall examine a sensory nerve, a motor nerve, and two mixed nerves.

A typical sensory nerve, the eighth cranial, is diagrammed in Fig. 15-2, *A.* It commences in a sense organ (in this instance, the membranous labyrinth) and terminates in the brain. Like all nerves, it is made up of nerve fibers. The cell bodies of sensory neurons are usually found in a ganglion on

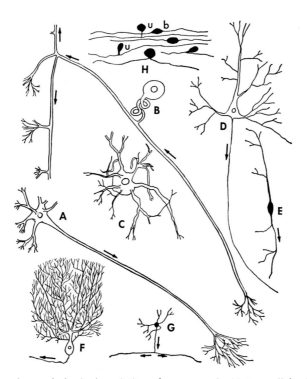

Fig. 15-1. Several morphological varieties of neurons. **A,** Motor cell body in the spinal cord, with the fiber utilizing the ventral root of the spinal nerve; **B,** dorsal root ganglion cell (sensory), with the fiber terminating at the left in the spinal cord; **C,** sympathetic ganglion cell; **D** and **E,** pyramidal and horizontal cells from the cerebral cortex; **F** and **G,** Purkinje and granular cells from cerebellum; **H,** a group of embryonic dorsal root ganglion cells in transition from bipolar, **b,** to unipolar, **u,** condition. **A, C, D, F,** and **G,** Multipolar neurons; **E,** bipolar; **B,** unipolar. Arrows indicate direction of impulse.

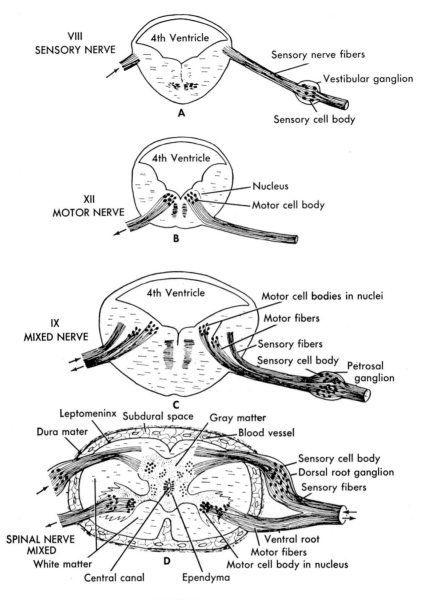

CHICKEN

Fig. 15-2. Sensory, motor, and mixed nerves. **A,** Sensory nerve with cell bodies in a ganglion. **B,** Motor nerve with cell bodies in a nucleus. **C,** Mixed nerve. **D,** Cross section of spinal cord showing roots of spinal nerve, with sensory cell bodies in the ganglion and motor cell bodies in the cord. Roman numerals designate specific cranial nerves.

the pathway of a nerve. A **ganglion** is a group of cell bodies outside the central nervous system. In lower vertebrates some of the cell bodies may be scattered along the nerve.

A typical motor nerve, part of the twelfth

cranial nerve of amniotes, is diagrammed in Fig. 15-2, *B.* The cell bodies of the motor neurons are found inside the central nervous system in a motor nucleus. Neurologically speaking, a **nucleus** is a group of cell bodies within the brain or spinal cord.

The motor fibers of the twelfth nerve terminate in striated muscle. There are almost no purely motor nerves in vertebrates, since nerves supplying striated muscles usually have sensory fibers for proprioception from the muscle (Fig. 10-1).

Typical mixed nerves containing both sensory and motor fibers are illustrated in Fig. 15-2, *C* and *D*. The cell bodies of the sensory neurons are in ganglia; those of the motor neurons are in motor nuclei. Most vertebrate nerves are mixed.

GROWTH AND DIFFERENTIATION OF THE NERVOUS SYSTEM

To achieve insight into the architecture of the adult nervous system, it is essential to know how the nervous system develops during embryonic life.

Neural tube. By the time the neural groove has been converted into a closed neural tube or shortly thereafter, the embryonic neural tube exhibits three zones: a **germinal layer** of actively mitotic cells, a **mantle layer** of cells proliferated from the germinal layer, and a **marginal layer** (Fig. 15-3). The germinal layer proliferates cells that add to the volume of the mantle layer until development is complete. The mantle layer cells sprout axons and dendrites to become neurons, or they develop into neuroglial cells, which are the connective tissue of the central nervous system. As the embryonic neurons (**neuroblasts**) of the mantle layer differentiate, their axonic sprouts grow centrifugally and produce and add to the marginal layer, which consists of nerve fibers. Because many of the axons become surrounded by a fatty myelin sheath, the marginal layer looks white when fresh and is called the **white matter.** The protoplasm of the cell bodies of the mantle layer causes this region to look gray, hence the name **gray matter.**

Many of the nerve fibers that grow into the marginal layer turn upward or downward in the cord or brain for short or long distances and synapse with neurons elsewhere in the central nervous system. The first of these intersegmental fibers to appear

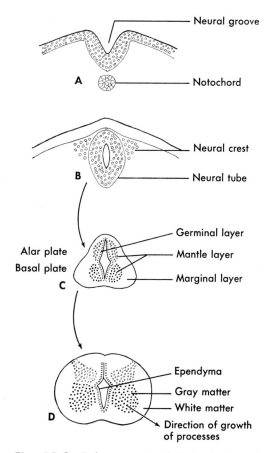

Fig. 15-3. Embryogenesis of the spinal cord. The alar plate contains association neurons, which will be in synapse with incoming (sensory) nerve fibers. The basal plate contains cell bodies of motor neurons, the axons of which grow in the direction of the arrow.

connect one segment with the next, a primitive condition. The later and longer ones become aggregated in ascending or descending bundles, or **fiber tracts,** each tract composed of functionally related fibers.

The embryonic cord and brainstem consist of **alar** and **basal plates** located above and below the sulcus limitans, respectively (Fig. 15-24). The alar plate becomes functionally associated with sensory (incoming) impulses, whereas the basal plate becomes primarily motor in function.

When all the nerve cells have been formed that will ever be formed in the cord or brain, the cells of the germinal layer cease to divide. Those that remain adjacent to the central canal become the **ependyma,** a connective tissue lining of the central canal.

All cell bodies and processes within the central nervous system are supported by **neuroglia,** a special connective tissue chiefly of ectodermal origin. The ependyma is one type of neuroglia and the only type found in cyclostomes. In higher vertebrates neuroglial cells become increasingly abundant and diversified.

Development of motor components of the nerves. A large number of the axons that sprout from neuroblasts in the basal plate grow out of, and away from, the neural tube (Fig. 15-3, *D*) to make contact with striated muscles. These become motor fibers of the cranial and spinal nerves. Because these fibers sprout from neuroblasts within the central nervous system, their cell bodies are within the adult brain or cord. Observe the location of the motor cell bodies in Fig. 15-2, *B* to *D*.

Some of the axons that sprout from neuroblasts in the basal plate grow out of the neural tube to make contact with neuroblasts in autonomic ganglia (Fig. 15-25). These axons are preganglionic fibers of the autonomic nervous system. The neuroblasts in the autonomic ganglia sprout postganglionic fibers that grow toward, and innervate, smooth muscles and glands. Thus, motor neurons have their cell bodies in the basal plate of the cord or brain with one exception: Postganglionic fibers of the autonomic system have their cell bodies in autonomic ganglia. Neuroblasts of the autonomic ganglia are migrants from neural crests, except those that, at least in amphibians, migrate outward from the basal plate.

Development of sensory components of the nerves. At the time the neural tube is forming, a longitudinal ribbon of neurectoderm separates on each side from the neural groove or tube. These neurectodermal ribbons become isolated parallel to the neural tube (Fig. 15-3, *B*). The ribbons soon become segmented to form a meta-

meric series of **neural crests** at the level of each body somite. Several neural crests also develop in the head lateral to the brain.

The cells of the neural crests have broad potentialities.[49] Some of them become neuroblasts that give rise to sensory neurons. In doing so, they pass through a bipolar stage (Fig. 15-1, *H*) in which one process grows toward, and into, the alar plate of the central nervous system, and the other process grows toward a sense organ. Thus, there is established an afferent neuronal connection between the sense organ and the cord or brain. Most of the embryonic bipolar cells later become unipolar in higher vertebrates (Fig. 15-1, *B*). Since each neural crest gives rise to a large number of sensory cell bodies, the result is usually a sensory ganglion on the pathway of the nerve. The cell bodies of sensory neurons therefore occur typically in sensory ganglia.

There are three exceptions to the generalization that sensory cell bodies develop from neural crests. The sensory cell bodies of the olfactory and optic nerves and those of proprioceptive fibers of the cranial nerves do not arise from neural crests.

The neuroblasts of olfactory nerve fibers develop from ectodermal cells in the olfactory epithelium. Their long processes grow into the nearest part of the brain, which is the olfactory bulb. Therefore, the cell bodies of the olfactory nerve are in the olfactory epithelium, and the nerve has no ganglion on its pathway, even though it is sensory.

Similarly, the neuroblasts that give rise to sensory fibers in the optic nerve are located in the embryonic retina, and the axons grow brainward along the optic stalk until they reach the optic chiasma. The cell bodies of the optic nerve therefore are in the retina, and there is no ganglion on the nerve. (In actuality, the retina is part of the brain, since it arises as an evagination from the diencephalon and never separates from it. The term "optic nerve" is therefore a misnomer.)

The cell bodies of proprioceptive fibers

in the cranial nerves do not arise from neural crests. Instead, neuroblasts within the embryonic brain sprout axons that grow out to the muscles. The cell bodies for proprioceptive fibers in the cranial nerves are therefore in nuclei inside the brain. The cell bodies of proprioceptive fibers of the spinal nerves arise from neuroblasts in the neural crests and hence are found in dorsal root ganglia.

SPINAL CORD

The spinal cord typically occupies the vertebral canal, surrounded by fat and protected, except in cyclostomes, by centra and neural arches. The protection given the cord in cyclostomes is shown in Fig. 1-11. The cord is surrounded in most fishes by a connective tissue membrane, the **meninx primitiva.** In some teleosts and in amphibians, reptiles, and some birds the embryonic meninx primitiva later forms an outer fibrous **dura mater,** and an inner vascular leptomeninx. In a few birds and in mammals the leptomeninx further differentiates into a web-like **arachnoid** layer and a **pia mater,** the latter intimately applied to the cord.

The spinal cord is considered to commence at the foramen magnum, but there is no landmark on the brain or cord precisely delimiting the two. Instead, a gradual transition from brain to cord extends over one or several body segments. The transition is more abrupt in higher forms.

The adult cord extends to the caudal end of the vertebral column only in vertebrates with abundant tail musculature, such as lower fishes, tailed amphibians, and reptiles. In other vertebrates the embryonic vertebral column elongates more rapidly than the spinal cord, with the result that at birth the cord is shorter than the column. In man the spinal cord terminates at the third lumbar vertebra. In frogs it ends anterior to the urostyle. In a few bony fishes the cord is actually shorter than the brain. It is only an inch or so in length in one fish several feet long.

When the cord is as long as the verte-

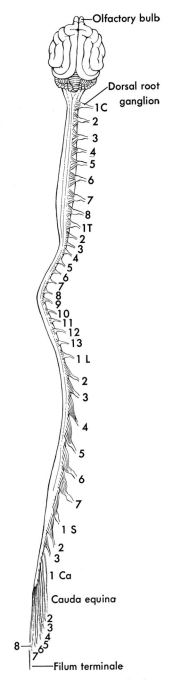

Fig. 15-4. Brain and spinal cord of a mammal (cat). The dura mater has been removed. **C,** Cervical spinal nerve; **T,** thoracic; **L,** lumbar; **S,** sacral, **Ca,** caudal spinal nerves. Note origin of spinal nerves by multiple rootlets, and enlargements of the cord in the cervical and lumbar regions.

bral column, each spinal nerve passes directly to the intervertebral foramen through which it emerges from the vertebral canal. If, however, the column subsequently elongates more than the cord, the spinal nerves then pass caudad within the vertebral canal to reach their respective foramina. As a result, the more caudal spinal nerves form a bundle of parallel nerves, the **cauda equina** (horsetail), within the vertebral canal (Fig. 15-4). Nonnervous elements (ependyma and meninges) of the foreshortened cord may continue farther caudad as a delicate strand, the **filum terminale.**

The spinal cord often exhibits cervical and lumbar enlargements at the level of the anterior and posterior appendages. The enlargements result from the large number of cell bodies and fibers innervating the appendage. When one pair of appendages is particularly muscular, such as the wings of birds with their pectoral muscles or the hind limbs of dinosaurs, the enlargement of the cord may be especially pronounced. Conversely, the spinal cord of turtles is particularly slender in the trunk because the thoracic and abdominal musculature is greatly reduced and sensory fibers to the shell are scanty. In many fishes the cord exhibits a swelling near the base of the caudal fin. The swelling marks the location of the urohypophysis (Fig. 17-6).

The cord is flattened in cyclostomes (Fig. 1-11), but tends to be cylindrical or quadrilateral in higher vertebrates. In general, the central canal within the cord is relatively large in lower forms and constricted in higher ones.

A cross section of a typical cord reveals the nuclei arranged in a definite pattern surrounding the central canal, where they comprise the gray matter (Fig. 15-5, high sacral). The nerve fibers occupy the periphery of the cord and constitute the white matter. The ascending and descending fibers are aggregated into functional groups of fiber tracts that interconnect one level of the cord with another or with the brain. Fibers for touch may constitute one tract,

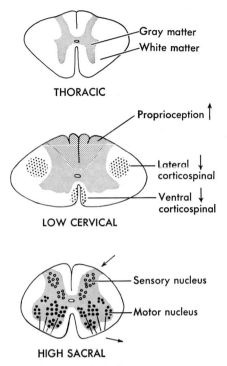

Fig. 15-5. Spinal cord of man in cross section at three levels, showing a few fiber tracts at the cervical level and a few nuclei at the sacral level. The cervical level is largest because it contains cell bodies associated with the anterior limb, all fibers ascending to the brain from lower levels, and all fibers descending from the brain to lower levels. The corticospinal tracts carry voluntary motor impulses from the cerebral cortex. The motor horn in the thoracic section is small because there are no limb muscles to be supplied at this level. Sensory nuclei contain the cell bodies of fibers that ascend toward the brain. They have synaptic relationships with the sensory fibers entering the cord via the dorsal root.

those for voluntary motor control another, and so forth. The fiber tracts of the cord are relatively few and simple in cyclostomes. They increase in number and complexity in higher forms.

SPINAL NERVES

Roots and ganglia. Except in cyclostomes, each spinal nerve arises from the cord by a dorsal and ventral root that unite to form the spinal nerve (Fig. 15-2). Each root is

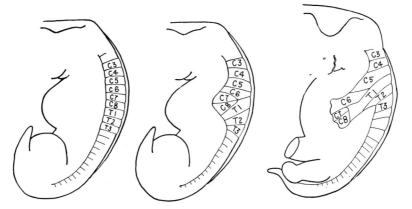

Fig. 15-6. Sensory innervation of the skin of the forelimb by successive spinal nerves. **C,** Cervical and, **T,** thoracic somites and associated nerves.

composed of a series of very short rootlets that unite close to their origin from the cord (Fig. 15-4). The dorsal root exhibits a ganglion and is predominantly sensory. The ventral root is strictly motor. There is considerable evidence that in the earliest vertebrates (1) the dorsal and ventral roots did not unite but continued independently to their destinations, (2) the dorsal roots were mixed, (3) dorsal root ganglia did not occur, and (4) primitive dorsal root ganglion cells were bipolar. These conclusions are based partly on study of the spinal nerves of the amphioxus and lower vertebrates.

The spinal nerves of the amphioxus consist of a series of dorsal nerves (mixed) alternating with ventral nerves (purely motor). The two do not unite. Each dorsal nerve arises from the cord at the level of a myoseptum and passes into the latter to be distributed to the skin (sensory) and to the viscera (sensory and motor). The cell bodies of the sensory fibers are bipolar and occur either within the spinal cord or scattered along the path of the dorsal nerve. There is no spinal ganglion. Each ventral nerve arises from the cord between two myosepta and penetrates the myomere, supplying motor fibers to the muscle.

The spinal nerves of cyclostomes are not much different. Dorsal and ventral roots alternate and remain independent in lam-

preys but unite in hagfishes. Some of the cell bodies of sensory fibers are for the first time aggregated in ganglia on the dorsal root, and some are located within the cord, as in the amphioxus. Most of the ganglion cells are bipolar. Visceral motor fibers occur in both roots. The ventral root is strictly motor.

Above cyclostomes, dorsal and ventral roots always unite. The dorsal root in teleosts still contains numerous visceral motor fibers in addition to sensory fibers, but in tetrapods most of these have been lost from the dorsal root. Most, if not all, of the sensory cell bodies occur in the dorsal root ganglion. These cells are bipolar in cartilaginous fishes; bipolar, intermediate, and unipolar in bony fishes; chiefly unipolar in amphibians; and entirely unipolar in amniotes (Fig. 15-1, *B*). The ventral root is motor (somatic and visceral), with cell bodies inside the cord.

Metamerism. A spinal nerve arises from each segment of the cord. These nerves are metamerically distributed to the skin and muscle of the body wall, neck, and tail. At the level where a fin or limb bud forms, they supply the skin (Fig. 15-6) and muscle (Fig. 10-7) of the appendage. The segmental distribution of spinal nerves to successive myomeres is best illustrated in fishes, since in them the metamerism of the body wall muscles is relatively undisturbed. In birds and mammals the metameric distribution to the abdominal muscles is obscured because of loss of myosepta or because of migration of embryonic muscle masses from their original segmental level. Muscle migration is illustrated in the nerve supply to the mammalian diaphragm, which, although it lies at the thoracolumbar junction, is a derivative of cervical myotomes. The phrenic nerve therefore receives contributions from several cervical spinal nerves—often C4, C5, and C6.

Vertebrates have almost as many spinal nerves as there are vertebrae, except at the end of the tail. Tadpoles have as many as forty pairs of spinal nerves and lose all but ten pairs at metamorphosis.

Rami and plexuses. Shortly after emerging from the vertebral canal, each typical spinal nerve divides into three mixed branches. A small **dorsal ramus** supplies the epaxial muscles and skin of the dorsum. A larger **ventral ramus** passes into the lateral body wall and supplies the hypaxial muscles and skin of the side and venter (Fig. 1-2). A third branch, the **white ramus communicans**, passes to a ganglion of the sympathetic trunk. It carries motor fibers to, and sensory fibers from, visceral organs (Fig. 15-25).

The ventral rami of successive spinal nerves often unite to form a plexus from which arise large nerve trunks. The chief plexuses are the brachial and pelvic, which supply nerves to the anterior and posterior appendages. The plexuses are relatively simple in anamniotes, since the metamerism of the muscles is only slightly interrupted by the fins. The plexuses become increasingly complicated in tetrapods, since the limbs have a large number of muscles derived from several successive somites (compare shark and mammal, Fig. 10-7). Autonomic plexuses occur in the visceral organs.

Occipitospinal nerves. In many fishes and amphibians several pairs of **occipitospinal nerves** arise from the transitional region between the brain and spinal cord just behind the vagal nerve. The first pair or two of these nerves may arise anterior to the foramen magnum. The nerves have no sensory roots and supply the hypobranchial musculature, including the tongue when present. The embryonic frog has an occipitospinal nerve in front of the one supplying the tongue, but it becomes suppressed during later embryonic development. The eleventh and twelfth cranial nerves of amniotes lack sensory roots and are derived in part from the ancient series of occipitospinal nerves found in fishes and amphibians.

Fiber components of spinal nerves. The nerve fibers in a typical spinal nerve are of four functional varieties listed in Table 15-1. Three of the varieties are referred to as **general** fibers (GSA, GVA, and GVE), to differentiate them from **special** types found only in cranial nerves (p. 370).

Table 15-1. Fiber components of spinal nerves

Components	Innervation
Sensory	
General somatic afferent fibers (GSA)	General cutaneous receptors for touch, pain, temperature, and pressure on skin
	Receptors on striated muscle cells and tendons (proprioceptive)
General visceral afferent fibers (GVA)	Viscera
Motor	
Somatic efferent fibers (SE)*	Myotomal muscle everywhere in body
General visceral efferent fibers (GVE)†	Smooth muscle, cardiac muscle, and glands; visceral fibers to skin are vasomotor, pilomotor (in mammals), or secretory (to skin glands) and they supply melanophores in many lower vertebrates

*The fibers to myotomal muscle are designated simply as SE, rather than as GSE (general somatic efferent) fibers, because there is no **special** somatic efferent component.

†Autonomic fibers.

BRAIN

Every vertebrate brain from fish to man is built in accordance with a single architectural plan. When the embryonic neural tube is first established, its anterior end is already enlarged and exhibits three primary brain vesicles, the forebrain, midbrain, and hindbrain (Fig. 4-7). The forebrain (prosencephalon) subsequently becomes subdivided into the **telencephalon** and **diencephalon.** The midbrain (**mesencephalon**) develops without further subdivision. The hindbrain (rhombencephalon) subdivides into the **metencephalon** and **myelenceph-**

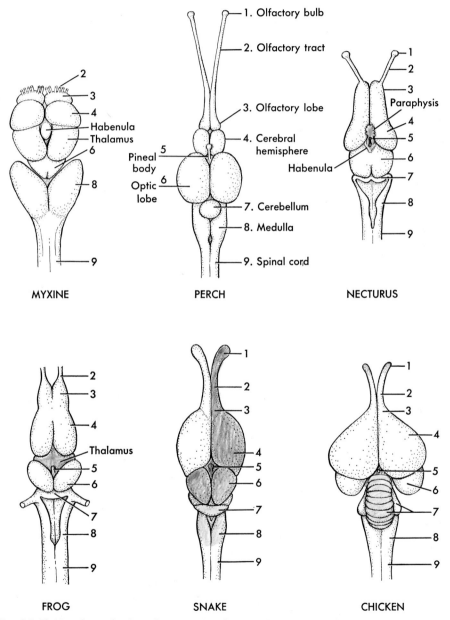

1. Olfactory bulb
2. Olfactory tract
3. Olfactory lobe
4. Cerebral hemisphere
5. Pineal body
6. Optic lobe
7. Cerebellum
8. Medulla
9. Spinal cord

Habenula
Thalamus

MYXINE   PERCH   NECTURUS

Paraphysis
Habenula

Thalamus

FROG   SNAKE   CHICKEN

Fig. 15-7. Vertebrate brains. The posterior choroid plexuses have been removed to expose the fourth ventricle. Suggestion: Color the telencephalon, diencephalon, mesencephalon, metencephalon, and myelencephalon, using a different color for each.

alon. Further differentiation of the five subdivisions during embryonic life involves extensive thickening of the walls in some locations and evagination or invagination in others, until the definitive brain has taken shape. The subdivisions are seen in their primitive relations in the brains of fishes (Fig. 15-7) but are readily demonstrable in all higher vertebrates, including man. The brain of the amphioxus has been discussed on p. 22.

Metencephalon and myelencephalon

The myelencephalon, represented chiefly by the medulla oblongata, merges imperceptibly with the spinal cord. The area of transition is characterized internally by gradual relocation of the ascending and descending fiber tracts (white matter), as a result of which the gray matter of the medulla is dispersed into interrupted columns or masses (nuclei).

The most conspicuous dorsal feature of the hindbrain is the cerebellum, a dorsal evagination of the metencephalon. The cerebellum functions in reflex control of the skeletal muscles. It receives input from the membranous labyrinth and feedback from the proprioceptive receptors in muscles, joints, and tendons. It discharges motor impulses that result in muscle tonus and bodily posture from fish to man. A dead fish falls onto its side and an unconscious bird or mammal collapses partly because the cerebellum is no longer bringing about synergistic contractions of those muscles essential to maintaining posture. The cerebellum is largest in those animals that have the greatest problem of maintaining posture and balance—particularly active or swimming forms. It is especially large in fishes, birds, and mammals, in which it may overlie both the medulla and the midbrain. It is relatively inconspicuous in amphibians. In cyclostomes it is not well developed and does not bulge from the brain. The cerebellum also receives regulatory impulses from the voluntary motor centers of the cerebral cortex and from other brain centers. The gray matter of the cerebellum is

on its surface, a condition not found in most parts of the brain.

Assisting the cerebellum as a reflex center for maintenance of equilibrium in some fishes, including elasmobranchs, are ruffle-like **restiform bodies** (Fig. 15-18) at the anterior lateral angles of the medulla. These, like the cerebellum, receive impulses from the membranous labyrinth. (The restiform bodies of fishes are not homologous with structures of the same name in higher forms.)

Another prominent feature of the anterior end of the medulla in a few fishes is the **vagal lobe** (Fig. 15-8). This marks the internal location of a visceral sensory nucleus (nucleus solitarius) that receives taste fibers from the seventh, ninth, and tenth cranial nerves in all vertebrates, including man. It bulges on the surface only in those fishes that have taste buds scattered over the surface of their body. The presence of the extra incoming fibers and association neurons causes the nucleus to be unusually large.

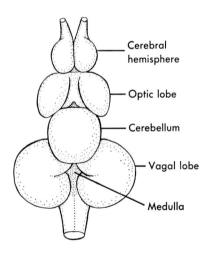

MODIFICATION FOR BOTTOM-FEEDING

Fig. 15-8. Brain of the buffalo fish *Carpiodes velifer*. Note unusual bulge (vagal lobe) on the alar plate of the medulla. It is a large nucleus and contains secondary cell bodies for taste. Here terminate the large number of incoming taste fibers characteristic of bottom-feeders.

The chief topographical features of the hindbrain laterally and ventrally are small swellings (indicating the presence of enlarged, underlying nuclei), elevated ridges, or transverse bands (indicating the presence of fiber tracts). These topographical markings are most prominent in mammals and least pronounced in fishes. Among mammalian markings are the **pyramids** (corticospinal tracts), carrying voluntary motor impulses from higher centers, and the **pons,** which contains fibers crossing from one side to the other (decussating) and connecting the cerebral hemispheres with the opposite side of the cerebellum (Fig. 15-9). Another transverse fiber tract of the hindbrain is the **trapezoid body,** which relays impulses for sounds. It is first demonstrable in frogs.

The cavity within the hindbrain is the fourth ventricle (Figs. 15-12 to 15-14). The cerebellum is part of its roof. The rest of the roof is a membranous **medullary velum,** a part of which hangs into the ventricle as the **choroid plexus** of the fourth ventricle.

Mesencephalon

The roof (**tectum**) of the mesencephalon exhibits a pair of prominent **optic lobes** in all vertebrates. These bulging gray masses serve partly as optic reflex centers that receive fibers from the retina. They the especially well developed in birds, which rely on visual stimuli for much of their information. A pair of **auditory lobes** are found caudal to the optic lobes in the roof of the mesencephalon commencing with reptiles. These gray masses are present in fishes and amphibians but are not large enough to bulge from the surface. The auditory lobes receive impulses from that part of the membranous labyrinth sensitive to vibratory stimuli—the lagena and its homologue the cochlea. With increased size of the cochlea, the auditory lobes enlarge. They also have other afferent connections.

The ventral portion of the mesencephalon consists of nuclear masses and of fiber tracts connecting lower and higher levels of the brain. In mammals these tracts

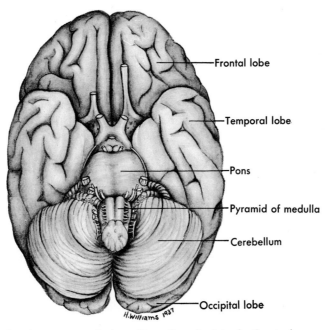

Fig. 15-9. Brain of man, ventral view. (From Francis: Introduction to human anatomy, ed. 5, St. Louis, 1968, The C. V. Mosby Co.)

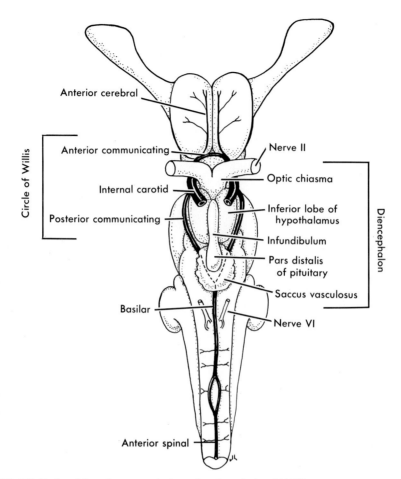

Fig. 15-10. Brain of *Squalus*, ventral view, showing circle of Willis.

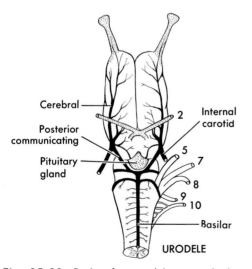

Fig. 15-11. Brain of a urodele, ventral view, showing blood supply. Note absence of circle of Willis.

become massive and are visible on the surface as the **cerebral peduncles.**

The ventricle of the midbrain is quite large in fishes and amphibians and extends dorsally into the optic lobes as optic ventricles. In higher vertebrates the optic lobes are not hollow, and the ventricle is restricted to a narrow **cerebral aqueduct** (Fig. 15-14, man).

Diencephalon

Optic chiasma (Figs. 15-10 and 15-11). The site of superficial origin of the optic nerves from the brain is the optic chiasma, a ventral landmark approximating the cephalic boundary of the diencephalon.

Pituitary gland (Figs. 15-10 and 15-11). Just caudal to the optic chiasma lies the pituitary body. It is attached by a stalk

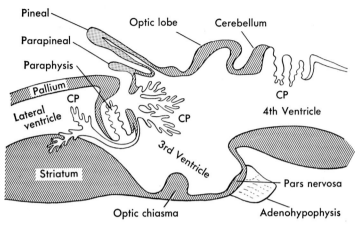

Fig. 15-12. Diencephalon and adjacent areas of a vertebrate brain, sagittal section. **CP,** Choroid plexus of lateral, third, and fourth ventricles. Based on the brain of a larval frog.

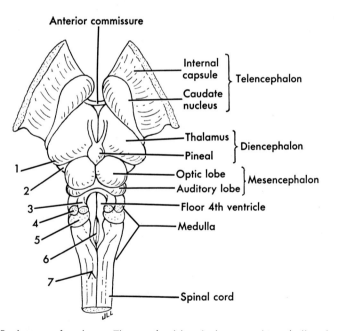

Fig. 15-13. Brainstem of a sheep. The cerebral hemispheres and cerebellum have been cut away to reveal the typical vertebrate structure. **1,** Location of lateral geniculate body of thalamus; **2,** medial geniculate body; **3 to 5,** anterior, middle, and posterior cerebellar peduncles, which carry fibers to and from the cerebellum (the peduncles had to be cut when the cerebellum was removed); **6,** hypoglossal trigone in the floor of the fourth ventricle and marking the site under which lies the hypoglossal nucleus; **7,** posterior funiculus containing ascending fibers for proprioception.

of brain tissue to the diencephalic floor. Its embryonic origin from two separate anlagen and its functions are discussed in Chapter 17.

Saccus vasculosus. Fishes, except cyclostomes, exhibit a thin-walled saccus vasculosus just caudal to, and often larger than, the pituitary. The sac apparently serves as a sense organ (p. 385) and is connected by fiber tracts with the cerebellum.

Hypothalamus. The portion of the diencephalon lying in the floor and ventrolateral walls of the third ventricle is the hypothalamus. The area contains a number of nuclei exerting partial control over the smooth muscles and glands of the body via the autonomic nervous system. In general the anterior part of the hypothalamus is associated with parasympathetic functions, whereas the caudal part is associated with sympathetic functions. The hypothalamus is also an important source of neurosecretions (Fig. 17-1).

Epithalamus (pineal and parapineal). The roof of the diencephalon is the epithalamus. It exhibits elevated gray masses, the habenulae (Fig. 15-7), marking the location of nuclei associated with olfaction. In addition, the roof of the primitive diencephalon gives rise to two prominent evaginations, the pineal body (**epiphysis**) and the parapineal (**parietal**) body (Fig. 15-12). The pineal has been the more persistent and is still present in almost all vertebrates.

The pineal body (Fig. 15-13) is an unpaired, elongated, club-shaped, knob-like, or, occasionally, thread-like organ attached by a stalk to the posterior roof of the diencephalon. In fishes and amphibians it sometimes projects upward through a foramen in the sagittal suture of the skull and has a knoblike tip lying under the skin of the head between the eyes. The brow spot of some species of frog marks the former location of the distal tip of the pineal body in the tadpole. In amniotes the pineal is buried between and beneath the enlarged cerebral hemispheres. No pineal develops in *Myxine,* and the embryonic

pineal of *Bdellostoma* becomes vestigial in adults. The pineal body is missing in crocodiles, armadillos, sirenians, and edentates, although embryonic vestiges occur in some of the mammals listed. It is microscopic in porpoises and small in birds and some mammals. In *Echidna*, rodents, ungulates, man, and other mammals it is relatively large.

The parapineal body has been widely distributed among vertebrates in the past but tends to have been lost in most modern forms. It is present in lampreys, but not in myxinoids, and in certain living, but mostly ancient, fishes such as *Amia*. It is absent in modern teleosts. It was widespread among ancient amphibians and is common in larval anurans, where it is known as the **frontal organ** (**stirnorgan**). It was present in ancient reptiles and is still found in *Sphenodon* and in several modern lizards, where it is more often called the parietal eye. It is vestigial or absent in other reptiles, in birds, and in mammals, although an anlage develops in bird embryos.

Both the pineal and parapineal bodies were photoreceptors at one time, and they continue to serve this role in a number of living vertebrates. The pineal also has an endocrine function in higher vertebrates, at least. The function of the pineal and parapineal bodies as photoreceptors is discussed in Chapter 16, and the endocrine role of the pineal is discussed in Chapter 17.

The pineal receives an afferent innervation from the superior cervical sympathetic ganglion via the **nervi conarii**. The significance of the innervation is not yet known. It may be related to the endocrine function of the gland.

Thalamus (Fig. 15-13). The largest subdivision of the diencephalon is the thalamus, a mass of nuclei surrounding the third ventricle. All sensory pathways ascending to the telencephalon synapse in one of the thalamic nuclei before continuing. The thalamus is small in lower vertebrates. It becomes increasingly prominent in higher forms and relays an increasing number of sensory impulses to the cerebral hemispheres. In mammals the thalamus is so enlarged that the left and right sides bulge into the ventricle and meet to form a gray commissure (middle commissure, or massa intermedia). The relay centers for sight and sound in the mammalian thalamus become swollen to form the lateral and medial geniculate bodies, respectively.

Third ventricle (Fig. 15-14). The cavity of the diencephalon is the third ventricle, continuous caudad with the cerebral aqueduct and cephalad with the lateral ventricles. The third ventricle is laterally com-

pressed, especially in higher vertebrates, by the thalamus. An optic recess of the ventricle extends toward the optic chiasma, and an infundibular recess extends into the pituitary stalk. The roof of the ventricle remains thin, becomes vascularized, and contributes to the choroid plexus of the third ventricle.

Telencephalon

The telencephalon consists of cerebral hemispheres and a rhinencephalon. In lower forms the rhinencephalon is as prominent as the cerebral hemispheres. In higher vertebrates the hemispheres increase in size, bulge forward over the rhinencephalon, and relegate this part of the brain to an inconspicuous anteroventral location. The change mirrors the increased role of the cerebral hemispheres in behavior.

Rhinencephalon. The rhinencephalon from fish to man consists of olfactory bulbs, olfactory tracts, and olfactory lobes (Fig. 15-20). The olfactory bulbs lie close to the olfactory epithelium. The two are separated by the olfactory capsule. Olfactory nerve fibers, with cell bodies in the olfactory epithelium, pass caudad through one or more foramina to synapse in the olfactory bulbs. The bulbs contain the cell bodies of the olfactory tracts. The tracts terminate chiefly in the olfactory lobes. These exhibit an ancient olfactory cortex, the **archipallium**. In mammals the olfactory lobes (now represented by the **hippocampal gyri**, or **pyriform area**) are dwarfed by the cerebral hemispheres. Fiber tracts connect the rhinencephalon with other parts of the forebrain.

Cerebral hemispheres. When one thinks of cerebral hemispheres, most likely what comes to mind are the enlarged cerebral hemispheres of mammals, with their thick cortex of gray matter (Figs. 15-15 to 15-17). But the cerebral cortex is a recent acquisition. The basic pattern upon which all vertebrate hemispheres are constructed and from which the hemispheres of mammals have evolved is seen in fishes.

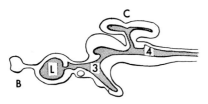

SHARK

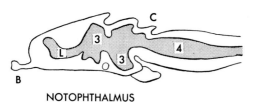

NOTOPHTHALMUS

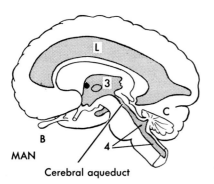

MAN

Cerebral aqueduct

Fig. 15-14. Sagittal brain sections showing the ventricles. **B,** Olfactory bulb; **C,** cerebellum; **L, 3,** and **4,** lateral, third, and fourth ventricles. The black foramen in the third ventricle in man is the interventricular foramen connecting the third and right lateral ventricles.

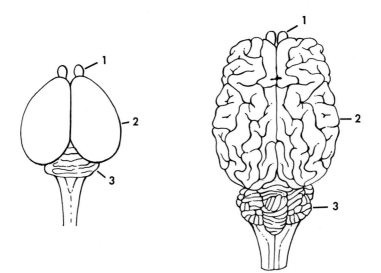

PLATYPUS   SHEEP

Fig. 15-15. Brain of a primitive mammal (platypus) lacking cortical gyri, and brain of sheep. **1,** Olfactory bulb; **2,** cerebral hemisphere, **3,** cerebellum.

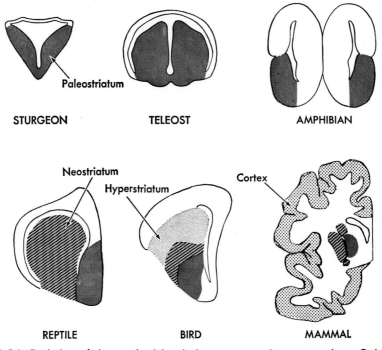

STURGEON   TELEOST   AMPHIBIAN

REPTILE   BIRD   MAMMAL

Fig. 15-16. Evolution of the cerebral hemispheres as seen in cross sections. Only the left hemisphere is shown in the lower figures. Gray indicates the paleostriatum. Reptiles and birds have added new nuclear masses (neostriatum and hyperstriatum). Mammals have developed a cortex. Note the old striatal complex (now called basal ganglia) still present in the mammal.

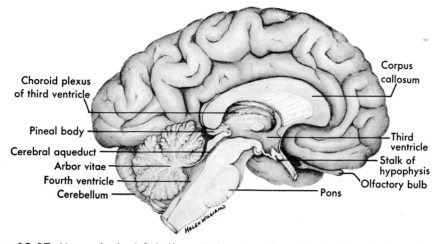

Fig. 15-17. Human brain, left half, sagittal section. (From Francis: Introduction to human anatomy, ed. 5, St. Louis, 1968, The C. V. Mosby Co.)

In fishes each hemisphere consists chiefly of a **corpus striatum,** often referred to as the paleostriatum because it is ancient *(paleo).* The striatum, the lateral ventricle, and the thin roof, or pallium, constitute the entire cerebral hemisphere (Fig. 15-16, sturgeon, teleost). The striatum consists of motor nuclei receiving incoming fibers, chiefly from the rhinencephalon. Fibers from the striatum enter ancient descending tracts that terminate in the motor nuclei of cranial and spinal nerves. Thus, olfactory stimuli result in reflex motor activity of many parts of the body, and the reflex is mediated by the striatum.

Sensory impulses from receptors other than the olfactory epithelium terminate mostly at levels lower than the hemispheres in fishes. Most of the optic nerve fibers terminate in the optic lobes and thalamus. Impulses from the membranous labyrinth end in the cerebellum and mesencephalon. Impulses from general cutaneous receptors (touch, pain, and temperature) ascend only to the thalamus.

The striatum of amphibians receives a larger number of sensory fibers projected forward from the thalamus than does that of fishes. The striatum therefore receives input from a wider spectrum of receptors. Nevertheless, the number of fibers relayed

from the thalamus to the striatum is small when compared with higher vertebrates, and the cerebral hemispheres of amphibians are still preempted by olfactory stimuli.

In specialized reptiles additional nuclei, constituting a **neostriatum** (Fig. 15-16), are added to the hemispheres, and these nuclei receive many more sensory fibers from the thalamus. A trace of cerebral cortex appears on the surface of the pallium of reptiles. Because of the added number of cell bodies and synapses, the hemispheres of reptiles are larger and bulge laterally and dorsally and, to a limited extent, backward over the diencephalon. Because of the increased flow of sensory information into the hemispheres, they now assume increased control of motor activity.

Bird hemispheres are essentially reptilian in structure and resemble the hemispheres of crocodilians in many respects. The striatum reaches a peak of development, and additional strata of nuclei, the **hyperstriatum,** are superimposed on the old striatum (Fig. 15-16). To the hyperstriatum come many sensory impulses which, after being relayed to the older striatum, result in stereotyped behavior such as nest-building, incubation of eggs, and care of the young. The avian cerebral cortex is better developed than that of reptiles, but almost complete ablation of the cortex has no effect on nesting, courtship, mating, or the rearing of young. In birds the olfactory lobes are greatly reduced in size, and smell is less important in influencing behavior.

In mammals the striatum and other nuclear masses, now called collectively the basal ganglia, continue to play an important, although not fully understood, role in the nervous system. But the cerebral cortex has become the most conspicuous part of the mammalian brain. As a result of the enormous upward, over, and backward growth of the pallium containing the cortex, the striatum, the diencephalon, and the midbrain are all hidden from dorsal view (Fig. 15-17). Removal of the overgrown cerebral hemispheres will reveal the primitive relationships of the rhinencephalon, striatum, thalamus, midbrain, metencephalon, and medulla (Fig. 15-13).

The mammalian cortex has at least four important functions. (1) It is the highest center to which sensory impulses may pass. These impulses give rise to sensations of a discriminative (epicritic) nature and contribute to the esthetic enjoyment of sensory stimuli. (2) It appears to be at least one location, so far as can be determined at present, where past experiences are stored as memory. (3) It is the center where all data, incoming or recalled, may be correlated, analyzed, and employed in making choices. (4) It is the highest center from which motor activity may be initiated. The cortex is, therefore, the "thinking" part of the brain. The manner in which the cerebral cortex is employed in the solution of human problems will determine the future fate of civilization insofar as it is under the control of man.

In many mammals, but not all, the cerebral cortex becomes so voluminous that it is folded into numerous ridges (gyri; sing., gyrus) and grooves (sulci; sing., sulcus). Under the cortex in the roof of the ventricles lies a broad transverse sheet of commissural nerve fibers, the corpus callosum, interconnecting the cortices of the two hemispheres (Fig. 15-17). The newborn opossum has no cerebral cortex; all cortical structures develop later.

Lateral ventricles. The cavities of the cerebral hemispheres are the lateral ventricles. These are continuous with the third ventricle via an interventricular foramen (Fig. 15-14). The location marks the original cephalic end of the embryonic neurocoel before evagination of the cerebral hemispheres took place. The roof of the ventricles includes in mammals the corpus callosum and the overlying cortex. The floor consists, in part, of the striatum and associated nuclei. Separating the left and right lateral ventricles is a thin, double-walled vertical partition, the **septum pellucidum.** The lateral ventricles, like the other brain ventricles, are lined by an ependyma that is extensively ciliated throughout life in many species, including man. The cilia may continue to beat for several hours after death.

Paraphysis. At many locations the roof of the vertebrate brain is very thin, non-nervous, and composed of ependyma or of choroid plexus and little else. At the caudal end of the telencephalon of fishes and some tailed amphibians, the roof tends to form a sac-like evagination, sometimes very large, called the paraphysis (Fig. 15-12). The vessels of the paraphysis receive a rich autonomic nerve supply, and a neurosecretory tract leads from the paraventricular nucleus of the hypothalamus to the base of the paraphysis. The functional significance of this sac is not known.

Choroid plexuses and cerebrospinal fluid. The cavities of the brain and spinal cord and certain of the submeningeal spaces are filled with a lymph-like cerebrospinal fluid secreted by choroid plexuses. A typical choroid plexus consists of the thin ependymal roof of the ventricle and of the pia mater (or leptomeninx in fishes), invaded by a rich vascular plexus. A choroid plexus hangs into the third and fourth ventricles and from the third ventricle extends forward into the lateral ventricles (Fig. 15-12).

Cerebrospinal fluid secreted into the

ventricles moves sluggishly caudad into the central canal of the spinal cord partly by ciliary action. From the fourth ventricle the fluid also passes into the submeningeal spaces via a pair of **lateral apertures of the fourth ventricle** under cover of the cerebellum. A median aperture is also present in many vertebrates. From the submeningeal spaces the cerebrospinal fluid passes outward along the roots of the cranial and spinal nerves for short distances. It also passes centrally along the nerve rootlets into the cord and brain and bathes each motor neuron. It seeps to the inner ear, where it contributes to the perilymph.

Cerebrospinal fluid is removed by lymph channels, especially along the roots of the spinal nerves. In higher vertebrates it is also removed by clusters of macroscopic arachnoid villi, which penetrate the dura mater and hang into the large venous sinuses of the brain.

Cerebrospinal fluid assists in protecting the central nervous system from concussion. It also exchanges metabolites with the tissues it bathes. In the saccus vasculosus of fishes the cerebrospinal fluid probably stimulates sensory neurons and initiates a train of impulses to the cerebellum.

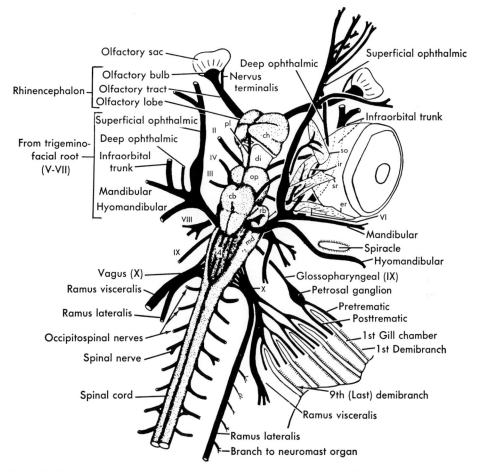

Fig. 15-18. Brain and cranial nerves of *Squalus*. **cb,** Cerebellum; **ch,** cerebral hemisphere; **di,** diencephalon; **er,** external rectus eyeball muscle; **ir,** internal rectus muscle; **md,** medulla; **op,** optic lobe; **pl,** pineal body; **rb,** restiform body; **so,** superior oblique muscle; **sr,** superior rectus muscle; **4,** fourth ventricle, roof removed; **II** to **X,** cranial nerves, pharyngeal branches omitted.

CRANIAL NERVES

The first ten cranial nerves of all vertebrates are distributed in accordance with a basic pattern. For convenience, these nerves may be grouped as follows: predominantly sensory nerves (I, II, and VIII), eyeball muscle nerves (III, IV, and VI), and branchiomeric nerves (V, VII, IX, and X). Amniotes have two additional pairs of nerves in the cranial series, which are purely or predominantly motor (XI and XII), and all vertebrates have a pair of terminal nerves anterior to the numbered series. The basic pattern of cranial nerve distribution is seen in sharks. Any variations in the distribution of these nerves in tetrapods result from adaptation to life on land.

The roots of cranial nerves do not come off the brain in a dorsal and ventral series. Nerves supplying myotomal muscles (III, IV, VI, and XII) have ventral roots only, and the branchiomeric nerves (V, VII, IX, X) and nerve XI have rootlets that are not in line with the rootlets of the spinal nerves. Instead, they come off the brain laterally and are called **lateral roots**.

Terminal nerve (Fig. 15-18). The terminal nerve emerges from the rhinencephalon close to the olfactory roots and passes to the olfactory epithelium. It appears to be vasomotor to the nasal mucosa. It is found in all vertebrates, including man.

Predominantly sensory cranial nerves*

Nerve I (olfactory). The cell bodies of the olfactory fibers are located in the olfactory epithelium, and the fibers extend caudad and perforate the olfactory capsule to terminate in the olfactory bulb (Fig. 15-19). In *Squalus acanthias* the olfactory epithelium lies so close to the bulb that an olfactory nerve cannot be distinguished as an anatomical entity. In other sharks (com-

---

*Nerves II and VIII, although functionally sensory, contain a number of efferent fibers from the brain to the deep layer of the retina or to the vestibular hair cells. These fibers apparently exert an inhibitory influence on the discharge of sensory impulses from the receptor.

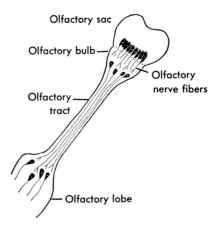

Fig. 15-19. Olfactory sac and rhinencephalon of a dogfish shark, showing location of cell bodies of sensory fibers.

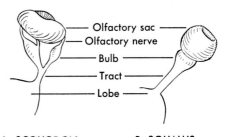

A. SCOLIODON   **B. SQUALUS**

Fig. 15-20. Rhinencephalon of two sharks. The olfactory sacs are not part of the brain. Olfactory nerve fibers connect the sac and bulb and form a discrete nerve in *Scoliodon* but not in *Squalus.*

pare Fig. 15-20, A and B), as in most vertebrates, one or more short bundles of olfactory fibers (**filia olfactoria**) extend between the olfactory epithelium and the bulb. These constitute collectively the olfactory nerve. In tetrapods a separate slip of the olfactory nerve innervates the vomeronasal organ.

In mammals the olfactory epithelium is in the upper part of the nasal passage, separated from the olfactory bulb by the cribriform plate of the ethmoid bone (derived from the olfactory capsule). The foramina in the cribriform plate (Fig. 8-6) transmit the filia olfactoria. When the brain of a vertebrate is lifted from the cranial cavity, the olfactory nerve bundles are torn from the olfactory bulbs and only stumps of the nerves remain attached to the brain.

Nerve II (optic). The cell bodies are found in the retina. The optic nerve emerges from the rear wall of the eyeball and extends to the optic chiasma, where the **optic tracts** commence. Except in mammals, all the optic nerve fibers decussate in the chiasma to enter the optic tract on the opposite side of the brain. In mammals only those fibers from the nasal side of the retina cross. As a result there is overlap of the visual fields for binocular vision.

Nerve VIII (vestibulocochlear). The eighth (formerly the auditory) nerve in lower vertebrates has two rami, an anterior and a posterior. The anterior ramus innervates the ampullae on the anterior vertical and horizontal semicircular canals, and the utriculus. The posterior ramus innervates the ampulla on the posterior vertical canal, the sacculus, and the lagena. Commencing with amphibians, the lagena enlarges and differentiates to become the cochlea for hearing. As a result, the posterior ramus becomes greatly enlarged. It is then called the **cochlear** nerve, but it continues to carry fibers from other parts of the membranous labyrinth. The anterior ramus is then called the **vestibular** nerve. Both branches enter the medulla and have a ganglion (**cochlear** or **vestibular**) on their pathway. The cochlear ganglion lies on the spiral membrane within the cochlea and hence is also known as the **spiral** ganglion. In some mammals (the rat and mouse, but not the bat or cat[108]) large cell bodies are also distributed along the entire length of the nerves, and their fibers extend into the medulla. These cell bodies are stimulated via synaptic endings from offshoots (collateral branches) of adjacent fibers whose cell bodies are in the ganglia. The mechanism presumably provides for reinforcement of the stimuli from the labyrinth.

Eyeball muscle nerves

Nerves III, IV, and VI (oculomotor, trochlear, and abducens). The third, fourth, and sixth nerves are distributed to the myotomal muscles of the eyeball and supply the superior oblique (IV), external rectus (VI), the four remaining extrinsic eyeball muscles (III), and certain other myotomal muscles of the eyes (Fig. 10-6, p. 221, and Table 10-1, p. 220). The eyeball muscle nerves resemble spinal nerves that have lost their dorsal roots. In addition to somatic motor fibers, the nerves contain sensory fibers for proprioception from the muscles innervated.

Nerve III arises from the ventral aspect of the mesencephalon. Nerve IV is the only nerve arising from the dorsal aspect of the brain (posterior end of the mesencephalon or anterior roof of the fourth ventricle) and one of the few nerves with motor fibers that decussate before emerging. Nerve VI emerges ventrally at the anterior end of the hindbrain. Nerves IV and VI are the smallest of the cranial nerves and therefore have the fewest fibers. The cell bodies of all the fibers in these nerves, both motor and proprioceptive, are in nuclei within the central nervous system. Hence the nerves exhibit no sensory ganglia.

Lampreys seem to lack an abducens nerve, but some authorities think it is represented by a small bundle that emerges from the hindbrain on the anterior surface of the trigeminal nerve.

AUTONOMIC FIBERS IN NERVE III. Nerve III contains visceral motor fibers. These terminate in the ciliary ganglion of the autonomic nervous system (Fig. 15-25). From the ganglion, postganglionic fibers pass to the sphincter pupillae and ciliary muscles of the iris diaphragm.

Branchiomeric nerves

One of the characteristics of vertebrates is the embryonic development of a series of visceral arches separated by pharyngeal pouches. The fate of the splanchnocranium and of the branchiomeric muscles of these arches has been discussed in earlier chapters. Whether the animal is to become fish or tetrapod, the muscles derived from the first arch are supplied by cranial nerve V;

those from the second arch, by nerve VII; from the third arch, nerve IX, and from succeeding visceral arches, nerve X. Since V, VII, IX, and X innervate branchiomeric muscles, they are called branchiomeric nerves.

The branchiomeric nerves are mixed nerves with important functions in addition to innervating branchiomeric muscles. Some of their motor fibers are components of the autonomic nervous system. Their sensory fibers supply several groups of sense organs, and each nerve includes proprioceptive fibers. The distribution of these nerves in *Squalus* illustrates the basic pattern. Alterations in tetrapods are chiefly the result of the elimination of gills and other adaptations to terrestrial life.

Nerve V (trigeminal). The fifth cranial nerve arises from the anterior end of the hindbrain and typically exhibits three divisions: **ophthalmic, maxillary,** and **mandibular.** In fishes the ophthalmic may be subdivided into superficial and deep ophthalmic nerves. All branches contain sensory fibers. Only the mandibular branch contains motor fibers.

SENSORY DISTRIBUTION. The fifth nerve is sensory to the head, including the teeth and anterior part of the tongue, for general cutaneous sensation (Fig. 15-21). The mandibular branch also contains proprioceptive fibers.

The cell bodies of all sensory fibers except proprioceptive are found in the **semi-lunar ganglion** unless, as in some lower vertebrates, the ophthalmic division has its own ganglion.

MOTOR DISTRIBUTION. The mandibular nerve supplies all muscles derived from the first pharyngeal arch. The predominant distribution is therefore to muscles of the jaws. However, mammals have a tensor tympani muscle attaching to the malleus (derived from the embryonic pterygoquadrate cartilage), which is therefore innervated by the fifth nerve. Table 10-4 gives the motor distribution of the fifth nerve.

Nerve VII (facial). The seventh cranial nerve arises from the anterior end of the hindbrain in close association with the fifth. In sharks the fifth and seventh nerves have a common (trigeminofacial) root.

SENSORY DISTRIBUTION. Nerve VII supplies the neuromast organs on the head of fishes and aquatic amphibians. With adaptation to land these fibers were lost. Branches also supply taste buds in the pharynx at the level of the second arch, any taste buds on the external surface of fishes, and taste buds on the anterior part of the tongue in tetrapods. The seventh nerve also contains proprioceptive fibers from the muscles and general sensory fibers from the endoderm of the second arch. The cell bodies of all sensory fibers, except proprioceptive, are in the **geniculate** (facialis) **ganglion.**

MOTOR DISTRIBUTION. The facial nerve is distributed to muscles of the second arch. These include the mimetic muscles of mam-

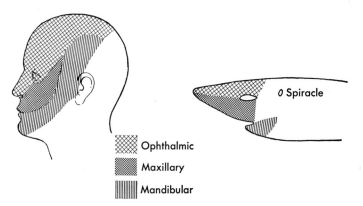

Ophthalmic
Maxillary
Mandibular

Fig. 15-21. Cutaneous distribution of the trigeminal nerve of vertebrates.

mals. Since the stapes is the dorsalmost hyoid cartilage, the stapedial muscle of mammals is also innervated by the seventh nerve. The seventh nerve also contains visceral motor fibers to the submandibular and sphenopalatine ganglia (Figs. 15-23 and 15-25). The former ganglion innervates the submandibular and sublingual salivary glands. The latter innervates the lacrimal gland and the mucous membrane of the nose.

Nerve IX (glossopharyngeal). In sharks the ninth nerve has three branches that are typical of branchiomeric nerves. These are **pretrematic** (sensory), **pharyngeal** (sensory), and **posttrematic** (mixed).* The pretrematic branch supplies the demibranch in the anterior wall of the first gill chamber for general sensation. The pharyngeal branch supplies the taste buds and general visceral receptors of the pharyngeal lining at the level of the third visceral arch. The posttrematic branch is sensory to the demibranch in the posterior wall of the first gill chamber, motor to the muscles of the third visceral arch, and proprioceptive from those muscles. Small branches of IX supply a short segment of the lateral-line canal at the junction of the head and trunk. Preganglionic fibers of the autonomic nervous system occur in IX. They have not been fully explored in all vertebrates, but in mammals they innervate the otic ganglion, from which postganglionic fibers pass to the parotid salivary glands. The ninth nerve arises from the medulla.

With loss of gills and of neuromast organs during adaptation to life on land, the ninth nerve lost many fibers. It continues to supply surviving taste buds in the third arch mucosa (on the posterior part of the tongue of mammals) and general receptors on the posterior part of the tongue and in the upper pharynx. Of the branchiomeric muscles of the third arch of fishes, only a stylopharyngeus remains in mammals.

The sensory cell bodies of the ninth nerve, except those for proprioception, are found in the **petrosal** ganglion of lower vertebrates, and in the **superior** (petrosal) and **inferior** ganglia of mammals.

Nerve X (vagus). The vagus arises from a series of rootlets along the lateral aspect of the medulla. The branchiomeric portion of the vagus in *Squalus* consists of a series of four trunks (Fig. 15-18) (more in elasmobranchs with more gill chambers), each of which exhibits a pretrematic, posttrematic, and pharyngeal branch distributed like the homologous branches of IX. The pretrematic branches supply the anterior walls of the last four gill chambers. Posttrematic branches supply the posterior walls of these chambers and are motor to arches IV to VII. It is therefore the chief respiratory nerve of fishes. The pharyngeal branches supply the pharyngeal epithelium for taste and general sensation.

In addition to the branchiomeric components, the vagus in *Squalus* has two other important trunks. The **ramus lateralis** is sensory to the lateral-line canal all the way to the tip of the tail, and the **ramus visceralis** supplies afferent and efferent visceral fibers to some of the coelomic viscera.

During the process of adapting to land the vagus lost those functions associated solely with life in the water but retained other functions. The prominent lateral-line branch disappeared. The sensory branches to the gill chambers were lost. However, sensory receptors in the mucosa of the pharynx, including surviving taste buds in the vicinity of the glottis, continue to be supplied by the vagus. Surviving also are the motor branches to those branchiomeric muscles of the fourth and successive arches that assumed new functions on land. These are chiefly the cricothyroid, cricoarytenoid, and thyroarytenoid muscles. Since much of

---

*The maxillary and mandibular branches of nerve V probably represent pretrematic and posttrematic branches of that nerve. (The ophthalmic is thought not to have belonged to V originally.) The pretrematic branch of VII is within the infraorbital trunk, and the posttrematic branch of VII (behind the spiracle) is the hyomandibular. The pharyngeal branch of VII in sharks is the palatine. The vagus of sharks exhibits four sets of pretrematics, posttrematics, and pharyngeals.

the distribution of the vagus has been lost in tetrapods, the ramus visceralis has become the major component of the nerve. It continues to supply the heart and viscera of the upper coelom.

In fishes as well as in tetrapods the vagus supplies preganglionic fibers to certain terminal ganglia of the autonomic system in the trunk (Fig. 15-25). Included in amniotes are autonomic fibers contributed by the internal ramus of the accessory nerve (Fig. 15-22). The vagus also contains proprioceptive fibers. And, through an unexplained circumstance, the vagus in mammals supplies the skin of part of the auricle of the ear.

The cell bodies of all sensory fibers, except those for proprioception, are found in one or more vagal ganglia such as the **lateralis** and **jugular** in fishes, or the **superior** (jugular) and **inferior** (nodose) **vagal ganglia** in mammals. In some elasmobranchs each of the four or more pretre-

matic and posttrematic branches to the gills has its own **epibranchial ganglion,** and it is likely that these branches were at one time four separate cranial nerves.

The innervation of the mucosa of the oral cavity and pharynx by nerves V, VII, IX, and X, in that sequence from mouth to esophagus and in all vertebrates from fish to man, demonstrates the negligible effects of life on land on the sensory innervation of the pharyngeal endoderm.

Accessory and hypoglossal nerves

Nerve XI (accessory). An independent accessory nerve is identified as an eleventh cranial nerve only among amniotes. It is purely motor. In mammals it has a series of medullary roots (bulbar roots), which arise from the medulla (bulb) of the brain immediately caudal to the roots of the vagus (Fig. 15-22). It also has a spinal root, which arises from a series of cervical rootlets with cell bodies in the spinal cord.

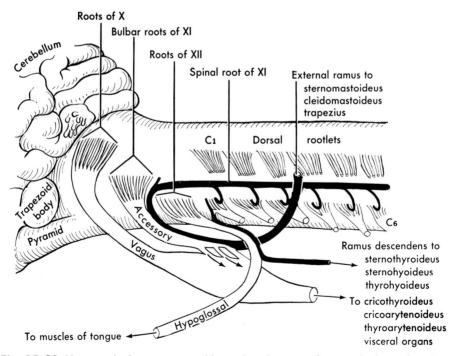

Fig. 15-22. Vagus, spinal accessory, and hypoglossal nerves of a cat. The hypoglossal roots (XII) in the medulla are in series with the ventral roots of the spinal nerves. Components in black are spinal nerve contributions. C_1, Dorsal rootlets of the first cervical spinal nerve; C_6, ventral rootlets of the sixth cervical spinal nerve.

The nucleus of origin of the spinal root occupies several segments of the cord—typically five or six in man, seven in horses, and fewer in many other mammals. The spinal rootlets typically merge into a common trunk, which passes cephalad close to the cord and enters the cranial cavity via the foramen magnum. Within the cranial cavity the spinal root joins the bulbar roots to form the eleventh cranial nerve. The nerve then passes through the jugular foramen in company with the ninth and tenth nerves. While close to the foramen the bulbar root fibers join the vagus to be distributed with the latter as the **internal ramus** of the accessory nerve. The internal ramus is composed of preganglionic fibers. The remainder of the eleventh nerve, composed of fibers of spinal origin, continues as the **external ramus** to the trapezius, sternomastoid, and cleidomastoid muscles.

Although an eleventh cranial nerve is not identified below amniotes, its components are actually present. In many fishes and amphibians the **last several rootlets of the vagus** unite to form a bulbar accessory nerve containing fibers supplying viscera and the branchiomeric trapezius (cucullaris) muscle. The addition in amniotes of cervical fibers supplying the sternomastoid and cleidomastoid muscles suggests that these two muscles may have myotomal as well as branchiomeric components. The eleventh cranial nerve is thus seen to be derived from cranial and occipitospinal nerves.

Nerve XII (hypoglossal). The twelfth cranial nerve of amniotes is motor, except for any proprioceptive fibers. It innervates the myotomal muscle of the tongue (genioglossus, styloglossus, hyoglossus, and lingualis, when present). It arises from the caudal end of the brain by a series of motor rootlets. Upon emerging from the hypoglossal foramen, the nerve is joined by branches from the first one or more cervical spinal nerves (Fig. 15-22). Some of the latter fibers are distributed via the main trunk of the hypoglossal to the geniohyoid muscle, but most of the spinal components

emerge from the hypoglossal to become the **ramus descendens.** This branch passes down the neck and supplies certain myotomal muscles of the hypobranchial series, including the omohyoid, sternothyroid, thyrohyoid, and sternohyoid muscles. (In birds the sternohyoid muscle becomes associated with the syrinx, which is therefore partly operated by the ramus descendens.)

That the hypoglossal is a cranial rather than a spinal nerve is dictated by the location of the foramen magnum. Like spinal nerves, it develops an embryonic dorsal root and ganglion (**Froriep's ganglion**). However, root and ganglion later disappear.

The twelfth nerve is a remnant of the occipitospinal series of fishes and amphibians, as may be deduced from the following facts: (1) The occipitospinal nerves have been reduced in number above fishes, whereas the cranial series has increased. (2) The adult hypoglossal nerve of amniotes, like occipitospinal nerves, lacks a dorsal root. (3) The occipitospinal nerves of lower vertebrates supply hypobranchial muscles derived from occipital somites, and the tongue is hypobranchial musculature of myotomal derivation. (4) Whereas the hypoglossal is the last cranial nerve in amniotes, it is the first spinal nerve in anamniotes. The foregoing and numerous other facts indicate that the twelfth cranial nerve represents a remnant of the occipitospinal series of lower vertebrates.

Sensory and motor innervation of the mammalian tongue (Fig. 15-23). The innervation of the mammalian tongue illustrates how a single organ may be served by numerous nerves, depending on the ontogenetic and phylogenetic history of its component parts. The mucosa of the anterior part of the tongue is ectodermal (stomodeal) and hence is innervated by the fifth nerve for cutaneous sensations. The taste buds on this part of the tongue are innervated by the seventh nerve, which supplies ectodermal taste buds whenever present in fishes. The mucosa on the posterior part of the tongue is innervated by the ninth nerve for both general sensation and

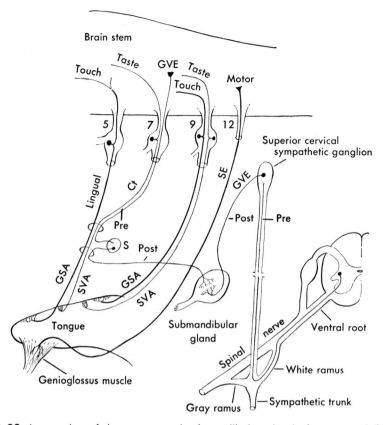

Fig. 15-23. Innervation of the tongue and submandibular gland of a mammal (based on cat and man). **5, 7, 9,** and **12,** Cranial nerves; **Ct,** chorda tympani; **Pre** and **Post,** preganglionic and postganglionic fibers of the autonomic nervous system; **S,** submandibular ganglion of the autonomic system. A key to the fiber components (**GSA, SE,** and so forth) is given in Tables 15-1 and 15-2.

taste because of the origin of this mucosa from the third visceral arch. The muscles are myotomal, hence are innervated by the twelfth nerve. Although four cranial nerves innervate the tongue, only three branches may be traced into it since the fibers from the seventh nerve (in the chorda tympani) enter the tongue as part of the lingual branch of the fifth nerve.

Fiber components of cranial nerves. In the discussion of spinal nerves it was pointed out that their fibers may be classified in four functional categories (GSA, GVA, SE, and GVE), each supplying specific types of general receptors or effectors (Table 15-1). One or more of these components may be found in most cranial nerves: SE in III, IV, VI, and XII; GVE

(autonomic fibers) in III, VII, IX, X, and XI; GSA (other than proprioceptive) chiefly in V; and GVA in VII, IX, and X.

In addition to the foregoing types of **general** fiber components, certain cranial nerves contain **special** components (Table 15-2), of which there are three types: special somatic afferent fibers from special somatic sense organs (retina, semicircular canals, cochlea, and neuromast organs) and found universally in nerves II and VIII and in VII, IX, and X when these supply neuromast organs; special visceral afferent fibers from special visceral sense organs (olfactory epithelium and taste buds) and found almost universally in I, VII, IX, and X; and special visceral efferent fibers sup-

Table 15-2. Special fiber components of cranial nerves*

Components	Innervation and nerve
Special somatic afferent fibers (SSA)	Special somatic receptors Retina (II) Membranous labyrinth (VIII) Neuromast organs (VII, IX, and X)
Special visceral afferent fibers (SVA)	Special visceral receptors Olfactory epithelium (I) Taste buds (VII, IX, and X)
Special visceral efferent fibers (SVE)	Branchiomeric muscle (V, VII, IX, X, and XI)

*In addition to these special components, cranial nerves other than I, II, and VIII have one or more of the general components listed in Table 15-1, p. 352.

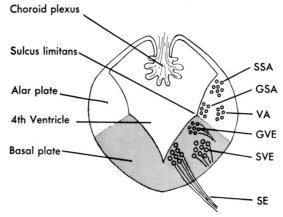

Fig. 15-24. Cross section of medulla. Sensory nuclei develop in the alar plate; motor nuclei, in the basal plate. The two plates are delimited by the sulcus limitans. A key to the fiber components is in Tables 15-1 and 15-2. **VA,** sensory nucleus for GVA and SVA fibers.

plying branchiomeric muscles and found universally in V, VII, IX, X, and XI.

Each of the foregoing nerve fibers (except those of nerves I and II) commences or terminates in the midbrain, hindbrain, or cord in a column of gray matter preempted by that specific component (Fig. 15-24).

AUTONOMIC NERVOUS SYSTEM
(Figs. 15-23 and 15-25)

The autonomic nervous system is that part of the nervous system that innervates glands and smooth and cardiac muscle. It consists chiefly of autonomic nerves, plexuses, and ganglia. It does not, however, constitute an anatomical entity; that is, it cannot be completely dissected away from the rest of the nervous system, since its components commence inside the central nervous system and emerge via cranial or spinal nerves. Traditionally, the system has been defined as a visceral motor system. However, sensory fibers from the organs innervated provide input to the central nervous system, and these fibers are sometimes included in the autonomic system.

Two motor neurons in series conduct the

impulse from the brain or cord to the typical effector. The first neuron of the chain has its cell body in the central nervous system, and the preganglionic fiber terminates in an autonomic ganglion. The second, or postganglionic, neuron has its cell body in the ganglion. Its fiber extends to the effector. Fibers from a ganglion to organs other than the skin usually aggregate to form numerous nerve strands, which accompany blood vessels to the organ. The celiac plexus consists of numerous strands of fibers radiating from the celiac ganglion.

The autonomic system is composed of the **thoracolumbar (sympathetic) system** emerging from the cord via thoracic and lumbar nerves and the **craniosacral (parasympathetic) system** emerging from the brain via cranial nerves III, VII, IX, X, and XI and from the cord via several sacral spinal nerves. Most visceral structures, except those in the skin, are supplied by postganglionic fibers from both these systems. The two systems act synergistically. The acceleratory or dilatory effects of one system counterbalance the inhibitory or constrictory effects of the other to bring about an appropriate state.

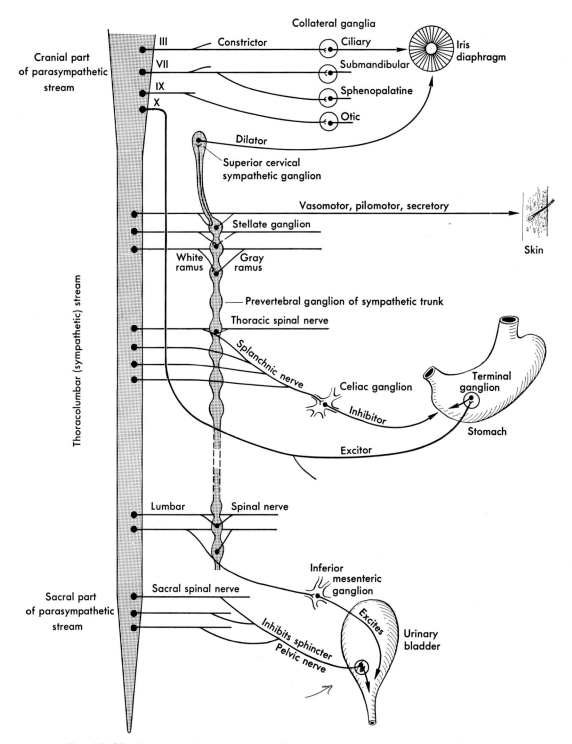

Fig. 15-25. Representative components of the autonomic nervous system of a mammal. Innervation of iris diaphragm, skin, stomach, and urinary bladder. Arrows emphasize dual control exerted elsewhere than in the skin by craniosacral and thoracolumbar streams. Preganglionic fibers are those with a cell body (black dot) in the central nervous system. Postganglionic fibers are those with a cell body in a ganglion.

The ganglia of the autonomic nervous system occur in three groups. **Prevertebral ganglia** lie close to the vertebral column and, except in some fishes, are connected to form a longitudinal sympathetic trunk. There is usually one prevertebral ganglion for each thoracic and lumbar nerve. Several ganglia also occur in the cervical region. Most of the prevertebral ganglia (except the cervical ganglia) are connected to the nearest thoracic or lumbar spinal nerve by a **white ramus communicans,** which conducts preganglionic fibers from the spinal nerve into the ganglion. The prevertebral ganglia contain the cell bodies of postganglionic fibers to the skin, head, and heart. A **gray ramus communicans,** absent in elasmobranchs, returns some of the postganglionic fibers to the spinal nerve to be distributed to the skin.

Collateral ganglia in the head are associated with cranial nerves III (**ciliary ganglion**), VII (**sphenopalatine** and **subman-** dibular **ganglia**), and IX (**otic ganglion**)— all supplying visceral organs within the head. Collateral ganglia in the trunk supply organs within the coelomic cavity. The **celiac** is a collateral ganglion of the trunk. It receives preganglionic fibers from the spinal cord via ventral roots, white rami communicantes, and the greater splanchnic nerve.

Terminal ganglia are embedded within the walls of the organ to be innervated. They occur at the preganglionic endings of the parasympathetic system. The cell bodies in these ganglia send short postganglionic fibers to the tissue to be innervated.

In all vertebrates the autonomic nervous system is primarily under involuntary control; that is, motor impulses are reflexly initiated by visceral afferent impulses. We are seldom aware of either the stimulus or the response.

Chapter summary

1. The neuron is the structural unit of the nervous system. A nucleus is a group of cell bodies within the central nervous system. Nuclei constitute the gray matter of the brain and cord. A ganglion is a group of cell bodies outside the central nervous system. The long process (axon) of a neuron is a nerve fiber. A bundle of nerve fibers inside the central nervous system is a fiber tract. Fiber tracts constitute the white matter of the brain and cord. A bundle of nerve fibers outside the central nervous system is a nerve.

2. The cell bodies of motor neurons occur within the brain or cord with one exception: the cell bodies of postganglionic fibers of the autonomic system occur in autonomic ganglia. Sensory cell bodies of higher vertebrates occur in sensory ganglia with three exceptions: the cell bodies of optic nerve fibers are in the retina; those of the olfactory nerve fibers are in the olfactory epithe- lium; and those of proprioceptive fibers of cranial nerves are in a nucleus within the brain.

3. Most spinal nerves exhibit sensory ganglia on the dorsal roots. The following cranial nerves exhibit sensory ganglia: V (semilunar), VII (geniculate), VIII (cochlear and vestibular), IX (petrosal; in mammals, inferior and superior), and X (lateral and jugular or, in mammals, inferior and superior).

4. The autonomic ganglia of the head and their associated nerves are ciliary (III), sphenopalatine (VII), submandibular (VII), and otic (IX).

5. Spinal nerves are metameric in origin and distribution. Most spinal nerves exhibit a dorsal and ventral root and three rami: dorsal, ventral, and white ramus communicans. Ventral rami unite to form simple or complicated plexuses.

6. There is considerable evidence that in the

earliest vertebrates the dorsal and ventral roots of spinal nerves did not unite, that dorsal roots were mixed and ventral roots were motor, and that the cell bodies of the sensory fibers were not aggregated in ganglia. These conditions are found in lampreys, except that some sensory cell bodies aggregate on the dorsal root. Above cyclostomes, dorsal and ventral roots always unite, ventral roots are motor, and dorsal roots are chiefly or wholly sensory.

7. Spinal nerves contain the following fiber components: GSA, SE, GVA, and GVE. Cranial nerves may contain one or more of the preceding and also one or more of the following components: SSA, SVA, and SVE.

8. Occipitospinal nerves lacking sensory roots and supplying the hypobranchial musculature arise between the vagus and the first typical spinal nerves. They are more numerous in lower vertebrates and are represented in part in amniotes by the cranial roots of nerves XI and XII.

9. Anamniotes have ten pairs of cranial nerves. Amniotes have twelve pairs. Nerves I, II, and VIII are purely sensory and supply special receptors of the head. Nerves III, IV, and VI supply the myotomal muscles of the eyeball. Nerves V, VII, IX, and X are branchiomeric and supply the jaws and gill arches in fishes and gill arch derivatives in tetrapods.

10. Cranial nerve V is the chief nerve for cutaneous sensation on the surface of the head and the anterior part of the oral cavity. Nerves VII, IX, and X supply neuromast organs and taste buds.

11. Nerve XI is derived phylogenetically, partly from X and partly from occipitospinal nerves. The internal ramus contains GVE fibers and is distributed with the vagus. The external ramus supplies the trapezius and sternocleidomastoid muscles.

12. Nerve XII represents one or more occipitospinal nerves and supplies the myotomal muscle of the tongue. Its ramus descendens, originating from the cord, supplies hypobranchial muscle of the neck.

13. A terminal nerve, accompanying nerve I, is sensory to the olfactory epithelium and motor to the smooth muscles of the blood vessels of the nasal mucosa.

14. The autonomic nervous system innervates smooth and cardiac muscles and glands. Preganglionic fibers of the craniosacral (parasympathetic) division emerge from the brain via cranial nerves III, VII, IX, X, and XI and from the sacral region of the cord via sacral spinal nerves. Preganglionic fibers of the thoracolumbar (sympathetic) division emerge from the cord via thoracic and lumbar spinal nerves.

15. Preganglionic fibers terminate in the prevertebral, collateral, or terminal ganglia of the head and trunk, from which postganglionic fibers pass to the visceral organs innervated. All such organs, except those of the skin, are supplied by fibers from both divisions. Visceral organs of the skin receive only sympathetic fibers that utilize gray rami communicantes and the cutaneous branches of spinal nerves.

16. A meninx primitiva occurs in some fishes. A dura mater and a leptomeninx develop in most vertebrates. In a few birds and in mammals the leptomeninx further differentiates into pia mater and arachnoid membranes.

17. The spinal cord often exhibits cervical and lumbar enlargements and, in fishes, a urohypophysis. When the cord is shorter than the vertebral column, the cord terminates in a filum terminale surrounded by a cauda equina.

18. Cerebrospinal fluid is secreted by choroid plexuses in the lateral, third, and fourth ventricles. The fluid fills the brain ventricles and the central canal of the spinal cord. It escapes to the meningeal spaces via foramina in the roof of the fourth ventricle.

19. The brain has three major subdivisions: prosencephalon (forebrain), mesencephalon (midbrain), and rhombencephalon (hindbrain). The forebrain consists of telencephalon and diencephalon. The hindbrain consists of metencephalon and myelencephalon. The more prominent structures associated with these parts are outlined on the facing page. Some occur only in lower or higher forms.

Telencephalon
  Rhinencephalon
    Olfactory bulbs
    Olfactory tracts
    Olfactory lobes
    Paleocortex on pallium
  Cerebral hemispheres
    Corpora striata or basal ganglia
    Corpus callosum
    Neocortex on pallium
    Paraphysis
  Lateral ventricles

Diencephalon
  Epithalamus
    Habenulae
    Pineal
    Parapineal (stirnorgan)
  Thalamus
  Hypothalamus
    Hypothalamic nuclei
    Optic chiasma
    Median eminence
    Infundibular stalk
    Pituitary
    Saccus vasculosus
    Mammillary bodies
  Third ventricle

Mesencephalon
  Optic lobes ⎫
          ⎬ Tectum
  Auditory lobes ⎭
  Cerebral peduncles
  Cerebral aqueduct

Metencephalon
  Cerebellum
  Trapezoid body
  Pons

Myelencephalon
  Medulla oblongata
  Restiform bodies
  Pyramids
  Fourth ventricle

Sense organs, or receptors, are transducers of energy. They transduce, or change, mechanical, electrical, thermal, chemical, or radiant energy into nerve impulses in an afferent nerve fiber. The energy constitutes a stimulus. Receptors like the cones of the eye are competent to respond to one kind of stimulus. Others, especially those scattered widely in the body, appear to respond to a broader spectrum of stimuli. **General receptors** are sense organs of wide distribution upon or within the body. **Special receptors** are those that are concentrated in small areas, particularly on the cephalic end.

Receptors that respond to environmental stimuli are **exteroceptors.** They inform the organism concerning the presence of an enemy, of food, or of a mate. One external stimulus, light, also regulates daily and seasonal cyclic changes in the central nervous system and endocrine organs. Receptors that respond to stimuli arising in the muscles, joints, and tendons and within the semicircular canals and utriculus are **proprioceptors.** These inform the central nervous system concerning the activity of the muscles, and provide input concerning the changing position of the animal in space. Receptors that respond to alterations in the internal environment are **interoceptors.** Interoceptors are also called **visceral receptors,** whereas exteroceptors and proprioceptors are **somatic receptors.**

The olfactory epithelium and the taste buds have a dual status. On the basis of the environmental source of their stimuli they are exteroceptive, but on the basis of the site of termination of their sensory pathways within the brain, and because they reflexly elicit visceral as well as somatic motor activity, they are classified as visceral receptors.

In this chapter somatic receptors will be introduced first. After these the visceral receptors will be discussed. In most instances we will consider briefly the morphology, morphogenesis, evolution, and biological roles of these organs, although not necessarily in that order.

SOMATIC RECEPTORS
General cutaneous receptors

The skin of all vertebrates contains free, (that is, naked) sensory endings that ramify among the epidermal cells everywhere on the surface, including the gills of fishes, and are stimulated by contact (Fig. 16-1). These are probably the oldest general cutaneous endings in vertebrates. In cyclostomes they are the only general endings in the skin. They give rise to what has been called protopathic sensation. This is a crude, poorly localized, probably phylogenetically ancient sensation. It is purely protective. It is not necessary that a fish know the texture of whatever touches him or whether the temperature of the object is 21° C. vs. 23° C. The mere fact of contact is sufficient to alert the brain to possible danger. The impulses ascend to the thalamus, where reflex motor activity is initiated to reorient the animal to avoid the stimulus.

In mammals free nerve endings entwine around the base of each hair. Touch a single hair on your arm and note the sensation. Localization of the stimulus is a function of the cerebral cortex.

In addition to free endings ramifying in the epidermis, tetrapods have acquired encapsulated bulb-like cutaneous endings in the dermis (Fig. 16-1). They consist of endings associated with epithelial-like cells and surrounded by a thin or thick connective tissue capsule. They may have evolved from free endings by the addition of capsular cells and by withdrawal into the dermal papillae or even deeper. They seem to have evolved along with the cerebral hemispheres and are associated with fine touch or temperature discrimination (epicritic

sensation). Birds and mammals have the largest number and variety of such corpuscles, including those on the beak of birds, at the end of the snout in mammals that rout in the soil, and on the external genitalia.

Neuromast organs

Cyclostomes, jawed fishes, larval amphibians, and adult aquatic amphibians such as *Necturus* and the underwater toad *Xenopus* have groups and linear series of epithelial receptors known as neuromasts. The lateral-line canal and cephalic canals of fishes and amphibians, ampullae such as those of Lorenzini, the closed vesicles of Savi found in electric rays, and the pit organs of fishes are variants or aggregates of neuromasts (Fig. 16-2, A). Although neuromasts in different locations on the animal show minor structural and, perhaps, functional variety, all are fluid-filled pits or ampullae. The least differentiated neuromasts (external, or naked, neuromasts) lie in the epidermis

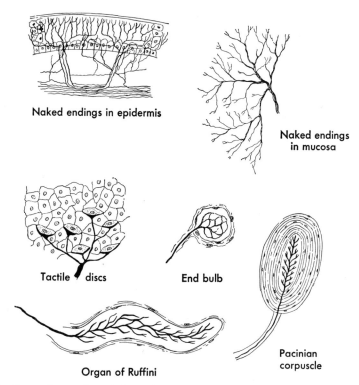

Naked endings in epidermis

Naked endings in mucosa

Tactile discs

End bulb

Organ of Ruffini

Pacinian corpuscle

Fig. 16-1. General somatic and visceral receptors.

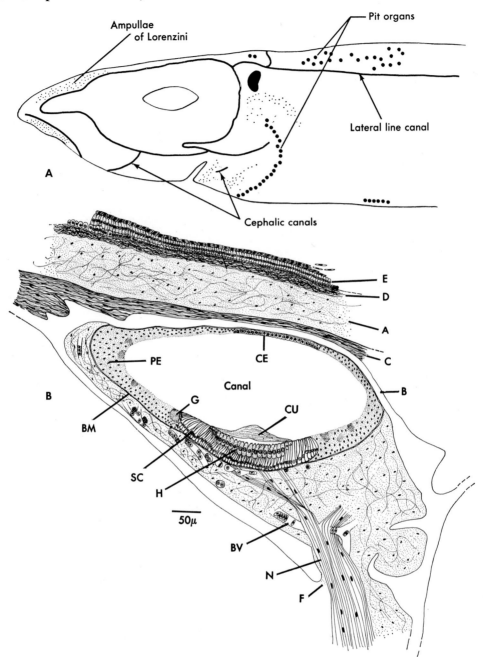

Fig. 16-2. A, Distribution of neuromast organs in skin of a shark. The precise distribution of the several types varies among the species. **B,** Lateral-line canal of bony fish in cross section at the level of a neuromast organ. The canal runs longitudinally under the skin embedded in dermal bone (B). E, epidermis; A, C, and D, dermis. The epithelial lining of the canal rests on a basement membrane (BM) and exhibits a cuboidal epithelium (CE), a pseudostratified epithelium (PE), goblet cells (G) and the neuromast organ. The latter is composed of sustentacular cells (SC), sensory cells (H), and a cupula (CU). The sensory nerve (N) innervating the neuromast organ penetrates the bone via a foramen (F). A blood vessel (BV) is shown approaching the receptor. (From Branson and Moore: Copeia, no. 1, p. 1, 1962.)

and open on the surface via pores (Fig. 16-3). Others lie under the skin in the wall of sunken canals, or even embedded in dermal bone (Fig. 16-2, *B*). The fish *Amia* has as many as 3,700 pores on the head alone!

The epithelium of a neuromast exhibits two cell types, a **sensory cell** and a **supportive (sustentacular) cell.** Bundles of sensory nerve fibers terminate in intimate association with the basal part of the sensory cells. At their apex the sensory cells exhibit one or more cilia, usually including a long one, which project into the neuromast chamber. Often, the cilia are embedded within a column of mucoid material, the **cupula,** secreted by the supporting cells. Movements of the fluid in the pit or canal displace the cupula and cilia. Displacement apparently releases a neurotransmitting agent, which activates sensory impulses in the nerve endings.

One of the most widely studied components of the neuromast system is the lateral-line canal system and its counterpart in the head, the cephalic canal system. In cyclostomes and amphibians the canals, consisting of linear series of neuromasts, are shallow grooves on the surface of the head and trunk. In dogfish sharks the canals are embedded in the trunk nearly to the caudal fin, where they emerge and continue caudad as open grooves. In *Chimaera*, another cartilaginous fish, the canals form elevated ridges clearly visible on the surface. In bony fishes the canals lie on the surface of, or are even embedded in, scales, which in some species are bony. Grooves in the dermal armor of the earliest vertebrates give evidence that the canal system is very ancient, and the arrangement of the grooves indicates that the system was once a network of superficial canals that covered the entire body. The network has been reduced in modern fishes by loss of many side branches and by interruption of the canals at one or more locations. In some modern fishes the canals are now restricted to the head.

In the salamander *Notophthalmus*, when the larva is metamorphosing into a red eft and migrating to land, the lateral-line system becomes buried under the proliferating stratum corneum. Later—several years later in some localities—when the eft returns to

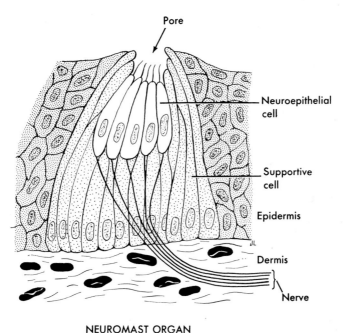

NEUROMAST ORGAN

Fig. 16-3. External (naked) neuromast in the epidermis of a teleost.

the water as a sexually mature newt, the stratum corneum is shed and the lateral-line is again exposed to the aquatic environment. Amphioxus has no lateral-line system.

The function of the neuromast system, including the ampullary types of neuromasts, has been the object of much early speculation and, subsequently, of extensive sophisticated research.[104] Two basic questions have been posed: What is the nature of the stimuli that evoke a response in the sensory cell? And what role does the information play in the behavior of the organism?

There is considerable agreement that at least parts of the neuromast system respond to water currents and to weak electric potentials. Water currents result from water flow past the organism, from disturbances in the water by passing prey, or from the locomotion of the organism in the dense watery medium. The information would enable an organism to orient itself appropriately in flowing water, to avoid enemies, and to participate in schooling, or it would provide feedback during locomotion similar to that from proprioceptors. The possibility that the system may have a kinesthetic function because of its location within the body wall cannot presently be ruled out.

With respect to electroreception, there is strong evidence that most animals produce enough electrical potentials when their muscles are contracting to make their presence detectable via neuromasts of sharks and rays at short range. Electroreception has been almost incontrovertibly demonstrated in the ampullae of Lorenzini. Salinity changes could also be a source of electric potential changes.

It has also been demonstrated experimentally that the canal system responds to compression waves in water. Sound waves are just such vibrations. However, investigators are hesitant to ascribe sound detection to the system because of the semantic problem of a definition of "sound"; and there is a possibility that waves of a length usually associated with sound may not be those that typically activate the sensory cells.

It is possible that the function of the system is a broad one and that its predominant biological role varies with habitat—swiftly flowing streams as opposed to the depths of the ocean, for example. Although the precise role of the system remains to be demonstrated, there is no doubt that the neuromast system, along with chemoreceptors and visual organs, constitutes a very important site of input of information into the central nervous systems of aquatic animals.

The neuromast organs arise from embryonic ectodermal placodes, which subsequently sink into or under the skin. All are innervated by cranial nerves VII, IX, or X. As will be seen later, the membranous labyrinth appears to be a highly modified canal system of neuromasts. For this reason, the neuromast organs and the membranous labyrinth are often referred to collectively as the acousticolateralis system.

Pit receptors in reptiles

Snakes and lizards have receptors in the form of pits that open to the surface between epidermal scales. These organs are of several varieties, the most common being **apical pits.** These are scattered over the surface of the body, especially on the trunk. As their name denotes, they lie at the apex (posterior free border) of the scales. There are usually one or two apical pits associated with each scale, but there may be as many as seven. In many instances a filamentous hair-like bristle projects from the pit. Since naked and encapsulated nerve endings in reptiles are buried under dense cornified scales, apical pits provide sites for the input of stimuli, probably tactile. Similar receptors (proto-triches; sing., protothrix; *thrix, tricho =* hair) have been described in some fishes and tailed amphibians.

A single pair of more specialized pit receptors is found on the head of crotalid snakes (snakes in the family Crotalidae) and on boas and their relatives. These specific receptors are called **pit organs.** In crotalid snakes the pits are directed forward, and may be several millimeters wide and twice as deep. Hence they are easily seen. For this reason, crotalids are called pit vipers. The pits of pit vipers are **loreal pits,** since they are located at the posterior end of the loreal scale (Fig. 5-20). (The loreal scale is found in the lore of reptiles, the location between the external naris and the eye). The pits in boas are slit-like and less obvious. Since they are associated with a labial scale, they are called **labial pits.**

Loreal and labial pits are thermal receptors that respond to radiant heat. Those in pit vipers are considerably more sensitive. Physiological and behavioral studies have shown the loreal pits to detect temperature changes of 0.001° C. at a distance of several feet. Therefore, they can detect the presence of objects, including warm-blooded animals on which they prey, if the object is only slightly warmer than the environment. Because of this sensitivity, they can locate and strike accurately at objects in the dark.

The pit receptors of reptiles bear a striking morphological resemblance to the external neuromasts of fishes. It is therefore an attractive hypothesis that pit receptors are evolutionary derivatives of neuromasts. On the other hand, their innervation by spinal nerves or by the fifth cranial nerve is not consistent with this hypothesis. Conceivably, they could represent instances of convergent evolution. Certainly, their relationship to similar receptors in other vertebrates is not clear.

Membranous labyrinth

All vertebrates exhibit a pair of membranous labyrinths embedded in the neurocranium lateral to the hindbrain. In fishes other than cyclostomes, each labyrinth consists of three **semicircular canals,** a **utriculus,** and a **sacculus** (Fig. 16-4). Tetrapods have, in addition, a cochlea. The anterior and posterior canals are vertical and lie at right angles to one another. The third canal lies in a horizontal plane. Each canal exhibits a small ampulla. Emerging dorsally from the sacculus or utriculus is an endolymphatic duct which usually terminates blindly in a small or large **endolymphatic sac** (Fig. 16-4, man). However, in elasmobranchs the endolymphatic ducts open onto the surface of the head via an endolymphatic pore. Lampreys have only the two vertical semicircular canals, and hagfishes have only the posterior vertical, although there are two ampullae. Membranous labyrinth is another name for **inner ear.**

The membranous labyrinth arises during embryonic life as an ectodermal placode directly in line with the placodes of the neuromast system when the latter are present. The placode, like the deep neuromasts, sinks under the skin to become a fluid-filled vesicle, or **otocyst** (Fig. 16-5). However, the otocyst develops into an organ of much greater complexity than the neuromasts. The mesenchyme surrounding the developing labyrinth gives rise to an

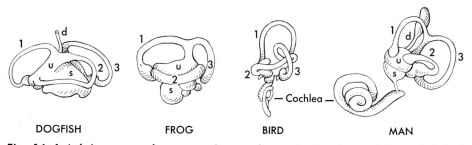

Fig. 16-4. Left inner ears of representative vertebrates. **1,** Anterior vertical canal; **2,** horizontal canal; **3,** posterior vertical canal; **d,** endolymphatic duct; **s,** sacculus; **u,** utriculus.

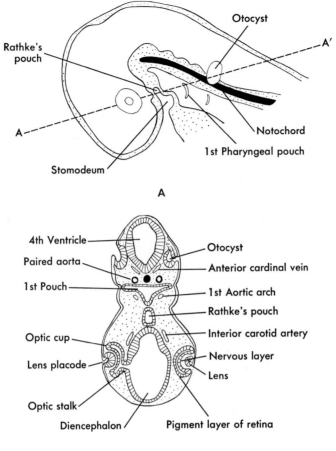

Fig. 16-5. Origin of inner ear (from otocyst), retina (from optic cup), and lens. **A,** Head region of embryo. **B,** Cross section of head at level of A to A'. Left side of **B** is slightly earlier than right side.

otic capsule. When development is complete, therefore, the membranous labyrinth occupies a similarly shaped skeletal labyrinth. The space between the membranous labyrinth and the skeleton is filled with perilymph, contributed partly by the cerebrospinal fluid that seeps in. The fluid within the labyrinth is endolymph. Within the endolymph are calcareous concretions, or **otoliths** (calcium carbonate in association with a protein in man). In some species there is a single large otolith that almost fills the chamber; in other species the otoliths are microscopic and very abundant, as in the dogfish shark. The size and shape of the otoliths (also called otoconia) vary with the species.

Elevated patches of sensory epithelia (cristae) occur in the ampullae, and others (maculae) occur in the sacculus and utriculus. The macula in the sacculus lies in the floor, whereas that in the utriculus lies in the wall in essentially a vertical plane. A depression in the floor of the sacculus, the **lagena,** contains another macula, which, in tetrapods, expands to become the cochlea, an organ for detecting sound. These receptor sites exhibit sensory cells, usually called **hair cells,** and **supportive cells.** Extending into the endolymph from the hair cells are cilia. Branching among the bases of the hair cells or ending in tiny "buttons" upon them are the sensory endings of cranial nerve VIII.

In all details the membranous labyrinth resembles a specialized neuromast organ. However, since a labyrinth occurred in the oldest vertebrates along with a neuromast system, the origin of the membranous labyrinth must remain conjectural.

Equilibratory function of the labyrinth. The function of the membranous labyrinth is to receive information concerning acceleration and deceleration of the head in motion, as in tilting the head or turning it (angular rotation). The endolymph acts as a plastic body and presses upon specific receptor sites. The information is transmitted to the brainstem and cerebellum, where reflex movements of the eyeballs are initiated so that the eyes are always looking in the same direction in which the head is directed. (It takes conscious effort to turn your head swiftly to the left while continuing to look to the right.) That the labyrinth controls eyeball movements can be demonstrated by spinning someone on a revolving chair rapidly for a minute or two and then stopping the chair abruptly. An observer will note that the eyeballs continue to exhibit rapid jerky side-to-front-to-side movements, which shortly cease.

Since the head is attached to the body—rigidly in fish and amphibia—information from the labyrinth also relates to the orientation of the trunk. Reflex motor impulses from the cerebellum result in contraction of the striated muscles of the neck, trunk, tail, and appendages appropriate for the welfare of the organism at any instant. The labyrinth therefore plays a role with the eyes and general proprioceptive receptors in reflexly maintaining bodily posture and, therefore, balance or equilibrium.

Auditory function of the labyrinth. The labyrinth has an auditory function as far down the phylogenetic scale as fishes. This function in fishes is subserved by a depression, the **lagena,** in the floor of the sacculus. The lagena has its own macula, which, at least in some fish, responds to longitudinal sinusoidal waves of the frequency of sound waves.

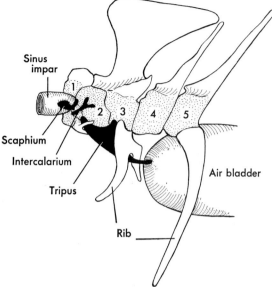

Fig. 16-6. Weberian ossicles (black) of a teleost. **1** to **5,** Centra of first five vertebrae. The sinus impar is an extension of the perilymphatic space.

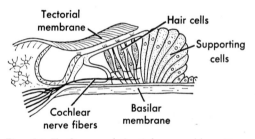

Fig. 16-7. Organ of Corti from cochlea. (From Tuttle and Schottelius: Textbook of physiology, ed. 15, St. Louis, 1965, The C. V. Mosby Co.)

In Cypriniformes, an order of teleosts that includes catfish, carp, and suckers, sound waves in the water evoke waves of similar frequency in the gas in the air bladder and these are transmitted by a series of **weberian ossicles** (Fig. 16-6) to the perilymphatic space and thence to the endolymph and lagena. Weberian ossicles are modified transverse processes of the first four (occasionally five) trunk vertebrae. The ossicles commence at the air bladder and end against the **sinus impar,** an extension of the perilymphatic space. In herring-like fish the air bladder has tubular extensions in direct contact with the sinus impar. It seems appropriate to say that these fish can "hear."

The lagena of ancestral fishes evolved in tetrapods into a spiral duct, the **cochlea.** With evolution of the cochlea, the number of sensory cells increased, and the sensory epithelium became the **organ of Corti** (Fig. 16-7). The branch of the eighth nerve that supplied the lagena became known as the cochlear nerve. The cochlea and its nerve, therefore, are modifications of structures in ancestral fishes, and are not completely new.

Middle ear of tetrapods. The most common route of conduction of sound waves to the labyrinth in tetrapods is from the eardrum (tympanic membrane) via the dorsal segment of the hyoid cartilage, variously known as the hyomandibula, columella, or stapes.* This cartilage or bone is located in the middle ear cavity (**cavum tympanum**). The middle ear cavity arises as an evagination of the first pharyngeal pouch, which grows toward the hyomandibula and commences to surround it. Independent erosion of the mesenchyme surrounding the hyomandibula also occurs. As a result, the hyomandibula, columella, or stapes becomes isolated in the middle ear cavity. In mammals the posterior tips of the embryonic upper and lower jaw cartilages are also encompassed by the expanding middle ear. Mammals, therefore, have three bones in the middle ear cavity—the malleus, incus, and stapes (Fig. 16-8). The columella or stapes in all vertebrates is attached to a **secondary tympanic membrane** stretched across the oval window in the wall of the otic capsule. Vibration of this membrane is transmitted to the perilymph and from there to the endolymph. The middle ear cavity remains in communication with the pharynx throughout life via the **eustachian tube.**

Although conduction of sound via a drum is the predominant route in tetra-

_____
*Although the terms are used here as synonyms, the columella and stapes may be derived in part from elements in addition to the hyomandibula.

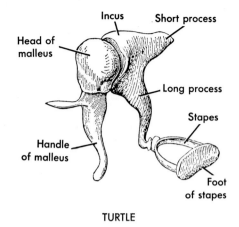

TURTLE

Fig. 16-8. Middle ear ossicles of a mammal. (From Tuttle and Schottelius: Textbook of physiology, ed. 15, St. Louis, 1965, The C. V. Mosby Co.)

pods, the drum is not always essential. Urodeles, apodans, a few anurans, and some lizards have no eardrum, no middle ear cavity, and even the columella is vestigial in urodeles. (Of course, aquatic urodeles have a lateral-line system, which perhaps compensates in part for lack of the auditory structures.) Since in the absence of a drum the columella may commence at the squamosal bone (urodeles), or at the quadrate (snakes, some lizards, and some extinct reptiles), sound conduction via bone in these organisms is enhanced if the floor of the buccal cavity, the lower jaw, or the anterior limbs are in contact with a dense substrate—water or earth. Certainly, the hyomandibular route to the membranous labyrinth was available long before fishes emerged onto land, since the hyomandibula articulates with the otic capsule as a primitive condition (Fig. 8-1).

Outer ear of tetrapods. The tympanic membrane is situated on the surface of the head in frogs and toads. In reptiles, birds, and mammals it is deeper in the head, situated at the end of an air-filled passageway, the **outer ear canal (external auditory meatus).** In mammals an appendage, the **pinna,** collects sound waves and directs them into the outer ear canal. The terms "outer ear" and "outer ear canal" should be considered synonyms.

Saccus vasculosus

Elasmobranch fishes, ganoids, and teleosts exhibit a saccular evagination of the thin floor of the diencephalon immediately behind the posterior lobe of the pituitary gland. Because the thin walls of the sac are highly vascular, the organ is known as the saccus vasculosus (infundibular organ). The epithelium lining the sac exhibits sustentacular cells and sensory cells, with cilia projecting into the cerebrospinal fluid within the sac. Passing from the sensory cells are nerve fibers that terminate in the hypothalamus and in other brain centers.

Evidently this is a sense organ, but what function does it perform? At present we have no experimental data. It has been assumed that the organ is stimulated by movements of the cerebrospinal fluid. It may be a depth detector, since the sac is best developed in deep-sea fish, less developed in freshwater forms, and least so in fishes that inhabit shallow water. No true saccus vasculosus is found in this location in cyclostomes, in lungfishes, or in any form above fishes, although the sac-like posterior lobe of urodeles has been called by this name. The term "saccus vasculosus" has been applied to a totally different structure in the amphioxus.

Light receptors

Light receptors, or photoreceptors, are sensitive to radiation in a narrow spectrum of wavelengths. The limits of the spectrum vary among species. Vertebrates have two sets of photoreceptors—the lateral (paired) eyes and the median (unpaired) eyes.

Lateral eyes. The receptor site of the lateral eye is the **retina,** a membrane rich in nervous tissue and synapses at the rear of a fluid-filled vitreous chamber (Fig. 16-9) of the eyeball. The retina arises as a bulbous evagination from the embryonic forebrain, which soon invaginates to become a double-walled **optic cup** (Fig. 16-5). The cup retains an attachment to the brain via the optic stalk. The layer of the optic cup that light first strikes becomes the nervous layer of the retina. Some of the cells differentiate to become rods and cones. Others give rise to bipolar cells, and still others to optic nerve cell bodies (Fig. 16-10). Optic nerve cell bodies (called ganglionic cells by the histologist) sprout long processes that grow along the optic stalk and into the brain. These processes constitute the optic nerve. From the description it can be seen that the retina arises from the brain and never becomes completely detached from it.

The embryonic mesenchyme surrounding the optic cup forms a pigmented **choroid coat** and a fibrous **sclerotic coat** (Fig. 16-9). The choroid coat is perforated anteriorly by a circular or slit-like aperture, the **pupil.** The part of the choroid coat surrounding the pupil is the **iris diaphragm.**

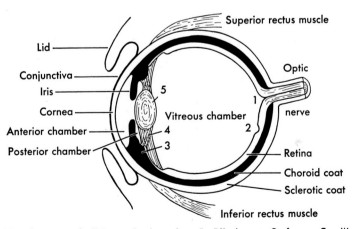

Fig. 16-9. Vertebrate eyeball in sagittal section. **1,** Blind spot; **2,** fovea; **3,** ciliary muscle; **4,** suspensory ligament; **5,** lens.

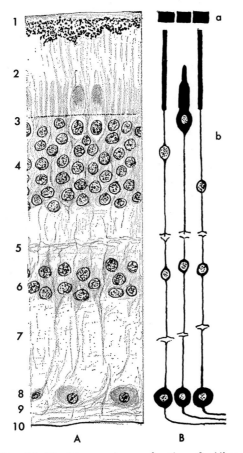

Fig. 16-10. Nervous layer of retina. **A,** Histological appearance; **B,** Diagram showing areas of synapse. The light enters the retina at layer **10** (base of picture) and passes through the other layers to the rods and cones. **1** and **a,** Pigmented epithelium; **2,** layer of rods and cones; **3,** external limiting membrane; **4,** nuclei of the rods and cones; **5,** outer molecular layer; **6,** layer of cell bodies of bipolar association fibers; **7,** inner molecular layer; **8,** layer of ganglion cells (cell bodies of optic nerve fibers); **9,** layer of optic nerve fibers; **10,** internal limiting membrane; **b,** rods and cones. (From Bevelander: Essentials of histology, ed. 5, St. Louis, 1965, The C. V. Mosby Co.)

Embedded in the diaphragm are two sets of intrinsic eyeball muscles (smooth muscles, except in reptiles and birds). Dilator muscles are radially arranged and increase the diameter of the pupil. Constrictors are arranged in circular fashion and by con-

tracting reduce the diameter of the pupil. Around the periphery of the iris diaphragm, except in fishes, is a ring of ciliary muscle known collectively as the **ciliary body.** Suspensory ligaments connect the ciliary body with the lens. Contraction or relaxation of the ciliary muscles alters the curvature of the lens in land-dwellers to accommodate for near or far vision. Although fishes lack ciliary muscles and therefore cannot alter the curvature of the lens, a special campanula in teleosts displaces the lens for accommodation for distance. The constrictors, dilators, and ciliary muscles are under control of the autonomic nervous system.

The sclerotic coat of the eyeball can be seen as the white of the eye. In front of the pupil the sclerotic coat is transparent and constitutes the **cornea.** Looking through the cornea, one sees the pigmented iris diaphragm around the pupil. The pupil appears black because there is no light emerging from the eyeball. Inserting on the sclerotic coat are the extrinsic muscles (rectus and oblique) of the eyeball. Piercing the sclerotic coat are blood vessels and nerves. Although the sclerotic coat is usually fibrous, it may be partly or wholly cartilaginous in fishes, and in birds and reptiles (except snakes and lizards) bony sclerotic plates develop within it (Fig. 6-3).

The lens arises as a thickened placode of surface ectoderm that sinks into position in the optic cup. It is induced to form by inductor substances released by the embryonic retina. That the retina serves as the inductor may be demonstrated by exchanging undifferentiated ectoderm from the thigh of an amphibian embryo with that at the site where a lens will form. A lens will then be induced in the tissue transplanted to the head but not in the potential lens ectoderm that was removed from the influence of the retina. Removal of the retina will result in no lens placode.

The chamber behind the lens (vitreous chamber) becomes filled with a jelly-like viscous refracting substance, the **vitreous humor.** Extending into the vitreous chamber of birds from a position near the entrance of the optic nerve is a pigmented and highly vascular fan-like plate, the pecten. Reptiles have a similar, though smaller, conical projection. What role the pecten plays is a matter of conjecture. The chambers between the lens and iris diaphragm (posterior chamber) and between the iris diaphragm and cornea (anterior chamber) are filled with **aqueous humor.**

The exposed surface of the eyeball—that portion which you can touch with your finger—is covered with transparent skin, known as the **bulbar conjunctiva.** This is directly continuous with the **palpebral conjunctiva** on the inner surface of the lids. The palpebral conjunctiva is continuous with the typical skin on the exposed surface of the lids.

Located in the orbit of most tetrapods are two epidermal glands (occasionally one) that keep the conjunctiva clean and moist. The **lacrimal gland** secretes a watery fluid. It is located behind the upper lid but is poorly developed in amphibians. The **harderian gland** is located ventrolaterally and secretes a more viscous fluid. It is lacking in some mammals, including man.

Vertebrates that live in dark recesses, such as cave salamanders, whales, moles, and certain fishes, often have degenerate eyes, or the lids may fail to open after birth. In hagfishes and even in certain mammals, the entire eyeball fails to differentiate.

Median eyes. The pineal and parapineal organs are ancient photoreceptors, which, like lateral eyes, arise as evaginations of the diencephalon. However, they arise as dorsal rather than lateral evaginations (Fig. 15-12). Median dorsal photoreceptors were present in all major groups of ancient fishes and in many ancient amphibians and reptiles. They are not rare, even today.

Both the pineal and parapineal bodies are light sensitive in lampreys. In these cyclostomes the pineal body ends in a hollow knob beneath an area of skin devoid of pigment between the paired eyes. The upper wall of the hollow knob consists of several layers of cells that form a lens. The lower wall contains sensory cells and, beneath these, ganglion cells with long processes that pass down the stalk to nerve centers in the right side of the diencephalon. Although the pineal eye is not as specialized as the lateral eyes and does not form an image, it has been shown to be sensitive to changes in light intensity. The parapineal eye of lampreys is essentially similar to the pineal eye in structure, and its descending fibers terminate in the left side of the diencephalon.

All tissues overlying the pineal body in some ganoids and some lower teleosts are translucent, and light striking the pineal can initiate responses of locomotor muscles and pigment cells. The integument overlying the pineal of at least some anuran tadpoles is likewise translucent, and electron micrographs indicate photoreceptor processes in the pineal of at least one adult tree frog.[106]

Above amphibians the pineal has the histological appearance of being chiefly glandular. Cells considered to be sensory, but not like the photoreceptor cells of the pineal of lower forms, are also present. The function of these sensory cells remains to be clarified.

The parapineal of reptiles, when present, is also overlaid by translucent tissues and serves as a third eye. It is often referred to as the **parietal eye.**

The evidence indicates that the pineal and parapineal bodies represent a pair of phylogenetically ancient dorsal eyes, and that one or the other has persisted in this capacity among numerous species until modern times, the parapineal having been less persistent. Additional discussions of the pineal and parapineal bodies will be found in Chapters 15 and 17.

Proprioceptors

Striated muscle cells, muscle tendons, and the bursas of joints are supplied with sensory endings that are stimulated when striated muscles contract. The sensation evoked by these stimuli is known as proprioception (*proprio* = self). It is also called kinesthesia and deep sensibility. Although some of the afferent impulses reach centers of consciousness—the cerebral cortex in mammals—many of them are shunted into more ancient centers, particularly the cerebellum. Here they provide feedback for the reflex modulation of motor impulses that are going to the muscles. The result of the reflex arc is coordinated muscular movements necessary for the maintenance of posture, for locomotion, for grasping, and for the performance of skilled activities of the appendages such as using tools and playing a piano. To demonstrate proprioception close your eyes, then extend your arm or leg, or "roll" your eyes. Your awareness of the change in position of the displaced part is an example of conscious proprioception.

One variety of proprioceptive ending on striated muscle is illustrated in Fig. 10-1. Pacinian corpuscles (Fig. 16-1) and other endings also serve as proprioceptors when situated in joints or tendons. Afferent fibers from muscles other than those in the head enter the spinal cord via spinal nerves, and their cell bodies are in dorsal root ganglia. Large, ascending fiber tracts in the cord (Fig. 15-5) transmit the impulses to the brain. Fibers from striated muscles in the head enter the brain via cranial nerves, and the cell bodies of these fibers are in the brain. Fishes seem to have no proprioceptors of a histological nature similar to those described above; however, the neuromast system, including the membranous labyrinth, perhaps compensates partly for their absence.

VISCERAL RECEPTORS
General visceral receptors

The mucosa of the internal tubes and organs of the body, the cardiac muscles, the smooth muscles including those of the blood vessels, and the capsules, mesenteries, and meninges of the viscera are supplied with general visceral afferent endings. These endings are mostly unencapsulated (Fig. 16-1), although pacinian corpuscles serve as visceral receptors in many visceral locations. General visceral receptors are stimulated mechanically by stretching or chemically by the presence of certain substances such as acid in the pyloric stomach. They are also stimulated by tactile and thermal stimuli in the pharyngeal and esophageal region, at least.

Most of the normal daily sensory input from general visceral receptors gives rise to no conscious sensation; stimulation results chiefly in reflex control of the smooth muscles and glands via visceral motor fibers of the autonomic nervous system. Because essentially the same monitoring of the internal environment must take place in all vertebrates whether in water or on land, general visceral receptors are not subject to so many selective pressures as are somatic receptors. Therefore, they vary relatively little from fish to man.

Cervical vascular monitors

Tetrapods have specialized chemoreceptors and pressure (*baro*) receptors intimately associated with the carotid vessels and the adult aortic arch. These receptors monitor the plasma oxygen content and the pressure of the arterial blood coming from the ventricle. Three such discrete receptors have been described in the cervical region of tetrapods. These are the **carotid body,** the **aortic body,** and the **carotid sinus.** Similar structures have not been described in fish.

The carotid body is a special chemoreceptor located close to the common carotid or internal carotid artery and receiving an arterial supply from a branch of one of these vessels. In turtles, birds, and mammals the carotid body is a discrete organ typically flattened against the wall of the carotid vessel. In lizards it is in the adventitia of the carotid artery. None has been described in snakes or crocodilians, but diffuse receptor tissue undoubtedly exists, since even in mammals the carotid body is sometimes broken into tiny nodules microscopic in size. Amphibia exhibit a structure called the carotid labyrinth, but its homologies are not known. No similar structure has been described in fishes.

The carotid body is usually an encapsulated organ consisting of an enormously rich network of sinusoidal spaces and several types of epithelioid cells richly supplied with sensory endings of the ninth cranial nerve. Some of the cells display a chromaffin reaction, but no amines have been demonstrated. The organ monitors the partial pressure of oxygen (P_{O_2}) in the blood passing through the sinusoidal spaces. Anoxia evokes a strong discharge of afferent impulses, which reflexly increase the respiratory rate. The carotid body arises from the third pharyngeal arch close to the future internal carotid artery.

The carotid sinus is a bulbous swelling of the common carotid or internal carotid artery close to the origin of the latter. The walls of the sinus exhibit localized thinning and are innervated by sensory endings of the ninth cranial nerve. The sinus serves as a baroreceptor that monitors arterial blood pressure. Lowered pressure evokes a sensory discharge that provides input to the cardiovascular regulatory center of the medulla. The carotid sinus nerve innervates both the sinus and the carotid body. In addition to fibers of the ninth nerve, it also contains vagal and sympathetic fibers.

The aortic body has been described only in mammals. It lies close to the arch of the aorta. It is histologically identical with the carotid body and has the same function, although it is innervated by the vagal nerve.

Olfactory organs

Olfactory epithelium. Olfactory organs are special visceral chemoreceptors. The epithelium lining the organ arises as a pair of ectodermal placodes (unpaired in cyclostomes), situated above the stomodeum. These placodes sink into the head to form a pair of **olfactory (nasal) pits** (Fig. 1-1). The ectodermal lining of the pits differentiates into a variety of sensory, secretory, and supportive cells. Some of the cells

sprout processes that grow toward and penetrate the olfactory bulb of the brain. These are the olfactory cells, of which there are an estimated 50 million in man. Their processes are the afferent fibers of the olfactory nerve. One theory of olfaction states that for a substance to act as an odorant, it must have a stereochemical configuration conforming to that of an initial binding substance on the surface membrane of the olfactory cell.[101]

In fishes the olfactory pits become surrounded by connective tissue and are thereafter known as **olfactory sacs.** These are blind sacs in all except lobe-finned fishes. A current of water into and out of the sacs is assured because each naris is divided into incurrent and excurrent apertures so situated that the forward motion of the fish propels a stream of water into one aperture and out the other. The mucosa containing the olfactory epithelium may exhibit many (schneiderian) folds, which increase the surface area of the epithelium. The olfactory cells monitor the water stream and are stimulated by odorants that may have their source in potential food, potential mates, or potential enemies. Primitively, olfactory stimuli resulted in reflex contraction of locomotor muscles, which propel the fish closer to, or farther from, the source of the odorant. Olfaction also reflexly stimulates the autonomic nervous system.

In tetrapods the olfactory pits push deep into the head to achieve outlets into the oral cavity or pharynx. The outlets are internal nares. The olfactory epithelium is confined to a portion of the lining of this newly established nasal passageway, so that it is appropriate in tetrapods to distinguish an olfactory epithelium and a respiratory epithelium. The olfactory epithelium contains the olfactory cell bodies just as in fish, but it monitors an airstream instead of a water stream. Odorants in the airstream dissolve on the moist olfactory epithelium and stimulate the olfactory cells. Among fishes only the Sarcopterygii have internal nares. Since Sarcopterygians are ancient fishes, internal nares are ancient structures.

Cyclostomes, unlike other vertebrates, develop a single olfactory sac with a single external naris. The olfactory sac is blind (Fig. 12-4), but a passageway may continue beyond the base of the sac.

The olfactory apparatus, including the epithelium and associated brain parts, is well developed in fishes and mammals and least developed in birds, which are therefore microsmatic; that is, they have a poor sense of smell. In some whales the olfactory nerves disappear during embryonic life. It seems logical that the ancestors of whales had a functional olfactory apparatus, but a mammal trying to inhale under water would drown! Loss of the olfactory nerve—a result of mutations—was therefore no disadvantage to the whale.

Vomeronasal (Jacobson's) organs. There has been a tendency among tetrapods, and even among some fishes, for a ventral segment of the olfactory epithelium to become more or less isolated from the nasal passageway, to the extent of becoming an independent olfactory organ in some species. The condition occurs in a rudimentary form in amphibians and reaches an evolutionary peak in lizards and snakes. The isolated olfactory area is named the vomeronasal organ because of its typical location above the vomer bone of the palate. It is also known as Jacobson's organ.

In urodeles the vomeronasal organs are simply a pair of deep grooves in the ventromedial floor of the nasal canal. (They are absent in perennibranchiates.) In anurans they are represented by blind sacs. In both instances the organs retain an opening to the nasal canal. In lizards and snakes the organs are tubular, lose their connection with the nasal canal, and achieve an opening into the anterior roof of the oral cavity. In these reptiles the two organs are moist chemoreceptive pockets into which the two tips of the forked tongue are thrust each time it darts out and back into the mouth; the organs monitor the chemicals that accumulate on the tongue each time it is thrust out.

Vomeronasal organs are vestigial or absent in turtles, crocodilians, and birds, but they have been transmitted to mammals. They are well developed in monotremes, marsupials, and in generalized insectivores, in which they are tubular structures above the false palate, retaining an opening into the nasal canal. At least some rodents also retain these structures, but in most other mammals they are vestigial and are often confined to embryos. In man the vomeronasal organs reach maximal size about the fifth month of intrauterine life and then regress.

The epithelium of the vomeronasal organ is supplied by a separate division of the olfactory nerve. It is also supplied by a branch of the fifth cranial nerve for general sensation. It is a chemoreceptor in all instances. Its relationship to the oral cavity and tongue in lizards is a reminder of the close functional relationship between olfaction and taste, a relationship that has been discussed by many authors and is aptly summarized by Noback.[111]

Organs of taste

Organs of taste (gustatory organs) occur in all vertebrates in the form of taste buds (Fig. 16-11). Each bud is a barrel-shaped fascicle of elongated taste cells and reserve taste cells (also called supportive cells) arranged in a pit around a central canal, which opens to the surface via a taste pore. Extending from each taste cell into the canal is a cilium. Surrounding the base of the cells in the bud and in contact with them are sensory nerve endings. Chemicals with an appropriate configuration in solution in the bud alter the chemistry of the taste cells, which then induce nerve impulses in the sensory endings. The impulses pass to the brain. The functional life of a taste cell is only 200 to 300 hours, or ten days, at which time they die from "wear and tear." They are replaced by reserve taste cells, which have been serving as supportive cells. New

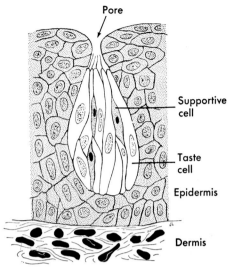

Pore

Supportive cell

Taste cell

Epidermis

Dermis

TASTE BUD

Fig. 16-11. Taste bud on the tongue of a monkey.

reserve cells are constantly proliferated from the germinal epithelium.

In fishes, taste buds are widely distributed in the roof, side walls, and floor of the oral cavity and pharynx, where they monitor the water stream to the gills. In bottom-feeders or scavengers such as catfish, carp, and suckers, taste buds are distributed over the entire surface of the body to the tip of the tail, providing these fish with additional avenues of informational input. They are abundant on the "whiskers" surrounding the mouth of catfish.

In tetrapods taste buds are less widely distributed in the body. They occur on the tongue of almost all tetrapods except birds, and they occur on the palate and in the pharynx of most tetrapods. They are most abundant in mammals, least abundant in birds. There are more taste buds in the human embryo than there are at age seven, as a result of failure to replace many of those that die during embryonic life and in infancy. There is evidence that the mammalian fetus makes some taste discriminations before birth.

Taste buds in all vertebrates are supplied by branches of cranial nerves VII, IX, and X, in that sequence, from mouth to esophagus. This arrangement is demonstrable in man, in which taste buds on the anterior surface of the tongue are supplied by nerve VII via the chorda tympani and lingual nerves, those on the posterior surface of the tongue by nerve IX, and those in the vicinity of the glottis by nerve X. Any taste buds on the external surface of the body of fishes are also supplied by nerve VII. The exaggerated size of the sensory nucleus receiving taste fibers in such fishes is illustrated in Fig. 15-8.

Chapter summary

1. Receptors are transducers of mechanical, electrical, thermal, chemical, or radiant energy. They may be classified as exteroceptors, proprioceptors, or interoceptors. General receptors are distributed over wide areas of the organism. Special receptors are concentrated in small areas or are paired or even median structures.

2. Exteroceptors and proprioceptors are somatic receptors. They inform the organism concerning the external environment, its orientation therein, and the activity of its skeletal muscles. The predominant reflex effect of input over these receptors is contraction of myotomal muscles in such patterns as to assure, insofar as possible, continued survival of the organism.

3. Interoceptors are visceral receptors. The predominant reflex effect of input over these receptors is stimulation of visceral efferent impulses in such a pattern as to assure, insofar as possible, maintenance of an appropriate internal environment.

4. General cutaneous receptors are exteroceptors located in the skin. The more primitive endings ramify among epidermal cells and give rise to protopathic sensations.

5. Neuromast organs are fluid-filled pits or

ampullae consisting of sensory and supportive cells and sensory terminals. They may open to the exterior (naked neuromasts) or be located in enclosed canals or vesicles. They occur only in fishes and aquatic amphibians. Electroreception and mechanoreception have been demonstrated. They are innervated by cranial nerves VII, IX, and X.

6. Snakes and lizards exhibit pit receptors of uncertain homology. Apical pits occur at the apex of body scales, and loreal and labial "pit organs" occur on the head of crotalid snakes and boas. Apical pits are probably mechanoreceptors. "Pit organs" are thermoreceptors for radiant heat.

7. The membranous labyrinth is a special somatic receptor consisting of canals, sacs, and ampullae lined with patches of sensory epithelia resembling neuromasts. The labyrinth is stimulated by mechanical displacement of the hairs of the sensory cells. The predominant effects of stimulation of the semicircular canals and utriculus are movements of the eyeball muscles and of the muscles that maintain equilibrium. The effect of stimulation of the sacculus of many fishes and of its evaginated derivative, the cochlear duct of tetrapods, is audition. Auditory stimuli reach the endolymph by bone conduction, chiefly weberian ossicles in cypriniform fishes and middle ear ossicles in tetrapods. An outer ear canal is found only in amniotes.

8. The saccus vasculosus is a midventral saccular evagination of the diencephalon caudal to the pituitary in fish. The epithelium exhibits sensory and supporting cells of unknown function.

9. There are two types of light receptors, paired lateral eyes and unpaired (median) pineal and parapineal bodies. Median eyes have a lens and sensory innervation but lack a retina and do not form an image. Pineal and parapineal bodies probably do not serve as light receptors above reptiles. The pineal has been more persistent than the parapineal.

10. General visceral receptors are unencapsulated endings, except for the pacinian corpuscles in visceral sites. Visceral receptors are mechanoreceptors or chemoreceptors, but some are subject to thermal and tactile stimulation.

11. Three special visceral receptors monitor the arterial blood of the aortic arch and carotid vessels of amniotes. Paired carotid bodies are chemoreceptors close to the carotid arteries that monitor the P_{O_2} of the blood. The carotid sinus is a baroreceptor on the pathway of the left and right carotid stream that monitors blood pressure. These two receptors are widely distributed among amniotes and are innervated by nerve IX. The unpaired aortic body lies close to the arch of the aorta, has a function identical to that of the carotid body, is innervated by the vagal nerve, and has been described in mammals only.

12. Organs for smell and taste are special visceral chemoreceptors. They include the olfactory sac of fishes, the olfactory epithelium of the nasal canal of tetrapods, the vomeronasal organs, and taste buds. Vomeronasal organs are independent olfactory organs innervated by a slip of the olfactory nerve. They are found in most tetrapods, although they are sometimes vestigial. Taste buds are distributed over the entire external surface of some fishes; otherwise, they are restricted to the oral cavity and pharynx. They are supplied by nerves VII, IX, and X in that sequence. They are found in all vertebrates but are least abundant in birds.

17

Endocrine organs

An endocrine organ produces one or more hormones. In this chapter we will discuss the thyroid, parathyroid, pituitary, and adrenal glands, Stannius corpuscles, pancreatic islands, and endocrine function of the gonads. In addition, because the nervous system produces a greater variety of internal secretions than any other tissue, we will be obliged to touch upon the pineal body, hypothalamus, and urohypophysis. We will conclude with a brief consideration of thymus tissue and of the ultimobranchial bodies.

NEUROENDOCRINE INTERRELATIONSHIPS

The hormones produced by the central nervous system are known collectively as **neurosecretions.** These regulate activity within other tissues, both endocrine and nonendocrine, assisting directly or indirectly in maintenance of reproductive cycles and of homeostasis. Neurosecretions are synthesized in the cell bodies of specialized neurons (neurosecretory neurons) of the brain and cord. The secretions pass from the cell body into the axon (neurosecretory fiber, Fig. 17-1) and travel along the axon as stainable droplets that accumulate at the axon terminals. There the neurosecretion is released into blood channels.

The majority of neurosecretions of vertebrate animals are produced in the hypothalamus. Some of these secretions are released into the hypophyseal portal vein (Figs. 17-1 and 17-2), which carries them to the part of the adenohypophysis commonly known as the anterior lobe. There they regulate the activity of pituitary cells that synthesize and release other hormones. Hypothalamic neurosecretions are also released from neurosecretory fibers into the pars nervosa of the pituitary, where they enter blood sinusoids to be transported to all parts of the body. The remote tissues affected by neurosecretions exert hormonal feedback onto the hypothalamus and pituitary, accelerating or decelerating the further synthesis and release of hypothalamic and hypophyseal secretions. The activities of the brain and other endocrine organs are thus integrated via internal secretions as well as via peripheral nerves.

Regulation of periodic, or rhythmical, activity of the neuroendocrine system, such as that associated with reproductive cycles, depends on the input of environmental stimuli into the brain via sense organs and afferent nerves (Fig. 17-3). These stimuli synchronize hypothalamic activity with daily (circadian) and seasonal environmental cycles. One effect is promotion of gametogenesis and of appropriate reproductive behavior (migration, territory defense, mating behavior, nest building, and care of the eggs and young) at the time of the year when environmental conditions are most suitable for the survival of offspring at birth. Such adaptive reflex neuroendocrine

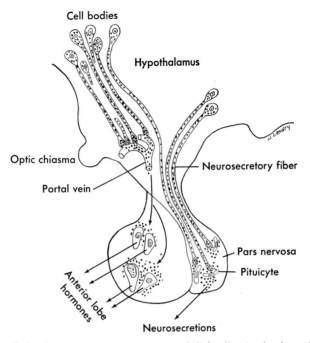

Cell bodies

Hypothalamus

Optic chiasma

Portal vein

Neurosecretory fiber

Pars nervosa

Pituicyte

Anterior lobe hormones

Neurosecretions

Fig. 17-1. Hypothalamic neurosecretory neurons. Cell bodies in the hypothalamus manufacture neurosecretions (black granules) that flow along the axons (neurosecretory fibers) and are discharged into the hypophyseal portal vein and into the pars nervosa of the pituitary.

arcs (receptor-brain-pituitary-gonad-behavior or metabolism) are the result of evolution. For an excellent discussion of neurosecretions, interestingly written, the authoritative work of the Scharrers[129] may be consulted.

Most hormones are combinations of amino acids (chiefly polypeptides, proteins, or nucleoproteins), or they are steroids. All these compounds are common products of biochemical synthesis not only among vertebrates, but also among invertebrates and plants as well. They represent chemical taxa that have evolved by radiation from simpler taxa, along with the evolution of the living organs that produce them. Steroids, for example, are found in yeast, in many green plants, and are universal in the animal kingdom. There is a greater variety in the lower animals than in higher ones. Inevitably some species of these chemical taxa have encountered, somewhere in the organism, cells whose metabolism was affected. In isolated instances the effect was

adaptive and had survival value for the organism. This biochemical species then supplemented the earlier role of the nervous system as an integrative mechanism. As a reminder of the days when the nervous system alone performed the regulative role, cells affected by hormones can also operate without the regulation of hormones, but the regulation has survival value.

Endocrine glands as well as their products have evolved. Some of them such as the adenohypophysis and the thyroid probably at one time were exocrine glands that secreted into the digestive tract. Evidence for this will be presented in the discussion to follow. The integrative functions of some of the hormones, prolactin, for example, have also evolved, and some of these changes will be noted. For the sake of organization we will discuss the endocrine organs according to their embryonic origins, commencing with those derived from ectoderm.

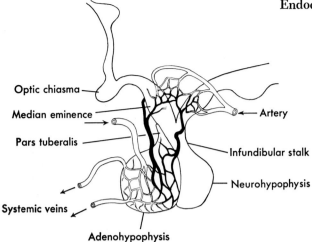

Fig. 17-2. The hypophyseal portal system (black) of mammals, schematic. Arrows indicate direction of blood flow.

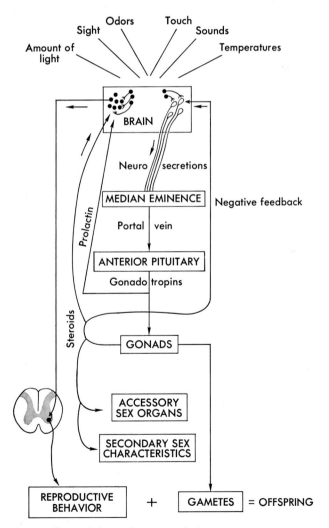

Fig. 17-3. Regulatory effects of the environment via hormones.

ENDOCRINE ORGANS DERIVED FROM ECTODERM
Pituitary gland

The pituitary gland, or hypophysis, lies underneath the diencephalon cradled, except in cyclostomes, in a depression in the sphenoid area of the skull. Because of its shape the depression is called the sella turcica (Turkish saddle). The gland consists of two major subdivisions with two different embryonic origins: a **neurohypophysis** derived from the floor of the diencephalon and an **adenohypophysis** derived from the roof of the stomodeum (Fig. 17-4). These parts and their gross subdivisions, with some common synonyms in parentheses, are as follows:

Neurohypophysis (pars neuralis)
  Median eminence
  Infundibular stalk
  Pars nervosa (posterior lobe)
Adenohypophysis (pars buccalis)
  Pars intermedia
  Pars distalis (pars glandularis or anterior lobe)
  Pars tuberalis
  Ventral (inferior) lobes of elasmobranchs

Neurohypophysis. The neurohypophysis is that part of the pituitary gland which arises from the floor of the diencephalon. Since the floor tends to evaginate ventrally, it contains a shallow or deeper recess of the third ventricle. The recess is prominent in amniotes, since in these forms the dien-cephalic floor is drawn out into an elongated stalk, the infundibulum. In fishes the part of the neurohypophysis in contact with the adenohypophysis exhibits little histological differentiation, but in lungfishes and tetrapods it has a specialized histoarchitecture and is called the pars nervosa.

Just behind the optic chiasma the diencephalic floor exhibits a swollen vascular area, the median eminence. The blood sinusoids of the median eminence are connected with the sinusoids of the adenohypophysis by the hypophyseal portal system (Fig. 17-2).

Both the median eminence and the pars nervosa are **neurohemal organs.** A neurohemal organ is a part of the nervous system containing endings of neurosecretory cells (in which the neurosecretion may be temporarily stored) and blood sinusoids, into which the neurosecretions are finally released. The neurosecretory substances that are released in the median eminence pass next to the adenohypophysis via the portal system. Those deposited in the pars nervosa are carried by systemic veins to the heart for distribution throughout the body. Neurohemal organs are not confined to vertebrates. The sinus gland of crustaceans and the corpus cardiacum of insects are neurohemal organs that receive neurosecretions from the brain, store them, and release them into vascular sinusoids.

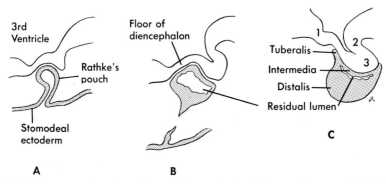

Fig. 17-4. Embryogenesis of amniote pituitary. **A,** Rathke's pouch stage. **B,** Isolation of adenohypophyseal anlage (gray) in contact with the floor of the diencephalon. **C,** Definitive pituitary consisting of adenohypophysis (gray) and neurohypophysis (white). **1,** Median eminence; **2,** infundibular stalk; **3,** pars nervosa. The subdivisions of the adenohypophysis are labeled.

The neurohypophysis produces no known hormones. It serves solely as a neurohemal organ for storage and release of neurosecretions manufactured in the hypothalamus. These neurosecretions are octapeptides with two chief effects. There are an antidiuretic effect (that is, a water-conserving effect), and a smooth muscle–stimulating effect (including stimulation of the smooth muscles of the blood vessels). Substitutions of one amino acid for another occur at three positions on the molecule; otherwise these polypeptides show no structural variation from cyclostomes to man. A substitution may alter considerably the biological activity of the molecule. Arginine vasotocin is probably the most primitive of these "neurohypophyseal hormones." It is the only one in cyclostomes and is found in all higher vertebrate classes, except mammals. The role of arginine vasotocin in fishes is not yet clear. In terrestrial vertebrates, arginine vasotocin serves to prevent excessive water loss from the organism. It causes the kidney tubules to reclaim water from the glomerular filtrate, and in toads, at least, it causes reclamation of water from the urinary bladder. In arid environments it causes absorption of water from the soil via the skin of toads. Arginine vasopressin, a mutant molecule, is the predominant antidiuretic hormone in mammals, including man. Another mutant, oxytocin in mammals, plays an important role in inducing uterine contraction during birth of the young and in causing contraction of the myoepithelial cells of the mammary glands, which results in the "letdown" of milk into the cisterns of the nipple when one nipple is stimulated by nursing.

Adenohypophysis. The adenohypophysis is the glandular (*adeno* = gland) portion of the pituitary. It arises as a bud of ectodermal cells from the roof of the stomodeum. In amniotes and in some lower fishes and selachians the bud is hollow and is known as Rathke's pouch. In other fishes and in amphibians the bud is solid. When the anlage of the adenohypophysis has made intimate contact over a broad area with the infundibular region of the diencephalon (the future neurohypophysis), the stalk between the stomodeal ectoderm and the adenohypophysis usually disappears. It remains as an open ciliated duct in *Calamoichthys* and *Polypterus* and leads to the buccal cavity, and in some reptiles and birds a solid strand remains. A vestige of the lumen of Rathke's pouch may remain in the adult gland as a cleft.

In cyclostomes the adenohypophysis arises off the stalk that gives rise to the olfactory sac. The origin of the adenohypophysis as an evagination of the stomodeum, or in association with the olfactory sac, and the persistence of a ciliated duct in some species suggest that the adenohypophysis at one time may have secreted into the buccal cavity. The ancients believed that the pituitary gland (*pituita* = phlegm) was the source of phlegm that falls into the throat. Although the ancients were wrong about the nature of the secretion, they may have been naively correct about the primitive site of secretion!

The adenohypophysis typically exhibits three regions—pars intermedia, pars distalis, and pars tuberalis (Fig. 17-5). The intermedia differentiates from that part of the stomodeal bud which comes in contact with the neurohypophysis, probably under the stimulus of the brain as an inductor. The pars distalis arises from most of the remainder of the anlage. The pars tuberalis is one of a pair of elongated extensions of the distalis, which grow along the floor of the diencephalon or up the infundibular stalk. No pars tuberalis forms in most fishes, but the rostral part of the distalis may represent potential tuberalis tissue that fails to elongate. Neither has a tuberalis been identified in certain snakes and lizards. In elasmobranchs the adenohypophysis bulges to form a fourth subdivision, the ventral (inferior) lobes.

PARS INTERMEDIA AND ITS HORMONES. The pars intermedia lies in intimate contact with the neurohypophysis or may even partially invade it or be invaded by it. However, in some species no recognizable

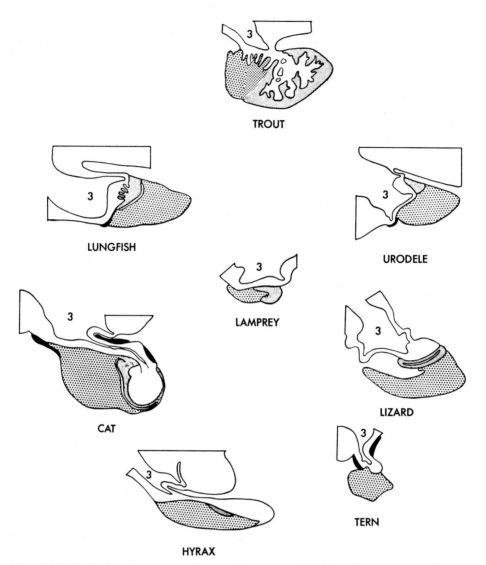

Fig. 17-5. Pituitaries of representative vertebrates, sagittal views. Dots represent the pars distalis, gray denotes the pars intermedia, and black denotes the pars tuberalis. The pars distalis of teleosts (trout) exhibits two cytological regions, a rostral part (large dots) and a proximal part (smaller dots). The neurohypophysis and associated infundibular region of the brain are shown in white. The infundibular region contains the infundibular recess, **3,** of the third ventricle.

intermedia develops. This is true in birds and some mammals such as cetaceans, elephants, manatees, armadillos, and beavers. In apes and man the intermedia grows smaller in size with age, after being relatively large in the embryo. This may be related to the function of intermedin, also known as melanophore-stimulating hor-

mone (MSH), a product of the intermedia. It causes the pigment granules in certain chromophores to disperse, thus darkening the skin. The phenomenon is characteristic only of cold-blooded animals.

The secretion of intermedin is a response to visual stimuli arising in the retina, transmitted to the brain via the optic nerve, thence through a series of intermediate brain centers over a pathway that has not yet been plotted, and then probably down the infundibular stalk via a fiber tract to the pars intermedia. Release of the hormone into the bloodstream brings about the color change in the skin.

HORMONES OF THE PARS DISTALIS. The pars distalis is known to produce six hormones from six morphological varieties of cells. Many of these hormones have been characterized according to their amino acid sequences. These hormones, along with their common synonyms and abbreviations, are as follows:

> Somatotropin (STH or growth hormone)
> Thyrotropin (thyroid-stimulating hormone or TSH)
> Adrenocorticotropin (ACTH)
> *Follicle-stimulating hormone (FSH)
> *Luteinizing hormone (LH)—in males known as interstitial cell-stimulating hormone (ICSH)
> Prolactin (lactogenic hormone)

STH, ACTH, and prolactin are proteins. Prolactin is the smallest molecule of the three. The others are glycoproteins, which contain a carbohydrate. All except STH are known to be under neurosecretory regulation by hypothalamic secretions. In the case of the control of prolactin, the effect of hypothalamic regulation is inhibitory. If the adenohypophysis is transplanted to another location in the animal, such as under the capsule of the kidney, only prolactin is produced in quantity, and it is produced without interruption. This is interpreted as indicating that the production of prolactin in this ectopic location is no longer under the inhibitory influence of the hypothalamus.

Somatotropin stimulates the synthesis of proteins from amino acids; hence it is a general growth-promoting hormone. When STH is administered experimentally, less nitrogen is excreted by the animal because amino acids that otherwise would be converted to urea are used in protein synthesis.

---
*These are gonadotropic hormones. They, along with TSH, are produced partly in the inferior lobes of the elasmobranch pituitary.

Muscle and cartilage respond to injections of STH by marked growth, and these both require protein. STH also exerts a control over carbohydrate and fat metabolism. When injected, it has a skin-darkening effect because the entire molecule of melanophore-stimulating hormone (α-MSH) is repeated as part of the amino acid sequence of the larger STH molecule.

Thyroid-stimulating hormone acts on the cells of the thyroid follicle and causes them to accumulate iodine, to synthesize thyroid hormone, which contains iodine, and to release thyroid hormone from the follicle to the circulatory channels.

Adrenocorticotropin regulates the cells of certain zones of the adrenal cortex and causes these cells to synthesize the adrenal cortical hormones other than aldosterone.

Follicle-stimulating hormone acts upon primary ovarian follicles and causes them to grow and to differentiate into graffian follicles containing mature ova. FSH also acts on the spermatogenic tubules of the testes and causes them to increase in size, diameter, and weight.

Luteinizing hormone induces luteinization of cells of the ovarian follicle before, and especially following, ovulation and causes the cells of the ruptured follicle to develop into a new endocrine body, the corpus luteum. LH is also necessary for ovulation to occur. In males, luteinizing hormone is better known as interstitial cell–stimulating hormone. It acts on the interstitial cells lying between the spermatogenic tubules and causes them to produce testosterone.

Prolactin was named for its effect on mammary tissue. Along with other hormones, it initiates the secretion of milk by the mammary gland. Prolactin has the same effect on the gland of the crop sac of pigeons. The crop sac secretes a nutritive "pigeon milk," which is regurgitated and fed to nestlings by the parent. When prolactin titers reach the level at which the epithelium of the gland will respond, milk is produced. In some teleost fishes, prolactin stimulates the secretion of mucus, which is eaten by the young for nourishment. It has also been implicated in the homeostatic adjustments necessary for saltwater platy-fish to survive in fresh water. In tailed amphibians this same hormone causes the immature land-dwelling eft to commence a migration to the ponds in which the animal will mate when the gonads have reached maturity and in which the female will deposit her eggs. Rising titers of prolactin in these efts herald the onset of sexual maturity. In all vertebrates from fish to man prolactin causes the organism to exhibit certain parental behavior patterns, including the building of nests (simple in most cold-blooded vertebrates and sometimes complex in warm-blooded ones); cleaning the area and the young; protection, turning, and incubation of eggs; and protection of the young, including transporting them about in some species. The diversity of actions of prolactin in the several vertebrate classes illustrates the effects of organic evolution on metabolic functions.

Urohypophysis

Many fishes have a neurohemal organ at the posterior end of the spinal cord consisting of large neurosecretory cells in the cord, which secrete into vascular sinusoids (Fig. 17-6). The spinal cord may evaginate to form a short or long stalk with the neurohemal organ at the end. The organ is comparable to the pars nervosa in that it receives the axon terminals of neurosecretory fibers, stores the secreted hormone(s), and releases it (them?) into the blood sinuses of the organ. The urohypoph-

ysis, or urophysis caudalis as it is also called, was first described in 1826, but only recently has it been studied extensively. Recent studies of absorption and excretion of salts have indicated that the neurosecretion probably regulates the transport of sodium ions into or out of the organism, according to the osmolarity of the organism and of the surrounding water.

Pineal body

The pineal body, an evagination of the roof of the diencephalon, produces a tryptamine derivative named **melatonin.** This hormone causes melanin granules to aggregate, thereby blanching the skin of ectotherms. The effect is opposite that of the melanophore-stimulating hormone of the pars intermedia.

Experimental evidence indicates that light affects the pineal body. The effect may be direct, when the pineal is located under translucent skin; otherwise, the effect is via afferent nerve fibers that enter the brain via the optic nerve. Light inhibits the synthesis of melatonin. Amphibian larvae with pineal glands intact become pale if placed in the dark. Pinealectomy

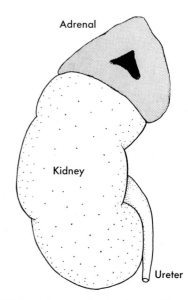

Fig. 17-7. Adrenal gland of man. The cortex, shown in gray, surrounds the medulla.

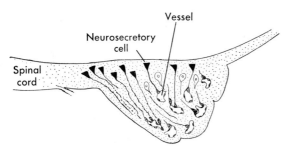

Fig. 17-6. Urohypophysis of a carp.

abolishes this response. Additional discussions of the pineal body will be found in Chapters 15 and 16.

Adrenal medulla and chromaffin tissue

The adrenal gland as most students envision it is a gland on the ventral surface of the kidney or near its cephalic pole. In mammals it consists of two components, a peripheral cortex and a central medulla (Fig. 17-7). Cortex and medulla are two entirely different glands spatially separated from each other in many fishes (lampreys, Chondrichthyes, and some teleosts) and more or less randomly interspersed among one another in most vertebrates, with the exception of mammals (Fig. 17-8). In a comparative study, therefore, we must discuss the adrenal as two separate glands, which, in actuality, it is. Only the chromaffin component (medulla) is ectodermal. It arises from neural crests.

If a solution of ferric chloride, potassium dichromate, or osmic acid is applied to the adrenal medulla, the cells will take on a greenish, yellowish, or brownish tint, according to which inorganic substance is applied. This is a histochemical response

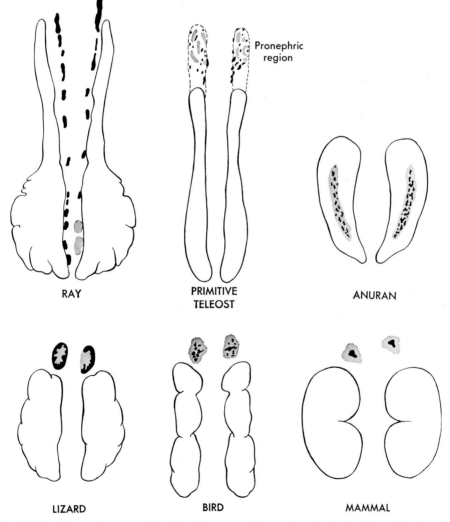

RAY

PRIMITIVE TELEOST

Pronephric region

ANURAN

LIZARD

BIRD

MAMMAL

Fig. 17-8. Chromaffin tissue (black) and interrenal tissue (gray) of selected vertebrates. The kidneys are shown in outline.

elicited by the presence of certain amines in the cells. The amines are products of cellular metabolism. The tinctorial response is called the chromaffin reaction, and a tissue giving the reaction is a chromaffin tissue. The adrenal medulla contains two amines, **epinephrine** (adrenaline) and **norepinephrine** (noradrenaline). Although a number of other tissues in vertebrates give the chromaffin reaction, only the adrenal medulla and its homologues in lower vertebrates can synthesize these hormones. To find the homologues of the mammalian medulla in fishes through birds, we must seek those groups of cells that give a chromaffin reaction and then test all such masses to see whether they contain the amines. By this technique it has been possible to locate the medullary homologue of the adrenal in lower vertebrates.

In fishes the medullary cells form masses of chromaffin tissue scattered along or close to the postcardinal vein (lamprey), or embedded in the connective tissue walls of the postcardinal veins (some teleosts), or close to the dorsal aorta or its segmental branches (lungfishes). Many of these masses tend to be segmentally arranged. In elasmobranchs the chromaffin masses are located between the mesonephroi, whereas in most teleosts they are near the cranial ends of the mesonephroi. Many of the masses are in association with veins that have penetrated the vestiges of the pronephroi and hence may be surrounded by lymphoid tissue.

In some teleosts the area occupied by the chromaffin tissue is also occupied by the adrenal cortical component (hereafter referred to as interrenal tissue), so that the chromaffin and the interrenal tissue are interspersed among one another close to the kidney. In most amphibians, reptiles, and birds the two components are interspersed. In lizards and some snakes the chromaffin tissue tends to aggregate, forming an almost complete capsule around the interrenal component. This condition is the reverse of that in mammals. Even in mammals there are species in which the inter-

renal tissue does not entirely surround the chromaffin component. In sea lions there is considerable interspersion, with small masses of cortical tissue scattered in the medulla and masses of medullary tissue scattered in the cortex. In the remaining mammals the interrenal tissue (cortex) completely encapsulates the chromaffin tissue.

The adrenal glands of anurans are flattened, elongated bodies located against the ventral surface of the kidney. In urodeles they are more diffuse and lie near its medial border. In amniotes the adrenal glands are typically located at or near the cephalic pole of the kidney.

Chromaffin tissue in mammals is not restricted to the adrenal medulla. Masses of chromaffin cells occur in close association with sympathetic ganglia, where they are called paraganglia, and in the ovary, testis, kidney, heart, and other visceral organs. That these cells synthesize epinephrine has not been demonstrated, and much work on the comparative physiology of the chromaffin tissue remains to be done. The heart of the hagfish *Myxine glutinosa* contains an abundance of chromaffin tissue containing catechol amines. Chromaffin tissue can be demonstrated in a number of invertebrates, and epinephrine is synthesized by a number of annelids, insects, and crustaceans.

The basic segmental arrangement in some fishes of the medullary component of the adrenal is understandable in the light of the embryonic origin of chromaffin cells. These cells are derived from neural crest ectoderm (neurectoderm), from which sympathetic ganglion cells are also derived. The postganglionic fibers of sympathetic ganglion cells release epinephrine and norepinephrine at their terminals. Thus, the postganglionic neurons of the autonomic system and the chromaffin cells of the adrenal medulla are homologous. This fact clarifies another unusual feature of the adrenal medulla—its innervation by a **preganglionic neuron** (Fig. 17-9). It is clear that the chromaffin cell of the medulla is

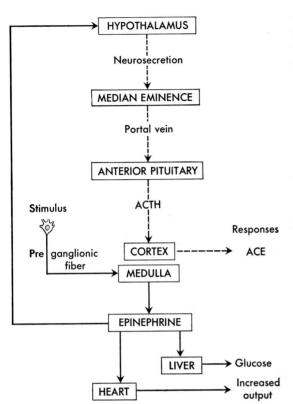

Fig. 17-9. Some regulatory functions of the adrenal medulla. When presented with a suitable nervous stimulus (left center), a preganglionic neuron of the sympathetic nervous system stimulates the adrenal medulla to release epinephrine. The latter elicits a broad spectrum of responses, three of which are indicated at the right.

a sympathetic neuron that has failed to sprout processes and that the medulla is, in essence, a sympathetic ganglion.

Epinephrine has a number of endocrine roles (Fig. 17-9). The most prominent are to increase the amount of blood sugar in times of sudden metabolic need and to stimulate increased production of adrenal cortical hormones in times of prolonged stress. Epinephrine causes a rise in blood sugar (glucose-6-phosphate) partly by stimulating glycogenolysis, the breakdown of glycogen, in the liver. Collectively, these and other effects are adaptive responses that tend to ensure survival in times of stress. Epinephrine also evokes dispersion

of pigment granules in melanophores. Norepinephrine is concerned chiefly with maintaining the tonus of the circulatory system through its vasoconstrictor effect.

ENDOCRINE ORGANS DERIVED FROM MESODERM
Interrenal bodies and the adrenal cortex

We have seen in the previous section that the adrenal gland is composed of two entirely different components—chromaffin and cortical tissue. The cortical tissue and its homologue, the interrenal bodies of fishes, are derived from mesoderm. Cortical cells of all vertebrates are budded off from localized thickenings of the coelomic mesothelium close to the gonadal ridge and from the underlying intermediate mesoderm. The cells assume a position near the major vessels in the vicinity of the kidney. In fishes they may form serial clusters or cords of interrenal tissue, or they may aggregate into one or two major masses, often with accessory masses extending cephalad or caudad. The major masses may be elongate, U-shaped, or oval, and they may wrap completely around a vein. Sometimes the distribution is quite diffuse. All such masses are referred to as interrenal bodies and are homologous with the adrenal cortex of mammals. The chromaffin bodies and interrenal bodies may be remote from one another, but more often they are interspersed. The term "cortex" is strictly applicable only in mammals, in which the interrenal tissue completely surrounds the chromaffin mass. Nevertheless, the term "cortical tissue" has gained acceptance to designate corticoid-producing tissue throughout the vertebrate groups. Even in mammals accessory cortical masses may remain apart from the adrenal gland.

The origin of the adrenal cortical tissue from the same mesothelium that gives rise to gonads is interesting because both cortical tissue and gonads produce steroid hormones, and these are the only sources of steroid hormones in the vertebrate body other than the mammalian placenta. Moreover, the cortical tissue can and, under appropriate conditions, does produce all the hormones synthesized by the male and female gonads. This indicates the similarity of the enzyme systems in the two tissues. Approximately fifty different steroids have been isolated from the adrenal cortex. Seven of those account for most of the activity of adrenal cortical extract (ACE).

Adrenal cortical extract has three chief effects. It regulates the excretion of Na$^+$ ions by stimulating the kidney tubules to reclaim sodium; it plays a role in maintaining glycogen reserves of the body by regulating the conversion of fats and proteins into glycogen (glyconeogenesis); and it produces androgens (male steroids) and estrogens (female steroids). The bearded lady is an example of what may happen when the adrenal cortex produces excessive quantities of androgens.

Those corticoids affecting primarily the resorption of Na$^+$ ions are called **mineralocorticoids.** They also affect the distribution of K$^+$ ions, since loss of sodium causes an efflux of potassium from the cells into the tissue spaces. The most potent mineralocorticoid is **aldosterone.** The corticoids with a predominantly glyconeogenic effect are referred to as **glucocorticoids.** Among these are **cortisone, cortisol,** and **corticosterone.** Species may produce one or two of these, but not all three. The hamster, for example, synthesizes cortisol and not corticosterone, whereas the rat is primarily a corticosterone secretor. Similar species differences occur among lower vertebrates. The production of ACE, except aldosterone, is regulated by ACTH from the adenohypophysis. Removal of the adrenal cortex in mammals results in death unless sodium is constantly supplied in the diet.

Corpuscles of Stannius

Embedded in the kidneys or attached to the mesonephric ducts of teleost and ganoid fishes are oval or spherical gland-like epithelioid bodies, the corpuscles of Stannius. In most teleosts there are two, but in *Amia* there are 40 to 50. In large salmon the corpuscles may reach the size of a green pea.[121] Although these may be mistaken morphologically for interrenal bodies, they arise as evaginations of the embryonic pronephric duct and hence are not homologous with adrenal cortical masses and have not been shown to be consistently responsive to ACTH. The cells contain granules that appear to be lipoprotein, but the nature of their secretions is unknown. Whether or not these corpuscles are endocrine organs has not yet been established. The cells are capable of converting one steroid into another in certain species,[124] but no steroids have been demonstrated in the corpuscles in over 5,000 fresh water eels.[126] Ablation of the corpuscles has been followed by demonstrable electrolyte changes within body fluids, but the effect did not appear to be via the kidneys. Ablation also caused a fall in arterial blood pressure in freshwater eels, and extracts of the corpuscles raised the pressure of rats as well as eels. There seems to be a functional relationship, direct or indirect, between the corpuscles and interrenal tissue, since ablation of the interrenal tissue stimulated the corpuscles and vice versa. These bodies may have some unknown relationship to electrolyte or water distribution within the body or to their excretion. Results of further studies of these organs are awaited with interest.

Gonads as endocrine organs

The gonads arise from the coelomic mesothelium as a pair of gonadal ridges medial to the kidneys (Fig. 14-16). In addition to gametes, the ovaries and testes of most vertebrates produce three types of steroid hormones—**estrogens, androgens,** and **progestogens.** Collectively, these hormones are essential for the successful propagation of the species.

Estrogens produced by ovarian follicles and androgens produced by the interstitial cells of the testes affect most prominently the accessory sex organs, including the reproductive tracts (Fig. 17-3). Differentiation of muellerian ducts to become uteri and oviducts in female embryos is partly an expression of the effects of estrogens, and failure of muellerian ducts to develop in males may be ascribed, in some lower vertebrates at least, to the dominance of androgens. These sex hormones are also responsible for secondary sex characteristics such as development of mammary glands in female mammals and the large muscles and skeleton in males. They also affect reproductive behavior, inducing a state of willingness to mate at a time most appropriate for fertilization to be assured.

Progesterone is synthesized by the ovaries and testes, though it is probably not released into the circulation by the testes. It is a precursor compound from which estrogens and androgens are formed. In female mammals it has achieved an independent role, that of maintaining the uterus in a progestational state—a state supporting pregnancy. By negative feedback to the hypothalamus, it also inhibits the formation of a new wave of ovarian follicles and hence delays the next ovulation. Progesterone in mammals is a product of the corpus luteum. This is the name given to an ovarian follicle after the cells have undergone luteinization (chemical and morphological change) under the influence of LH.

In addition to steroid hormones, the mammalian ovary produces relaxin, a peptide. Relaxin is produced during pregnancy and softens the ligaments of the pubic symphysis before birth, thus enlarging the birth canal for easier delivery of the fetus.

ENDOCRINE ORGANS DERIVED FROM ENDODERM

The endoderm of the foregut gives rise to the pancreatic islets and to the thyroid and parathyroid glands. The pharyngeal pouch endoderm also gives rise to a series of epithelioid organs with no known endocrine function, but which will be discussed under thymus and ultimobranchial bodies.

Pancreatic islets

In addition to cells that secrete digestive enzymes, the pancreas typically contains islands of endocrine tissue that secrete two hormones, **insulin** and **glucagon**. These, along with other hormones of the body, regulate blood sugar levels. Present knowledge indicates that both insulin and glucagon release is regulated by the titers of sugar in the blood passing through the pancreas, rather than by another hormone or by innervation from the nervous system.

Insulin is a protein produced by the beta cells of the islets. It was the first protein to be characterized regarding the precise sequential order of its amino acids (1954). Insulin regulates glycogenesis in the liver (the conversion of glucose into glycogen) and influences the utilization of glucose in living cells. When insulin levels are low, glucose accumulates in the blood and may spill over into the urine, a condition known as sugar diabetes, or diabetes mellitus.

Glucagon is a small protein consisting of a straight chain of twenty-nine amino acid residues. It is believed, on the basis of fairly strong histological and experimental evidence, to be produced by the alpha cells of the islets. Alpha cells are especially abundant in reptiles and birds, whereas some urodeles apparently lack these cells. Glucagon promotes glycogenolysis (the breakdown of glycogen into glucose) in the liver by increasing the active phosphorylase needed to catalyze the reaction. Because it brings about a rise in blood sugar, it is also known as the **hyperglycemic factor** (HGF).

Pancreatic islands occur in the pancreas of most vertebrates. In cyclostomes, however, the islands are separate and form a compact mass at the blind end of the intestinal cecum (lamprey) or encircling the bile duct at its entrance into the intestine (*Myxine*). In some bony fishes and in a few snakes the islands occur as nodules outside of the pancreas. The endocrine islands have no connection with the lumen of the alimentary canal.

It is probable that all pancreatic cells, endocrine and exocrine alike, were at one time part of the intestinal epithelium and that only later did a part of the gut wall evaginate to establish a separate pancreas. The duodenal epithelium still contains endocrine cells that produce **secretin,** which stimulates the exocrine portion of the pancreas. Since the liver, which is the chief target organ of the pancreatic hormones, is also an evagination from the intestine, it is possible that insulin, glucagon, and secretin were all produced by the intestinal lining and were effective locally at one time. Later, when the liver and pancreas separated from the gut, the hormones could reach their target organs only via the circulatory system.

Thyroid gland

At a very early stage in vertebrate evolution certain cells of the pharyngeal floor acquired the capacity to accumulate diffusible iodides from the body fluids and to convert these into molecular iodine. These iodide-capturing cells may have arisen by alteration of some of the cells of a primitive endostyle. The endostyle of the larval lamprey (ammocoete) is able to accumulate

iodine (Fig. 17-10), and this ability is increased by injection of mammalian TSH. In the adult lamprey the iodine-accumulating cells lie beneath the pharyngeal floor between the second and fifth pharyngeal pouches.

The thyroid gland of all higher vertebrates, like the endostyle, arises as an unpaired evagination from the midventral pharyngeal floor, usually at the approximate level of the second pharyngeal pouches (Figs. 17-11 and 17-13). After the distal tip of the thyroid evagination has reached its adult location, the tissue organizes into thyroid follicles consisting of a cuboidal epithelium surrounding a central reservoir. The central reservoir contains a colloid, and into it is secreted the captured iodides. The embryonic stalk of cells attaching the thyroid to the pharyngeal floor usually disappears, isolating the thyroid in its adult site. However, a duct remains in the elasmobranch *Chlamydoselache.* It perforates the basihyal cartilage and empties into the pharyngeal floor. In other species of fishes a solid stalk may persist. In mammals a small pit, the foramen cecum, on

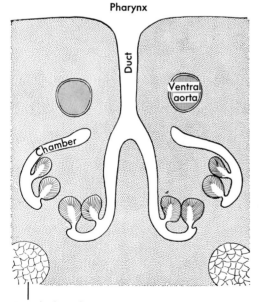

Fig. 17-10. Subpharyngeal gland (endostyle) of a larval lamprey. All chambers empty into the duct. At metamorphosis the duct will close.

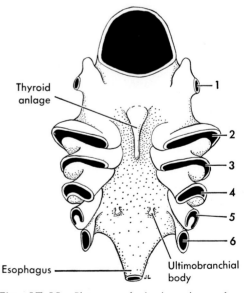

Fig. 17-11. Pharynx of shark embryo, from below. **1,** Spiracle; **2** to **6,** gill slits. (After Camp.[37])

the caudal surface of the tongue marks the site of the embryonic evagination. Remnants of the stalk may persist in mammals as a cyst-like thyroglossal duct, which occasionally requires surgical removal in man.

In hagfishes and most teleosts there is no organized thyroid. The many follicles are scattered singly or in small groups along the ventral aorta and along some of the afferent branchial arteries. They may even accompany the arteries into the base of the gills. In a few teleosts, however, the follicles are not dispersed but form a median compact mass or two masses in tandem, between the bases of the first gill pouches.

Except in hagfishes and teleosts the unpaired thyroid anlage develops into a single compact gland, or a pair of glands, which become encapsulated by connective tissue. Elasmobranchs, like a few teleosts, have a median thyroid located at the ante-

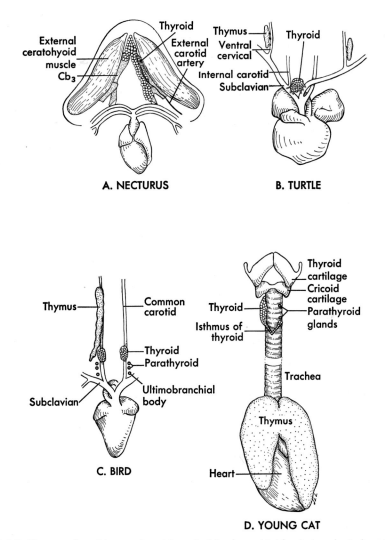

Fig. 17-12. Thymus, thyroid, parathyroid, and ultimobranchial body in selected vertebrates. The thymus of *Necturus* lies in the angle between the posterior ends of the masseter and external ceratohyoid muscles and is not illustrated. **Cb₃**, The ceratobranchial cartilage of the third pharyngeal arch.

rior bifurcation of the ventral aorta. Unpaired thyroids are also characteristic of snakes, turtles, some lizards, and *Echidna,* an egg-laying mammal. Most other adult vertebrates have paired thyroid glands. However, the two glands in most lizards and in mammals remain connected across the midline by an **isthmus** of thyroid tissue, and in these instances the gland is sometimes referred to as "bilobed."

In amphibians the glands lie under cover of the mylohyoid muscle and usually close to the anterior branchial cartilages in the floor of the pharynx (Fig. 17-12, *A*). In amniotes the glands migrate caudad varying distances from the pharynx, taking a position close to the common carotid arteries (from which they receive a rich arterial supply) and trachea (Fig. 17-12, *B* to *D*). In reptiles and birds they lie just anterior to the systemic arch. In mammals they usually lie at the level of the thyroid cartilage or just caudal to it, although in *Echidna* they reach the thorax. The gland was named "thyroid" because of its position in mammals close to the thyroid (shield-shaped) cartilage.

The capacity to accumulate iodine is not restricted to the thyroid gland or even to animals. Certain marine algae have this capacity, and in vertebrates a number of tissues accumulate iodine in quantities greater than that of the body fluids. These tissues include the notochord, gastric glands, chloride-secreting cells of gills, and numerous other extrathyroid tissues. Only thyroid cells, however, are known to have the capacity to unite the iodine with tyrosine to form the thyroid hormones, **thyroxin** and **triiodothyronine.** Molecular iodine displaces hydrogen on the tyrosine molecule to form iodotyrosine. Further enzymatic action results in the formation of the hormones. These unite with a globulin present in the colloid to form the storage product, thyroglobulin. Later, when needed, the hormone is split from the thyroglobulin molecule by proteolytic digestion. The synthesis of the hormone is stimulated by thyroid-stimulating hormone from the adeno-hypophysis. (Whether or not this is true in cyclostomes has not yet been established.)

Although it may seem as though we know a great deal about thyroid hormone, the fact is that our knowledge is still very incomplete. We are not even sure whether the hormone is synthesized in the thyroid cell or in the colloid. We know little about its passage across cell membranes, and we do not know precisely what it does for the organism. At present we have only a long list of observations of effects resulting from experimentation. The overt effect is an increase in the rate at which the organism carries on cellular respiration.

It is sufficient in this text to point out that thyroid hormone is essential in a large number of tissues if normal metabolic rates are to be maintained; that it is particularly essential in young animals that are undergoing growth, differentiation, and maturation, including metamorphosis in amphibians; that it is related, directly or indirectly, to the function of the mitochondria. There has been speculation that thyroid hormone may function within the mitochondrion, regulating the addition of phosphorus to adenosine diphosphate (ADP) and thus converting this substance into adenosine triphosphate (ATP). ATP, produced in the Krebs cycle, is the source of energy for all of the cell's activities. Therein may lie a clue regarding why the thyroid is of such importance in so many tissues in vertebrate animals.

Parathyroid glands

The parathyroid glands, consisting of one to several pairs, are so named because in mammals they lie beside (*para*) the thyroid gland (Fig. 17-12, *D*). Parathyroid glands have not been identified in fishes. The parathyroid anlagen of tetrapods usually arise as endodermal outgrowths from the tips of pharyngeal pouches II, III, and IV. Although in a few reptiles all three pairs differentiate and function, most tetrapods have only two pairs, since the pouch II anlagen usually fail to mature

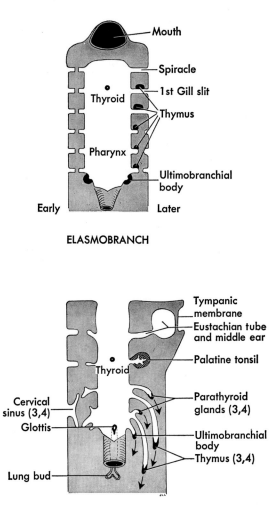

ELASMOBRANCH

MAMMAL

Fig. 17-13. Pharyngeal derivatives of shark and mammal. Left sides are shown earlier in ontogeny than the right sides.

hormone regulating the distribution of calcium and phosphate between bone, tissue fluids, and blood. Parathormone also retards the excretion of calcium by stimulating calcium resorption from the glomerular filtrate by the kidney tubules. Because removal of the parathyroid glands results in unregulated excretion of calcium, most parathyroidectomized tetrapods develop muscular spasms, which may end in tetany. Feeding calcium orally or intravenously results in a dramatic remission of the symptoms. Parathormone secretion seems to be regulated by the level of calcium in the bloodstream passing through the gland. Low levels evoke release of parathormone, which brings about a release of calcium from the reserve in the bones.

OTHER PHARYNGEAL EPITHELIOID BODIES

The walls of most of the embryonic pharyngeal pouches exhibit dorsal and ventral thickenings that differentiate into various organs known collectively as epithelioid bodies (epithelioid because some of the cells retain a resemblance to epithelial cells). Such bodies usually, but not invariably, separate from the walls of the pouches and sink into the surrounding mesenchyme. In many cases the thickenings take the form of pouch-like evaginations, which, by further growth, finally occupy sites remote from their origin. In addition to parathyroids, the several thymus masses and the ultimobranchial bodies are epithelioid organs derived from pharyngeal pouches.

Thymus

The thymus masses are lymphoid organs. Phylogenetically, the thymus appears to have been derived from each of the embryonic pharyngeal pouches (Fig. 17-13, elasmobranch). In the cyclostome *Petromyzon*, thymic tissue is said to differentiate in all seven pouches. In other fishes and in tailed amphibians thymus anlagen appear, with variations, in the walls of all pouches except the first. Transient anlagen have also

(Fig. 17-13). In urodeles one pair is typical. Some amniotes have only one pair, which develops from pouch III or IV, depending on the species. Among mammals two pairs are common in cats, guinea pigs, rabbits, and man, whereas in rats, pigs, seals, and some others there is only one adult pair. Accessory parathyroid tissue is not uncommon, and as many as twelve parathyroid masses have been described from human cadavers.

The parathyroid glands produce **parathormone**, a polypeptide. It is the principal been described in the wall of the first (spiracular) pouch in elasmobranchs, and in caecilians the first six pouches participate. There is a tendency for the thymus to be more restricted in origin in higher forms, so that in most reptiles, birds, and mammals, pouches III and IV are the sole contributors of this tissue. In a few mammals (rat and mouse) pouch III is the sole source of thymus, and in a larger number of mammals, including man, contributions from pouch IV are variable. In at least one species of frog, pouch II is the sole origin. Historically, however, the thymus is of multiple origin.

There is a tendency for the successive thymus anlagen to unite during development so that there may be fewer adult thymus masses than there were embryonic anlagen (Fig. 17-12, B to D). In fishes, except for elasmobranchs, all anlagen more or less fuse to form a single, elongated gland lying above the branchial chambers. In at least one elasmobranch (Heptanchus cinereus) ducts lead from the first six lobes of the thymus into the pharynx, in young animals at least. In amphibians and reptiles the masses may fuse or remain separate. The single gland on each side in frogs usually lies just behind the tympanic membrane. In reptiles and birds the thymus may consist of a series of large nodes in the neck, seven in chicken, extending caudad as far as the thyroid. In mammals the young thymus is usually represented by a relatively large bilobed mass in the thoracic cavity dorsal to the sternum and mostly anterior to the heart, constituting part of the mediastinum. It becomes infiltrated by fat and more difficult to identify after puberty.

Interest in the thymus as an endocrine organ has alternately waxed and waned over a span of many years as bits of evidence for an endocrine role have been reported in the literature, only to fail the test of reproducibility. Among the facts that had to be taken into account is that the thymus is largest in young mammals and undergoes fatty infiltration as the animal attains sexual maturity. Recent studies by immunologists have provided one explanation. The thymus is large just before birth and early in life because at that time it is producing and liberating into the blood lymphocytes that pass to the spleen and to lymph nodes and which, thereafter throughout life, serve as the source of antibodies.

In birds the role of the thymus as the source of cells that will produce antibodies is supplemented by an organ that arises as an evagination of the cloaca. This organ, the **bursa of Fabricius,** is somewhat like the thymus in structure and disappears completely at sexual maturity. The appendix of the rabbit has a similar function.

Ultimobranchial bodies

Epithelioid bodies growing from the caudal walls of the last pharyngeal pouches have been called ultimobranchial bodies or postbranchial bodies (Figs. 17-11 and 17-13). They do not seem to have an endocrine function. This might be expected in view of their origin from the last pharyngeal pouches, since the last visceral arch may also lack certain components. When serially arranged ancient structures become modified, the anteriormost and posteriormost components of the series seem to be affected first. This tendency is illustrated by the kidney tubules, the somatic efferent nerves of the body, the gill apparatus, the aortic arches, and the vertebrae, for example.

Despite the fact that the ultimobranchial bodies seem to make little contribution to body function, they often remain in adult tetrapods as prominent structures. They may differentiate in the direction of thyroid follicles, but that these follicles can produce the thyroid hormone has not been shown. In many mammals they actually become incorporated in the thyroid lobes and sometimes constitute as much as half of the total thyroid mass. For this reason, early embryologists often referred to these bodies as "lateral thyroid" glands. Epithelioid bodies in the ventral wall of the last pair of gill pouches in cyclostomes and in the roof of the pericardial cavity of fishes (**suprapericardial bodies**) probably represent the ultimobranchial bodies in these forms. The suprapericardial bodies seem to respond by hyperplasia to thyroid hormone and to parathyroid hormone.

Chapter summary

1. Hormones are metabolic products of tissues located in restricted areas of organisms. They exert a regulatory effect over the metabolism of nearby or remote cells of a different nature.

2. Neurosecretions are hormones produced in the cell bodies of secretory neurons and released into circulatory channels at axon terminals. The terminals of the neurosecretory axons and the associated vascular channels constitute a neurohemal organ. The median eminence, pars nervosa of the hypophysis, and urohypophysis of fishes are neurohemal organs.

3. Endocrine organs derived from ectoderm include the adenohypophysis, pineal, and adrenal medullary tissue. The hormones produced by ectoderm are combinations of amino acids.

4. Endocrine organs derived from mesoderm are the gonads and the cortical component of the adrenal, including the interrenal bodies of fishes. These produce steroid hormones.

5. Endocrine organs derived from endoderm are the pancreatic islets, thyroid, and parathyroid glands. These produce peptides and proteins.

6. The hypophysis (pituitary) consists of a neurohypophysis derived from the diencephalon and of an adenohypophysis derived from the roof of the stomodeum. The parts of the neurohypophysis are the median eminence, infundibular stalk, and pars nervosa, the latter beginning with lungfish. The adenohypophysis typically consists of a pars distalis, pars intermedia, and pars tuberalis. The pars distalis produces TSH, ACTH, FSH, LH (ICSH), prolactin, and STH.

7. Three endocrine tissues produce chromatophore-regulating hormones. The pars intermedia produces intermedin, which causes pigment granule dispersion and darkening of the skin of ectotherms. Melatonin from the pineal and epinephrine from the adrenal medulla cause pigment granules to aggregate, thus blanching the skin.

8. Three endocrine organs play prominent roles in regulating the levels of salt in the body. The interrenal bodies (adrenal cortex of mammals) regulate sodium and potassium with aldosterone; the parathyroids regulate calcium and phosphate levels with parathormone; and the urohypophysis may regulate the transport of sodium ions into and out of the organism.

9. Interrenal bodies arise from the coelomic mesoderm and remain separate in many fishes, are interspersed among chromaffin tissue in most amphibians, reptiles, and birds, and form an adrenal cortex in mammals. Parathyroids are derivatives of a series of pharyngeal pouches and are lacking in fish.

10. Corpuscles of Stannius are derivatives of the pronephric duct that superficially resemble interrenal bodies. They occur in ray-finned fishes and may affect electrolyte distribution or water balance.

11. Three endocrine organs play prominent roles in regulating blood sugar levels. The pancreas produces insulin and glucagon, the adrenal cortex produces glucocorticoids, and the adrenal medulla produces epinephrine. Insulin lowers blood sugar levels. The others tend to raise the levels. STH also elevates blood sugar levels by an unknown route.

12. The midventral pharyngeal floor in all vertebrates evaginates to produce an organ, some of the cells of which accumulate iodine. Although arising as an unpaired median evagination, the adult thyroid is usually paired. Its hormones, thyroxin and triiodothyronine, increase metabolic rates and the utilization of oxygen. Thyroid hormone is especially essential in young

animals that are undergoing maturation.

13. The gonads of both sexes produce androgens and estrogens, with the former predominating in males and the latter in females. The ovary also produces progesterone, which affects the function of the adult female reproductive tract. It also produces relaxin, which softens the pubic ligaments before birth.

14. The thymus and ultimobranchial bodies are derivatives of a series of successive pharyngeal pouches, but they produce no known hormone. The thymus involutes at sexual maturity.

Appendix

Abridged classification of the vertebrates

An abridged classification has been provided so that the reader may readily determine the relationships of animals he meets in the text. If he consults it at once when an unfamiliar group of vertebrates is first encountered, the student may be pleased to find he is gaining a working knowledge of vertebrate relationships without resorting to rote memorization.

Authorities are not in complete accord in matters of classification. This is not necessarily unhealthy, since it fosters continual inquiry into the appropriateness of existing schemes. For example, Agnatha are sometimes considered a superclass, or even a subphylum[4] in recognition of the chasm that separates agnathous fish from the more recent jawed fish, which are then placed in the superclass or subphylum Gnathostomata. Since no single classification is universally employed, the primary consideration in devising this one was to achieve simplicity while remaining in harmony with a majority.

All classes are included, but entire subclasses have been omitted if all members are extinct and no representative has been cited in the text. All orders containing living members are included except under Osteichthyes and Aves. Inclusion of the many taxa in those groups would not be in accord with the purpose of the abridgment. Extinct forms are indicated by an asterisk(*).

PHYLUM CHORDATA

Subphylum I. Urochordata
Subphylum II. Cephalochordata
Subphylum III. Vertebrata

SUPERCLASS PISCES

Class Agnatha. Jawless fishes
  *Subclass Ostracodermi. Extinct armored agnathostomes
  Subclass Cyclostomata.
       Order Myxinoidea (Myxiniformes). Hagfishes *(Myxine, Bdellostoma)*
       Order Petromyzontia. Lampreys *(Petromyzon, Lampetra)*
       Three additional extinct orders
*Class Placodermi. Armored Paleozoic gnathostome fishes
    *Order Acanthodii. Acanthodians
    *Order Arthrodira. Arthrodires *(Coccosteus)*

*Order Antiarchi. *Bothriolepis*

*Order Stegoselachii.

Class Chondrichthyes. Cartilaginous fishes

  Subclass Elasmobranchii. Naked gill slits

    *Order Cladoselachii. Primitive Paleozoic sharks *(Cladoselache)*

    Order Selachii. Sharks *(Carcharhinus, Hexanchus, Heptanchus, Mustelus, Scoliodon, Squalus)*

    Order Batoidea, Sawfishes, skates, rays *(Torpedo, Raia, Dasyatis)*

  Subclass Holocephali. Gill slits covered by an operculum *(Chimaera)*

Class Osteichthyes. Higher bony fishes

  Subclass Sarcopterygii (Choanichthyes). Lobe-finned fishes with internal nares

    Order Crossopterygii. Chiefly Paleozoic

      *Suborder Rhipidistia. Probable ancestors to amphibians *(Eusthenopteron)*

      Suborder Coelacanthini. Specialized crossopterygians, internal nostrils absent *(Latimeria* sole living crossopterygian)

    Order Dipnoi. Lungfishes *(Lepidosiren, Epiceratodus* [= *Neoceratodus*], *Protopterus* sole living genera)

  Subclass Actinopterygii. Ray-finned fishes

    Superorder Chondrostei. Cartilaginous ganoids, chiefly Paleozoic *(Polypterus, Calamoichthys, Acipenser* [sturgeon], *Polyodon* [spoonbill])

    Superorder Holostei. Bony ganoids, dominant Mesozoic fishes *(Amia* [bowfin], *Lepidosteus* [gar] sole living genera)

    Superorder Teleostei. Recent bony fishes; 95% of all living fishes

      Order Clupeiformes. Herring-like fishes, salmon, etc.

      Order Cypriniformes. Exhibit weberian apparatus (goldfish, carp, catfish, buffalo fish)

      Order Anguilliformes. Eels *(Electrophorus)*

      Order Gadiformes. Codfishes, etc. *(Gadus)*

      Order Perciformes. Perch-like fishes *(Perca)*

      And up to 35 additional living orders

SUPERCLASS TETRAPODA

Class Amphibia. Highest anamniotes

  *Superorder Labyrinthodontia. Stem amphibians, precursors of reptiles *(Seymouria,* sometimes classified as a cotylosaur reptile)

  Superorder Salientia

    *Order Proanura. Probable precursors of Anura

    Order Anura. Frogs, toads *(Rana, Hyla, Bufo, Xenopus, Triprion)* (18 families)

    Order Caudata (Urodela). Tailed amphibians

      Family Hynobiidae. Five genera in Asia

      Family Proteidae. *Necturus*

      Family Amphiumidae. *Amphiuma*

      Family Cryptobranchidae. *Cryptobranchus*

      Family Salamandridae. Notophthalmus (= *Diemictylus*)

      Family Ambystomatidae. *Ambystoma*

      Family Plethodontidae. *Plethodon, Desmognathus*

     Family Sirenidae. *Siren, Pseudobranchus* sole genera
    Order Apoda (Gymnophiona). Caecilians
  Class Reptilia. Lowest amniotes; mostly extinct
   Subclass Anapsida.
     *Order Cotylosauria. Stem reptiles
     Order Chelonia. Turtles and tortoises *(Chrysemys, Trionyx)*
   *Subclass Synapsida. Mammal-like reptiles *(Cynognathus)*
   *Subclass Synaptosauria. Includes plesiosaurs (large marine reptiles)
   *Subclass Parapsida. Marine reptiles *(Ichthyosaurus)*
   Subclass Lepidosauria.
     Order Rhynchocephalia. *Sphenodon* sole living genus, one species
     Order Squamata
      Suborder Lacertilia (Sauria). Lizards, horned toads *(Anolis, Helo-*
        derma, Chirotes)
      Suborder Serpentes (Ophidia). Snakes
   Subclass Archosauria. Ruling reptiles
     *Order Pterosauria. Flying reptiles
     *Order Saurischia. Ancestors of birds; includes many dinosaurs *(Tyran-*
        nosaurus)
     *Order Ornithischia. Dinosaurs with bird-like pelvis *(Triceratops)*
     Order Crocodilia. Crocodiles, alligators
  Class Aves. Vertebrates with feathers
   *Subclass Archaeornithes. Earliest birds (*Archaeopteryx, Archaeornis* sole
       genera)
   Subclass Neornithes. All other birds
    *Superorder Odontognathae. *Ichthyornis, Hesperornis*
    Superorder Paleognathae. Ratites (emus, kiwis, cassowaries, rheas, os-
        triches, *moas)
    Superorder Neognathae. Carinates
     Order Columbiformes. Doves *(Columba)*
     Order Pelecaniformes. Pelicans
     Order Anseriformes. Ducks, geese, other waterfowl
     Order Falconiformes. Hawks, eagles, vultures
     Order Galliformes. Chickens, grouse, quail
     Order Psittaciformes. Parrots, paroquets
     Order Passeriformes. Perching birds—up to 64 families, including
       songbirds
    And 15 other orders
  Class Mammalia. Vertebrates with hair
   Subclass Prototheria. Egg-laying mammals
     Order Monotremata. Spiny anteaters *(Echidna)*, duck-billed platy-
       puses *(Ornithorhynchus)*
   Subclass Theria. Give birth to young and have mammary glands; all except
       the Marsupialia have a chorioallantoic placenta
     Order Marsupialia. Yolk sac serving as placenta (opossum *[Didel-*
       phis], kangaroo, Tasmanian wolf, phalangers)
     Order Insectivora. Moles, shrews, hedgehogs
     Order Dermoptera. Gliding lemurs *(Cynocephalus = Galeopithecus)*
     Order Chiroptera. Bats
     Order Primates

Suborder Prosimii. Lemurs *(Tarsius)*
Suborder Anthropoidea. Monkeys, apes, man
  Superfamily Ceboidea. New world monkeys (*Cebus, Ateles, Alouatta* [howler])
  Superfamily Cercopithecoidea. Old world monkeys (rhesus monkey [*Macaca*])
  Superfamily Hominoidea. Apes and man
    Family Pongidae. Apes (orangutan)
    Family Hominidae. Man
      Australopithecus africanus
      Homo erectus. Java man
      Homo neanderthalensis (sometimes cited as *Homo sapiens neanderthalensis*)
      Homo sapiens. Modern and *Cro-Magnon man
Order Edentata. Sloths, armadillos, South American anteaters
Order Pholidota. Scaly anteaters, or pangolins *(Manis)*
Order Lagomorpha. Rabbits, hares
Order Rodentia. Rats, mice, chipmunks, beavers, porcupines, woodchucks, squirrels, chinchillas, guinea pigs, hamsters, grisons, pocket gophers, marmots, coypus
Order Cetacea. Whales, dolphins, porpoises
Order Carnivora. Cats (*Felis, Lynx*, etc.), dogs, bears, otters, badgers, skunks, weasels, racoons, hyenas, seals, sea lions, walruses
Order Tubulidentata. Aardvarks
Order Proboscidea. Elephants,* mastodons
Order Hyracoidea. Conies *(Hyrax)*
Order Sirenia. Dugong, manatees
Order Perissodactyla. Odd-toed ungulates (horses, rhinoceros, tapirs)
Order Artiodactyla. Even-toed ungulates
  Suborder Suina. Pigs, hippopotamuses
  Suborder Ruminantia.
    Family Camelidae. Camels, llamas
    Family Cervidae. Deer
    Family Giraffidae. Giraffes
    Family Antilocapridae. American pronghorn antelope sole species
    Family Bovidae. Cattle, sheep, goats, gnu, gazelles, *Bison*, ibex, antelope (except pronghorns)
    Family Tragulidae. Chevrotains

Literature cited and selected readings

Publications cited by number in the text will be found in this section, along with selected readings ranging from short papers to comprehensive works with extensive bibliographies. Attention is also directed to the growing list of symposia in the *American zoologist*. Among recent topics are vertebrate locomotion, feeding mechanisms, metamorphosis, evolution of amphibia, olfaction, and the vertebrate ear. The student is encouraged to browse among the journals in the library and to examine the tables of contents not only in current issues, but also in early ones to discover past and current trends in research in vertebrate morphology.

GENERAL

1. Barrington, E. J. W.: The biology of Hemichordata and Protochordata, San Francisco, 1965, W. H. Freeman & Co., Publishers.
2. Bensley, B. A.: Practical anatomy of the rabbit, Philadelphia, 1938, P. Blakiston's Son & Co., Inc.
3. Bishop, S. C.: The salamanders of New York, New York State Museum Bulletin no. 324, Albany, 1941.
4. Brodal, A., and Fänge, R., editors: The biology of Myxine, Oslo, 1963, Norway Universitetsforlaget.
5. Cook, M. J.: The anatomy of the laboratory mouse, New York, 1965, Academic Press, Inc.
6. Cope, E. D.: The crocodilians, lizards, and snakes of North America, Washington, D. C., 1900, Government Printing Office.
7. Daniel, J. F.: The elasmobranch fishes, Berkeley, 1934, University of California Press.
8. Ditmars, R. L.: Reptiles of the world, New York, 1933, The Macmillan Co.
9. Ecker, A.: The anatomy of the frog, New York, 1889, The Macmillan Co. (Translated from German by George Haslam.)
10. Francis, E. T. B.: The anatomy of the salamander, London, 1934, Oxford University Press.
11. Gilbert, P. W., Mathewson, R. F., and Rall, D. P., editors: Sharks, skates, and rays, Baltimore, 1967, The Johns Hopkins Press.
12. Goodrich, E. S.: Studies on the structure and development of vertebrates, London, 1930, The Macmillan Co., Ltd. (Reprinted by Dover Publications, Inc., New York, 1958.)
13. Grassé, P.-P., editor: Traité de zoologie. Anatomie, systématique, biologie (17 vols.), Paris, 1948-1967, Masson et Cie.
14. Gray, P., editor: The encyclopedia of the biological sciences, New York, 1961, Reinhold Publishing Corp.
15. Gray, P., editor: The dictionary of the biological sciences, New York, 1967, Reinhold Publishing Corp.
16. Harmer, S. F., and Shipley, A. E., editors: The Cambridge natural history (11 vols.), London, 1895-1909, The Macmillan Co., Ltd.
17. Hartman, C. G., and Straus, W. L., Jr., editors: The anatomy of the rhesus monkey, Baltimore, 1933, The Williams & Wilkins Co.
18. Hyman, L. H.: Comparative vertebrate anatomy, Chicago, 1942, University of Chicago Press.
19. Jensen, D.: The hagfish, Scientific American **214:**82, Feb., 1966.
20. Miller, M. E., Christensen, G. C., and Evans, H. E.: Anatomy of the dog, Philadelphia, 1964, W. B. Saunders Co.
21. Parker, T. J., and Haswell, W. A.: Textbook

of zoology, vol. II (revised by C. Forster-Cooper), ed. 7, London, 1960, The Macmillan Co., Ltd.

22. Pennak, R. W.: Collegiate dictionary of zoology, New York, 1964, The Ronald Press Co.

23. Peters, J. A.: Dictionary of herpetology, New York, 1964, Hafner Publishing Co., Inc.

24. Prosser, C. L., and Brown, F. A.: Comparative animal physiology, Philadelphia, 1961, W. B. Saunders Co.

25. Reighard, J., and Jennings, H. S.: Anatomy of the cat, New York, 1935, Henry Holt & Co., Inc.

26. Romer, A. S.: Vertebrate paleontology, Chicago, 1945, University of Chicago Press.

27. Romer, A. S.: The vertebrate story, Chicago, 1958, University of Chicago Press.

28. Rugh, R.: The mouse—its reproduction and development, Minneapolis, 1968, Burgess Publishing Co.

29. Sisson, S., and Grossman, J. D.: The anatomy of the domestic animals, Philadelphia, 1938, W. B. Saunders Co.

30. Slijper, E. J.: Whales, New York, 1962, Basic Books, Inc.

31. Taylor, W. T., and Weber, R. J.: Functional mammalian anatomy (with special reference to the cat), New York, 1951, D. Van Nostrand Co., Inc.

32. Templeman, W.: The life history of the spiny dogfish, Department of Natural Resources, Res. Bull. no. 15, St. John's, Newfoundland, 1944.

33. Thomson, Sir A. L., editor: A new dictionary of birds, New York, 1964, McGraw-Hill Book Co.

34. Young, J. Z.: The life of vertebrates, New York, 1962, Oxford University Press.

DEVELOPMENTAL ANATOMY

35. Amoroso, E. C.: Comparative anatomy of the placenta, Annals of the New York Academy of Sciences **75**:855, 1959.

36. Bone, Q.: The assymmetry of the larval amphioxus, Proceedings of the Zoological Society of London **130B**:289, 1958.

37. Camp, W. E.: The development of the suprapericardial (postbranchial, ultimobranchial) body in Squalus acanthias, Journal of Morphology **28**:369, 1917.

38. Conklin, E. G.: The embryology of amphioxus, Journal of Morphology **54**:69, 1932.

39. DeHaan, R. L., and Ursprung, H., editors: Organogenesis, New York, 1965, Holt, Rinehart & Winston, Inc.

40. Fleischmajer, R., and Billingham, R. E.: Epithelial-mesenchymal interactions, Baltimore, 1968, The Williams & Wilkins Co.

41. Hamilton, W. J., Boyd, J. D., and Mossman,

H. W.: Human embryology, Baltimore, 1962, The Williams & Wilkins Co.

42. Matthews, L. H.: The evolution of viviparity in vertebrates, Memoirs of the Society for Endocrinology, no. 4, Cambridge, 1955, Cambridge University Press.

43. Nelsen, O. E.: Comparative embryology of the vertebrates, New York, 1953, Blakiston Division, McGraw-Hill Book Co.

44. Patten, B. M.: Embryology of the pig, ed. 3, New York, 1948, The Blakiston Co.

45. Patten, B. M.: Early embryology of the chick, New York, 1951, The Blakiston Co.

46. Patten, B. M.: Foundations of embryology, New York, 1964, McGraw-Hill Book Co.

47. Rugh, R.: The frog—its reproduction and development, New York, 1951, The Blakiston Co.

48. TeWinkel, L. E.: Observations on later phases of embryonic nutrition in Squalus acanthias, Journal of Morphology **73**:177, 1943.

49. Willier, B. H., editor: Analysis of development, Philadelphia, 1955, W. B. Saunders Co.

50. Witschi, E.: Development of vertebrates, Philadelphia, 1956, W. B. Saunders Co.

SKELETON AND MUSCULATURE

51. Auffenberg, W.: A review of the trunk musculature in the limbless land vertebrates, American Zoologist **2**:183, 1962.

52. Ayers, H., and Jackson, C. M.: Morphology of the Myxinoidei. I. Skeleton and musculature, Journal of Morphology **17**:185, 1901.

53. Baldauf, R. J.: Contributions to the cranial morphology of Bufo w. woodhousei (Girard), The Texas Journal of Science **7**:275, 1955.

54. Branson, B. A.: Comparative cephalic and appendicular osteology of the fish family Catostomidae. Part I. Cycleptus elongatus (Lesueur), The Southwestern Naturalist **7**:81, 1962.

55. Byerly, T. C.: The myology of Sphenodon punctatum, University of Iowa Studies in Natural History, (First Series, no. 98) **11**:no. 6, 1925.

56. De Beer, G. R.: The development of the vertebrate skull, Oxford, 1937, The Clarendon Press.

57. Evans, F. G.: The morphology and functional evolution of the atlas-axis complex from fish to mammals, Annals of the New York Academy of Sciences **39**:29, 1939.

58. Grundfest, H.: Electric fishes, Scientific American **203**(4):115, 1960.

59. Harrington, R. W., Jr.: The osteocranium of the American cyprinid fish, Notropis bifrenatus, with an annotated synonymy of

teleost skull bones, Copeia, no. 4, p. 267, 1955.

60. Harris, J. P., Jr.: The skeleton of the arm of Necturus, Field and Laboratory **20**:78, 1952.

61. Holtzer, H.: Aspects of chondrogenesis and myogenesis. In Synthesis of molecular and cellular structure, New York, 1961, The Ronald Press Co., Inc.

62. Huber, E.: Evolution of facial musculature and cutaneous field of trigeminus. Parts I and II, Quarterly Review of Biology **5**:133, 389, 1930.

63. James, P. S. B. R.: Comparative osteology of the ribbon fishes of the family Trichiuridae from Indian waters, with remarks on their phylogeny, Journal of the Marine Biological Association of India **3**:215, 1961.

64. Keynes, R. D.: Electric organs. In Brown, M. E., editor: The physiology of fishes, New York, 1957, Academic Press, Inc.

65. Manville, R. C.: Bregmatic bones in North American lynx, Science **130**:1254, 1959.

66. Priddy, R. B., and Brodie, A. F.: Facial musculature, nerves and blood vessels of the hamster in relation to the cheek pouch, Journal of Morphology **83**:149, 1948.

67. Reed, C. A., and Schaffer, W.: Evolutionary implications of cranial morphology in the sheep and goats, American Zoologist **6**:565, 1966.

68. Romer, A. S.: Osteology of the reptiles, Chicago, 1956, University of Chicago Press.

69. Salih, M. S., and Kent, G. C.: The epaxial muscles of the golden hamster, Anatomical Record **150**:319, 1964.

70. Trueb, L.: Unusual dermal modifications of the skull of the hylid frog Triprion petasatus, American Zoologist **5**:691, 1965.

71. Wells, L. J.: Development of the human diaphragm and pleural sacs, Carnegie Institution of Washington Contributions to Embryology **35**:107, 1954.

RESPIRATORY SYSTEM AND SKIN

72. Atz, J. W.: Narial breathing in fishes and the evolution of internal nares, Quarterly Review of Biology **27**:367, 1952.

73. Elias, H., and Bortner, S.: On the phylogeny of hair, American Novitates, no. 1820, p. 1, 1957.

74. Fox, D. L.: Coloration of animals. In Gray, P., editor: The encyclopedia of the biological sciences, New York, 1961, Reinhold Publishing Corp.

75. Jones, F. R. H., and Marshall, N. B.: The structure and functions of the teleostean swimbladder, Biological Reviews **28**:16, 1953.

76. Krejsa, R. J.: On the supposed participation of the ectoderm in the ontogenesis of teleost scales, American Zoologist **7**:774, 1967.

77. Locy, A., and Larsell, O.: The embryology of the bird's lung, American Journal of Anatomy **20**:1, 1916.

78. Montagna, W., and Lobitz, W. C., editors: The epidermis, New York, 1964, Academic Press, Inc.

79. Parker, G. H.: Animal color changes and their neurohumours, London, 1948, Cambridge University Press.

80. Parker, M. V.: The amphibians and reptiles of Reelfoot Lake and vicinity, with a key for the separation of species and subspecies, Journal of the Tennessee Academy of Sciences **14**:72, 1939.

CIRCULATORY SYSTEM

81. De Ryke, W.: The development of the renal portal system in Chrysemys marginata Belli (Gray), University of Iowa Studies in Natural History (New Series, no. 88) **11**: no. 3, 1925.

82. Easton, T. W.: Two venous anomalies of commercial laboratory specimens, Turtox News **45**(2):66, 1967.

83. Fox, M. H., and Goss, C. M.: Experimentally produced malformations of the heart and great vessels in rat fetuses. Transposition complexes and aortic arch abnormalities, American Journal of Anatomy **102**:65, 1958.

84. Millen, J. E., Murdaugh, H. V., Bauer, C. B., and Robin, E. D.: Circulatory adaptation to diving in the freshwater turtle, Science **145**:591, 1964.

85. Nandy, K., and Blair, C. B.: Double superior venae cavae with completely paired azygos veins, Anatomical Record **151**:1, 1965.

86. Ogren, H., and Mitchen, J.: Tracing oxygenated and unoxygenated blood through the organs of a frog by means of radioisotopes, Turtox News **45**(5):130, 1967.

87. Roofe, P. G.: The endocranial blood vessels of Amblystoma tigrinum, Journal of Comparative Neurology **61**:257, 1935.

88. Rusznyák, I., Földi, M., and Szabó, G.: Lymphatics and lymph circulation: physiology and pathology, ed. 2, New York City, 1967, Pergamon Press, Inc.

89. Seib, G. A.: On the azygos vein in Pithecus (Macacus) rhesus, Anatomical Record **51**: 285, 1932.

90. Struthers, P. H.: The aortic arches and their derivatives in the embryo porcupine (Erethizon dorsatus), Journal of Morphology and Physiology **50**:361, 1930.

91. Yoffey, J. M., and Courtice, F. C.: Lymphatics, lymph, and lymphoid tissue, Cambridge, 1956, Harvard University Press.

92. Zucchero, P. J.: An anomaly of the right

subclavian artery of the cat, Turtox News **42:**(1):14, 1964.

URINOGENITAL SYSTEM

93. Altschule, M. D.: The change(s) in the mesonephric tubules of human embryos ten to twelve weeks old, Anatomical Record **46:** 81, 1930.

94. Baker, C. J., and Taylor, W. W.: The urogenital system of the male Ambystoma, Journal of the Tennessee Academy of Sciences **39:**1, 1964.

95. Fox, H.: The amphibian pronephros, Quarterly Review of Biology **38:**1, 1963.

96. Jaffee, O. C.: Morphogenesis of the pronephros of the Leopard frog (Rana pipiens), Journal of Morphology **95:**109, 1954.

97. McCrady, E., Jr.: The development and fate of the urinogenital sinus in the opossum, Didelphys virginiana, Journal of Morphology **66:**131, 1940.

98. Schmidt-Nielsen, B.: Comparative morphology and physiology of excretion. In Moore, J. A., editor: Ideas in modern biology, Garden City, 1965, Natural History Press.

99. Zuckerman, S., editor: The ovary, vol. I, New York, 1962, Academic Press, Inc.

NERVOUS SYSTEM AND SENSE ORGANS

100. Adams, W. E.: The comparative morphology of the carotid body and carotid sinus, Springfield, Ill., 1958, Charles C Thomas, Publisher.

101. Amoore, J. E., Johnston, J. W., Jr., and Rubin, M.: The stereochemical theory of odor, Scientific American **210**(2):42, 1964.

102. Ariëns Kappers, C. U., Huber, G. C., and Crosby, E. C.: The comparative anatomy of the nervous system of vertebrates, including man, New York, 1936, The Macmillan Co. (Republished by Hafner Publishing Co., Inc., 1960.)

103. Branson, B. A., and Moore, G. A.: The lateralis components of the acoustico-lateralis system in the sunfish family Centrarchidae, Copeia, no. 1, p. 1, 1962.

104. Cahn, P. H., editor: Lateral-line detectors, Bloomington, 1967, Indiana University Press.

105. Denison, R. H.: The origin of the lateral-line sensory system, American Zoologist **6:** 369, 1966.

106. Eakin, R. M., Quay, W. B., and Westfall, J. A.: Cytological and cytochemical studies on the frontal and pineal organs of the tree frog, Hyla regilla, Zeitschrift für Zellforschung **59:**663, 1963.

107. Eakin, R. M., and Westfall, J. A.: Fine structure of photoreceptors in Amphioxus, Journal of Ultrastructure Research **6:**531, 1962.

108. Harrison, J. M., Warr, W. B., and Irving, R. E.: Second order neurons in the acoustic nerve, Science **138:**893, 1962.

109. Hassler, R., and Stephen, H., editors: Evolution of the forebrain, New York, 1967, Plenum Publishing Corp.

110. Kitay, J. I., and Altschule, M. D.: The pineal gland, Cambridge, 1954, Harvard University Press.

111. Noback, C. R.: The human nervous system, New York, 1967, McGraw-Hill Book Co.

112. Norris, H. W., and Hughes, S. P.: The cranial, occipital, and anterior spinal nerves of the dogfish, Squalus acanthias, Journal of Comparative Neurology **31:**293, 1920.

113. Olson, E. C.: The middle ear—morphological types in amphibians and reptiles, American Zoologist **6:**399, 1966.

114. Szabo, T.: Sense organs of the lateral line system in some electric fishes of the Gymnotidae, Mormyridae and Gymnarchidae, Journal of Morphology **117:**229, 1965.

115. Tester, A. L., and Kendall, J. I.: Cupulae in shark neuromasts: composition, origin, generation, Science **160:**772, 1968.

116. Tumarkin, A.: On the evolution of the auditory conducting apparatus: a new theory based on functional considerations, Evolution **9:**221, 1955.

117. van Bergeijk, W.: Evolution of the sense of hearing in vertebrates, American Zoologist **6:**371, 1966.

ENDOCRINE ORGANS

118. Comparative aspects of parathyroid function (Symposium), American Zoologist **7:**882, 1967.

119. Enami, M.: The morphology and functional significance of the caudal neurosecretory system of fishes. In Gorbman, A., editor: Comparative endocrinology, New York, 1959, John Wiley & Sons, Inc.

120. Frazer, J. F. D.: The sexual cycles of vertebrates, London, 1959, Hutchinson University Library.

121. Gorbman, A., and Bern, H. A.: A textbook of comparative endocrinology, New York, 1962, John Wiley & Sons, Inc.

122. Harris, G. W., and Donovan, B. T., editors: The pituitary gland (3 vols.), Berkeley, 1966, University of California Press.

123. Holmgren, U.: Neurosecretion in teleost fishes: the caudal neurosecretory system, American Zoologist **4:**37, 1964.

124. Idler, D. R., and Freeman, H. C.: Steroid transformation by corpuscles of Stannius of the Atlantic cod (Gadus morhua L.), Journal of the Fisheries Research Board of Canada **23:**1249, 1966.

125. Imai, K., Stanley, J. G., Fleming, W. R., and

Bern, H. A.: On the suggested ionoregulatory role of the teleost caudal neurosecretory system, Proceedings of the Society for Experimental Biology and Medicine 118:1102, 1965.

126. Jones, I. C., Henderson, I. W., Chan, D. K., and Rankin, J. C.: The role of the adrenal cortex and the corpuscles of Stannius in electrolyte balance of the eel, General and Comparative Endocrinology 5:669, 1965.

127. Leach, W. J.: The endostyle and thyroid gland of the brook lamprey, Ichthyomyzon fossor, Journal of Morphology 65:549, 1939.

128. Lynn, W. G., and Wachowsky, H. E.: The thyroid in cold-blooded vertebrates, Quarterly Review of Biology 26:123, 1951.

129. Scharrer, E., and Scharrer, B.: Neuroendocrinology, New York, 1963, Columbia University Press.

130. Young, W. C., editor: Sex and internal secretions, vols. I and II, Baltimore, 1961, The Williams & Wilkins Co.

EVOLUTION

131. Barrington, E. J. W.: Hormones and vertebrate evolution, Experientia 18:201, 1962.

132. Campbell, B.: Human evolution—an introduction to man's adaptations, Chicago, 1966, Aldine Publishing Co.

133. Carter, G. S.: Structure and habit in vertebrate evolution, Seattle, 1967, University of Washington Press.

134. Clark, W. E. L.: The fossil evidence for human evolution, Chicago, 1964, University of Chicago Press.

135. Colbert, E. H.: Evolution of the vertebrates, New York, 1955, John Wiley & Sons, Inc.

136. Eaton, T. H.: Origin of tetrapod limbs, American Midland Naturalist 46:245, 1951.

137. Gregory, W. K., and Raven, H. C.: Studies on the origin and early evolution of paired fins and limbs, Annals of the New York Academy of Sciences 42:273, 1941.

138. Gunter, G.: Origin of tetrapod limbs, Science 123:495, 1956.

139. Laughlin, W. S.: Eskimos and Aleuts: their origins and evolution, Science 142:633, 1963.

140. Mayr, E.: Animal species and evolution, Cambridge, 1963, Harvard University Press.

141. Mayr, E.: Evolution at the species level. In Moore, J. A., editor: Ideas in modern biology, Garden City, 1965, Natural History Press.

142. Napier, J.: The evolution of the hand, Scientific American reprint no. 140, San Francisco, 1962, W. H. Freeman & Co., Publishers.

143. Nursall, J. R.: Swimming and the origin of paired appendages, American Zoologist 2:127, 1962.

144. Parsons, T. S., and Williams, E. E.: The relationships of the modern amphibia: a reexamination, Quarterly Review of Biology 38:26, 1963.

145. Romer, A. S.: The early evolution of fishes, Quarterly Review of Biology 21:33, 1946.

146. Schmalhausen, I. I.: The origin of terrestrial vertebrates (translated from the Russian by Leon Kelso), New York, 1968, Academic Press.

147. Simpson, G. G.: Horses; the story of the horse family in the modern world and through sixty million years of history, New York, 1951, Oxford University Press.

148. Szarski, H.: Origin of the amphibia, Quarterly Review of Biology 37:189, 1962.

149. Williams, G. C.: Adaptation and natural selection, Princeton, 1966, Princeton University Press.

MISCELLANEOUS

150. Bock, W. J.: Experimental analysis of the avian passive perching mechanism, American Zoologist 5:681, 1965.

151. Butler, P. M.: Studies of the mammalian dentition, Proceedings of the Zoological Society of London 109B:1, 1939.

152. Carson, R. L.: Odyssey of the eel. In Beebe, W., editor: The book of naturalists, New York, 1945, Alfred A. Knopf.

153. Cole, F. J.: A history of comparative anatomy, London, 1944, The Macmillan Co., Ltd.

154. Cracraft, J.: Comments on homology and analogy, Systematic Zoology 16:355, 1967.

155. Darnell, R. M.: Nocturnal terrestrial habits of the tropical gobioid fish Gobiomorus dormitor, with remarks on its ecology, Copeia, no. 3, p. 237, 1955.

156. Greenway, P.: Body form and behavioral types in fish, Experientia 21:489, 1965.

157. Gregory, J. T.: The jaws of the cretaceous toothed birds, Ichthyornis and Hesperornis, The Condor 54:73, 1952.

158. Hildebrand, M.: Anatomical preparations, Berkeley, 1968, University of California Press.

159. Miller, R.: The morphology and function of the pharyngeal organs in the clupeid, Dorosoma petenense (Gunther), Chesapeake Science 5:194, 1964.

160. Nomina Anatomica, ed. 3, Amsterdam, 1966, Excerpta Medica Foundation.

161. Nomina Anatomica Veterinaria, Department of Anatomy, Ithaca, N. Y., 1968, State Veterinary College.

162. Oguri, M.: Rectal glands of marine and fresh-water sharks; comparative histology, Science 144:1151, 1964.

163. Savile, D. B. O.: Gliding and flight in the vertebrates, American Zoologist 2:161, 1962.

164. Schmidt-Nielsen, K.: Physiology of salt glands, American Journal of Physiology **204:**264, 1963.

165. Schumacher, E.: The last paradises, New York, 1967, Natural History Press.

166. Simpson, G. G.: The biological nature of man, Science **152:**472, 1966.

167. Snyder, R. C.: Adaptations for bipedal locomotion of lizards, American Zoologist **2:**191, 1962.

168. Vertebrate adaptations. Readings from Scientific American with introductions by Norman K. Wessells, San Francisco, 1952-1968, W. H. Freeman & Co., Publishers.

Index